Advances in Neurobiology
Volume 35

covers basic research in neurobiology and neurochemistry. It provides in-depth, book-length treatment of some of the most important topics in neuroscience including molecular and pharmacological aspects. The main audiences of the series are basic science researchers and graduate students as well as clinicians including neuroscientists (neurobiologists and neurochemists) and neurologists. *Advances in Neurobiology* is indexed in PubMed, Google Scholar, and the Thompson Reuters Book Citation Index.

Patrick L. Kerr • Cristian Sirbu • John M. Gregg
Editors

Endogenous Opioids

From Basic Science to Biopsychosocial Applications

Springer

Editors
Patrick L. Kerr
West Virginia University School of Medicine-Charleston
Charleston, WV, USA

Cristian Sirbu
Center for Cancer Research
Charleston Area Medical Center
Charleston, WV, USA

John M. Gregg
Department of Surgery
VTCSOM
Blacksburg, VA, USA

ISSN 2190-5215 ISSN 2190-5223 (electronic)
Advances in Neurobiology
ISBN 978-3-031-45492-9 ISBN 978-3-031-45493-6 (eBook)
https://doi.org/10.1007/978-3-031-45493-6

This Springer imprint is published by the registered company Springer Nature Switzerland AG
The registered company address is: Gewerbestrasse 11, 6330 Cham, Switzerland

PLK
This work is dedicated…
To my wife, Kim, and our children, Jackson and Ella, who have supported my work on this book every way throughout this journey. I am eternally grateful for your love and support.
To my mom and dad, who have supported me at every step of my journey, professionally and personally.
To the memory of my sister, Sarah, who was taken from this world far too soon during the completion of this work. May the knowledge and wisdom in this book somehow prevent more lives from being lost in the same way.

JMG
Dedicated to my loving, talented and globally extended family.

CS
Dedicated to my family.

Preface

A new era and significant hope for opioid research started in 1973 with the discovery of opiates brain binding sites, identification of multiple opioid receptors, and soon after the identification and cloning of endogenous ligands binding to these receptors. For the first time after millennia of opiate compounds oscillation between glorification and demise, the scientific community was able to shed light into the complexity of the endogenous opioid system. The 1990s enhanced this promise, with two important developments: the cloning of canonical opioid receptors and the development of genetic animal models allowing the identification of their specific functional roles. A decade into the twenty-first century, the crystal structure of the opioid receptors in inactive and active forms was deciphered, and, for the first time, the scientific community is hopeful that the "holy grail" of identifying an opioid compound with morphine potency and no tolerance, emotional side effects, or respiratory depression is within reach. In an era of unprecedented public health challenges from the COVID-19 pandemic, which has only compounded the existing opioid crisis, there is an urgency for translating the massive volume of knowledge about the endogenous opioid system into lifesaving interventions.

This book (*Endogenous Opioids: From Basic Science to Biopsychosocial Applications*) is motivated by this urgency and the desire of our contributors to close the loop between the exciting basic science findings and much needed clinical applications. Our volume takes a decidedly unique trajectory to synthesize basic science, behavioral science, and social science related to endogenous opioids into a unified and useful whole. We have diligently undertaken our mandate to present an integrated "biopsychosocial" model, as our title denotes.

In the space between the basic and translational science of endogenous opioids are adjacent spaces occupied by behavioral, affective, and social science. Insights about the endogenous opioid as a critical evolutionary survival system with paramount importance in establishing homeostasis when environmental and internal challenges arise provide a glimpse into the complex molecular processes involved. The awareness that endogenous opioids are key players that act as neurotransmitters, neuromodulators, and hormones across the nervous, endocrine, and immunological systems has broad implications for understanding normal physiological processes involved in stress and pain responses, natural immunity, social bonding, emotions, reward, feeding, reproduction, exercise, or placebo response. In pathological conditions such as cardiovascular disease, metabolic and eating disorders, and cancer, the involvement of endogenous opioid system shapes the progression of the disease and the prospect of a cure. Additionally, challenges arise when the same opioid receptors are exposed to opiates and the delicate balance of the endogenous opioid system is disrupted. Each chapter was written with the desire of synthesizing the basic science findings of the endogenous opioid system and their application to both healthy and pathological physiological processes.

In 2023, we were celebrating five decades of research since the discovery of the endogenous opioid system, and we are highly indebted to the contributions of opioid research giants and internationally renowned laboratories that added critical pieces of knowledge to the field. Prestigious journals (i.e., *Peptides*) publish summaries of endogenous opioid research outlining thousands of studies every year. When we started planning this book, we were aware of the tremendous challenges of reviewing an intimidating body of knowledge, trading across controversial topics and outlining new directions in the field. To our delight, the book contributors brilliantly addressed these challenges, bringing to life novel and exciting chapters that we hope will stimulate new ideas and conversations in the field. In such a large and complex field, we cannot claim that this work is completely comprehensive, nor was that the intention. Instead, the current volume is an attempt at reducing the distance between basic and translational research by outlining the tremendous importance of endogenous opioids across the continuum from normal to pathological processes. We believe that the distance between basic and translational science for endogenous opioids is reduced with the biopsychosocial bridge represented by the diverse yet interrelated chapters in this book.

We are in the midst of an "omics" era. Our knowledge of the molecular aspects of the endogenous opioid system is evolving at light speed. Analytical methods for big data generate hypotheses that were unfathomable just a decade ago. New clinical and scientific tools allow research findings on endogenous opioids to be built upon in unparalleled rapid succession. As we look to the journey ahead of meaningfully applying this knowledge, we hope that basic scientists and clinicians alike will find this book useful in their efforts to build much needed translational bridges.

Charleston, WV, USA Patrick L. Kerr
Charleston, WV, USA Cristian Sirbu
Blacksburg, VA, USA John M. Gregg
May 1, 2023

Contents

About the Editors

Patrick L. Kerr, PhD, completed his Master's and doctoral degrees in clinical psychology at the University of North Dakota. He completed his pre-doctoral internship training at the West Virginia University School of Medicine-Charleston campus and Charleston Area Medical Center. Dr. Kerr is a clinical psychologist, specializing in the treatment of severe psychopathology, suicidality, and traumatic stress. He is currently an Associate Professor in the Department of Behavioral Medicine and Psychiatry at West Virginia University School of Medicine-Charleston. He serves as Director of the WVU Behavioral Science and Psychopathology Research Division, and as Director of the WVU Dialectical Behavior Therapy Services Program.His main lines of research and academic work emphasize common mechanisms of severe psychiatric disorders, emotion regulation, suicide risk, trauma, and the psychobiological mechanisms of psychopathology.

Cristian Sirbu, DDS, PhD, PsyD, is a Clinical Associate Professor at West Virginia University School of Medicine-Charleston campus and Charleston Area Medical Center (CAMC) and a Research Scientist at the CAMC Center for Cancer Research. He completed his dental and psychology doctoral degrees in Romania and a clinical psychology doctoral degree at Marshall University.His scholarly and clinical works are focused on assessment and treatment of anxiety, mood disorders, and chronic pain across multiple populations and the enhancement of psychosocial interventions using pharmacological and technology-based approaches. He is interested in immunological mechanisms of psychopathology and the implementation of patient-reported outcomes in oncology.

John M. Gregg, DDS, MS, PhD, is a retired oral and maxillofacial surgeon and who served in academic positions at multiple institutions during his career. Dr. Gregg's academic career has included appointments as Professor of Surgery at University of North Carolina, Virginia Tech University, and Virginia Commonwealth University. During his academic and professional training, he completed five degrees as well as clinical residency in Oral and Maxillofacial surgery at the University of Michigan. Dr. Gregg's preclinical research was the first to demonstrate that peripheral injury of rodent trigeminal nerves may produce neuroanatomic pathoses in the transganglionic and central spinal trigeminal complex. He continues his active program of clinical research post-retirement, with an emphasis on mechanisms of neuropathic pain and microsurgical management of trigeminal nerve injuries.

Introduction to the Volume: The Journey Ahead

Patrick L. Kerr, John M. Gregg, and Cristian Sirbu

Abstract The endogenous opioid system (EOS) is complex. The line of research contributing to our current body of knowledge about this system is diverse, as are the ways in which endogenous opioids affect human health and behavior. This chapter serves as an introduction to the edited volume. It includes commentary about the current public discourse related to opioids, the rationale for this book, and the unique contributions of each chapter within this volume.

Keywords Endogenous opioids · Opioid research

The Present Moment

Between 2020 and 2022, the world changed in fundamental ways. The SARS-CoV-2 pandemic and its resulting illness (COVID-19) reshaped nearly every aspect of human life globally in some way. Along the way, COVID-19 became a leading cause of death, and a resulting drop in years of life expectancy was attributed to it. COVID-19 also complicated other leading causes of death that had been increasing; most notably, and relevant to this volume, deaths that were attributable to opioid poisoning or "overdose". While COVID-19 reached the rank of third leading cause of death for 2021, it was followed closely by unintentional injuries as the fourth leading cause of death (Ahmad et al., 2022). Ahmad noted about these data that "Unintentional injury deaths were largely driven by drug overdose deaths, and likely contributed to the increased death rate in younger populations." (p. 600). As

P. L. Kerr (✉)
West Virginia University School of Medicine-Charleston, Charleston, WV, USA
e-mail: pkerr@hsc.wvu.edu

J. M. Gregg
Department of Surgery, VTCSOM, Blacksburg, VA, USA

C. Sirbu
Center for Cancer Research, Charleston Area Medical Center, Charleston, WV, USA

P. L. Kerr et al. (eds.), *Endogenous Opioids*, Advances in Neurobiology 35,
https://doi.org/10.1007/978-3-031-45493-6_1

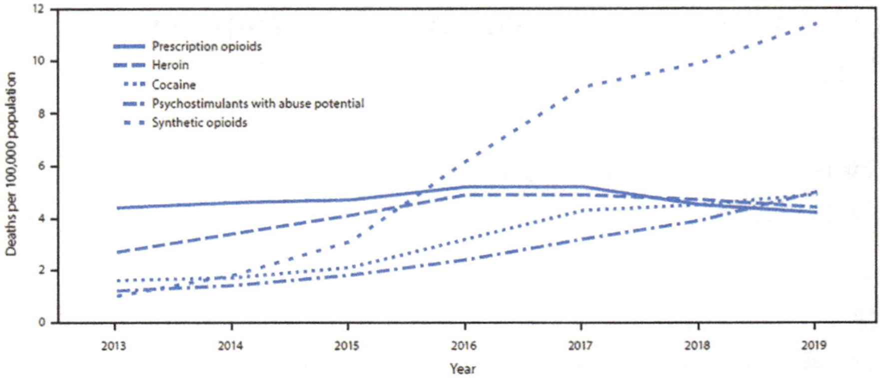

Source: National Vital Statistics System, Mortality File. https://wonder.cdc.gov/

* Rate per 100,000 population age-adjusted to the 2000 U.S. standard population using the vintage year population of the data year.

† Deaths were classified using the *International Classification of Diseases, Tenth Revision*. Drug overdoses are identified using underlying cause-of-death codes X40–X44 (unintentional), X60–X64 (suicide), X85 (homicide), and Y10–Y14 (undetermined).

§ Drug overdose deaths, as defined, that involve natural and semisynthetic opioids (T40.2) or methadone (T40.3).

¶ Drug overdose deaths, as defined, that involve heroin (T40.1).

** Drug overdose deaths, as defined, that involve cocaine (T40.5).

†† Drug overdose deaths, as defined, that involve psychostimulants with abuse potential (T43.6).

§§ Drug overdose deaths, as defined, that involve synthetic opioids other than methadone (T40.4).

¶¶ Because deaths might involve more than one drug, some deaths are included in more than one category. In 2019, 6.3% of drug overdose deaths did not include information on the specific type of drug(s) involved.

Fig. 1 Age-adjusted rates[*] of drug overdose deaths[†] involving prescription opioids, [§]heroin, [¶]cocaine, [**]psychostimulants with abuse potential,[††] and synthetic opioids other than methadone[§§,¶¶]—United States, 2013–2019. (Adapted from Mattson et al. (2021))

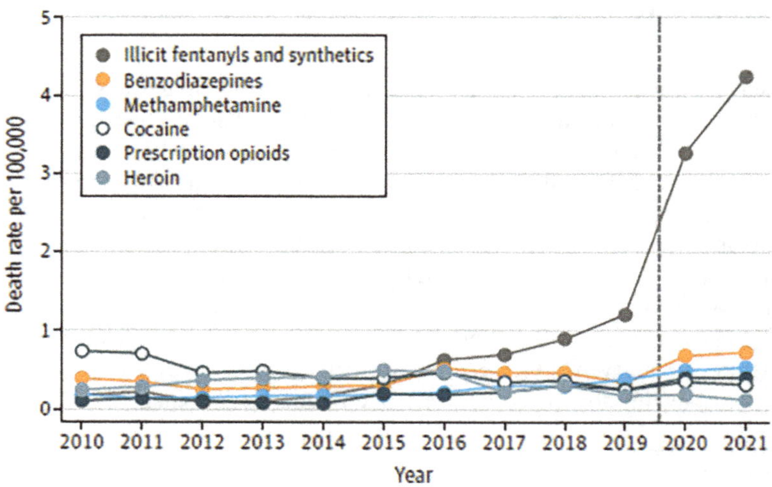

Fig. 2 Adolescent overdose deaths, 2010–2021: Overdose mortality among adolescents by substance type. (Adapted from Friedman et al. (2022))

the data in Figs. 1 and 2 make clear, the majority of these drug poisoning deaths in both adults and adolescents were driven by opioids and increasingly by a sharp rise in the availability of synthetic opioids (Friedman et al., 2022; Mattson et al., 2021).

However, these data do not reflect a new phenomenon. In November 2018, data from the US Centers for Disease Control and Prevention (Murphy et al., 2018) reported two alarming data points: that life expectancy in the United States decreased between 2016 and 2017; and that drug overdose deaths were a significant contributing factor to this. Moreover, overdose deaths from synthetic opioids, which increased by 45% in that time period, appear to be leading this trend. Since that initial report, the trend has continued and has only been exacerbated by the COVID-19 pandemic (Hedegaard et al., 2021; see Fig. 3). Nearly 4 years later in 2022, the CDC (2022) reported further declines in life expectancy from 2020 (77.4 years) to 2021 (76.4 years), with approximately one in six deaths attributable to accidental/unintentional injuries, and at least half of those deaths attributable to unintentional drug poisonings (i.e., "overdoses"). It is possible that the intersecting pandemics of the COVID-19 pandemic and unrelenting rates of opioid use disorder will constitute one of the most deadly syndemics in human history. Only time and data will tell.

These reports, along with the panoply of media coverage they have spawned, have painted a stark picture of the ways in which opioids have wreaked havoc on human lives. However, missing from these reports and the media descriptions of them is any discussion of two fundamental questions: Why? and How? These questions necessitate science and not speculation.

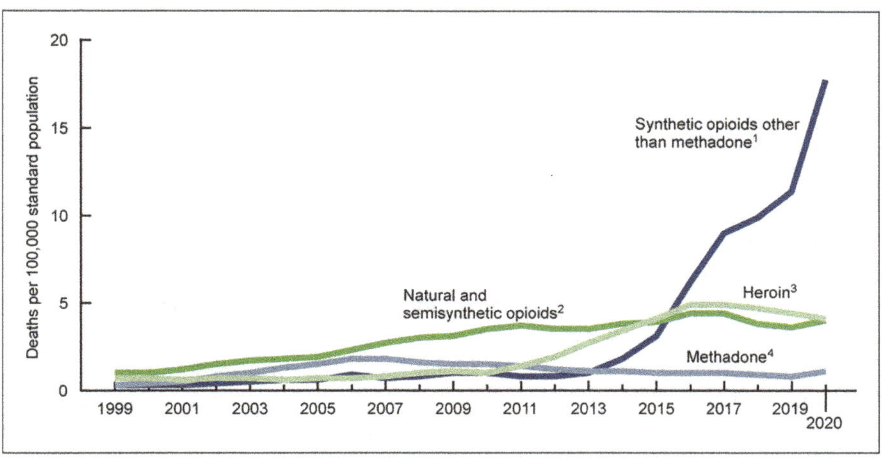

[1]Significant increasing trend from 1999 through 2020, with different rates of change over time, $p < 0.05$.
[2]Significant increasing trend from 1999 to 2010, and stable trend from 2010 through 2020, $p < 0.05$.
[3]Significant increasing trend from 2005 to 2016, with different rates of change over time, and significant decreasing trend from 2016 through 2020, $p < 0.05$.
[4]Significant increasing trend from 1999 to 2006, with different rates of change over time, significant decreasing trend from 2006 through 2017, and stable trend from 2017 through 2020, $p < 0.05$.
NOTES: Drug overdose deaths are identified using the *International Classification of Diseases, 10th Revision* (ICD–10) underlying cause-of-death codes X40–X44, X60–X64, X85, and Y10–Y14. Drug overdose deaths involving selected drug categories are identified by specific multiple-cause-of-death codes: any opioid, T40.0–T40.4 and T40.6; heroin, T40.1; natural and semisynthetic opioids, T40.2; methadone, T40.3; and synthetic opioids other than methadone, T40.4. Deaths involving more than one opioid category (such as a death involving both methadone and a natural or semisynthetic opioid) are counted in both categories. The percentage of drug overdose deaths that identified the specific drugs involved varied by year, ranging from 75%–79% from 1999 through 2013 and increasing from 81% in 2014 to 94% in 2020. Access data table for Figure 4 at: https://www.cdc.gov/nchs/data/databriefs/db428-tables.pdf#4.
SOURCE: National Center for Health Statistics, National Vital Statistics System, Mortality.

Fig. 3 Age-adjusted drug overdose death rates involving opioids, by type of opioid: United States, 1999–2020. (Adapted from Hedegaard et al. (2021))

Endogenous Opioids: The Journey Ahead

Our goal in developing this book is to provide a thoughtful and scientifically based exploration of the multifaceted world of opioids, with a distinct emphasis on endogenous opioids, e.g., endorphins, enkephalins, and dynorphins. At the time of this writing, the term "opioid" has achieved household name status, commonly connected to the phrases "opioid epidemic" or "opioid crisis". Data from Google Trends alone suggest a five- to ten-fold increase in interest based on search patterns worldwide over the past decade. There is no shortage of interest in this topic. With the most recent data from CDC described above, there is likely to be enduring interest for some time.

Despite this increase in interest, the general public's understanding of opioids, and especially endogenous opioids, has not kept pace. There are understandable reasons for this. References to opioids, especially in popular culture, often focus on medications (e.g., hydrocodone) or substances to which people become addicted (e.g., heroin), which are exogenous opioids. Rarely is there any discussion of what happens when those molecules interact with the body. Moreover, there is even less consideration of the innate opioids naturally produced by the body—the native system engaged by exogenous opioids. It is impossible to explain or understand one without the other; yet, attempts to do so happen all too often. Failing to consider the dynamic nature of opioids, "inside and out," so to speak, can have alarming consequences at the regulatory, policy, societal, and community levels, all of which can lead to devastating consequences at the individual level. Consider misguided regulation that results from a misunderstanding of the science of opioid use disorder or drug policy based more on subjective moral reasoning than scientific reasoning, e.g., the blockade or outlawing harm reduction programs. Consider community investment in prevention and intervention programs that "should" work based on a misunderstanding of the underpinnings of opioids and opioid use disorder. As the data above indicate, any obstacles to effective intervention can cost lives.

At the same time, there are health conditions, from eating disorders, to breast cancer, to depression, to obesity, in which opioids play a role, but are often not considered—either in clinical treatment or clinical research. There are places in the system of healthcare services in which endogenous opioids play a role through placebo (and nocebo) effects, but to which are rarely if ever attended. Debates about the rising increase in Cesarean sections and increasing reliance on induction of labor in childbirth abound, yet rarely are the important roles of endogenous opioids mentioned. Managing pain, especially chronic pain, is a common corollary discussion of the "opioid crisis," yet the discussion of what actually happens in the body of someone with chronic pain being managed with exogenous opioids is nearly absent. In sum, the discussion of endogenous opioids is a boat we are missing across many territories.

This book is aimed at starting the conversations we should be having about opioids. We have conceptualized the range of this book to be from basic science to clinical applications, realizing that there cannot be one without the other. The research literature on the basic science of opioids—their structure, their metabolism, their physiological transactions—is fascinating in its own right; it is even more compelling when considered in the context of human interactions and clinical interventions. Throughout the chapters of this book, our contributors have interwoven discussions of these basic components of endogenous opioids in their respective topics.

Neither opioids nor opioid research is new in any sense. Opioids were discovered, and quickly put to use, millennia ago. We have had an understanding of the effects of opioids for centuries, which has shaped human history in fascinating ways. Indeed, across several eras and generations of human progenitors, opioids have contributed to cross-cultural relationships, the establishment of intercontinental trade routes, warfare, revolutions in pain management, and, more recently, public health crises and epidemics.

What is new is our nuanced understanding of why and how opioids yield their effects vis-à-vis endogenous opioid activity. The historical emphasis on exogenous opioids and opiates has led to an almost complete neglect of endogenous opioids. This neglect is problematic in that, by ignoring what is happening in the brain and body, efforts to reduce the adverse effects of exogenous opioids (whether at the public health, legislative, public policy, or intervention levels) have been ill-informed and myopic. Opioids are a two-sided coin, and both sides must be given equal weight. However, the relationship between these two sides is of paramount importance, as it portends an opportunity for translational science. The distinct bias in media, entertainment, and even in research toward exogeny overlooks the dynamic roles that opioids play in human lives. Those roles have a broad range: from how we respond biologically to falling in love, to how we respond physiologically to falling down; from how we develop substance use disorders, to how we could optimize treatments; from how we manage depression through engaging in exercise, to how we engage with others in helpful ways; from the pain we experience from fear, to fear we experience from pain. Endogenous opioids are ubiquitous and are part of what makes humans human. We believe that the chapters within this volume reflect long-missing pieces of the dialogue about opioids.

Why the Road Turns: Reasons the Research Focus Has Changed Over Time

There are understandable reasons that research on endogenous opioids has languished, and that potentially productive lines of research have turned to other targets. Most notably, endogenous opioids are difficult to study with reliability and precision. While improved technology has made headway in this direction, the rapid catabolism of endogenous opioid molecules has historically made them difficult to locate and image. They are perhaps the "yeti" of neurophysiology, albeit with far more empirical evidence for their existence. Scientific investigations of endogenous opioids require technological capability, resources, and persistence, all of which may prove difficult to allocate during an era in which funders often emphasize short-term return on investments. Thus, scientific efforts have turned to other biological targets that are ostensibly lower hanging fruit.

Fortunately, research on endogenous opioids has persisted. However, examination of the recent history of endogenous opioid research reveals that studies of this system have been distributed across subgenres of science, and that findings were largely "siloed" for decades. Nonetheless, there is clear evidence of an interest in endogenous opioid research across disciplines. Such evidence derives from the range of journals and scientific disciplines represented in the endogenous opioid literature as a whole. The datasets in different disciplines are disparate and disconnected, which is a disservice to the applied value that integrating these lines of research would have. This book is aimed at breaking down those siloes, and each of our authors has contributed to this mission by integrating literature reviews, or presenting translational science approaches, within their respective chapters.

A Synthesis of Parallel Paths

This volume represents an effort to develop a cross-disciplinary lens on a trans-systemic phenomenon. Our chapter contributors extend scientific branches to the edges of what is known about endogenous opioids, while at times proposing innovative next steps to advance our knowledge base. We have conceptualized this work as a synthesis of parallel paths. Each line of research is distinct in focus, methodology, and applicability. However, all share a common purpose—advancement of the human condition through an understanding of endogenous opioids. In the last two centuries, opioids have traversed from causes of warfare to sources of human welfare. This is an encouraging metamorphosis indeed.

No single field has produced all the data on endogenous opioids; all the data there are come from nearly all of biological science. In science, there is strength in unity: strength to replicate findings, strength to translate findings into applications, and strength to test translational applications of basic science. From its inception, this volume has been aimed at strengthening this field of research to make the thousands of papers that have been published on the endogenous opioid system as meaningful and impactful as possible. One volume cannot accomplish all the unification that is needed, but we saw importance in taking this necessary step toward doing so.

Organization and Structure of the Volume

Endogenous Opioids: From Basic Science to Biopsychosocial Applications spans three broad themes, which are interwoven throughout the book: the roles of endogenous opioids in health-related functioning; the roles of endogenous opioids in pathology; and clinical applications and interventions of endogenous opioid science. Across the themes within this book, the authors provide thorough syntheses of research demonstrating the many ways in which endogenous opioids serve as vehicles for preservation of health. Our volume launches with a flagship chapter on neurobiology by Tache and colleagues, providing an extensive exploration of the molecular structure of endogenous opioid peptides, and their multitude of receptor types. The chapter by Barenz and colleagues outlines the role that endogenous opioids have in facilitating pleasant emotional states, which are critical to human functioning and health. Rusu describes a creative model that places endogenous opioids at the heart of altruistic behaviors associated with volunteering. Capitalizing on cardiovascular research, Dr. Cristina Sirbu describes the cardioprotective effects of endogenous opioids. Dr. Alan Goldfarb and colleagues make an empirically sound case for the involvement of endogenous opioids, and especially endorphins, in hypoalgesia associated with physical exercise. The chapter by Felicione and colleagues explores the role of endogenous opioids in pain and fear conditioning. Spanning the full range of pleasure to pain, Marjan Khajehei galvanizes a new lens on reproductive health in her review of the involvement of endorphins in childbirth and sexual functioning.

Some chapters address endogenous opioids with a more pathocentric focus, with chapters identifying opioid-related mechanisms involved in pathology. Dr. Lindsay Acree makes an impactful contribution to this discussion by describing the molecular intricacies of opioid use disorder. Sessle's chapter on craniofacial pain places endogenous opioids on the proverbial hook and in the hotseat for evaluation of their contributions. Chapters by Flores, Stephano, and Albu turn the lens on endogenous opioids as prime biological suspects in disordered eating.

Our book also shines a hopeful light on the contributions endogenous opioids make to alleviating suffering. The contribution by Pettrey and colleagues updates our current knowledge base about the endogenous opioids as a mechanism through which exercise may alleviate depression. A far-reaching explanation of placebo and nocebo effects is presented in the chapter by Kerr and Gregg, elucidating the ways in which these well-established effects could be harnessed. Dr. David Nguyen's chapter poses thoughtful questions about how the body's own defenses can be weaponized against breast cancer and proposes several novel experiments that would clearly advance our ability to capitalize on these processes to save lives. The chapter by Hancock and colleagues bookends Dr. Nguyen's chapter, offering a model of endogenous opioids' contribution to recovery from breast cancer, and diving deeply into a different yet complementary dimension of the breast cancer literature. Finally, Dr. Karin Westlund-High and her team describe their preclinical research examining enkephalin as an intervention for temporomandibular joint pain, with encouraging preliminary findings.

For now, we turn to the journey ahead, and we begin down the winding road through research past, present, and planned.

References

Ahmad, F. B., Cisewski, J. A., & Anderson, R. N. (2022). Provisional mortality data—United States, 2021. *Morbidity and Mortality Weekly Report, 71*(17), 597–600.

Centers for Disease Control and Prevention, National Center for Health Statistics. (2022, August 31). *Life expectancy in the U.S. dropped for the second year in a row in 2021.* Retrieved from: https://www.cdc.gov/nchs/pressroom/nchs_press_releases/2022/20220831.htm

Friedman, J., Godvin, M., Shover, C. L., Gone, J. P., Hansen, H., & Schriger, D. L. (2022). Trends in drug overdose deaths among US adolescents, January 2010 to June 2021. *JAMA, 327*(14), 1398–1400.

Hedegaard, H., Miniño, A. M., Spencer, M. R., & Warner, M. (2021). *Drug overdose deaths in the United States, 1999-2020. NCHS Data Brief, no 428.* National Center for Health Statistics.

Mattson, C. L., Tanz, L. J., Quinn, K., Kariisa, M., Patel, P., & Davis, N. L. (2021). Trends and geographic patterns in drug and synthetic opioid overdose deaths—United States, 2013–2019. *Morbidity and Mortality Weekly Report, 70*(6), 202.

Murphy, S. L., Xu, J. Q., Kochanek, K. D., & Arias, E. (2018). *Mortality in the United States, 2017. NCHS data brief, No 328.* National Center for Health Statistics.

The Foundational Science of Endogenous Opioids and Their Receptors

Simona Tache, Patrick L. Kerr, and Cristian Sirbu

Abstract The function of endogenous opioids spans from initiating behaviors that are critical for survival, to responding to rapidly changing environmental conditions. A network of interconnected systems throughout the body characterizes the endogenous opioid system (EOS). EOS receptors for beta-endorphin, enkephalin, dynorphin, and endomorphin underpin the diverse functions of the EOS across biological systems. This chapter presents a succinct yet comprehensive summary of the structure of the EOS, EOS receptors, and their relationship to other biological systems.

Keywords Endogenous opioids · Systems · Peptides · Beta endorphins · Dynorphins · Enkephalin

Introduction

Opioids were in use long before they were understood, and opioid receptors were being engaged by exogenous opioids far prior to their identification and description in scientific literature. Like many other facets of human history, the history of opioids predates the history of science. For millennia, human societies used morphine (derived from poppy plants, or mixed in tinctures of paregoric, theriaca, and others) to treat ailments such as pain or diarrhea, as well as to induce pleasure. It would take

S. Tache
Iuliu Haţieganu University of Medicine and Pharmacy, Cluj-Napoca, Romania

P. L. Kerr (✉)
West Virginia University School of Medicine-Charleston, Charleston, WV, USA
e-mail: pkerr@hsc.wvu.edu

C. Sirbu
Center for Cancer Research, Charleston Area Medical Center, Charleston, WV, USA

P. L. Kerr et al. (eds.), *Endogenous Opioids*, Advances in Neurobiology 35,
https://doi.org/10.1007/978-3-031-45493-6_2

hundreds of years, and the development of the scientific method, before the molecule responsible for these ostensibly miraculous effects would be isolated in the early nineteenth century through work by Friedrich Sertuner (Krishnamurthi & Rao, 2016) and by Derosne (1803). Belying the most nascent conceptualization of this discovery, Derosne (1803) originally referred to morphine as "the salt of opium." Another century would pass before the molecular structure of morphine was characterized in the early 1920s by the Nobel laureate chemist Robert Robinson and his research group (Cahn & Robinson, 1926).

A full account of the history of endogenous opioids research is beyond the scope of this chapter, but is summarized in Fig. 1 by Benyhe et al. (2015). Endogenous opioids were first discovered in the mid-1970s, when opioid receptors were discovered and located within the central nervous system and the peripheral tissue. As depicted by Benyhe et al. (2015), progress in identifying and then understanding endogenous opioids, their receptors, their functions, and their structure, has been gradual. This scientific work has been punctuated by occasional breakthroughs and

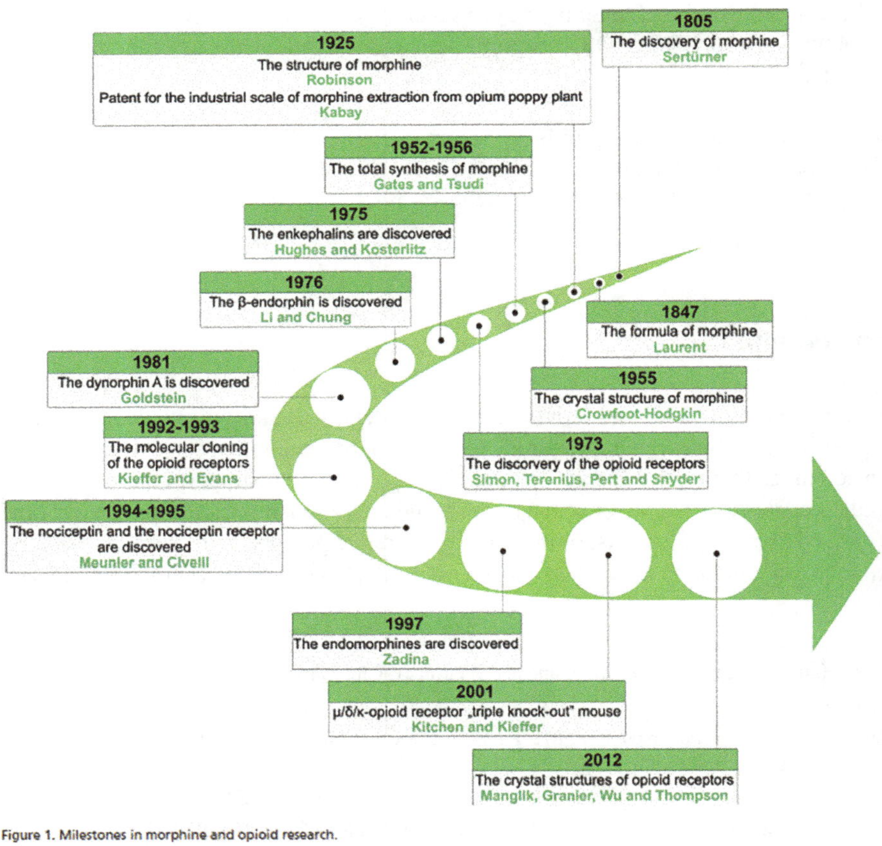

Fig. 1 Milestones in morphine and opioid research. (Adapted from Benyhe et al. (2015))

discoveries. It is worth noting that morphine itself represented a significant discovery in the field of organic chemistry—it was the first alkaloid ever identified.

Together, over centuries of research, the aforementioned work has laid the foundation for this chapter. The goal of this chapter is not to explicate all of the myriad nuances, debates, and controversies of this line of scientific research; some of these are addressed in subsequent chapters in this volume. For a more thorough catalogue of such nuances, the reader is directed to the most recent series edition on opioids and behavior by Bodnar (2022), as well as to the references for the contemporary and historical work by prior researchers noted in the introduction to this book. Conversely, our aim in this chapter is to straightforwardly describe the present state of knowledge regarding the structures, functions, and interactions of endogenous opioids and their receptors.

Basic Scientific Foundations of Endogenous Opioids

Foundations 1: Nomenclature, Terminology, and Typology

At the outset of this chapter, some basic terminology must be established. First, although often erroneously used interchangeably, *opioids* and *opiates* are different. The term *opioid* (which includes the suffix –oid, meaning resembling or like) refers to any chemical (natural, semisynthetic, or fully synthetic) that engages opioid receptors in the body and brain, and which attenuates or eliminates pain signals. The related term *opiate* (with the suffix –ate meaning to act on in a specific way), refers specifically and exclusively to naturally occurring opioids, e.g., morphine, heroin, and codeine. Thus, opioids include natural, exogenous substances such as opium, morphine and codeine; synthetic and semisynthetic substances; and natural endogenous substances.

In the nearly five decades since the discovery of the first endogenous opioids in the mid-1970s, researchers have made significant progress in identifying a full range of EOS molecules and receptors. The currently known endogenous opioids include: β-endorphin, enkephalin, dynorphin, and endomorphin. An additional "orphan" peptide (nociception/Orphanin) was also identified relatively recently; because of its actions, it is sometimes conceptualized as an "antiopioid" (Cox et al., 2015). See Tables 1, 2, 3, 4 and 5.

Enkephalins

Enkephalins (derived from the Greek word *enkephalos,* or "brain") are smaller peptides involved in the regulation of analgesia in the body. Two types of enkephalins have been identified in the human body. These are the pentapeptides methionine [met-]enkephalin and leucine [leu-]enkephalin). Met-enkephalins are endogenous

Table 1 Main location of opioid peptides

Opioid peptides	Distribution
Enkephalins	Brain: hypothalamus, brain stem, periaqueductal gray Spinal cord: substantia gelatinosa Nerve endings in the gastrointestinal tract and carotid body Retina
β-Endorphine	Brain: hypothalamus, thalamus, brain stem (nucleus of solitary tract) Pituitary gland Gastrointestinal tract Lung, placenta, retina
Dynorphins	Brain: hypothalamus, periaqueductal gray, rostroventral medulla Spinal cord: substantia gelatinosa Posterior pituitary gland (Dynorphin 1–8) Duodenum (Dynorphin 1–17)
Endomorphins	Brain: cortex, amygdale, thalamus, hypothalamus, striatum Brain stem (nucleus accumbens, nucleus of solitary tract) Spinal cord Dorsal root ganglia Histaminergic neurons

opioid pentapeptides, a type of neurotransmitter found naturally in the brain, with molecular formula $C_{27}H_{35}N_5O_7S$. Leu-enkephalins are endogenous opioids pentapeptides, with molecular formula $C_{26}H_{17}N_5O_7$. These peptides are synthesized as part of larger precursor molecules (discussed below) called proenkephalin (PENK). PENK was first identified decades ago in the adrenal medulla (Birch & Christie, 1986; Quach et al., 1984). Each PENK molecule contains four met-enkephalins, one leu-enkephalin, one octapeptide, and one heptapeptide.

In addition to CNS and PNS distribution noted in subsequent sections of this chapter, PENK expression has been found in various tissues in humans. These include heart, smooth and skeletal muscle, kidney and intestinal tissues. Enkephalins are metabolized primarily by two peptidases. The first is enkephalinase A, which splits the Gly-Phe bond. The second is enkephalinase B, which splits the Gly-Gly bond. Aminopeptidase splits the Tyr-Gly bond and contributes to their metabolism. From PENK are synthesized other potent opioid peptides, such as peptide E,F and BAM P 22,20,12, and nonopioid peptides such as synenkephalin, peptide I, and peptide B are also described (Rossier, 1993).

Both the mechanisms of action and the functional roles of enkephalins are diverse. Enkephalins act through both MOPs and DOPs. Through these pathways, enkephalins serve a variety of functions. These include functioning as neurotransmitters in the brain, pain regulation and modulation, cardiac and respiratory functions, immunological functions, ischemic tolerance (e.g., myocardial ischemia and/or angina), mediating the effects of alcohol, and altering emotional responses. Thus, the ways in which enkephalin activity affects the brain, body, behavior, emotion, and cognition goes well beyond the concept of pain that may most immediately come to mind when one thinks of endogenous opioids.

Table 2 Opioid peptides, their precursors and location, and structures

Opioid peptides	Precursor	Location	Structures
Met-enkephalin	Proenkephalin (PENK)	Adrenal medulla Brain	Tyr-Gly-Gly-Phe-Met_5
Leu-enkephalin			Tyr-Gly-Gly-Phe-Leu_5
Octapeptide			Tyr-Gly-Gly-Phe-Met-Arg-Gly-Leu_8
Heptapeptide			Tyr-Gly-Gly-Phe-Met-Arg-Phe_7
β Endorphin	Proopiomelanocortin (POMC)	Anterior and intermediate pituitary gland lobes Brain	Tyr-Gly-Gly-Phe-Met-Thr-Ser-Glu-Lys-Ser-Gln-Thr-Pro-Leu-Var-Thr-Leu-Phe-Lys-Asn-Ala-Ile-Val-Lys-Asn-Ala-His-Lys-Lys-Gly-Glu_{31}
Dynorphin 1–8 (B)	Prodynorphin (PDYN	Posterior pituitary gland Hypothalamus Duodenum Hypothalamus Gastrointestinal tract Hypothalamus Gastrointestinal tract	Tyr-Gly-Gly-Phe-Leu-Arg-Arg-Ile_8
Dynorphin 1–17 (A)			Tyr-Gly-Gly-Phe-Leu-Arg-Arg-Ile-Arg-Pro-Lys-Leu-Lys-Trp-Asp-Asn-Gln_{17}
α-Neoendorphin			Tyr-Gly-Gly-Phe-Leu-Arg-Lys-Tyr-Pro-Lys_{10}
β-Neoendorphin			Tyr-Gly-Gly-Phe-Leu-Arg-Lys-Tyr-Pro_9
Endomorphins Endomorphins 1	Unknown	Brain Upper brainstem	Tyr-Pro-Trp-Phe-NH_2
Endomorphin 2		Spinal cord Lower brain stem	Tyr-Pro-Phe-Phe-NH_2

Endorphins

Endorphins (an English neologism derived from a combination of the words "endogenous" and "morphine") are endogenous opioid polypeptides, composed of 31 amino acids. They are produced in the pituitary gland and in the hypothalamus in vertebrates during stress, strenuous exercise, pain, and orgasm-induced excitement. Four types of endorphins are produced in the human body: alpha (α), beta (β), gamma (γ), and sigma (σ). Each of these has different numbers and types of amino acids in their molecules, ranging from 16 to 31 amino acids in each molecule.

These peptides are synthesized from the precursor molecule proopiomelanocortin (POMC). POMC is a large precursor molecule found in anterior and intermediate lobes of the pituitary gland in the brain. POMC stimulates β-endorphins. β-endorphins are the most powerful endogenous opioid peptides and are found in the neurons of the CNS (hypothalamus, pituitary gland, limbic system) and PNS.

The mechanisms of action for endorphins are generally well-established (McNally & Akil, 2002). Endorphins act through opiate receptors; β endorphins have the highest affinity for μ_1 opioid receptor, a slightly lower affinity for μ_2 and σ opioid receptors, and a low affinity for the κ_1 opioid receptors.

Table 3 Opioid receptors location and responses by stimulation

Receptor	Central nervous system location	Endogenous ligands	Effects produced by stimulation
μ	Brain (laminae III and IV of the cortex, thalamus, periaqueductal gray) Spinal cord (substantia gelatinosa)	Endomorphins B-Endorphin Dynorphins Enkephalins	Supraspinal analgesia Physical dependence Action site of morphine Respiratory depression Euphoria Reduced gastrointestinal motility Meiosis Increased secretion of growth hormone and prolactin
κ	Brain (hypothalamus, periaqueductal gray, claustrum) Spinal cord (substantia gelatinosa)	Dynorphin A	Spinal analgesia Sedation Meiosis Inhibition of antidiuretic hormone release and diuresis
δ	Brain (pontine nucleus, amygdale, olfactory bulbs, deep cortex)	Enkephalins β-Endorphin Dynorphins	Analgesia Euphoria Physical dependence

Table 4 Effector systems coupled through G proteins to opioid receptors

Receptor type	Effector
μ, δ, κ	Inhibit adenylyl cyclase and ↓cAMP
μ, δ	Increase K^+ conductance
Δ	Decrease Ca^{2+} conductance
κ	Stimulate IP_3

Similar to enkephalins, endorphins have diverse roles in health and pathophysiology. First, endorphins act as neurotransmitters in CNS and PNS. Some data indicate they may function protectively to prevent obesity (see Flores & Zuniga, Chap. 18 of this volume), and diabetes, potentially via metabolic control assisting in glucose and lipid homeostasis. Considering the emotional effects of endorphins, as described by Pettrey and colleagues (chapter "Physical Exercise as an Intervention for Depression: Evidence for Efficacy and Mu-Opioid Receptors as a Mechanism of Action", this volume), endorphins may also play a role in mitigating psychiatric diseases, especially depression. As discussed by Felicione and colleagues (chapter "Pain, Fear, Anxiety, and Stress: Relations to the Endogenous Opioid System", this volume), endorphins may downregulate anxiogenic factors. In a similar vein as that observed for enkephalin, and as discussed by Acree (chapter "Endogenous and Exogenous Opioids: Role in Substance Use Disorders", this volume), endorphins may mediate the effects of alcohol and other substances. Endorphins are also involved in the release of many sex hormones, which is discussed in careful detail by Khajehei (chapter "Endorphins, Sexuality, and Reproduction", this volume). As noted early on in research on endorphins, these peptides appear to be involved in immunological functioning (Fischer & Falke, 1984; Panerai & Sacerdote, 1997).

Table 5 Opioid receptors and their ligands

Receptor	Endogenous ligands	Selective agonist	Exogenous ligands antagonist	Ligands agonist	Nonselective ligands antagonist
μ or previous IUPHAR name OP_3 IUPHAR name MOP (MOR)	Endorphins Endomorphins	DAMGO Methadone Fentanyl Dermorphin	CTOP	Levorphanol Etorphine	Naloxone Naltrexone B-funaltrexamine
κ or kappa previous IUDHAR name OP_2 current IUDHAR name KOP(KOR)	Dynorphin A	Spiradoline $U_{50,488}$	Nor-BNI	Levorphanol Etorphine EKC	Naloxone Naltrexone
δ or delta previous IUPHAR: OP_1 name current IUPHAR name DOP(DOR)	Enkephalins	DPDPE Deltorphin DSLET	Naltrindole NTB BNTX-7	Levorphanol Etorphine	Naloxone Naltrexone

Adapted from Koneru et al. (2009)
BNTX-7 benzylidenenaltroxone, *EKC* etylketocyclazosine, *NBT* benzofuran ethyl, *Nor-BNI* nor-binaltrophimine, *DAMGO* [D-Ala2,MePhe4,Gly(ol)5] enkephalin, *DPDPE* [D-Pen 2,D-Pen 5] enkephalin, *DSLET* O[D-Ser2,Leu5] enkephalin -Thr 6, *CTOP* D-Phe-Cys-Tyr-D-Trp-Orn-Thr-Pen-Thr-NH_2, *OP* opioid receptors, *MOR* μ opioid receptors, *KOR* κ opioid receptors, *DOR* δ opioid receptors

Perhaps most well-known of all mechanisms of action and outcomes is that endorphins act as analgesic factors that modulate/regulate pain.

Dynorphins

Dynorphins are produced in different parts of the brain, including the hypothalamus, hippocampus, midbrain, medulla, and pons. They are also found in the spinal cord. Dynorphins have 13 amino acid length protein. These peptides are synthesized from the precursor prodynorphin (PDYN). Proprotein convertase (PC 2) cleaves PDYN in active peptides: Dynorphin A and Dynorphin B. Dynorphins primarily exert their effects through the KOP receptors, and secondarily through MOP and DOP receptors.

Dynorphins act as modulators of pain responses. Research also implicates these opioid peptides in maintenance and return to homeostasis agent through appetite control, which may also account for their proposed role in weight maintenance. There is some evidence of dynorphins' role core biological processes related to basic survival, including in circadian rhythm regulation, as well as biothermal regulation.

Endomorphins

Endomorphins are a relatively new addition to the family of endogenous opioids. Early work by the research group spearheaded by James Zadina and Lazslo Hackler (Hackler et al., 1997; Zadina et al., 1997) first identified and characterized these peptides. Zadina et al. (1999) state that the two new peptides (Tyr-Pro-Trp-Phe-NH2 and Tyr-Pro-Phe-Phe-NH2) were endogenous peptides that had a high and selective affinity for MOP receptors. Therefore, they were given the names endomorphin-1 and -2 (EM1 and EM2).

EM1 and EM2 are C-terminally amidated tetrapeptides, with a high affinity and specificity for MOP receptors. As noted above, the N-terminal message sequence of EM1 is defined by two Tyr and Trp. These pharmacophoric amino acid residues are necessary for recognition of MOP receptors. The N-terminal sequence for EM2 is similar to EM1, except that it includes Phe (instead of Trp, as in EM1). For both EM1 and EM2, the N-terminal sequence composition also includes a spacer (Pro), which links the pharmacophoric residues.

Endomorphins yield diverse effects, and may regulate a variety of physiological functions and associated behaviors. These include both sedative and arousal behaviors; pain perception; responses to stress, reward, arousal and vigilance; autonomic, cognitive, neuroendocrine and limbic homeostasis. Additionally, EM1 and EM2 may modulate the secretion of multiple neurotransmitters, including dopamine (e.g., Bagosi et al., 2006; Ukai & Lin, 2002a), norepinephrine (e.g., Al-Khrasani et al., 2003; Rialas et al., 1998), serotonin (e.g., Tao & Auerbach, 2002), and acetylcholine (e.g., Patel et al., 1999; Ukai & Lin, 2002b). Finally, endomorphins may modulate the release of the neurohormones oxytocin and vasopressin (Doi et al., 2001). For an extensive discussion of these roles of endomorphins, we direct readers to the excellent authoritative review by Fichna et al. (2007).

Opioid Receptor Types and Subtypes

In addition to the discovery of the endogenous opioid ligands themselves, the discovery of receptors for these molecules reflected a significant advancement in understanding pain signaling, and disruptions in pain signaling, including chronic pain and opioid use disorder. The opioid receptors, together with the endogenous opioids, form the so-called endogenous opioid system. This system produces diverse effects and functions throughout the body (see Fig. 2) via wide distribution throughout the CNS (see Fig. 3)

The opioid receptors are characterized by a heptahelical structure, with 400 aa in μ, 372 aa in δ, and 380 aa in κ. These receptors belong to the super-family of the G-protein-coupled receptors (GPCRs). The GPCRs are located in the cell membrane and most of them are activated by molecules outside the cell to trigger signal, transduction pathways inside the cell. The binding of opioid peptides to these receptors initiates biochemical events and various effects.

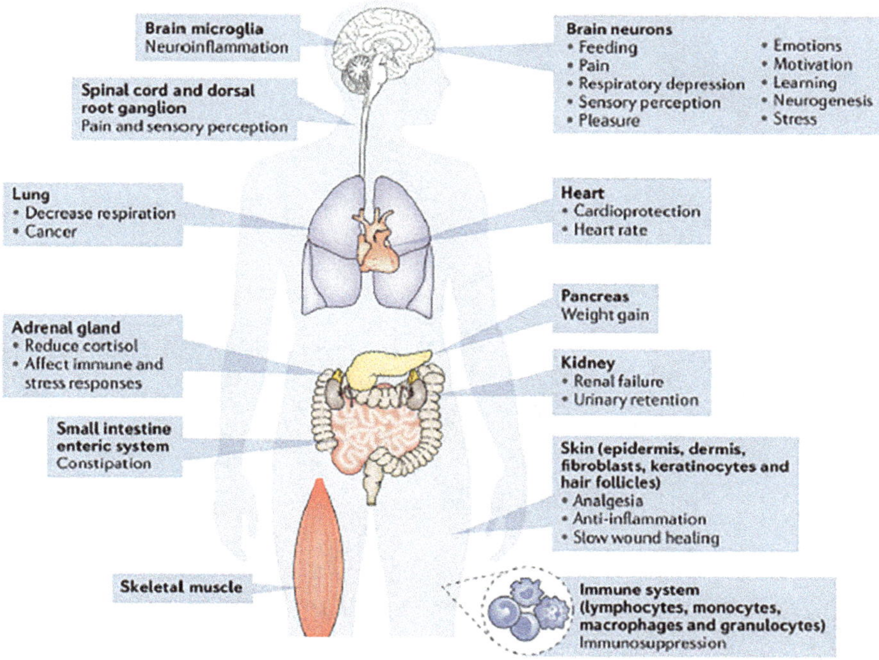

Fig. 2 Opioid actions throughout the body. (Adapted from Kibaly et al. (2019). Used with permission)

As characterized by Simon and Gioannini (1993), methods for separation and purification of binding sites of opioid receptors consist of:

- Solubilization
- Physical separation
- Affinity cross-linking
- Partial purification
- Purification to homogeneity

The three G-protein-coupled opioid receptors are mu (μ; MOP), kappa (κ; KOP), delta (δ; DOP), and nociceptin (NOP). As discussed by Cox et al. (2015), there has been some debate about whether NOP/Orphanin is appropriate for classification in the endogenous opioid nomenclature. However, research has demonstrated that all four of these receptors share a highly similar crystalline structure (e.g., Granier et al., 2012; Manglik et al., 2012; Thompson et al., 2012; Wu et al., 2012; see Fig. 4a, b).

MOP receptors are characterized by their high affinity for morphine. Endogenous ligands for MOP receptors are Endomorphine-1 and Endomorphin-2. β-Endorphins, Enkephalins and Dynorphins bind to μ receptors but with lower affinity.

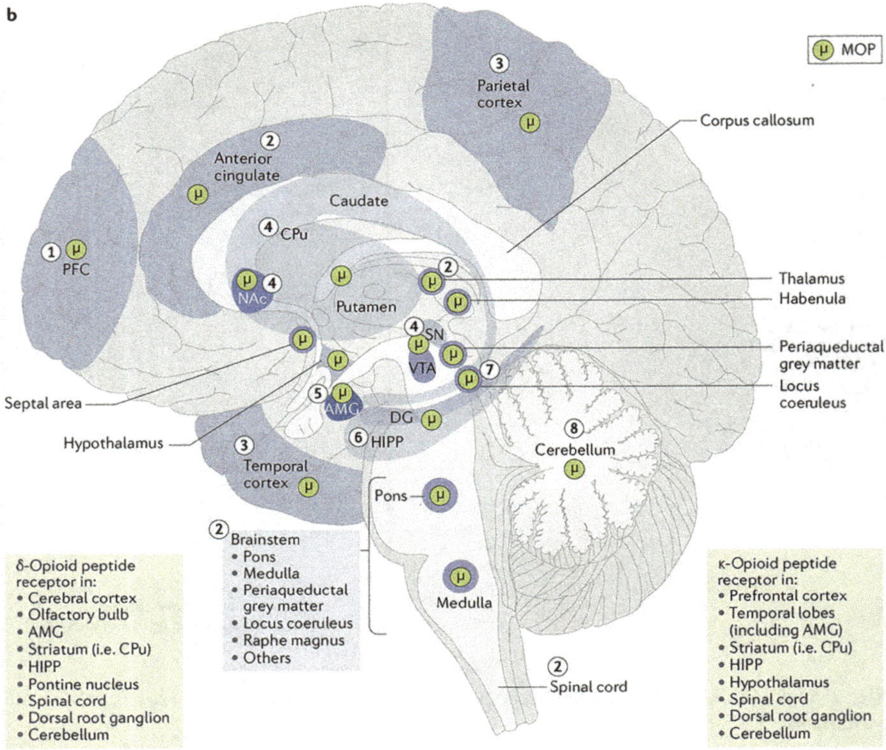

Fig. 3 Opioid actions in brain functions. (Adapted from Kibaly et al. (2019). Used with permission)

Subtypes of MOP receptors have been identified and characterized as follows:

- $\mu_{1:}$ with higher affinity for morphine; mediates supraspinal analgesia; selectively blocked by naloxone
- μ_2: with lower affinity for morphine, mediates spinal analgesia, respiratory depression and constipation.

Regarding KOP, several laboratories have emphasized two populations of KOP receptors. These can be differentiated based on their affinity for benzeneacetamides as: $U_{50,488}$; $U_{69,583}$; and PD 117302. They are also displaceable by KOP agonists and KOP antagonists as follows:

- κ1: sensitive sites
- κ2: insensitive sites

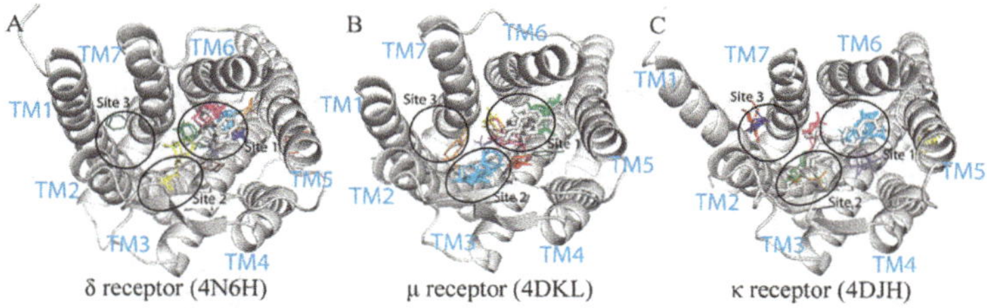

Fig. 2. FTMAP results using δ (A), μ (B), and κ (C) receptor crystal structures. Only chains A of the receptors, excluding non-receptor atoms, were uploaded to the FTMAP web server for fragment docking calculation. Receptors are shown in silver cartoon representation, ligands in the crystal structures are shown in silver sticks, and FTMAP probe clusters are shown as colored lines. The FTMAP probe clusters are colored in the order of decreasing size as follows: green, cyan, magenta, yellow, red, blue, purple, orange, dark green, and chocolate.

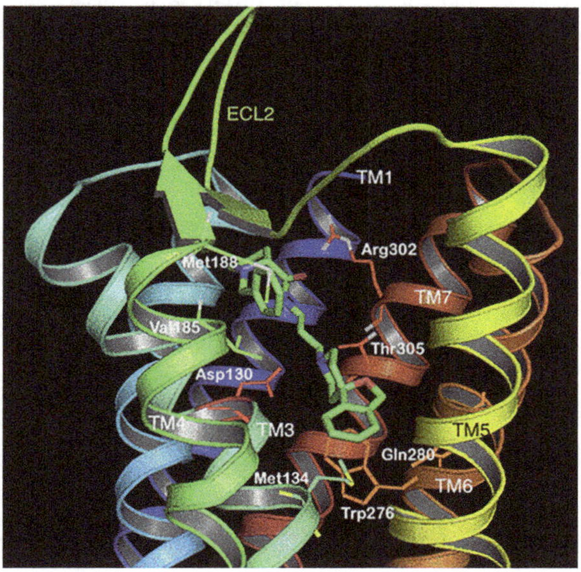

Fig. 4 Crystal structures of MOP, KOP, DOP (4a), and NOP (4b) receptors. (Shang and Filizola (2015). Used with permission). (**a**) FTMAP results using δ(**A**), μ(**B**), and κ(**C**) receptor crystal structures. Only chains A of the receptors, excluding non-receptor atoms, were uploaded to the FTMAP web server for fragment docking calculation. Receptors are shown in silver cartoon representation, ligands in the crystal structures are shown in silver sticks, and FTMAP probe clusters are shown as colored lines. The FTMAP probe clusters are colored in the order of decreasing size as follows: green, cyan, magenta, yellow, red, blue, purple, orange, dark green, and chocolate. (**b**) Molecular model of the NOP receptor crystal structure bound to NOP antagonist C-24 (green) (PDB ID: 4EA3). The TM helices are colored in seven different colors and labeled. The ECL2 loop, between TM4 and TM5, is shown in green. Side chains of amino acids interacting with the antagonist are shown as sticks and labeled. (Toll et al. (2016). Used with permission)

Opioid Receptor Distribution

MOPs are the most widely distributed of the EOS receptors, found throughout multiple structures within the central nervous system (CNS), including the amygdala (basolateral nuclear complex; Zhang et al., 2015), cerebral cortex, thalamus, periaqueductal gray (PAG), and rostral ventromedial (Wang et al., 2019). KOPs are found in a more limited range of structures: the hypothalamus and PAG. DOPs are found in the basal ganglia, pontine nucleus, and amygdala. Finally, NOPs are distributed widely throughout the brain Although there is some neuroanatomical overlap, each receptor type also occupies unique locations within the CNS. This may account for the different functions that each has been observed to serve. Specifically, MOP functions as part of a reward system, while KOP and DOP are associated with aversive states (i.e., antireward). This likely accounts for the association of MOP with the experience of pleasant affective states (e.g., euphoria), and KOP and DOP with unpleasant affective states (e.g., dysphoria). Figure 5 describes opioid engagement in the amygdala.

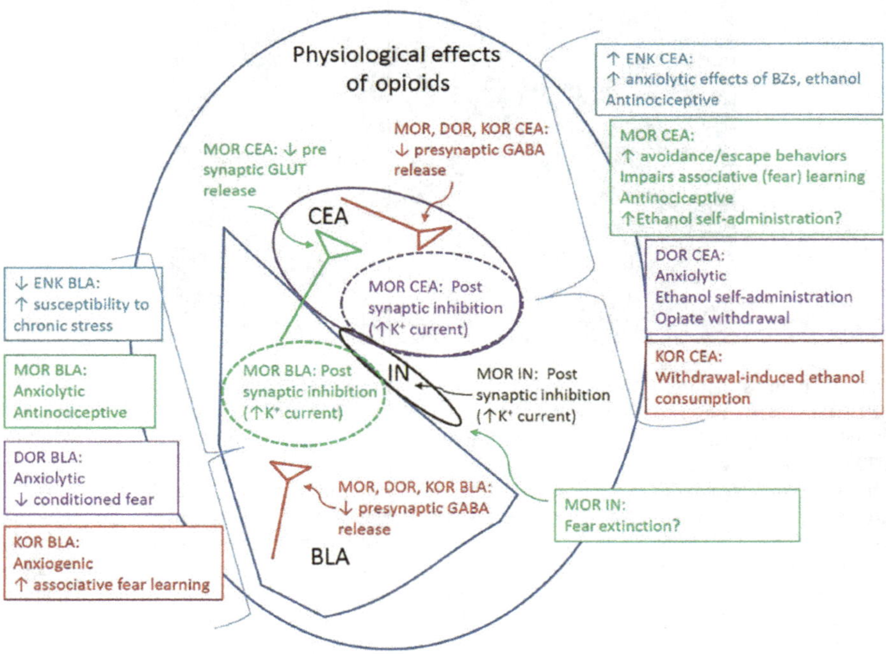

Fig. 5 Overview of opioid-mediated effects in the amygdala. Central figure shows main electrophysiological presynaptic and postsynaptic effects induced by activation of mu (MOR), delta (DOR), or kappa (KOR) receptors in the basolateral/lateral amygdala (BLA), central amygdala (CEA), or intercalated nuclei (IN). Primary functional effects of receptor activation in these regions are summarized, implicating the role of amygdala opioids in nociception, stress and anxiety-related responses, associative learning and conditioned fear, ethanol consumption or withdrawal, and opiate dependence. GLUT, glutamate. (From Wilson and McDonald (2020). Used with permission)

Foundations 2: Organic Chemistry of Endogenous Opioids

A discussion of the organic chemical foundations of endogenous opioids is warranted. Endogenous opioids (see section "Foundations 1: Nomenclature, Terminology, and Typology"), are classified as a type of compound called a peptide. Peptides are biomolecules comprised of 2–50 chains of amino acids (i.e., polypeptides). Polypeptide chains with more than 50 polypeptides linked together form the structure of proteins. At the time of this writing, there are 20 known amino acids (see Fig. 6).

Endogenous opioid peptides are small molecules that are produced in the central nervous system and in various glands such as the pituitary and adrenal glands. The traditional opioid peptides are endorphins, enkephalins and dynorphins (with the late addition of nociception/orphanin), all of which all share the Tyr-Gly-Gly-Phe amino acid sequence at the N terminus. Opioid peptides are also derived from other sources, including:

- Bovine milk for β-casomorphin-5 and -7
- Human milk for β-casomorphin-5 and -7
- Human blood for hemorphin-4 and -7
- Frog skin for dermorphin and dermenkephalin
- Bovine brain for endomorphin-1 and -2.

These peptides produce the same effects as classic alkaloid opiates, such as morphine and heroin. Furthermore, natural opioid peptides may subdivide again in typical and atypical opioid peptides; their synthetic derivatives may be subdivided in typical and atypical opioid peptide analogues.

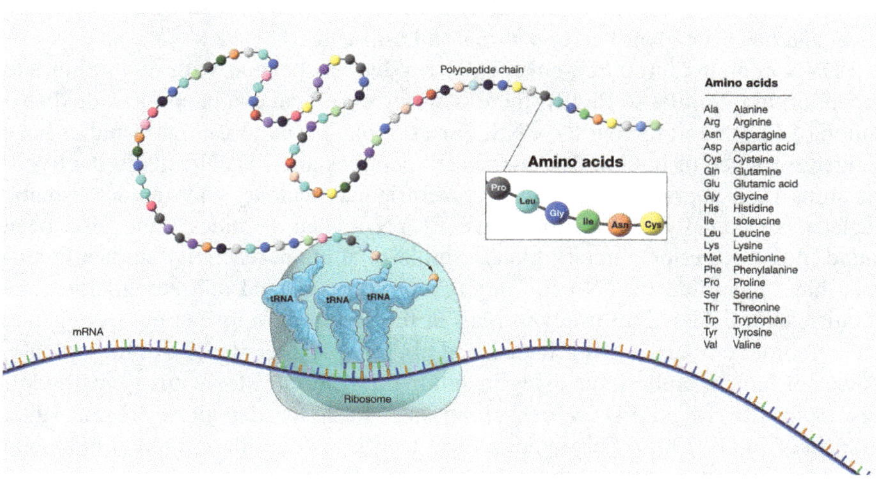

Fig. 6 Relationship between amino acids, peptides, and polypeptide chains. (National Human Genome Research Institute, 2022)

In his seminal contribution to endogenous opioid research, Teschemacher (1993) differentiated between "atypical" opioid peptides and those considered "typical." Teschemacher articulated that "typical" opioid peptides are derived from three precursor molecules only (i.e., proenkephalin, prodynorphin, and proopiomelanocortin), and all have the same N-terminal amino acid sequence (i.e., Tyr-Gly-Gly-Phe). Conversely, "atypical" opioid peptides are derived from a variety of parent proteins, Tyr residue is obligatory for their N-terminal amino acid sequence; and N-terminal extension of the sequence beyond the Tyr residue may occur (Teschemacher, 1993).

Production of Endogenous Opioids

All subtypes of endogenous opioids (see section "Foundations 1: Nomenclature, Terminology, and Typology") are derived from three precursor proteins. Enkephalin is derived from proenkephalin (PENK). Dynorphin is derived from prodynorphin (PDYN). Beta-endorphin is derived from proopiomelanocortin (POMC).

Proenkephalin was first identified in the adrenal medulla, and it is the precursor for Met-enkephalin and Leu-enkephalin in the brain. Each proenkephalin molecule contains four Met-enkephalin, one Leu-enkephalin, one octapeptide, and one heptapeptide. Neurobiological studies have yielded strong empirical evidence establishing wide distribution of PENK throughout the central nervous system, including the brain and spinal cord. This evidence has included immunoreactivity of subtypes of enkephalin (the pentapeptides methionine [met-]enkephalin and leucine [leu-] enkephalin) within several structures of the CNS, including the hypothalamus, amygdala, cerebral cortex, locus coeruleus, nucleus accumbens, caudate putamen, substantia nigra, and spinal cord among others (Cahill et al., 2021; Fricker et al., 2020). Furthermore, evidence of wide distribution of proenkephalin in the CNS is derived from the observation of PENK mRNA expression in several CNS structures, and has been identified in neurons and astrocytes.

PDYN contains three Leu-enkephalin residues associated with dynorphin and neoendorphin. Similar to PENK, there is strong empirical evidence for wide distribution of PDYN throughout the CNS. For example, immunoreactivity studies have yielded evidence of both PDYN-associated peptides and PDYN cells in the hypothalamus (i.e., supraoptic nucleus, paraventricular nucleus, and suprachiasmatic nucleus, specifically). Immunoreactive PDYN-related peptides have also been found in the posterior pituitary gland. Similarly, immunoreactivity studies in animals have identified PDYN cells throughout corticostriatal and amygdalostriatal circuits, with evidence of prodynorphin in the striatum, hippocampus, amygdala, and cerebral cortex (e.g., Khachaturian et al., 1985; Lin et al., 2006). Importantly, studies of human brain tissue using in situ hybridization histochemistry methodology have also found PDYN cells in tissues from these regions (Hurd, 1996; Nikoshkov et al., 2005). This indicates that the PDYN distribution found in animal models is likely consistent with human CNS distribution.

The wide distribution of PDYN has led to the hypothesis that it is multifunctional (Berezniuk & Fricker, 2011). Adding credence to this proposed multifunctional model, evidence of PDYN mRNA has been found in several CNS locations, including within the corticostriatal circuit, hippocampus, spinal cord, and the parabrachial and supraoptic nuclei (Lein et al., 2007). As discussed later in this volume, the potential multifunctionality of PDYN, as well as dynorphin, may account for its proposed role in a variety of pathology, including psychopathology (Tejeda et al., 2012).

POMC was identified in the anterior and intermediate lobes of pituitary gland and in the brain. POMC contains β-endorphin, a polypeptide of 31 amino and residues that have Met-enkephalin at its amino terminal.

Interactions of the Opioid Peptides

Endogenous opioid peptides interact with other neurotransmitter systems. Specifically, these molecules have complex interactions with cholinergic, dopaminergic and GABAergic pathways in the central nervous system (Wood, 1993). Within the cholinergic system, endogenous opioid peptides interact with the septo-hippocampal cholinergic pathway and the nucleus basalis-cortical cholinergic pathway. The dopaminergic pathways with which endogenous opioid peptides interact include the nigrostriatal, mesolimbic, and mesocortical pathways (Wood, 1993).

Endogenous Opioid Peptides: Functions and Interrelationships

Understanding the processing of endogenous opioid peptides is prerequisite to understanding their function. Endogenous opioid peptides are natural ligands for opioid receptors. These peptides are modified during posttranslational processing, which can involve the addition of various chemical groups to the peptides, including sugar molecules (i.e., glycosylation), acetyl groups (i.e., acetylation), phosphate groups (i.e., phosphorylation), and methyl groups (i.e., methylation).

Endogenous opioid peptides function as hormones and as neuromodulators. Endogenous opioid peptides that act as hormones are secreted into circulation by glands and delivered to a variety of distant target tissues. Endogenous opioid peptides that act as neuromodulators are produced and secreted by neurons and act in the brain and spinal cord to modulate the action of other neurotransmitter.

Endogenous opioid peptides are often produced together with other neurotransmitter molecules. Froehlich (1997) cites examples of such coproduction, including neurons in the carotid body that secrete both enkephalin and dopamine; and peripheral nervous system neurons in rodent species (e.g., rats), which simultaneously secrete norepinephrine and enkephalin.

Opioid peptides can alter the release of other neurotransmitters. They inhibit the release of acetylcholine, dopamine and norepinephrine in the brain and the release of serotonin and gamma-aminobutyric acid in the brain. The opioid peptides act as neuromodulators, charging the release of other neurotransmitters in the CNS (brain and spinal cord) and periphery, which may contribute to complex behaviors.

Opioid receptors types (μ, δ, κ) are present on both pre- and postsynaptic membranes of opioid and opioid-target neurons. Opioid neurotransmission appears to influence the processes of the central nervous system: nociception; cardiovascular regulation; extrapyramidal locomotor activity; food intake; gastrointestinal tract; sexual behavior; learning and memory; social defeat.

Because of their complex, integrated relationships with several systems in the CNS and PNS, opioid peptides are implicated either directly or indirectly in the pathophysiology of neurological and psychiatric disorders, including Alzheimer's, Parkinson's, Huntington's disease, stroke, epilepsy, brain and spinal cord injury, substance use disorders, eating disorders, depression, anxiety disorders, and schizophrenia (see Tejeda et al., 2012). The relationships between endogenous opioids and many of these conditions are addressed in subsequent chapters in this volume.

Conclusion

In this chapter, we have provided a cursory introduction to some of the foundational scientific underpinnings and principles of the endogenous opioid system. An entire book could not do justice to an area of research characterized by such enormity, much less a single chapter. Tremendous progress has been made in understanding the structures and functions of endogenous opioids and their receptors. This progress has been achieved in a relatively short duration of time, and only through immense scientific effort. The findings from the scientific work described in this brief chapter have facilitated the clinical and translational applications that have only recently been realized. Those applications and their connections to this foundational science are territories canvassed throughout the rest of this volume.

References

Al-Khrasani, M., Elor, G., Abbas, M. Y., & Rónai, A. Z. (2003). The effect of endomorphins on the release of 3Hnorepinephrine from rat nucleus tractus solitarii slices. *Regulatory Peptides, 111(1–3), 97–101.*

Bagosi, Z., Jászberényi, M., Bujdosó, E., Szabó, G., & Telegdy, G. (2006). The effects of endomorphins and diprotin A on striatal dopamine release induced by electrical stimulation—an in vitro superfusion study in rats. *Neurochemistry International, 49*(7), 665–668.

Benyhe, S., Zádor, F., & Ötvös, F. (2015). Biochemistry of opioid (morphine) receptors: Binding, structure and molecular modelling. *Acta Biologica Szegediensis, 59*(suppl. 1), 17–37.

Berezniuk, I., & Fricker, L. D. (2011). Endogenous opioids. In G. Pasternak (Ed.), *The opiate receptors* (2nd ed., pp. 93–120). Humana Press.

Birch, N. P., & Christie, D. L. (1986). Characterization of the molecular forms of proenkephalin in bovine adrenal medulla and rat adrenal, brain, and spinal cord with a site-directed antiserum. *Journal of Biological Chemistry, 261*(26), 12213–12221.

Bodnar, R. J. (2022). Endogenous opiates and behavior: 2020. *Peptides, 151*, 170752.

Cahill, C., Tejeda, H. A., Spetea, M., Chen, C., & Liu-Chen, L. Y. (2021). Fundamentals of the dynorphins/kappa opioid receptor system: From distribution to signaling and function. In

L. Y. Liu-Chen & S. Inan (Eds.), *The kappa opioid receptor* (pp. 3–21). Springer International Publishing.

Cahn, R. S., & Robinson, R. (1926). CXXVI.—The morphine group. Part IV. A new oxidation product of codeine. *Journal of the Chemical Society (Resumed), 129*, 908–912.

Cox, B. M., Christie, M. J., Devi, L., Toll, L., & Traynor, J. R. (2015). Challenges for opioid receptor nomenclature: IUPHAR review 9. *British Journal of Pharmacology, 172*(2), 317–323.

Derosne, J. F. (1803). Natural product chemistry for drug discovery. *Annali di Chimica, 45*, 257.

Doi, N., Brown, C. H., Cohen, H. D., Leng, G., & Russell, J. A. (2001). Effects of the endogenous opioid peptide, endomorphin 1, on supraoptic nucleus oxytocin and vasopressin neurones in vivo and in vitro. *British Journal of Pharmacology, 132*(5), 1136–1144.

Fichna, J., Janecka, A., Costentin, J., & Do Rego, J. C. (2007). The endomorphin system and its evolving neurophysiological role. *Pharmacological Reviews, 59*(1), 88–123.

Fischer, E. G., & Falke, N. E. (1984). β-endorphin modulates immune functions. *Psychotherapy and Psychosomatics, 42*(1–4), 195–204.

Fricker, L. D., Margolis, E. B., Gomes, I., & Devi, L. A. (2020). Five decades of research on opioid peptides: Current knowledge and unanswered questions. *Molecular Pharmacology, 98*(2), 96–108.

Granier, S., Manglik, A., Kruse, A. C., Kobilka, T. S., Thian, F. S., Weis, W. I., & Kobilka, B. K. (2012). Structure of the δ-opioid receptor bound to naltrindole. *Nature, 485(7398), 400–404.*

Hackler, L., Zadina, J. E., Ge, L. J., & Kastin, A. J. (1997). Isolation of relatively large amounts of endomorphin-1 and endomorphin-2 from human brain cortex. *Peptides, 18*(10), 1635–1639.

Hurd, Y. L. (1996). Differential messenger RNA expression of prodynorphin and proenkephalin in the human brain. *Neuroscience, 72*(3), 767–783.

Khachaturian, H., Lewis, M. E., Haber, S. N., Houghten, R. A., Akil, H., & Watson, S. J. (1985). Prodynorphin peptide immunocytochemistry in rhesus monkey brain. *Peptides, 6*, 155–166.

Kibaly, C., Xu, C., Cahill, C. M., Evans, C. J., & Law, P. Y. (2019). Non-nociceptive roles of opioids in the CNS: Opioids' effects on neurogenesis, learning, memory and affect. *Nature Reviews Neuroscience, 20*(1), 5–18.

Koneru, A., Satyanarayana, S., & Rizwan, S. (2009). Endogenous opioids: Their physiological role and receptors. *Global Journal of Pharmacology, 3*(3), 149–153.

Krishnamurti, C., & Rao, S. C. (2016). The isolation of morphine by Serturner. *Indian Journal of Anaesthesia, 60*(11), 861.

Lein, E. S., Hawrylycz, M. J., Ao, N., Ayres, M., Bensinger, A., Bernard, A., et al. (2007). Genome-wide atlas of gene expression in the adult mouse brain. *Nature, 445*(7124), 168–176.

Lin, S., Boey, D., Lee, N., Schwarzer, C., Sainsbury, A., & Herzog, H. (2006). Distribution of prodynorphin mRNA and its interaction with the NPY system in the mouse brain. *Neuropeptides, 40*(2), 115–123.

Manglik, A., Kruse, A. C., Kobilka, T. S., Thian, F. S., Mathiesen, J. M., Sunahara, R. K., ... & Granier, S. (2012). Crystal structure of the μ-opioid receptor bound to a morphinan antagonist. *Nature, 485(7398), 321–326.*

McNally, G. P., & Akil, H. (2002). Opioids peptides and their receptors: Overview and function in pain modulation. In L. D. Kenneth, D. Charney, J. T. Coyle, & C. Nemeroff (Eds.), *Neuropsychopharmacology: The fifth generation of progress* (pp. 35–46). American College of Neuropsychopharmacology.

National Human Genome Research Institute. (2022). *Glossary of genetic and genomic terms*. Accessed from: https://www.genome.gov/genetics-glossary/Amino-Acids

Nikoshkov, A., Hurd, Y. L., Yakovleva, T., Bazov, I., Marinova, Z., Cebers, G., et al. (2005). Prodynorphin transcripts and proteins differentially expressed and regulated in the adult human brain. *The FASEB Journal, 19*(11), 1543–1545.

Panerai, A. E., & Sacerdote, P. (1997). β-Endorphin in the immune system: A role at last? *Immunology Today, 18*(7), 317–319.

Patel, H. J., Venkatesan, P., Halfpenny, J., Yacoub, M. H., Fox, A., Barnes, P. J., & Belvisi, M. G. (1999). Modulation of acetylcholine release from parasympathetic nerves innervating guinea-pig and human trachea by endomorphin-1 and-2. *European Journal of Pharmacology, 374*(1), 21–24.

Quach, T. T., Tang, F., Kageyama, H., Mocchetti, I., Guidotti, A., Meek, J. L., et al. (1984). Enkephalin biosynthesis in adrenal medulla. Modulation of proenkephalin mRNA content of cultured chromaffin cells by 8-bromo-adenosine 3′, 5′-monophosphate. *Molecular Pharmacology, 26*(2), 255–260.

Rialas, C. M., Fimiani, C., Bilfinger, T. V., Salzet, M., & Stefano, G. B. (1998). Endomorphin-1 and-2 inhibit human vascular sympathetic norepinephrine release: lack of interaction with mu 3 opiate receptor subtype. *Zhongguo yao li xue bao= Acta Pharmacologica Sinica, 19(5), 403–407.*

Rossier, J. (1993). Biosynthesis of enkephalins and proenkephalin-derived peptides. In *Opioids (pp. 423–447). Springer Berlin Heidelberg.*

Shang, Y., & Filizola, M. (2015). Opioid receptors: Structural and mechanistic insights into pharmacology and signaling. *European Journal of Pharmacology, 763*, 206–213.

Simon, E. J., & Gioannini, T. L. (1993). Opioid receptor multiplicity: Isolation, purification, and chemical characterization of binding sites. In H. Akil & E. J. Sion (Eds.), *Handbook of experimental pharmacology, volume 104: Opioids I* (pp. 3–26). Springer.

Tao, R., & Auerbach, S. B. (2002). Opioid receptor subtypes differentially modulate serotonin efflux in the rat central nervous system. *Journal of Pharmacology and Experimental Therapeutics, 303*(2), 549–556.

Tejeda, H. A., Shippenberg, T. S., & Henriksson, R. (2012). The dynorphin/κ-opioid receptor system and its role in psychiatric disorders. *Cellular and Molecular Life Sciences, 69*(6), 857–896.

Teschemacher, H. (1993). Atypical opioid peptides. In *Opioids (pp. 499–528). Springer Berlin Heidelberg.*

Thompson, A. A., Liu, W., Chun, E., Katritch, V., Wu, H., Vardy, E., ... & Stevens, R. C. (2012). Structure of the nociceptin/orphanin FQ receptor in complex with a peptide mimetic. *Nature, 485(7398), 395–399.*

Toll, L., Bruchas, M. R., Cox, B. M., & Zaveri, N. T. (2016). Nociceptin/orphanin FQ receptor structure, signaling, ligands, functions, and interactions with opioid systems. *Pharmacological Reviews, 68*(2), 419–457.

Ukai, M., & Lin, H. P. (2002a). Endomorphins 1 and 2 induce amnesia via selective modulation of dopamine receptors in mice. *European Journal of Pharmacology, 446*(1–3), 97–101.

Ukai, M., & Lin, H. P. (2002b). Involvement of μ1-opioid receptors and cholinergic neurotransmission in the endomorphins-induced impairment of passive avoidance learning in mice. *Behavioural Brain Research, 129*(1–2), 197–201.

Wang, W., Zhong, X., Li, Y., Guo, R., Du, S., Wen, L., ... & Wei, X. H. (2019). Rostral ventromedial medullamediated descending facilitation following P2X7 receptor activation is involved in the development of chronic postoperative pain. *Journal of Neurochemistry, 149(6), 760–780.*

Wilson, M. A., & McDonald, A. J. (2020). The amygdalar opioid system. In *Handbook of behavioral neuroscience* (Vol. 26, pp. 161–212). Elsevier.

Wood, P. L. (1993). Interrelationships of opioid, dopaminergic, cholinergic, and GABAergic pathways in the central nervous system. In *Opioids (pp. 625–643). Springer Berlin Heidelberg.*

Wu, H., Wacker, D., Mileni, M., Katritch, V., Han, G. W., Vardy, E., ... & Stevens, R. C. (2012). Structure of the human κ-opioid receptor in complex with JDTic. *Nature, 485(7398), 327–332.*

Zadina, J. E., Hackler, L., Ge, L. J., & Kastin, A. J. (1997). A potent and selective endogenous agonist for the μ-opiate receptor. *Nature, 386*(6624), 499–502.

Zadina, J. E., Martin-Schild, S., Gerall, A. A., Kastin, A. J., Hackler, L., Ge, L. J., & Zhang, X. (1999). Endomorphins: Novel endogenous μ-opiate receptor agonists in regions of high μ-opiate receptor density. *Annals of the New York Academy of Sciences, 897*(1), 136–144.

Zhang, J., Muller, J. F., & McDonald, A. J. (2015). Mu opioid receptor localization in the basolateral amygdala: An ultrastructural analysis. *Neuroscience, 303, 352–363.*

Interactions Between Endogenous Opioids and the Immune System

Wei Du

Abstract The endogenous opioid system, which consists of opioid receptors and their ligands, is widely expressed in the nervous system and also found in the immune system. As a part of the body's defense machinery, the immune system is heavily regulated by endogenous opioid peptides. Many types of immune cells, including macrophages, dendritic cells, neutrophils, and lymphocytes are influenced by endogenous opioids, which affect cell activation, differentiation, proliferation, apoptosis, phagocytosis, and cytokine production. Additionally, immune cells also synthesize and secrete endogenous opioid peptides and participate peripheral analgesia. This chapter is structured into two sections. Part one focuses on immunoregulatory functions of central endogenous opioids; and part two describes how opioid peptide-containing immune cells participate in local analgesia.

Keywords Endogenous opioids · Cytokines · Immunology · Opioid receptors

Introduction

Opioid System

The endogenous opioid system consists of opioid receptors and their ligands that are expressed abundantly in both the central and peripheral nervous systems, and can also be found in many other peripheral organs and tissues, including the immune system (Plein & Rittner, 2018).

The endogenous opioid system modulates a variety of physiological and pathophysiological functions, not only limited to emotions and analgesia but also immune reactions to stress, injury, and disease. Sharing many overlapping neurobiological mechanisms, systemically administered exogenous opioids are among the strongest

W. Du (✉)
Clinical Sciences Research, CAMC Institute for Academic Medicine, Charleston, WV, USA
e-mail: wei.du@camc.org

P. L. Kerr et al. (eds.), *Endogenous Opioids*, Advances in Neurobiology 35,
https://doi.org/10.1007/978-3-031-45493-6_3

analgesics for managing chronic and cancer-related pain. However, several severe undesirable side effects coexist; such as addiction, respiratory suppression, nausea, sedation, and psychotic symptoms. Significant preclinical and human research has shown that a critical side effect of exogenous opioid therapy is immunosuppression (Roy et al., 2011).

Opioid Receptors and Endogenous Opioids

Opioid receptors include classic mu (μ, MOP), delta (δ, DOP), kappa (κ, KOP), and nociceptin/orphanin FQ (NOP) receptors (Cox et al., 2015). These receptors are involved in mediating analgesic effects, while NOP plays a role in both analgesia and hyperalgesia/antiopioid hypersensitivity actions. Opioid receptors belong to the seven-transmembrane, G-protein-coupled receptors superfamily, encoded by four genes (Evans et al., 1992; Kieffer et al., 1992; Mestek et al., 1995; Mogil & Pasternak, 2001; Simonin et al., 1995), the structures of which have been revealed by crystal analysis (Che et al., 2018; Granier et al., 2012; Huang et al., 2015; Thompson et al., 2012). Classical opioid receptors are blocked by the antagonist naloxone and activated by exogenous opiates such as morphine, and its semisynthetic or synthetic analogs, or by endogenous opioid peptides, including endorphin, enkephalins, dynorphins, and endomorphins. Each type of these endogenous opioids is derived from different precursors and displays different affinity to opioid receptors (Goldstein et al., 1981; Hughes et al., 1975; Kakidani et al., 1982; Kangawa et al., 1981; Li et al., 1976; Nakanishi et al., 1979; Noda et al., 1982; Zadina et al., 1997). NOP precursors have not yet been identified, however, they are not blocked by naloxone. Their specific endogenous agonist is nociception/orphanin FQ (Meunier et al., 1995; Reinscheid et al., 1995). A summary of this information is presented in Table 1. Much of the understanding of receptor functions in differing

Table 1 Major endogenous opioid peptides

Family	Precursor	Peptides	Receptors and affinity
Endorphins	Proopiomelanocortin (POMC)	α-endorphin β-endorphin γ-endorphin	μ, δ μ = δ
Enkephalins	Proenkephalin (PENK)	Met-enkephalin Leu-enkephalin	δ, μ δ > μ
Dynorphins	Prodynorphin (PDYN)	Dynorphin A Dynorphin B α-neoendorphin β-neoendorphin	κ, μ, δ κ > μ, δ
Endomorphins	Unknown	Endomorphin-1 Endomorphin-2	μ
Nociceptin/Orphanin FQ	Prepronociceptin	Nociceptin	Nociceptin receptor

organs and tissues has been aided by the availability of receptor gene knockout mice models (Fürst et al., 2020; Gavériaux-Ruff, 2013).

Immune System

The immune system is the biological defense barrier for the body against invading pathogens and injuries; it consists of several organs and many types of cells. Immune system functions are divided into innate and adaptive immunity, based on how the cells react to pathogens. Generally, innate immune responses occur rapidly but without development of specific cell memories. The innate immune system includes several cellular components: neutrophils, macrophages, monocytes, dendritic cells (DCs), natural killer (NK) cells, complements, physical barriers (such as gastrointestinal barrier), and some secreted enzymes. Adaptive immunity, in contrast, is highly specific and capable of developing memory responses. The adaptive immune system includes T and B lymphocytes and the circulating antibodies that have been produced by B cells.

It is well established that leukocytes contribute to pain generation during inflammatory disease through mediators such as cytokines and sensitized nociceptive signaling pathways. Both innate and adaptive immune systems are involved in generating pain mediators (Calvo et al., 2012). Immune cells also communicate directly with peripheral neurons to control pain by interacting with the endogenous opioid system (Machelska, 2007; Stein & Machelska, 2011). About 30–40% of immune cells that infiltrate around injured nerve sites express endogenous opioid peptides and participate in local analgesia (Labuz et al., 2009).

Part 1: Immunoregulatory Functions of Endogenous Opioids

Opioid Receptors Expressed by Immune Cells

Classical opioid receptors MOP, DOP, and KOP and nociception/FQ receptors are expressed in both rodent and human immune cells (Boyadjieva et al., 2004; Lopker et al., 1980; Stefano et al., 1992; Williams et al., 2007). Some literatures shows variable results of specificity of receptors expression on different cell types (Al-Hashimi et al., 2016; Machelska & Celik, 2020; Wick et al., 1996; Williams et al., 2007). The expression levels of peripheral opioid receptors are relatively low compared to those in the nervous system and are modulated by either pathophysiologic conditions or pharmacological treatments (Gunji et al., 2000; Jaume et al., 2007; Toskulkao et al., 2010).

Both exogenous and endogenous opioids appear to affect functions of the immune system; suppressing the overall immune response and increasing the susceptibility to diseases such as infections. Indeed, in animal studies, morphine and fentanyl are associated with a high incidence of infections and cancer, including mechanisms that impair the function of macrophages, natural killer (NK) cells and T cells, or weaken the intestinal barrier (Plein & Rittner, 2018). In addition, opiate and opioid peptides were believed to affect immune system indirectly, through opioid effects of the brain (Pomorska et al., 2014). Opioid receptors are identified at multiple types of immune cells, suggesting a direct interaction between opioid receptor ligands and immune cells (Bidlack, 2000; Bidlack et al., 2006; McCarthy et al., 2001; Sharp, 2006; Sharp et al., 1998; Sibinga & Goldstein, 1988). Most evidence points to immunomodulatory effect of opioids on specific immune cells, rather than overall suppression. The influence of opioids on immune cells includes many aspects of cell status, such as cell proliferation, maturation, cell signaling, chemotaxis, chemokine receptor expression, phagocytosis, and cytokine production profiles (Finley et al., 2008; McLaughlin et al., 2015; Ninkovic & Roy, 2013; Plein & Rittner, 2018; Sacerdote, 2006; Salzet et al., 2000; Sharp, 2006). It is now well recognized, therefore, that opioid receptors participate in immune functions, involving both innate and adaptive immune system.

The mechanisms of how endogenous opioid peptides regulate immune system functions are not conclusive yet. We have cited several recent studies that attempt to explain how different opioid peptides affect different immune cells. Most of these studies, however, are based on in vivo and ex vivo experiments and are lacking in controlled human research.

Endorphins

Endorphins activate both μ and δ opioid receptors, with similar affinity. Many of the endorphin studies have concentrated on β-endorphin (βE) and their immunostimulatory effects.

βE showed stimulatory effects on unfractionated leukocyte suspensions with increased IL-1β and decreased IL-8 production, with or without the presence of lipopolysaccharide (LPS) in vitro (Gein et al., 2009). A similar study using the same cell population, also demonstrated that βE could enhance IL-1β production, but not IL-6 nor TNFα levels (Gein et al., 2007). Both of these studies revealed this stimulatory effect on IL-1β secretion and could not be blocked by naloxone or naltrindole, but IL-8 inhibitory functions could be blocked by naloxone or using calcium-deficient medium (Nandhra & Carson, 2000). Using human articular chondrocytes obtained from patients whom underwent total knee arthroplasty, N Andjelkov et al. found that TNFα and IL-1β levels in supernatants from cell culture significantly increased after stimulation with βE, and this stimulatory effect was blocked by naltrexone (Andjelkov et al., 2005). The different reaction to naltrexone blockade also indicated there might be alternative pathways between immune cells and other cells, also suggesting that the interactions between opioid peptides and immune cells might involve not only u and δ receptor binding sites but also nonopioid βE receptors (Kovalitskaya et al., 2011; Navolotskaya et al., 2004).

Other studies with differing cell population showed similar stimulatory results. In an early study of neurohumoral interactions, researchers found that βE could inhibit spontaneous IL-6 secretion by spleen macrophages (Straub et al., 1998). βE could induce $CD4^+$ T cells to secrete IL-4, and macrophages to secrete IL-1 and IL-6 (van den Bergh et al., 1993). Synthesized peptides corresponding to βE fragments also showed immune stimulatory reactions in peritoneal macrophages as well as spleen T and B cells (Kovalitskaya et al., 2011).

In conclusion, studies of endorphin influences on immune system support its primary stimulatory functions, but lack of in vivo and human research evidence.

Enkephalins

Endogenous enkephalins, including leu- and Met-enkephalin, mainly bind to δ opioid receptor, and are suggested to play important roles in connecting neuroendocrine and immune system and regulate various function of cells from both innate and adaptive immune system, and potentially influence tumorigenesis. The evidence and proposed mechanisms are discussed as follows:

1. Enhanced antigen presentation. DCs play a critical role in antigen presentation and regulation of subsequent T-cell responses (Li et al., 2012; Liu et al., 2012; Shan et al., 2011). Studies have shown that met-enkephalins induce bone marrow-derived (BM) DCs polarized toward myeloid (mDC) subtypes rather than plasmacytoid (pDC) subtypes, facilitating Th1 responses. Met-enkephalin exposure markedly induces DC maturation and positively modulate the pathway between DCs and T cells. Treatments with met-enkephalin also lead to upregulation of MHC II and costimulatory molecules (CD86 and CD40) in BM-derived DCs. Met-enkephalin also produced more proinflammatory cytokines (IL-12 and TNFα), which are essential signals for T-cell activation. T cells primed by these enkephalin-treated DCs showed enhanced proliferation, cytotoxicity, and antitumor activity (Li et al., 2012; Liu et al., 2012; Shan et al., 2011).
2. Modulate T-cell function. Met-enkephalin alone or combined with IL-2 or IFNγ, promotes $CD4^+$ T-cell expansion, CD4 molecule expression, and IL-2 production (Shan et al., 2011). However, other studies have demonstrated that tumor cells, such as colorectal carcinoma cells, may secrete met-enkephalin to inhibit $CD4^+$ T-cell proliferation and induce apoptosis to significantly affect cancer progression (Ohmori et al., 2009). Treatments with met-enkephalin in $CD8^+$ T cell have shown increased proliferation and enhanced cytotoxicity to tumor cells in vivo and in vitro; these antitumor effects could be blocked by naltrexone, indicating the involvement of classic opioid receptors. These met-enkephalin-mediated T-cell functional changes are associated with upregulated Ca^{2+} influx into cytoplasm and translocation of transcription factor NFAT2 into nucleus (Li et al., 2014).
3. Regulate macrophage/microglia polarization. In an animal tumor model, met-enkephalin downregulates expression of CD206 and arginase-1 (protumor M2

phenotype markers), while significantly upregulating the expression of CD64, MHC II, and nitric oxide (antitumor M1 phenotype markers) synthase in tumor-associated macrophages (TAMs) in vivo (Chen et al., 2012). Moreover, in vitro study of BM-derived macrophages showed increased tumoricidal activity after treatments with met-enkephalin (Chen et al., 2012). Microglia are the most important brain resident immune cells, with the identical origin with BM-derived macrophage during embryo development (Ginhoux et al., 2010). It is hypothesized that met-enkephalin treatments of microglia could produce similar tumoricidal affects by inducing polarization toward an M1 phenotype (Xu et al., 2016).

4. Alteration of NK cell activation. Studies with synthetic enkephalin analogs or δ-selective peptides showed that NK cell activity was greatly enhanced by peptides such as DPDPE, even at very low concentrations. NK cell cytotoxicity and IL-2 production significantly increase after peptides treatment, although some research has shown opposite effects (House et al., 1996; Mehrotra et al., 2002).
5. Tumor immunology. Met-enkephalin research has pointed to antitumor effects, therefore, the therapeutic potential of met-enkephalin has been extensively investigated. Two major mechanisms have been hypothesized: the first mechanism focuses on the influence of pro- and antitumor immune cells such as T cells, macrophages, DCs, and NK cells discussed above. The second mechanism points to the evidence that met-enkephalins directly inhibit tumor cell growth. However, to date neither hypothesis is conclusively evidence-base supported (Tuo et al., 2020).

Enkephalins clearly alter immune responses in various immune cells, exerting either stimulatory or inhibitory effect on tumor growth. However more in vivo experiments and human clinical investigations are needed to clarify the interactions of enkephalins and endogenous opioid complexes and determine how enkephalins affect cancer.

Dynorphins

Dynorphins comprise a family of opioid peptides with high affinity for κ opiate receptors. These receptors have been found expressed on immune cells similar to that found in the central nervous system (Gein, 2014). Dynorphin A may affect immune cell functions either by activating κ opiate receptors mainly expressed on macrophage and T cells, through nonopioid dynorphin-binding intracellular site, or on designated NMDA receptors expressed on T cells (Zainullina et al., 2011). Recent research with patients with major depression has shown that dynorphins regulate immune cell proliferation, activate cytokine production, and display both pro- and anti-inflammatory activities (Al-Hakeim et al., 2020; Bidlack, 2000).

Using a mouse model, dynorphin and related opioid peptides enhanced macrophage activities, increased production of IL-1, increased macrophage-mediated tumoricidal activity, superoxide production, and oxidative burst. Most of these effects were blocked by naloxone, indicating that classical κ opiate receptors were involved. Additionally, early research with dynorphin A showed a dose-dependent increase in the phagocytosis of peritoneal macrophages, an effect not reversed by naloxone (Foster & Moore, 1987; Hagi et al., 1994).

More recent studies have demonstrated that κ opiate receptors regulate the maintenance of immune cell homeostasis, and that overstimulation of these receptors may change cytokine levels and alter antibody production (Pomorska et al., 2014). However, it has been shown that κ opiate receptors do not affect NK cell activity and might suppress B-cell proliferation after LPS stimulation (Hsueh et al., 1996).

Endomorphins

Endomorphins activate μ opioid receptors, which are expressed on immune cells. Most studies of endomorphins have shown that immunosuppressive activity similar to that seen with morphine (Anton et al., 2008; Azuma & Ohura, 2002a). The evidence and proposed mechanisms are discussed as follows:

1. Humoral immunity. Endomorphin 1 (EM-1) and endomorphin 2 (EM-2) inhibit antibody formation at ultra-low concentrations, and this immunomodulatory effect can not be blocked by naloxone, indicating a mechanism that does not involve opioid receptors (Anton et al., 2008).
2. Innate immune system. EM-2 alters macrophage function, by decreasing TNFα, IL-10, and IL-12 production. Moreover, there is evidence that EM-2 suppresses phagocytosis, chemotaxis, and superoxide anion production by macrophages (Azuma & Ohura, 2002a). Both EM1 and EM2 suppress LPS-induced IL-10 and IL-12 production in the human monocytic cell line THP-1, which is differentiated to macrophage-like cells (Azuma & Ohura, 2002b). Endorphins have also showed regulatory effects on to neutrophils. EM1 and EM-2 inhibit the production of superoxide anions by PMA-stimulated neutrophils in a concentration-dependent manner, and this effect is blocked by the selective μ opioid receptor antagonist naltrindole. In contrast, in unstimulated neutrophils, EM1 and EM2 treatment significantly enhances superoxide anion production (Azuma et al., 2000). EM1 and EM2 also delays neutrophil apoptosis and activates the phosphoinositide 3-kinase pathway, determined by phosphorylation of BAD (pro-apoptotic protein). Unlike macrophage responses to endomorphin treatments, phagocytosis of opsonized *E. coli* by neutrophils was not changed by endomorphins treatments (Azuma et al., 2002).
3. Antigen presentation. There is evidence that EM-1 alters antigen presentation, affecting cell maturation and function. In vitro studies showed that peripheral blood mononuclear cells (PBMCs) derived DCs treated by EM-1 would upregulate the expression of surface molecules CD86, CCR7, and CD36, while decrease the cytokine production and TLR4 expression. EM-1-treated DCs also showed decreased ability to stimulate T-cell proliferation, which can be blocked by naloxone (Liu et al., 2018).

In summary, opioid peptides exert major regulatory effects on the immune system. Different endogenous opioid peptides perform differently regarding different immune cell types and disease models, although more in vivo and human studies are needed. These studies seem to indicate that immune cells are not the one that only regulated by opioid peptides, but also may use alternative endogenous peptides pathways to perform physiological and pathophysiological function.

Part 2: Immune Cells Secrete Endogenous Opioids to Participate in Local Analgesia

Celik et al. (2016) have demonstrated that activation of opioid receptors on leukocytes initiates a subsequent secretion of endogenous opioid peptides, including met-enkephalin, βE, and dynorphin A, which, in turn, affect local neuronal receptors and bring about analgesia (Celik et al., 2016). Using the chronic constriction injury of the sciatic nerve-induced neuropathic pain mouse model, it was found that wild-type, opioid peptide- and receptor-expressing leukocytes could induce analgesia, while receptor-lacking leukocytes could not. Ex vivo experiments show that exogenous opioids could induce leukocytes to produce opioid peptides, and this peptides release was mediated by G protein-PLC-IP3 receptors–intracellular Ca^{2+} pathway (Celik et al., 2016). This leukocytes-induced local analgesia could help explain the decreased tolerance to peripheral morphine analgesia that develops in the presence of infiltrated immune cells in inflamed synovia after human knee surgery (Stein et al., 1996).

Endogenous Opioid Peptides Expressed in Immune Cells Affect Local Analgesia

Endogenous opioids peptides are synthesized and released by immune cells in response to different stimulations (such as CRF, certain chemokines, and inflammation) and situations (such as stress, trauma; Cabot et al., 1997; Likar et al., 2007; Maddila et al., 2017; Mousa et al., 2004; Rittner et al., 2006). Gene expression of precursors POMC, PENK, and PDYN, as well as critical enzymes needed for processing to opioid peptides are expressed in immune cells, and localization of endogenous opioid peptides have been found in T and B lymphocytes, macrophage, and granulocytes in peripheral inflamed tissue in animal models (Chadzinska et al., 2005; Maddila et al., 2017; Mousa et al., 2004; Przewlocki et al., 1992; Sitte et al., 2007) and humans (Mousa et al., 2007). After release from immune cells, these endogenous opioid peptides function locally by activating neuronal opioid receptors to regulate analgesia (Hua, 2016; Stein & Machelska, 2011; Stein et al., 2003). These leukocytes-induced, opioid peptides-mediated analgesic actions have been identified in mouse/rat inflammation (Machelska et al., 2004; Rittner et al., 2001;

Schafer et al., 1994), neuropathic (Labuz et al., 2009, 2010), and osteosarcoma (Baamonde et al., 2006) pain models, and also shown in patients after surgery (Likar et al., 2007; Stein et al., 1993). During inflammation, peripheral sensory and sympathetic nerve fibers augment the expression of ICAM-1 on vascular endothelial cells and recruit opioid-containing immune cells to inflammatory sites to promote analgesia (Mousa et al., 2010). However, endogenous peripheral antinociception effect was not augmented by recruited opioid-containing leukocytes (Brack et al., 2004). The secretion of nociception/OFQ has also been found in lymphoid tissues and leukocytes (Fiset et al., 2003; Pampusch et al., 2000), although these findings lack confirmation by controlled clinical studies.

Endogenous Opioids Secreted by Immune Cells Affect Local Analgesia

It has been shown with in vivo and ex vivo experiments that, compared to unstimulated M0 and proinflammatory M1, anti-inflammatory M2-polarized macrophages produce and secrete higher amount of endogenous opioid peptides, including met-enkephalin, βE, and dynorphin, thereby contributing to local analgesia. Perineural transplantation of M2 macrophage could significantly reduce neuropathy-induced mechanical hypersensitivity, and this analgesic effect could be blocked by perineural injection of opioid receptor antagonist naloxone methiodide (Pannell et al., 2016). Additionally, DCs activated by LPS and other Toll-like receptor ligands can induce the expression and secretion of endomorphins EM1 and EM2, and subsequently suppress T-cell proliferation through activation of μ opioid receptors (Yang et al., 2012).

Mucosal effector $CD4^+$ T cells in gastrointestinal system, which have been identified as colitogenic, could produce endogenous opioids to counteract inflammation-associated visceral pain in inflammatory bowel disease in animal models (Basso et al., 2015; Boue et al., 2014). Notably, in this model, macrophages and epithelial cells did not show the ability to produce opioid peptides and colitogenic $CD4^+$ T cells that are the major contributors to the local analgesic effect.

Labuz et al. (2010) showed that T cells accumulated at nerve injury sites, expressing βE and releasing locally in response to corticotropin-releasing factors (CRF), to activate local neuronal opioid receptors and reduce neuropathy-induced mechanical hypersensitivity (Labuz et al., 2010). Similar effects on pain hypersensitivity were seen when severe combined immunodeficiency (SCID) mice without T cells showed benefit from adoptive transfer WT mice-derived T cells.

In summary, immune cells participate in peripheral analgesia by synthesizing and releasing endogenous opioids in response to varied pathophysiological conditions and stimulations.

Conclusions

The nervous and immune systems are in constant communication to regulate our physiological and pathological processes. Endogenous opioid peptides are critical mediators during these processes. Immune cells not only express opioid receptors to receive signals and modify responses to different pathogens, but also release opioid peptides to facilitate analgesia during disease. These discoveries indicate that targeting opioid-secreting immune cells could potentially promote endogenous pain control and offer novel therapeutic opportunities.

References

Al-Hakeim, H. K., Zeki Al-Fadhel, S., Al-Dujaili, A. H., & Maes, M. (2020). In major depression, increased kappa and mu opioid receptor levels are associated with immune activation. *Acta Neuropsychiatrica, 32*(2), 99–108. https://doi.org/10.1017/neu.2019.47

Al-Hashimi, M., McDonald, J., Thompson, J. P., & Lambert, D. G. (2016). Evidence for nociceptin/orphanin FQ (NOP) but not micro (MOP), delta (DOP) or kappa (KOP) opioid receptor mRNA in whole human blood. *British Journal of Anaesthesia, 116*(3), 423–429. https://doi.org/10.1093/bja/aev540

Andjelkov, N., Elvenes, J., Martin, J., & Johansen, O. (2005). Opiate regulation of IL-1beta and TNF-alpha in cultured human articular chondrocytes. *Biochemical and Biophysical Research Communications, 333*(4), 1295–1299. https://doi.org/10.1016/j.bbrc.2005.06.035

Anton, B., Leff, P., Calva, J. C., Acevedo, R., Salazar, A., Matus, M., Pavon, L., Martinez, M., Meissler, J. J., Adler, M. W., Gaughan, J. P., & Eisenstein, T. K. (2008). Endomorphin 1 and endomorphin 2 suppress in vitro antibody formation at ultra-low concentrations: Anti-peptide antibodies but not opioid antagonists block the activity. *Brain, Behavior, and Immunity, 22*(6), 824–832. https://doi.org/10.1016/j.bbi.2008.02.004

Azuma, Y., & Ohura, K. (2002a). Endomorphin-2 modulates productions of TNF-alpha, IL-1beta, IL-10, and IL-12, and alters functions related to innate immune of macrophages. *Inflammation, 26*(5), 223–232. https://doi.org/10.1023/a:1019766602138

Azuma, Y., & Ohura, K. (2002b). Endomorphins 1 and 2 inhibit IL-10 and IL-12 production and innate immune functions, and potentiate NF-kappaB DNA binding in THP-1 differentiated to macrophage-like cells. *Scandinavian Journal of Immunology, 56*(3), 260–269. https://doi.org/10.1046/j.1365-3083.2002.01128.x

Azuma, Y., Wang, P. L., Shinohara, M., & Ohura, K. (2000). Immunomodulation of the neutrophil respiratory burst by endomorphins 1 and 2. *Immunology Letters, 75*(1), 55–59. https://doi.org/10.1016/s0165-2478(00)00274-1

Azuma, Y., Ohura, K., Wang, P. L., & Shinohara, M. (2002). Endomorphins delay constitutive apoptosis and alter the innate host defense functions of neutrophils. *Immunology Letters, 81*(1), 31–40. https://doi.org/10.1016/s0165-2478(01)00335-2

Baamonde, A., Lastra, A., Juarez, L., Garcia-Suarez, O., Meana, A., Hidalgo, A., & Menendez, L. (2006). Endogenous beta-endorphin induces thermal analgesia at the initial stages of a murine osteosarcoma. *Peptides, 27*(11), 2778–2785. https://doi.org/10.1016/j.peptides.2006.07.004

Basso, L., Bourreille, A., & Dietrich, G. (2015). Intestinal inflammation and pain management. *Current Opinion in Pharmacology, 25*, 50–55. https://doi.org/10.1016/j.coph.2015.11.004

Bidlack, J. M. (2000). Detection and function of opioid receptors on cells from the immune system. *Clinical and Diagnostic Laboratory Immunology, 7*(5), 719–723. https://doi.org/10.1128/cdli.7.5.719-723.2000

Bidlack, J. M., Khimich, M., Parkhill, A. L., Sumagin, S., Sun, B., & Tipton, C. M. (2006). Opioid receptors and signaling on cells from the immune system. *Journal of Neuroimmune Pharmacology, 1*(3), 260–269. https://doi.org/10.1007/s11481-006-9026-2

Boue, J., Basso, L., Cenac, N., Blanpied, C., Rolli-Derkinderen, M., Neunlist, M., Vergnolle, N., & Dietrich, G. (2014). Endogenous regulation of visceral pain via production of opioids by colitogenic CD4(+) T cells in mice. *Gastroenterology, 146*(1), 166–175. https://doi.org/10.1053/j.gastro.2013.09.020

Boyadjieva, N. I., Chaturvedi, K., Poplawski, M. M., & Sarkar, D. K. (2004). Opioid antagonist naltrexone disrupts feedback interaction between mu and delta opioid receptors in splenocytes to prevent alcohol inhibition of NK cell function. *Journal of Immunology, 173*(1), 42–49. https://doi.org/10.4049/jimmunol.173.1.42

Brack, A., Rittner, H. L., Machelska, H., Shaqura, M., Mousa, S. A., Labuz, D., Zollner, C., Schafer, M., & Stein, C. (2004). Endogenous peripheral antinociception in early inflammation is not limited by the number of opioid-containing leukocytes but by opioid receptor expression. *Pain, 108*(1–2), 67–75. https://doi.org/10.1016/j.pain.2003.12.008

Cabot, P. J., Carter, L., Gaiddon, C., Zhang, Q., Schafer, M., Loeffler, J. P., & Stein, C. (1997). Immune cell-derived beta-endorphin. Production, release, and control of inflammatory pain in rats. *The Journal of Clinical Investigation, 100*(1), 142–148. https://doi.org/10.1172/JCI119506

Calvo, M., Dawes, J. M., & Bennett, D. L. (2012). The role of the immune system in the generation of neuropathic pain. *Lancet Neurology, 11*(7), 629–642. https://doi.org/10.1016/S1474-4422(12)70134-5

Celik, M. O., Labuz, D., Henning, K., Busch-Dienstfertig, M., Gaveriaux-Ruff, C., Kieffer, B. L., Zimmer, A., & Machelska, H. (2016). Leukocyte opioid receptors mediate analgesia via Ca(2+)-regulated release of opioid peptides. *Brain, Behavior, and Immunity, 57*, 227–242. https://doi.org/10.1016/j.bbi.2016.04.018

Chadzinska, M., Starowicz, K., Scislowska-Czarnecka, A., Bilecki, W., Pierzchala-Koziec, K., Przewlocki, R., Przewlocka, B., & Plytycz, B. (2005). Morphine-induced changes in the activity of proopiomelanocortin and prodynorphin systems in zymosan-induced peritonitis in mice. *Immunology Letters, 101*(2), 185–192. https://doi.org/10.1016/j.imlet.2005.05.009

Che, T., Majumdar, S., Zaidi, S. A., Ondachi, P., McCorvy, J. D., Wang, S., Mosier, P. D., Uprety, R., Vardy, E., Krumm, B. E., Han, G. W., Lee, M. Y., Pardon, E., Steyaert, J., Huang, X. P., Strachan, R. T., Tribo, A. R., Pasternak, G. W., Carroll, F. I., Stevens, R. C., Cherezov, V., Katritch, V., Wacker, D., & Roth, B. L. (2018). Structure of the nanobody-stabilized active state of the kappa opioid receptor. *Cell, 172*(1–2), 55–67. https://doi.org/10.1016/j.cell.2017.12.011

Chen, W., Liu, J., Meng, J., Lu, C., Li, X., Wang, E., & Shan, F. (2012). Macrophage polarization induced by neuropeptide methionine enkephalin (MENK) promotes tumoricidal responses. *Cancer Immunology, Immunotherapy, 61*(10), 1755–1768. https://doi.org/10.1007/s00262-012-1240-6

Cox, B. M., Christie, M. J., Devi, L., Toll, L., & Traynor, J. R. (2015). Challenges for opioid receptor nomenclature: IUPHAR review 9. *British Journal of Pharmacology, 172*(2), 317–323. https://doi.org/10.1111/bph.12612

Evans, C. J., Keith, D. E., Jr., Morrison, H., Magendzo, K., & Edwards, R. H. (1992). Cloning of a delta opioid receptor by functional expression. *Science, 258*(5090), 1952–1955. https://doi.org/10.1126/science.1335167

Finley, M. J., Happel, C. M., Kaminsky, D. E., & Rogers, T. J. (2008). Opioid and nociceptin receptors regulate cytokine and cytokine receptor expression. *Cellular Immunology, 252*(1–2), 146–154. https://doi.org/10.1016/j.cellimm.2007.09.008

Fiset, M. E., Gilbert, C., Poubelle, P. E., & Pouliot, M. (2003). Human neutrophils as a source of nociceptin: A novel link between pain and inflammation. *Biochemistry, 42*(35), 10498–10505. https://doi.org/10.1021/bi0300635

Foster, J. S., & Moore, R. N. (1987). Dynorphin and related opioid peptides enhance tumoricidal activity mediated by murine peritoneal macrophages. *Journal of Leukocyte Biology, 42*(2), 171–174. https://doi.org/10.1002/jlb.42.2.171

Fürst, S., Zádori, Z. S., Zádor, F., Király, K., Balogh, M., László, S. B., Hutka, B., Mohammadzadeh, A., Calabrese, C., Galambos, A. R., Riba, P., Romualdi, P., Benyhe, S., Timár, J., Schmidhammer, H., Spetea, M., & Al-Khrasani, M. (2020). On the role of peripheral sensory and gut mu opioid receptors: Peripheral analgesia and tolerance. *Molecules, 25*(11). https://doi.org/10.3390/molecules25112473

Gavériaux-Ruff, C. (2013). Opiate-induced analgesia: Contributions from mu, delta and kappa opioid receptors mouse mutants. *Current Pharmaceutical Design, 19*(42), 7373–7381. https://doi.org/10.2174/138161281942140105163727

Gein, S. V. (2014). Dynorphins in regulation of immune system functions. *Biochemistry (Mosc), 79*(5), 397–405. https://doi.org/10.1134/S0006297914050034

Gein, S. V., Gorshkova, K. G., & Tendryakova, S. P. (2007). Role of beta-endorphin in the regulation of proinflammatory cytokine production by peripheral blood monocytes in vitro. *Bulletin of Experimental Biology and Medicine, 143*(2), 214–217. https://doi.org/10.1007/s10517-007-0054-6

Gein, S. V., Gorshkova, K. G., & Tendryakova, S. P. (2009). Regulation of interleukin-1beta and interleukin-8 production by agonists of mu and delta opiate receptors in vitro. *Neuroscience and Behavioral Physiology, 39*(6), 591–595. https://doi.org/10.1007/s11055-009-9166-y

Ginhoux, F., Greter, M., Leboeuf, M., Nandi, S., See, P., Gokhan, S., Mehler, M. F., Conway, S. J., Ng, L. G., Stanley, E. R., Samokhvalov, I. M., & Merad, M. (2010). Fate mapping analysis reveals that adult microglia derive from primitive macrophages. *Science, 330*(6005), 841–845. https://doi.org/10.1126/science.1194637

Goldstein, A., Fischli, W., Lowney, L. I., Hunkapiller, M., & Hood, L. (1981). Porcine pituitary dynorphin: Complete amino acid sequence of the biologically active heptadecapeptide. *Proceedings of the National Academy of Sciences of the United States of America, 78*(11), 7219–7223. https://doi.org/10.1073/pnas.78.11.7219

Granier, S., Manglik, A., Kruse, A. C., Kobilka, T. S., Thian, F. S., Weis, W. I., & Kobilka, B. K. (2012). Structure of the delta-opioid receptor bound to naltrindole. *Nature, 485*(7398), 400–404. https://doi.org/10.1038/nature11111

Gunji, N., Nagashima, M., Asano, G., & Yoshino, S. (2000). Expression of kappa-opioid receptor mRNA in human peripheral blood lymphocytes and the relationship between its expression and the inflammatory changes in rheumatoid arthritis. *Rheumatology International, 19*(3), 95–100. https://doi.org/10.1007/s002960050110

Hagi, K., Uno, K., Inaba, K., & Muramatsu, S. (1994). Augmenting effect of opioid peptides on murine macrophage activation. *Journal of Neuroimmunology, 50*(1), 71–76. https://doi.org/10.1016/0165-5728(94)90216-x

House, R. V., Thomas, P. T., & Bhargava, H. N. (1996). A comparative study of immunomodulation produced by in vitro exposure to delta opioid receptor agonist peptides. *Peptides, 17*(1), 75–81. https://doi.org/10.1016/0196-9781(95)02051-9

Hsueh, C. M., Chen, S. F., Huang, H. J., Ghanta, V. K., & Hiramoto, R. N. (1996). Activation of mu-opioid receptors are required for the conditioned enhancement of NK cell activity. *Brain Research, 737*(1–2), 263–268. https://doi.org/10.1016/0006-8993(96)00740-8

Hua, S. (2016). Neuroimmune interaction in the regulation of peripheral opioid-mediated analgesia in inflammation. *Frontiers in Immunology, 7*, 293. https://doi.org/10.3389/fimmu.2016.00293

Huang, W., Manglik, A., Venkatakrishnan, A. J., Laeremans, T., Feinberg, E. N., Sanborn, A. L., Kato, H. E., Livingston, K. E., Thorsen, T. S., Kling, R. C., Granier, S., Gmeiner, P., Husbands, S. M., Traynor, J. R., Weis, W. I., Steyaert, J., Dror, R. O., & Kobilka, B. K. (2015). Structural insights into micro-opioid receptor activation. *Nature, 524*(7565), 315–321. https://doi.org/10.1038/nature14886

Hughes, J., Smith, T. W., Kosterlitz, H. W., Fothergill, L. A., Morgan, B. A., & Morris, H. R. (1975). Identification of two related pentapeptides from the brain with potent opiate agonist activity. *Nature, 258*(5536), 577–580. https://doi.org/10.1038/258577a0

Jaume, M., Laffont, S., Chapey, E., Blanpied, C., & Dietrich, G. (2007). Opioid receptor blockade increases the number of lymphocytes without altering T cell response in draining lymph nodes *in vivo*. *Journal of Neuroimmunology, 188*(1–2), 95–102. https://doi.org/10.1016/j.jneuroim.2007.06.013

Kakidani, H., Furutani, Y., Takahashi, H., Noda, M., Morimoto, Y., Hirose, T., Asai, M., Inayama, S., Nakanishi, S., & Numa, S. (1982). Cloning and sequence analysis of cDNA for porcine beta-neo-endorphin/dynorphin precursor. *Nature, 298*(5871), 245–249. https://doi.org/10.1038/298245a0

Kangawa, K., Minamino, N., Chino, N., Sakakibara, S., & Matsuo, H. (1981). The complete amino acid sequence of alpha-neo-endorphin. *Biochemical and Biophysical Research Communications, 99*(3), 871–878. https://doi.org/10.1016/0006-291x(81)91244-4

Kieffer, B. L., Befort, K., Gaveriaux-Ruff, C., & Hirth, C. G. (1992). The delta-opioid receptor: Isolation of a cDNA by expression cloning and pharmacological characterization. *Proceedings of the National Academy of Sciences of the United States of America, 89*(24), 12048–12052. https://doi.org/10.1073/pnas.89.24.12048

Kovalitskaya, Y. A., Nekrasova, Y. N., Sadovnikov, V. B., Zolotarev, Y. A., & Navolotskaya, E. V. (2011). Immunostimulating effect of the synthetic peptide octarphin corresponding to beta-endorphin fragment 12-19. *Biochemistry (Mosc), 76*(5), 596–604. https://doi.org/10.1134/S0006297911050105

Labuz, D., Schmidt, Y., Schreiter, A., Rittner, H. L., Mousa, S. A., & Machelska, H. (2009). Immune cell-derived opioids protect against neuropathic pain in mice. *The Journal of Clinical Investigation, 119*(2), 278–286. https://doi.org/10.1172/JCI36246

Labuz, D., Schreiter, A., Schmidt, Y., Brack, A., & Machelska, H. (2010). T lymphocytes containing beta-endorphin ameliorate mechanical hypersensitivity following nerve injury. *Brain, Behavior, and Immunity, 24*(7), 1045–1053. https://doi.org/10.1016/j.bbi.2010.04.001

Li, C. H., Chung, D., & Doneen, B. A. (1976). Isolation, characterization and opiate activity of beta-endorphin from human pituitary glands. *Biochemical and Biophysical Research Communications, 72*(4), 1542–1547. https://doi.org/10.1016/s0006-291x(76)80189-1

Li, W., Meng, J., Li, X., Hua, H., Yiming, M., Wang, Q., Wang, E., & Shan, F. (2012). Methionine enkephalin (MENK) improved the functions of bone marrow-derived dendritic cells (BMDCs) loaded with antigen. *Human Vaccines & Immunotherapeutics, 8*(9), 1236–1242. https://doi.org/10.4161/hv.21128

Li, W., Chen, W., Herberman, R. B., Plotnikoff, N. P., Youkilis, G., Griffin, N., Wang, E., Lu, C., & Shan, F. (2014). Immunotherapy of cancer via mediation of cytotoxic T lymphocytes by methionine enkephalin (MENK). *Cancer Letters, 344*(2), 212–222. https://doi.org/10.1016/j.canlet.2013.10.029

Likar, R., Mousa, S. A., Steinkellner, H., Koppert, W., Philippitsch, G., Stein, C., & Schafer, M. (2007). Involvement of intra-articular corticotropin-releasing hormone in postoperative pain modulation. *The Clinical Journal of Pain, 23*(2), 136–142. https://doi.org/10.1097/01.ajp.0000210954.93878.0d

Liu, J., Chen, W., Meng, J., Lu, C., Wang, E., & Shan, F. (2012). Induction on differentiation and modulation of bone marrow progenitor of dendritic cell by methionine enkephalin (MENK). *Cancer Immunology, Immunotherapy, 61*(10), 1699–1711. https://doi.org/10.1007/s00262-012-1221-9

Liu, C. M., Yang, T. H., Huang, M., Zhou, C., Li, Y. H., & Li, Z. H. (2018). Effect of endomorphin-1 on maturation and expression of TLR4 in peripheral blood dendritic cells induced by high glucose. *Zhongguo Shi Yan Xue Ye Xue Za Zhi, 26*(3), 886–893. https://doi.org/10.7534/j.issn.1009-2137.2018.03.043

Lopker, A., Abood, L. G., Hoss, W., & Lionetti, F. J. (1980). Stereoselective muscarinic acetylcholine and opiate receptiors in human phagocytic leukocytes. *Biochemical Pharmacology, 29*(10), 1361–1365. https://doi.org/10.1016/0006-2952(80)90431-1

Machelska, H. (2007). Targeting of opioid-producing leukocytes for pain control. *Neuropeptides, 41*(6), 355–363. https://doi.org/10.1016/j.npep.2007.06.001

Machelska, H., & Celik, M. O. (2020). Opioid receptors in immune and glial cells-implications for pain control. *Frontiers in Immunology, 11*, 300. https://doi.org/10.3389/fimmu.2020.00300

Machelska, H., Brack, A., Mousa, S. A., Schopohl, J. K., Rittner, H. L., Schafer, M., & Stein, C. (2004). Selectins and integrins but not platelet-endothelial cell adhesion molecule-1 regulate opioid inhibition of inflammatory pain. *British Journal of Pharmacology, 142*(4), 772–780. https://doi.org/10.1038/sj.bjp.0705837

Maddila, S. C., Busch-Dienstfertig, M., & Stein, C. (2017). B lymphocytes express Pomc mRNA, processing enzymes and beta-endorphin in painful inflammation. *Journal of Neuroimmune Pharmacology, 12*(1), 180–186. https://doi.org/10.1007/s11481-016-9715-4

McCarthy, L., Wetzel, M., Sliker, J. K., Eisenstein, T. K., & Rogers, T. J. (2001). Opioids, opioid receptors, and the immune response. *Drug and Alcohol Dependence, 62*(2), 111–123. https://doi.org/10.1016/s0376-8716(00)00181-2

McLaughlin, P. J., McHugh, D. P., Magister, M. J., & Zagon, I. S. (2015). Endogenous opioid inhibition of proliferation of T and B cell subpopulations in response to immunization for experimental autoimmune encephalomyelitis. *BMC Immunology, 16*, 24. https://doi.org/10.1186/s12865-015-0093-0

Mehrotra, S., Prajapati, R. K., Haq, W., & Singh, V. K. (2002). Immunomodulation by biphalin, dimeric synthetic opioid peptide, and its analog. *Immunopharmacology and Immunotoxicology, 24*(1), 83–96. https://doi.org/10.1081/iph-120003405

Mestek, A., Hurley, J. H., Bye, L. S., Campbell, A. D., Chen, Y., Tian, M., Liu, J., Schulman, H., & Yu, L. (1995). The human mu opioid receptor: Modulation of functional desensitization by calcium/calmodulin-dependent protein kinase and protein kinase C. *The Journal of Neuroscience, 15*(3 Pt 2), 2396–2406. https://www.ncbi.nlm.nih.gov/pubmed/7891175

Meunier, J. C., Mollereau, C., Toll, L., Suaudeau, C., Moisand, C., Alvinerie, P., Butour, J. L., Guillemot, J. C., Ferrara, P., Monsarrat, B., et al. (1995). Isolation and structure of the endogenous agonist of opioid receptor-like ORL1 receptor. *Nature, 377*(6549), 532–535. https://doi.org/10.1038/377532a0

Mogil, J. S., & Pasternak, G. W. (2001). The molecular and behavioral pharmacology of the orphanin FQ/nociceptin peptide and receptor family. *Pharmacological Reviews, 53*(3), 381–415. https://www.ncbi.nlm.nih.gov/pubmed/11546835

Mousa, S. A., Shakibaei, M., Sitte, N., Schafer, M., & Stein, C. (2004). Subcellular pathways of beta-endorphin synthesis, processing, and release from immunocytes in inflammatory pain. *Endocrinology, 145*(3), 1331–1341. https://doi.org/10.1210/en.2003-1287

Mousa, S. A., Straub, R. H., Schafer, M., & Stein, C. (2007). Beta-endorphin, Met-enkephalin and corresponding opioid receptors within synovium of patients with joint trauma, osteoarthritis and rheumatoid arthritis. *Annals of the Rheumatic Diseases, 66*(7), 871–879. https://doi.org/10.1136/ard.2006.067066

Mousa, S. A., Shaqura, M., Brendl, U., Al-Khrasani, M., Furst, S., & Schafer, M. (2010). Involvement of the peripheral sensory and sympathetic nervous system in the vascular endothelial expression of ICAM-1 and the recruitment of opioid-containing immune cells to inhibit inflammatory pain. *Brain, Behavior, and Immunity, 24*(8), 1310–1323. https://doi.org/10.1016/j.bbi.2010.06.008

Nakanishi, S., Inoue, A., Kita, T., Nakamura, M., Chang, A. C., Cohen, S. N., & Numa, S. (1979). Nucleotide sequence of cloned cDNA for bovine corticotropin-beta-lipotropin precursor. *Nature, 278*(5703), 423–427. https://doi.org/10.1038/278423a0

Nandhra, T. S., & Carson, R. J. (2000). beta-endorphin inhibits the production of interleukin-8 by human chorio-decidual cells in culture. *Molecular Human Reproduction, 6*(6), 555–560. https://doi.org/10.1093/molehr/6.6.555

Navolotskaya, E. V., Kovalitskaya, Y. A., Zolotarev, Y. A., Kolobov, A. A., Kampe-Nemm, E. A., Malkova, N. V., Yurovsky, V. V., & Lipkin, V. M. (2004). Characteristics of non-opioid beta-endorphin receptor. *Biochemistry (Mosc), 69*(4), 394–400. https://doi.org/10.1023/b:biry.0000026194.10324.22

Ninkovic, J., & Roy, S. (2013). Role of the mu-opioid receptor in opioid modulation of immune function. *Amino Acids, 45*(1), 9–24. https://doi.org/10.1007/s00726-011-1163-0

Noda, M., Teranishi, Y., Takahashi, H., Toyosato, M., Notake, M., Nakanishi, S., & Numa, S. (1982). Isolation and structural organization of the human preproenkephalin gene. *Nature, 297*(5865), 431–434. https://doi.org/10.1038/297431a0

Ohmori, H., Fujii, K., Sasahira, T., Luo, Y., Isobe, M., Tatsumoto, N., & Kuniyasu, H. (2009). Methionine-enkephalin secreted by human colorectal cancer cells suppresses T lymphocytes. *Cancer Science, 100*(3), 497–502. https://doi.org/10.1111/j.1349-7006.2008.01073.x

Pampusch, M. S., Serie, J. R., Osinski, M. A., Seybold, V. S., Murtaugh, M. P., & Brown, D. R. (2000). Expression of nociceptin/OFQ receptor and prepro-nociceptin/OFQ in lymphoid tissues. *Peptides, 21*(12), 1865–1870. https://doi.org/10.1016/s0196-9781(00)00332-6

Pannell, M., Labuz, D., Celik, M. O., Keye, J., Batra, A., Siegmund, B., & Machelska, H. (2016). Adoptive transfer of M2 macrophages reduces neuropathic pain via opioid peptides. *Journal of Neuroinflammation, 13*(1), 262. https://doi.org/10.1186/s12974-016-0735-z

Plein, L. M., & Rittner, H. L. (2018). Opioids and the immune system - friend or foe. *British Journal of Pharmacology, 175*(14), 2717–2725. https://doi.org/10.1111/bph.13750

Pomorska, D. K., Gach, K., & Janecka, A. (2014). Immunomodulatory effects of endogenous and synthetic peptides activating opioid receptors. *Mini Reviews in Medicinal Chemistry, 14*(14), 1148–1155. https://doi.org/10.2174/1389557515666150101095237

Przewlocki, R., Hassan, A. H., Lason, W., Epplen, C., Herz, A., & Stein, C. (1992). Gene expression and localization of opioid peptides in immune cells of inflamed tissue: Functional role in antinociception. *Neuroscience, 48*(2), 491–500. https://doi.org/10.1016/0306-4522(92)90509-z

Reinscheid, R. K., Nothacker, H. P., Bourson, A., Ardati, A., Henningsen, R. A., Bunzow, J. R., Grandy, D. K., Langen, H., Monsma, F. J., Jr., & Civelli, O. (1995). Orphanin FQ: A neuropeptide that activates an opioidlike G protein-coupled receptor. *Science, 270*(5237), 792–794. https://doi.org/10.1126/science.270.5237.792

Rittner, H. L., Brack, A., Machelska, H., Mousa, S. A., Bauer, M., Schafer, M., & Stein, C. (2001). Opioid peptide-expressing leukocytes: Identification, recruitment, and simultaneously increasing inhibition of inflammatory pain. *Anesthesiology, 95*(2), 500–508. https://doi.org/10.1097/00000542-200108000-00036

Rittner, H. L., Labuz, D., Schaefer, M., Mousa, S. A., Schulz, S., Schafer, M., Stein, C., & Brack, A. (2006). Pain control by CXCR2 ligands through Ca2+–regulated release of opioid peptides from polymorphonuclear cells. *The FASEB Journal, 20*(14), 2627–2629. https://doi.org/10.1096/fj.06-6077fje

Roy, S., Ninkovic, J., Banerjee, S., Charboneau, R. G., Das, S., Dutta, R., Kirchner, V. A., Koodie, L., Ma, J., Meng, J., & Barke, R. A. (2011). Opioid drug abuse and modulation of immune function: Consequences in the susceptibility to opportunistic infections. *Journal of Neuroimmune Pharmacology, 6*(4), 442–465. https://doi.org/10.1007/s11481-011-9292-5

Sacerdote, P. (2006). Opioids and the immune system. *Palliative Medicine, 20*(Suppl 1), s9–s15. https://www.ncbi.nlm.nih.gov/pubmed/16764216

Salzet, M., Vieau, D., & Day, R. (2000). Crosstalk between nervous and immune systems through the animal kingdom: Focus on opioids. *Trends in Neurosciences, 23*(11), 550–555. https://doi.org/10.1016/s0166-2236(00)01642-8

Schafer, M., Carter, L., & Stein, C. (1994). Interleukin 1 beta and corticotropin-releasing factor inhibit pain by releasing opioids from immune cells in inflamed tissue. *Proceedings of the National Academy of Sciences of the United States of America, 91*(10), 4219–4223. https://doi.org/10.1073/pnas.91.10.4219

Shan, F., Xia, Y., Wang, N., Meng, J., Lu, C., Meng, Y., & Plotnikoff, N. P. (2011). Functional modulation of the pathway between dendritic cells (DCs) and CD4+T cells by the neuropeptide: Methionine enkephalin (MENK). *Peptides, 32*(5), 929–937. https://doi.org/10.1016/j.peptides.2011.01.033

Sharp, B. M. (2006). Multiple opioid receptors on immune cells modulate intracellular signaling. *Brain, Behavior, and Immunity, 20*(1), 9–14. https://doi.org/10.1016/j.bbi.2005.02.002

Sharp, B. M., Roy, S., & Bidlack, J. M. (1998). Evidence for opioid receptors on cells involved in host defense and the immune system. *Journal of Neuroimmunology, 83*(1–2), 45–56. https://www.ncbi.nlm.nih.gov/pubmed/9610672

Sibinga, N. E., & Goldstein, A. (1988). Opioid peptides and opioid receptors in cells of the immune system. *Annual Review of Immunology, 6*, 219–249. https://doi.org/10.1146/annurev.iy.06.040188.001251

Simonin, F., Gaveriaux-Ruff, C., Befort, K., Matthes, H., Lannes, B., Micheletti, G., Mattei, M. G., Charron, G., Bloch, B., & Kieffer, B. (1995). kappa-Opioid receptor in humans: cDNA and genomic cloning, chromosomal assignment, functional expression, pharmacology, and expression pattern in the central nervous system. *Proceedings of the National Academy of Sciences of the United States of America, 92*(15), 7006–7010. https://doi.org/10.1073/pnas.92.15.7006

Sitte, N., Busch, M., Mousa, S. A., Labuz, D., Rittner, H., Gore, C., Krause, H., Stein, C., & Schafer, M. (2007). Lymphocytes upregulate signal sequence-encoding proopiomelanocortin mRNA and beta-endorphin during painful inflammation in vivo. *Journal of Neuroimmunology, 183*(1–2), 133–145. https://doi.org/10.1016/j.jneuroim.2006.11.033

Stefano, G. B., Melchiorri, P., Negri, L., Hughes, T. K., Jr., & Scharrer, B. (1992). [D-Ala2]deltorphin I binding and pharmacological evidence for a special subtype of delta opioid receptor on human and invertebrate immune cells. *Proceedings of the National Academy of Sciences of the United States of America, 89*(19), 9316–9320. https://doi.org/10.1073/pnas.89.19.9316

Stein, C., & Machelska, H. (2011). Modulation of peripheral sensory neurons by the immune system: Implications for pain therapy. *Pharmacological Reviews, 63*(4), 860–881. https://doi.org/10.1124/pr.110.003145

Stein, C., Hassan, A. H., Lehrberger, K., Giefing, J., & Yassouridis, A. (1993). Local analgesic effect of endogenous opioid peptides. *Lancet, 342*(8867), 321–324. https://doi.org/10.1016/0140-6736(93)91471-w

Stein, C., Pfluger, M., Yassouridis, A., Hoelzl, J., Lehrberger, K., Welte, C., & Hassan, A. H. (1996). No tolerance to peripheral morphine analgesia in presence of opioid expression in inflamed synovia. *The Journal of Clinical Investigation, 98*(3), 793–799. https://doi.org/10.1172/JCI118852

Stein, C., Schafer, M., & Machelska, H. (2003). Attacking pain at its source: New perspectives on opioids. *Nature Medicine, 9*(8), 1003–1008. https://doi.org/10.1038/nm908

Straub, R. H., Dorner, M., Riedel, J., Kubitza, M., Van Rooijen, N., Lang, B., Scholmerich, J., & Falk, W. (1998). Tonic neurogenic inhibition of interleukin-6 secretion from murine spleen caused by opioidergic transmission. *The American Journal of Physiology, 274*(4), R997–R1003. https://doi.org/10.1152/ajpregu.1998.274.4.R997

Thompson, A. A., Liu, W., Chun, E., Katritch, V., Wu, H., Vardy, E., Huang, X. P., Trapella, C., Guerrini, R., Calo, G., Roth, B. L., Cherezov, V., & Stevens, R. C. (2012). Structure of the nociceptin/orphanin FQ receptor in complex with a peptide mimetic. *Nature, 485*(7398), 395–399. https://doi.org/10.1038/nature11085

Toskulkao, T., Pornchai, R., Akkarapatumwong, V., Vatanatunyakum, S., & Govitrapong, P. (2010). Alteration of lymphocyte opioid receptors in methadone maintenance subjects. *Neurochemistry International, 56*(2), 285–290. https://doi.org/10.1016/j.neuint.2009.10.013

Tuo, Y., Tian, C., Lu, L., & Xiang, M. (2020). The paradoxical role of methionine enkephalin in tumor responses. *European Journal of Pharmacology, 882*, 173253. https://doi.org/10.1016/j.ejphar.2020.173253

van den Bergh, P., Rozing, J., & Nagelkerken, L. (1993). Beta-endorphin stimulates Ia expression on mouse B cells by inducing interleukin-4 secretion by CD4+ T cells. *Cellular Immunology, 149*(1), 180–192. https://doi.org/10.1006/cimm.1993.1146

Wick, M. J., Minnerath, S. R., Roy, S., Ramakrishnan, S., & Loh, H. H. (1996). Differential expression of opioid receptor genes in human lymphoid cell lines and peripheral blood lymphocytes. *Journal of Neuroimmunology, 64*(1), 29–36. https://doi.org/10.1016/0165-5728(95)00144-1

Williams, J. P., Thompson, J. P., McDonald, J., Barnes, T. A., Cote, T., Rowbotham, D. J., & Lambert, D. G. (2007). Human peripheral blood mononuclear cells express nociceptin/

orphanin FQ, but not mu, delta, or kappa opioid receptors. *Anesthesia & Analgesia, 105*(4), 998–1005. table of contents. https://doi.org/10.1213/01.ane.0000278865.11991.9d

Xu, X., Gao, Y., Wen, L., Zhai, Z., Zhang, S., Shan, F., & Feng, J. (2016). Methionine enkephalin regulates microglia polarization and function. *International Immunopharmacology, 40*, 90–97. https://doi.org/10.1016/j.intimp.2016.08.037

Yang, X., Xia, H., Chen, Y., Liu, X., Zhou, C., Gao, Q., & Li, Z. (2012). Inducible expression of endomorphins in murine dendritic cells. *Neural Regeneration Research, 7*(35), 2811–2817. https://doi.org/10.3969/j.issn.1673-5374.2012.35.009

Zadina, J. E., Hackler, L., Ge, L. J., & Kastin, A. J. (1997). A potent and selective endogenous agonist for the mu-opiate receptor. *Nature, 386*(6624), 499–502. https://doi.org/10.1038/386499a0

Zainullina, L. F., Yamidanov, R. S., Vakhitov, V. A., & Vakhitova, Y. V. (2011). NMDA receptors as a possible component of store-operated Ca(2)(+) entry in human T-lymphocytes. *Biochemistry (Mosc), 76*(11), 1220–1226. https://doi.org/10.1134/S0006297911110034

The Opioid Growth Factor in Growth Regulation and Immune Responses in Cancer

Gerald R. Hankins and Robert T. Harris

Abstract It has become apparent that endogenous opioids act not only as neurotransmitters and neuromodulators, but have multiple functions in the body. Activation of the opioid system by opiate drugs is associated with a risk of cancer development through direct stimulation of tumor cell proliferation and through immunosuppression. In contrast, the endogenous peptide opioid [Met^5]-enkephalin, now commonly referred to as Opioid Growth Factor (OGF), negatively regulates cell proliferation in a wide number of cells during development, homeostasis, and neoplasia. This action is mediated through the opioid growth factor receptor, originally designated the zeta (ζ) opioid receptor. Further, contrary to the traditional notion of opiates as immunosuppressive, endogenous OGF has been shown to possess a number of positive immunomodulatory properties and may provide a beneficial effect in cancer by augmenting the activity of cells involved in both innate and acquired immunity. Taken together, the evidence supports consideration of opioid peptides such as OGF as new strategies for cancer therapy.

Keywords Opioid growth factor · Immunology · Cancer · Endogenous opioids

Introduction

Endogenous opioid peptides serve not only as neurotransmitters and neuromodulators, but have multiple functions in the body. Among these is the regulation of cell proliferation (Fichna & Janecka, 2004; Chen et al., 2008). Systematic identification of the opioid peptides involved in this regulation has depended on both the investigation of development and regeneration of normal tissues and on effects on tumorous cells and tissues. This has been reviewed by Zagon et al. (2002). Multiple studies indicated that [Met^5]-enkephalin (MENK, structurally

G. R. Hankins (✉) · R. T. Harris
Department of Biology, West Virginia State University, Institute, WV, USA
e-mail: ghankins@wvstateu.edu

P. L. Kerr et al. (eds.), *Endogenous Opioids*, Advances in Neurobiology 35,
https://doi.org/10.1007/978-3-031-45493-6_4

Try-Gly-Gly-Phe-Met) had the widest and, generally, greatest effect on cell proliferation. [Met^5]-enkephalin is 1 of 11 proteolytic products of the preproprotein encoded by the proenkephalin gene (HUGO nomenclature PENK, also known as preproenkephalin, proenkephalin-A, enkephalin A, Peptide F, PENK-A, and PE) which, in humans, is cytogenetically located at 8q12.1. Because of its widespread distribution, function in negative growth regulation and lack of elicitation of dependence, tolerance or withdrawal, [Met^5]-enkephalin has been given the alternative name Opioid Growth Factor (OGF), which will be the name most frequently used in this chapter (Zagon & McLaughlin, 1993). However this peptide, like other secreted signaling peptides, has multiple functions, including functions in apoptosis and immune system regulation (Zagon & McLaughlin, 2003; Shan et al., 2011; Zhao et al., 2016; Liang et al., 2016; Li et al., 2018; Wang et al., 2018a; Ma et al., 2020).

In early studies on neuroblastomas by Ian Zagon, Patricia McLaughlin, and collaborators, selective ligands for the classical opioid receptors (μ, δ, κ, ε, and σ) were not found effective in inhibiting cell proliferation in vitro and growth of tumors transplanted into immunodeficient mice. These investigations led to the characterization a saturable and specific high affinity receptor-binding site for [Met^5]-enkephalin binding in S20Y murine neuroblastoma cells which they proposed to be on a previously identified opioid receptor that was termed the zeta (ζ) opioid receptor (Zagon et al., 1989, 1990, 1993, 1991; Zagon & McLaughlin, 1993). Cloning of the receptor, first in rat, and later in mouse and human, indicates a protein that is unrelated to those of the classical opioid receptors, which are G protein-coupled receptors (Zagon et al., 1999b, 2000c, d). The proteins are between 580 and 677 amino acids long, depending on the species, are highly conserved among the species in the N-terminal end, and have three nuclear localization signals. The proteins lack the G protein-coupled receptor structures including the characteristic seven transmembrane domains, and a C-terminal tail that would interact with trimeric G proteins. The zeta opioid receptor name has largely been dropped in favor of the receptor being called the opioid growth factor receptor (OGFr). In humans, the gene (HUGO nomenclature OGFR) is located at 20q13.3, a subtelomeric location. The first part of this chapter focuses on the more well-defined actions of OGF through OGFr in control of cell proliferation. The later part reviews actions of OGF in immunomodulation where, when defined, the actions appear to occur through classical opioid receptors.

Tissue Distribution of OGF and Its Uptake by Cells

OGF is a secreted peptide that acts in an autocrine or paracrine fashion and is widely distributed and found an all tissues that have been evaluated (Zagon et al., 2003c). Internalization of OGF has been evaluated in COS-7 cells, an African Green Monkey derived cell line that lack classical opioid receptors. These cells do express both OGF and OGFr. Exogenous OGF suppresses cell replication in log-phase cultures while cell replication is increased by the opioid receptor antagonist naltrexone.

Treatments with other opioid agonists or antagonists do not have any significant effect on COS-7 cell proliferation (Zagon et al., 2003c). Using OGF labeled with 5,6-tetramethylrhodamine, Cheng et al. (2010a) demonstrated that OGF is internalized by clathrin-mediated endocytosis independent of endosomal or Golgi pathways.

Cellular Localization of OGFr and Entry into the Nucleus After OGF Binding

The first demonstration of the residence of OGFr on the nuclear envelope was in mouse keratinocytes, where OGF inhibits DNA synthesis dependent on the circadian rhythm (Zagon et al., 1994b, 2005b). Further, the regulation of cell proliferation by OGFr is dependent on functional nuclear localization signals. Using a human squamous cell carcinoma of the head and neck cell line, Cheng et al. demonstrated that upon binding by OGF, OGFr is bound at two of its three nuclear localization signals by the soluble transport factor karyopherin β (kapβ or KPNB1) (Cheng et al., 2009b). The complex then enters the nucleus by facilitated diffusion through the pore complex, dependent on action of the small GTPase RAN for the directionality of diffusion (Cheng et al., 2010a, b). This is illustrated in Fig. 1 (McLaughlin & Zagon, 2012). Only karyopherin β of the more than 20 karyopherin β family members was evaluated, so the function, if any, of other members of the family in OGFr transport is unknown. However siRNA knockdown of either karyopherin β or RAN was sufficient to increase cell proliferation. At low doses, binding of the opioid receptor antagonist naltrexone to OGFr can result in the naltrexone–OGFr complex signaling in a similar fashion as well as producing a feedback loop leading to more OGF and OGFr production (Wang et al., 2019). Although expression of p16 and p21 are known to be upregulated after entry of the OGF-OGFr complex into the nucleus, it is not clear whether this is due to direct interaction of the complex with promoter or enhancer regions. OGFr has a functional nuclear export signal and export is dependent on exportin 1 (XPO1 or exp1, also known as chromosome region maintenance 1 or CRM1), the most common mechanism for protein export from the nucleus (Kren et al., 2016).

OGF and OGFr Are Expressed During Normal Vertebrate Development and Function to Regulate Cell Proliferation

Early studies indicated that endogenous opioid signaling systems function in development and results of treatment of rodents with the opioid receptor antagonist naltrexone led Ian Zagon and Patricia McLaughlin to postulate that opioids can act as growth factors (Zagon & McLaughlin, 1983a, b, 1984). Subsequent studies have revealed OGF is a widespread regulator of cell proliferation during mammalian

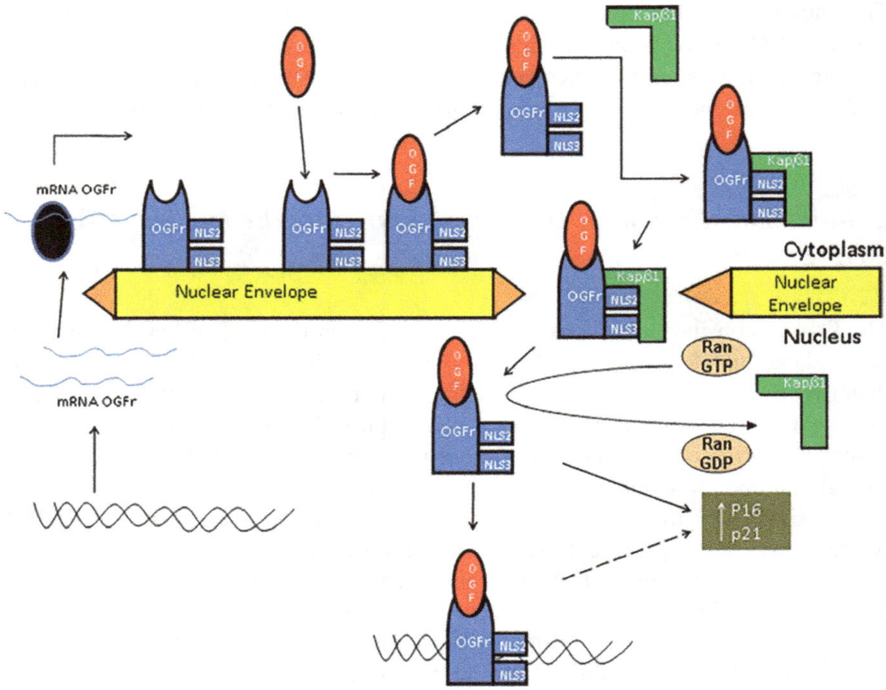

Fig. 1 Model of localization of nucleocytoplasmic trafficking of OGFr and function inside of the nucleus. Upon binding by OGF, OGFr is bound at two of its three nuclear localization signals by the soluble transport factor karyopherin β then enters the nucleus by facilitated diffusion through the pore complex, dependent on action of the small GTPase RAN. Expressions of p16 and p21 are known to be upregulated after entry of the OGF-OGFr complex into the nucleus; however, it is not clear whether this is due to direct interaction of the complex with promoter or enhancer regions. (From: McLaughlin and Zagon (2012))

development that is found in tissues derived from all three of the basic embryonic derivatives of gastrulation: ectoderm, mesoderm and endoderm (Keshet et al., 1989; Kew & Kilpatrick, 1990; Polakiewicz & Rosen, 1990; Zagon & McLaughlin, 1991; Zagon et al., 1995, 1997b). Whole embryo studies in rat embryos demonstrate decreased DNA synthesis in skin, spinal cord, cerebellum, vertebra, adrenal cortex, heart and liver after treatment with OGF (Zagon et al., 1999c). Treatment with naltrexone resulted in increased DNA synthesis in cerebrum, skin, spinal cord, rib, adrenal cortex, heart, liver, tongue, and intestine. The same studies found immunoreactivity against both OGF and OGFr in rat embryo brain, rib, and tongue and in skin, muscle, and intestine from a 31-week human fetus.

The initial studies of the presence of and regulation by OGF during development were in the central nervous system (Zagon & McLaughlin, 1987, 1991). The most extensive studies of the OGF-OGFr system in normal tissues have been in the rat CNS where the developmental regulation of location and timing of gene expression of both preproenkephalin and OGFr have been documented (Harlan et al., 1987;

Isayama & Zagon, 1991; Isayama et al., 1995, 1996; Zagon et al., 1994a; Zagon & McLaughlin, 2004; Osborne et al., 1991; Palmer et al., 1982). Receptor blockage with the antagonist naltrexone during development in rats results in increased brain size and cell numbers (Zagon & McLaughlin, 1983a).

In the cardiovascular system, preproenkephalin gene expression was first identified in the adult rat heart (Howells et al., 1986). During development, the message is high prior to birth, decreases prior to weaning, then increases, and reaches its highest levels in adulthood (Springhorn & Claycomb, 1989; Boluyt et al., 1993; McLaughlin & Allar, 1998; McLaughlin & Wu, 1998). Contrary to the levels of message, the levels of [Met^5]-enkephalin are very low compared to those in newborn rats (Barron et al., 1995; Boluyt et al., 1993; Howells et al., 1986; McLaughlin & Allar, 1998; Springhorn & Claycomb, 1989). Secretion of the peptide from adult myocytes has been described; however, at this time, it is not clear whether the peptides are acting in other heart cells (Springhorn & Claycomb, 1992). During development, if naltrexone is used to block the receptor rats show higher DNA synthesis and whole heart and ventricular weight increases (McLaughlin, 1994). This increase in heart size is similar to that seen in the brain of naltrexone treated animals. Similar developmental expression has also been observed in the rat aorta where both OGF and OGFr are detected in the mesenchymal wall by day E16 and in smooth muscle cells, fibroblasts and endothelial cells by day E20. As in the heart, OGF message levels decreased during the post natal period but rose again at adulthood (Wu et al., 1998; Zagon et al., 1996a). Both OGF and OGFr present endothelial and mesenchymal cells in chick chorioallantoic membrane. Exogenous OGF has an inhibitory effect on angiogenesis that is abrogated by naloxone in this system (Blebea et al., 2000).

Developmental expression has also been studied in the bone and uterus (Környei et al., 2003; Rosen et al., 1991, 1995; Vértes et al., 1996). Primary rat calvaria cells and the murine calvaria-derived cell line, MC 3T3/E1, express proenkephalin A mRNA and OGF. In rat development, there are high levels of PENK message prior to birth and shortly after, but the levels decrease after 6 days postpartum. Similarly, the message decreases as preosteoblasts begin to differentiate in vitro, whether spontaneously or in response to the bone-differentiating agents TGFβ or 1,25-dihydroxyvitamin D3.

Proliferation of uterine cells from 7-day-old rats was reduced by treatment with either exogenous OGF or another opioid peptide; [D-Met^2-Pro^5]-enkephalinamide was reduced whether the cells were unstimulated or stimulated with epidermal growth factor. However, OGF, but not [D-Met^2-Pro^5]-enkephalinamide, reduced proliferation of uterine cells from 14 day old rats. OGF and most other opioid peptides had no effect on the proliferation of estradiol or epidermal growth factor stimulated uterine cells taken from 21 to 60 day old rats; however [D-Met^2-Pro^5]-enkephalinamide and [D-Ala^2, N-Me-Phe^4, Gly^5-ol]-enkephalin did decrease proliferation of these cells (Környei et al., 2003). Overall, the data that OGF is active in controlling uterine cell proliferation early in development while some other opioid peptides act later.

In adults, OGF-OGFr regulation is involved in control of cell renewal in tissues including cornea and other epidermis, fibroblasts and vascular smooth muscle (Klocek et al., 2010; McLaughlin et al., 2012; Nissen & Kragballe, 1997; Zagon et al., 2003a, b, 2005b, c, 1997b, 1998, 2000b, 1994b, 1996b). Although outside the scope of this review, it should be noted that this has led to interest in inhibiting OGF-OGFr activity to promote regeneration and wound repair as well as in negative regulation of pathological growth besides cancer (Immonen et al., 2014; Zagon et al., 2003a).

OGF and OGFr and Regulation of Cell Proliferation in Cancer

Control of Cell Proliferation in Cancer Cells In Vitro by OGF-OGFr

The widespread function of the OGF-OGFr axis in controlling cell proliferation in normal development and regeneration suggests that this growth factor-receptor system would also be involved in tumorigenesis. Not surprisingly, its evaluation in tumors and tumor cell lines has been concomitant with the study of the system in normal development. Expression of OGF and OGFr is widespread in cancer cells and cell proliferation control by OGF-OGFr has been documented in over 50 tumor cell lines. These are derived from a diverse variety of human tumors which include astrocytomas (including glioblastoma), breast cancers, colon cancers, esophageal cancer, leukemia, hepatocellular carcinoma, lung cancers, melanomas, myeloma, neuroblastoma, ovarian cancers, pancreatic cancer, renal cancer, sarcomas, stomach cancer, squamous cell carcinoma of the head and neck, thyroid cancers, and uterine cancer (Avella et al., 2010; Bisignani et al., 1999; Donahue et al., 2009; McLaughlin et al., 1999, 2000, 2007, 2009; Rosen et al., 1991; Zagon et al., 2009, 1999a, 2000c; Wang et al., 2018a, b; Wu et al., 2018; Zheng et al., 2019).

There are several lines of evidence for the regulation of proliferation of cancer cells in vitro by the OGF-OGRr axis:

First, OGF and OGFr are widely, if not ubiquitously, expressed in human cancer cells. Expression has been evaluated at both the RNA level, and at the protein level by immunohistochemistry (Avella et al., 2010; Bisignani et al., 1999; Donahue et al., 2009; McLaughlin et al., 2009; Rosen et al., 1991; Zagon et al., 2009, 1996a). Further, both OGF and OGFr can be detected in both the cytoplasm and in local areas of the nucleus, which is consistent with previously reported results in normal cells (Avella et al., 2010; Donahue et al., 2009; McLaughlin et al., 2009; Zagon et al., 2009).

Second, exogenous OGF inhibits cell proliferation while the antagonist naltrexone can block the inhibition of proliferation by exogenous OGF, and in some cells may increase cell proliferation when cells are treated with the antagonist alone (Avella et al., 2010; Bisignani et al., 1999; Donahue et al., 2009; McLaughlin et al., 2009; Wang et al., 2016; Zagon et al., 2009, 1996a). However, low-dose naltrexone can decrease cell proliferation through a mechanism similar to OGF after binding OGFr and feedback can also lead to increased production of OGF and OGFr (Li et al., 2018; Wang et al., 2019; Ma et al., 2020).

Third, the inhibition of cell proliferation is reversible when media with exogenous OGF is replaced by media without exogenous OGF (Avella et al., 2010; Bisignani et al., 1999; Donahue et al., 2009; McLaughlin et al., 2009; Zagon et al., 1996a).

Fourth, neutralizing antibodies against OGF can block the inhibition of cell proliferation (Bisignani et al., 1999; McLaughlin et al., 2009; Zagon et al., 2009).

Fifth, siRNA knock-down of OGFr blocks the inhibitory action of OGF on cell proliferation, while OGFr overexpression reduces DNA synthesis and cell numbers (Avella et al., 2010; McLaughlin et al., 2007, 2009; Wu et al., 2018; Zagon et al., 2009).

Sixth, silencing of OGFr blocks the inhibitory action on cell proliferation of the opioid receptor antagonist, naltrexone. This has been demonstrated via siRNA knock-down of OGFr expression (Avella et al., 2010; Zagon et al., 2009). Recently, OGFr expression has been shown to be upregulated at both the RNA and protein levels by knockdown of the long noncoding RNA HOX transcript antisense intergenic RNA (HOTAIR) with concomitant inhibition of proliferation of hepatocellular carcinoma, HeLa, MCF-7 breast cancer, and HEK293T cells (Wu et al., 2018). The effect of HOTAIR on OGFr expression does not appear to be direct, and the indirect mechanism is not clear at this time. Another long noncoding RNA that affects OGFr expression is LINC00673. LINC00673 is upregulated in epithelial ovarian cancer cell lines and tissue and induces proliferation, migration and invasion of the cells and inhibits apoptosis in vitro. Knockdown of this LINC results in increased expression of OGFr RNA both in vitro and in tumor xenografts, while OGFr knockdown can ablate the effects of it on cell proliferation, migration and invasion (Zheng et al., 2019). It would be interesting to investigate whether HOTAIR expression has any effect on that of LINC00673.

Induction of Apoptosis in Cancer Cells by OGF-OGFr

When apoptosis and necrosis are examined, activation or inhibition of the OGF-OGFr axis has not shown any effect in most studies whether examined by TUNEL assay or trypan blue exclusion (Avella et al., 2010; Bisignani et al., 1999; Donahue et al., 2009; McLaughlin et al., 2009; Zagon et al., 2009, 1996a). Although naltrexone is an antagonist to the classical opioid receptors (μ, δ and κ) as well as OGFr, knock-down of OGFr led to an increase in cell numbers compared to treatment with naltrexone alone, suggesting that OGFr, not other naltrexone target receptors, was responsible for the effect of naltrexone treatment (Zagon et al., 2009). Further, while exogenous OGF inhibits growth of follicular thyroid carcinoma cells, ovarian cancer cells and hepatocarcinoma cells, treatment with specific agonists to the μ, δ and κ opioid receptors do not result in significant changes in cell proliferation (Avella et al., 2010; Donahue et al., 2009; McLaughlin et al., 2009). However, some studies have implicated OGF in induction of apoptosis through the μ and δ opioid receptors. In contrast to the earlier reports, more recent studies, largely from Fengping Shan's research group have implicated signaling through OGFr in induction of apoptosis. Nuclear factor of activated T cells (NFAT-1) activation by OGF was shown to result in apoptosis in rat C6 glioma cells (Lu et al., 2018). Exogenous OGF treatment of human gastric cancer cells increased OGFr expression, apoptotic cell numbers and mRNA levels of the proapoptotic factor, BAX, while downregulating mRNA for the antiapoptotic genes BCL-2 and survivin (Wang et al., 2018b). Also, protein levels of BAX, cleaved caspase 3, and cleaved PARP increased in the treated cells while protein levels of BCL-2 decreased. Decrease in PI3K, pAKT, and mTOR proteins along with BCL-2 indicates that OGR may be inhibiting signaling through the PI3K/AKT/mTOR pathway. Knock-down of OGFr reversed the effects of OGF on BAX, BCL-2, PI3K, pAKT, and mTOR (Wang et al., 2018a). Similar effects were observed in human colorectal cancer cells after treatment with low-dose naltrexone, indicating that the same apoptotic pathway may be acting after low-dose naltrexone treatment (Ma et al., 2020).

Control of Normal and Cancer Cell Proliferation by OGF-OGFr Acts Through Cyclin-Dependent Kinase Inhibitors p16INK4a and p21WAF1/CIP1

FACS analysis of human pancreatic (BxPC-3), colon (HT-29), and head and neck cancer (CAL-27) demonstrated that OGF, acting through OGFr, increases the proportion of cells in the G0/G1 phase of the cell cycle while reducing the proportion in the S, G2 and M phases. On the other hand, naltrexone decreased the proportion of cells in G0/G1 while increasing the proportion in S, G2 and M (Zagon et al., 2000a). This suggests that OGFr activation inhibits DNA synthesis by preventing passage through the G1 restriction point of the cell cycle. In BxPC-3 cells, OGF decreased phosphorylation of the retinoblastoma protein (RB1) and cyclin-dependent kinase 2 (Cdk2) activity and increased levels of cyclin-dependent kinase inhibitor 1A (CDKN1A, also known as $p21^{CIP1}$ and $p21^{WAF1}$), while naloxone abolished the increased expression of CDKN1A (Cheng et al., 2008). siRNAs to CDKN1A blocked the effects of OGF on BxPC-3 cells and two other pancreatic cancer cell lines, PANC-1 and Capan-2. CDKN1A inhibits the activity of Cdk2 which when complexed with cyclin E hyperphosphorylates RB1 leading to the passage of the cell beyond the G1 restriction point (Figs. 2 and 3, Weinberg, 2014). A similar study by the same group but on human head and neck squamous cell carcinoma cells also found decreased phosphorylation of RB1 after OGF treatment, but correlated with decreased Cdk4 rather than Cdk2 activity. Consistent with this latter observation was the increased expression of cyclin-dependent kinase inhibitor 2A (CDKN2A, also known as $p16^{INK4A}$) (Cheng et al., 2007). CDKN2A inhibits the activity of Cdk4, which when complexed to cyclin D. phosphorylates RB1. Without this first phosphorylation of RB1, hyperphosphorylation by Cdk2 doesn't occur.

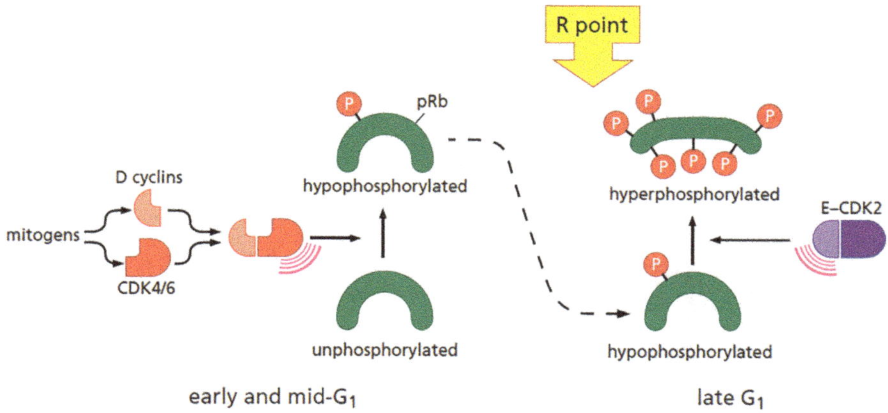

Fig. 2 Phosphorylation of the retinoblastoma tumor suppressor by cyclin-dependent kinases. Cyclin-dependent kinases complexed to cyclins phosphorylate and deactivate the retinoblastoma tumor suppressor protein, resulting in progression of the cell cycle beyond the G1 restriction point. (From Weinberg (2014). Copyright © 2014 by Garland Science, Taylor & Francis Group, LLC. Used by permission of W. W. Norton & Company, Inc.)

In either case, OGF-OGFr acts to inhibit cell cycle progression beyond the G1 restriction point by increasing transcription of CDKN2A or CDKN1A. Expressions of both of these cyclin-dependent kinase inhibitors were upregulated along with that of OGFr in hepatocellular carcinoma cells by HOTAIR knockdown (Wu et al., 2018). In two human ovarian cancer cell lines that have been investigated, knockdown of CDKN2A was required to block the effects of OGF on SKOV-3 cells, while knockdown of either CDKN2A or CDKN2B was sufficient for blocking the effects in OVCAR-3 cells (Donahue et al., 2009). This latter result was similar to that observed for normal cells. Knockdown of either of either CDKN2A or CDKN2B was sufficient for blocking the effects of OGF in human epidermal keratinocytes, umbilical vein endothelial cells, dermal fibroblasts and mesenchymal stem cells (Cheng et al., 2009c). The difference between the normal and most cancer cells may be due to deficiencies in the cyclin-dependent kinase inhibitors that are often observed in cancer cells, but this needs to be correlated with the specific cancer cells that have been investigated above. OGF treatment had no effects on protein expression of other cyclin-dependent kinase inhibitors examined: CDKN2B ($p15^{INK4B}$), CDKN2C ($p18^{INK4C}$), CDKN2D ($p19^{INK2D}$), and CDKN1B ($p27^{KIP1}$) (Fig. 4) (Cheng et al., 2009c; Weinberg, 2014). It is not clear at this point whether the OGF-OGFr complex interacts with the CDKN2A and CDKN2B promoters, or whether the mechanism is less direct.

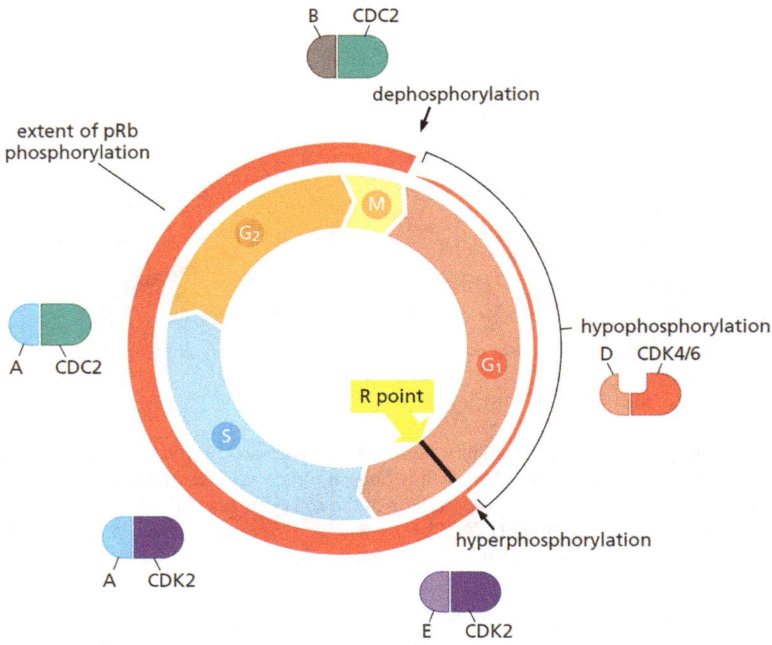

Fig. 3 The retinoblastoma tumor suppressor, pRB, remains hyperphosphorylated through mitosis. New phosphorylation is required to proceed beyond G1. (From Weinberg (2014). Copyright © 2014 by Garland Science, Taylor & Francis Group, LLC. Used by permission of W. W. Norton & Company, Inc.)

OGF-OGFr Expression in Human Surgical Cancer Specimens

Both OGF and OGFr are commonly found in cancerous tumors, although reports vary regarding the relative amounts compared to normal tissue. In lung cancer patients, Krajnik et al. observed colocalization of OGF, proenkephalin, and enzymes that involved in processing proenkephalin into OGF, prohormone convertases 1 and 2 and carboxypeptidase E. However, their study evaluated expression of the delta opioid receptor and not OGFr (Krajnik et al., 2010). Levin et al. found immunoreactivity for both OGF and OGFr in seven specimens of head and neck squamous cell carcinoma (Levin et al., 1997). In three hepatocellular specimens, immunohistochemical staining indicated no difference in staining between tumor and surrounding tissue was observed in three normal specimens (Avella et al., 2010). The limits of semiquantitative densitometry for stained tissue may have contributed to the results. However, a comparison by semiquantitative densitometry among specimens from 16 ovarian cancers, 27 benign ovarian cysts and normal human ovarian surface epithelial cells from postmenopausal women undergoing hysterectomy and oophorectomy found decreased expression of both OGF and OGFr in ovarian cysts compared to normal human ovarian surface epithelial cells. Expression of OGF and OGFr was further reduced in the cancers. Binding assays indicated only one-fifth as

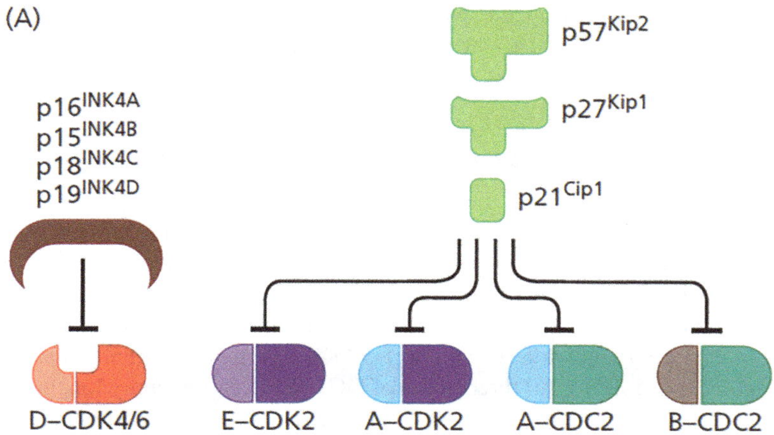

Fig. 4 Cyclin-dependent kinase inhibitors block cyclin-dependent kinase activity. These include the activity two whose expression is increased by OGF-OGFr: p16^{INK4A}, which blocks activation of cyclin-dependent kinases 4/6 that function in pRB hypophosphorylation and p21^{CIP1}, which blocks activity of cyclin-dependent kinase 2 that functions in pRB hyperphosphorylation. (From Weinberg (2014). Copyright © 2014 by Garland Science, Taylor & Francis Group, LLC. Used by permission of W. W. Norton & Company, Inc.)

many OGFr-binding sites in the cancers as in the cysts (Fanning et al., 2012). In an examination of 16 nonmedullary thyroid cancers, both OGF and OGFr were present. Normal thyroid tissue was not investigated, but 16 goiter samples were used to compare. OGRr binding capacity for [^{3}H]-[Met^5]-enkephalin was numerically reduced from that of patients with various goiters although the difference was not statistically significant (Goldenberg et al., 2008). In another study comparing cancers with noncancerous tissue from other patients, tissue from 64 squamous cell carcinomas of the head and neck were compared to tissue from 49 uvulopalatoplasty or tonsillectomy patients. OGFr protein levels from the cancers were one fifth those of normal epithelium while OGF binding was only one-ninth. mRNA levels were comparable, indicating that the protein differences were due, at least in part, to posttranscriptional mechanisms (McLaughlin et al., 2003b). Taken together, the results indicate that OGFr may be considered a tumor suppressor, although it is not clear how much of the difference is due to changes to the gene. One missing piece until recently was inactivating mutations in the gene. Out of 13 missense mutations in OGFr in the Somatic Mutations in Cancer (COSMIC) database, Kren et al. characterized two regarding their effect on nuclear trafficking of OGFr and effect on cell proliferation (Kren et al., 2016). Cells transfected with plasmid expression vectors for both mutated OGFr demonstrated lack of response to exogenous OGF and also to naltrexone. Cells transfected with OGFr with a mutation (R444H) in the ligand binding domain had a decrease in OGFr nuclear/cytoplasmic ratio and in DNA synthesis while those transfected with OGFr with a mutation (S378I) in a potential phosphorylation did not show these latter effects. The number of mutations of OGFr in COSMIC database has since expanded significantly, so further examination will likely demonstrate more that effect OGFr function.

Preclinical Studies of Human Tumors in Animal Models

Several studies by Zagon, McLaughlin and their colleagues have investigated the inhibition of human cancers after subcutaneous transplantation into athymic (nude) mice. HT-29 human colon cancer cells were subcutaneously injected into young male NCr-nu mice, followed one day later by daily injections of sterile water, OGF (0.5, 5, or 25 mg/kg), naloxone (10 mg/kg) or OGF (5 mg/kg) plus naloxone (10 mg/kg) (Zagon et al., 1996b). All three doses of OGF led to significantly longer periods of latency (measured by time to palpable, visible and measurable tumor), decreased tumor volume for mice with tumors, and reduction of mice with tumors on days 21 and 50 after tumor cell injection. While over 13/14 of the control mice developed tumors over the course of the experiment, only 18/42 of the OGF-treated mice did. No significant differences were observed among the groups receiving different doses of OGF. Naloxone abrogated the response. OGFr, immunoreactivity in the HT-29 tumors was observed in the cytoplasm, but not the nucleus. Subcutaneous injections of BxPC-3 human pancreatic cancer cells were followed by injections of 5 mg/kg OGF three times daily resulted in a longer period of latency, however 12/13 of the OGF treated mice had tumors 30 days after injection compare to 10/10 of the non-OGF-treated mice did) (Zagon et al., 1997a). Although binding of radiolabeled OGF was detected in nuclear homogenates from the tumor, the binding capacity was reduced to less than half of control levels. Results of transplantation of CAL-27 human squamous cell carcinomas into young male BALB/c nu/nu mice were intermediate between those described above. Mice were given daily intraperitoneal injections of saline, OGF (10 mg/kg), naltrexone, or both OGF and naltrexone. OGF treated mice had delayed appearance of detectable tumor, reduced tumor size and lower tumor incidence (24/29 OGF treated mice compared to 21/21, 27/27 and 20/20 in saline control, OGF/naltrexone and naltrexone respectively; McLaughlin et al., 2003a, b). As before, OGFr was presence in the tumors, but decreased numbers of receptors. Taken together, these studies indicate that exogenous OGF may reduce tumor onset and/or incidence, but that established tumors show a reduction in the numbers and/or function of the receptor. Evaluation of OGF and OGFr in triple negative human breast tumors (MDA-MB-231) in young female athymic nu/nu mice found downregulation of both in large tumors compared to small tumors (Worley et al., 2015). Reduction of OGFr expression in large tumors has also been reported for human squamous cell carcinomas of the head and neck (McLaughlin & Zagon, 2006). Interestingly, OGFr expression in MDA-MB-231 and MCF-7 breast cancer cells is reduced in confluent cells in vitro compared to log-phase cells (Worley et al., 2015). Therefore it is not clear how much of the downregulation in tumors is due to selection of cells that express less OGFr and how much is due to physiological conditions.

Naltrexone enters cells rapidly through passive diffusion; this makes it relatively easy to intermittently block OGFr (Cheng et al., 2009a). While continuous blockage of OGFr by naltrexone results in increased cell proliferation, intermittent low doses can result in decreased proliferation (McLaughlin & Zagon, 2015). Human ovarian cancer cell (SKOV-3) tumors in young female athymic nu/nu mice treated with daily injections of either OGF (10 mg/kg) or low-dose naltrexone (0.1 mg/kg) were reduced in size compared to those of saline treated mice, with reduction of both tumor cell proliferation and tumor angiogenesis (Donahue et al., 2011a). Both treatments increased the OGF in tumors, but only low-dose naltrexone increased OGFr protein. In a companion study, low-dose naltrexone reduced DNA angiogenesis, angiogenesis, and tumor progression in mice with established ovarian tumors (Donahue et al., 2011b). Further, low-dose naltrexone increased the inhibitory effect of cisplatin and reduced the weight loss from cisplatin treatment. Combination of low-dose naltrexone with taxol did not increase the inhibitory effect in this study. Shan and colleagues have demonstrated inhibition of in vitro proliferation of gastric and colorectal cancer cells by low-dose naltrexone (Ma et al., 2020; Wang et al., 2018a). Xenograft tumor size was also demonstrated in the same studies. However, Suzuki et al. used 0.3 mg/kg of methylnaltrexone to block OGF-OGFr signaling and release human diffuse gastric cancer cells from cell cycle arrest so they would be sensitive to docetaxel (Suzuki et al., 2015). Multiple gastric cancer cell lines along with PANC-1 pancreatic cells were evaluated in vitro. A highly peritoneal-seeding gastric cancer cell line (60As6) was inoculated into young female C.B17/Icr-scid mice and allowed to establish for 7 days before treatment with saline, methylnaltrexone, docetaxel or methylnaltrexone and docetaxel. The combination treatment extended median survival time from 56 days for docetaxel alone to 74 days for methylnaltrexone and docetaxel and decreased abdominal pain (based on scoring of hunching behavior).

Combination therapy using OGF and standard chemotherapeutic agents has also been evaluated in xenografts. Xenografts of human squamous cell carcinoma cells (SCC-1) in young male athymic mice followed by injections of OGF (10 mg/kg daily) and paclitaxel (8 mg/kg every other day) resulted in significantly higher body weights and longer survival than for mice receiving only paclitaxel (Jaglowski et al., 2005). This appears to be largely due to moderation of the toxicity of paclitaxel as survival with the combination was not significantly improved over treatment with OGF alone or saline. Combination therapy of OGF (10 mg/kg daily) and gemcitabine (120 mg/kg every third day) after injection of MIA PaCa-2 pancreatic adenocarcinoma cells resulted in increased latency time for tumor appearance (mean of 10.1 days for saline, 10.7 days for OGF, 11.1 days for gemcitabine, and 19.5 days for OGF and gemcitabine) and decreased tumor volumes compared to saline or single-agent treatment (Zagon et al., 2005a). More recently, Sikong et al. evaluated combination of exogenous OGR and cisplatin on xenografts of human hepatocellular carcinoma in athymic mice. After 28 days of treatment, mice treated with both OGF and cisplatin had lower tumor volumes than those of mice treated with only OGF or cisplatin alone while tumors in the saline treated controls where the highest (Sikong et al., 2019). Immunohistochemical staining indicated higher levels of

OGFr, p16^{INK4a}, p21^{Cip1}, and p53 in the tumors from mice treated with both OGF and cisplatin than with cisplatin alone.

Taken together, the work to date indicates that OGF alone may be useful as an anticancer agent. However, either OGF or naltrexone in common with other therapeutic agents may be more effective. The relative effectiveness will likely prove to be dependent on the state of OGF and OGFr in the target tumors and how widespread latent cells are.

Clinical Trials Targeting OGF-OGFr Regulation of Cell Proliferation

Phase I and phase II clinical trials have been initiated on the use of OGF as an anticancer therapy has indicated low toxicity of OGF but mixed results on effectiveness. Results of two trials in pancreatic cancer patients have been published in primary form and in a review (Smith et al., 2004, 2010; Zagon & McLaughlan, 2014). A phase I trial found that the maximum tolerated dose was 250 μg/kg weekly with intravenous injection with hypotension being the dose-limiting toxic effect (Smith et al., 2004). Subcutaneous treatment was limited to 50 μg/kg twice daily because of limited solubility of OGF in the relatively small volumes that could be effectively delivered by this route. During the chronic phase of the study, ten patients were treated via the intravenous route and sex via subcutaneous injection. Treatment by both routes increased the plasma OGF levels. Interestingly, plasma levels of OGF have been found to be higher in pancreatic cancer patients and this has been suggested as a diagnostic indicator of the disease (Smith et al., 2000). Two of the patients treated via the intravenous route had transient resolution of hepatic metastases and one of these also experiences significant decrease in size of the primary tumor. The survival of these two patients was 18 and 23 months post diagnosis as compared to the range of 2–23 months for the 10 patients in the intravenous treatment group. In a phase II trial by the Smith et al., 24 advanced pancreatic cancer patients for whom standard chemotherapy had failed were treated with intravenous 250 μg/kg OGF weekly (Smith et al., 2000). The median survival time was 65.5 days as compared to 21 days for 166 patients meeting the same inclusion criteria but opting for hospice rather than further treatment. Multiple quality of life measurements showed either no change or improvement. Although plasma OGF levels increased during the study and in spite of the function of the OGF-OGFr axis in normal homeostasis, no significant changes were observed in hematologic parameters. One obvious extension would to be clinical trials of combination treatments. A phase I study combining OGF and gemcitabine for pancreatic cancer was designed that would be a logical extension of the animal study described in the previous section (Milton S. Hershey Medical Center, 2008; Zagon et al., 2005a, b, c). Unfortunately, problems with the IRB led to termination of recruitment.

Other clinical studies on OGF for cancer therapy are either unpublished or have been terminated. Results of a phase I study on unresectable hepatocellular carcinoma have not been posted (University of Missouri-Columbia, 2008). Recruitment for a phase I study for solid tumors was terminated by Innovive Pharmaceuticals (Innovive Pharmaceuticals, 2005). A phase II study on treatment of head and neck tumors was discontinued when the treatment was decided to be ineffective (Milton S. Hershey Medical Center, 2009).

Effect of OGF on Regulation of the Immune System

It is now apparent that the effects of endogenous opioids extend well beyond that of functioning as powerful centrally acting analgesics. Indeed, opioids have been clearly shown to have effects on multiple functions of the body including that of the immune system (Liang et al., 2016; Zhao et al., 2016).

It should thus not be surprising that the peptide opioid [Met^5]-enkephalin is now referred to commonly as Opioid Growth Factor. As is quite common with the names of signaling molecules such as circulating hormones, even this contemporary term fails to adequately represent the multifunctional nature of this signaling peptide.

It has been long known that activation of the opioid system by opiate drugs is associated with a risk of cancer development. An active area of research in fact is directed toward development of drugs that provide adequate pain management while minimizing the negative impact of these side effects. In addition to direct stimulation of tumor cell proliferation, opiates such as morphine may induce tumorigenesis indirectly through inhibition of the antitumor cytokine IL-2 and stimulation of the immunosuppressive cytokine IL-10. Thus, given the traditional notion of opiates as immunosuppressive, is may be surprising that the endogenous opioid OGF has been shown to possess a number of positive immunomodulatory properties (Liang et al., 2016; Zhao et al., 2016).

Opioids Produce Effects in the Periphery

The demonstration of peripheral sources of OGF, such as the adrenal glands and the landmark discovery of binding sites for opioids on the surface of T cells in 1979 paved the way for research that has demonstrated multiple effects of OGF on immune function (Tischler et al., 1983; Wybran et al., 1979). Not long after this work, it was found that endorphins and proenkephalin A-derived peptides including OGF are produced and released by peripheral human blood lymphocytes (Smith et al., 1985; Sibinga & Goldstein, 1988). A decade after the discovery of these receptors, it was shown that an analogue of OGF, D-[Ala2]methionine enkephalinamide (MEA) was able to inhibit the growth PYB6 fibrosarcoma tumors by 72% (Srisuchart et al., 1989). Of major significance was the finding that MEA was not directly toxic to PYB6 tumor cells but instead worked by selectively enhancing the lymphoproliferative response to a T-cell mitogen.

It is quite clear that the various opiate drugs and endogenous opioids do not share the same immune profile. Although numerous studies have demonstrated that these molecules are often immunomodulatory, they produce differential effects on immune cells which may potentially alter interactions among the cells and ultimately lead to markedly different overall immune system responses. Some opioids seem to have no effects on immune function, whereas others tend to be either immunosuppressive or immunostimulatory. This is probably due to a combination of effects, including direct effects on immune cells, indirect effects which involve production and secretion of immunomodulatory mediators that influence the activity of other immune cells, and perhaps a global in vivo effect of the molecules at the whole organism level.

Despite the apparent contradictory findings in the literature, there is growing evidence that the endogenous peptide OGF may enhance the activity of the body's immediate, nonspecific responses which are provided by its innate immunity. Further, OGF appears to also have the potential to activate the body's second layer of protection which is provided by the adaptive arm of the immune system. When considering the potential significance of data regarding the in vitro effects of opioids such as OGF, it is important to remember that the reported specificities of ligands are relative and not absolute. In other words, at higher levels of concentration, a specific molecule may bind to receptors other than the one to which it is said to be "selective." This is in fact a well-known phenomenon involving opioid binding to the classical delta, kappa, and mu receptors (Zagon et al., 2002; Plotnikoff et al., 1997). Secondly, it is important to note that the concentrations of opioid peptides used in different studies are highly variable. As soluble, low-molecular-weight signaling molecules/hormones and cytokines, through binding to specific receptors on target cells, play a vital role in immune function. Owing to the remarkably high affinity of these receptors, cytokines can produce biological affects at very low concentrations. It should thus not be surprising that experiments using different cytokine concentrations show highly variable and sometime even opposite effects.

OGF Has Multiple Immunomodulatory Effects on T-Cell Function

The existence of receptors for OGF (and morphine) on the surface of T lymphocytes was demonstrated nearly four decades ago (Wybran et al., 1979). Research that followed uncovered a complex set of effects on T-cell immune function elicited by OGF (Liang et al., 2016; Tischler et al., 1983). While some of the early studies in this area suggested that OGF enhances the function of T cells, others found a suppression of T-cell responses. To date, however, a majority of the research involving

T cells seems to support the notion of using OGF as a component of cancer immunotherapy.

Lymphocytes are mononuclear leukocytes that are involved in both humoral and cell-mediated immunity and are classified broadly as B cells and T cells. Maturing mainly in the thymus (or tonsil), T cells are further classified as the long-lived memory cells that target future infections and effector cells, which comprise a superset of T cells that respond immediately to a stimulus. It is the three types of effector cells, cytotoxic (CD8+) T cells, helper (CD4+) T cells, and regulatory T cells that have been the subject of numerous studies involving the influence of opioids on immunity. It is clear that the functions all three types of effector cells can be modulated by OGF.

Increase in the Frequency and Activity of Cytotoxic T Cells and Helper T Cells

A cytotoxic T cell (also commonly known as a CTL, T_C, killer T cell, etc.) is a T lymphocyte that possesses the machinery to recognize and kill host cells that are viral infected, damaged, or cancerous. Since the cell relies on the presence of the transmembrane glycoprotein CD8 which serves as a coreceptor for the T-cell receptor (TCR), these cells are referred to formally as $CD8^+$ T cells. A helper T (T_H) cell is a lymphocyte that serves to stimulate other immune cells such as cytotoxic T cells, B cells, and macrophages to engage in an immune response. Mature T_H cells express the surface protein CD4 and are thus often referred to as $CD4^+$ T cells. The ability of helper T cells to activate antigen-specific effector cells such as cytotoxic T cells and to recruit cells of the innate immune system such as macrophages is central to the development of an immune response.

Some of the earliest investigation of OGF and immunity involved efforts to determine the extent of rosette formation in cultured T cells (Miller et al., 1983, 1984; Murgo et al., 1985; Plotnikoff et al., 1987; Wybran et al., 1979). The agglutination of cells such as lymphocytes to erythrocytes forms a flower-like pattern which is referred to as a rosette and the degree of rosetting is indicative of the number of active cells that are present. OGF was shown to significantly enhance active T-cell rosette formation by human peripheral blood lymphocytes from both normal subjects and lymphoma patients (Miller et al., 1983, 1984; Murgo et al., 1985; Plotnikoff et al., 1987; Wybran et al., 1979). It was thus suggested in these early studies that OGF could have a significant role in enhancing cell-mediated immunity and T-cell function.

Perhaps the most controversial area of OGF related immunomodulation involves the extent and direction of OGF involvement in T-cell proliferation. In contrast to early studies showing enhances T-cell activity, some researchers have reported that T-cell proliferation may in some situations by suppressed by OGF (Zagon et al., 2011; Shahabi & Sharp, 1995). Zagon's group examined the effects of OGF on splenic-derived murine lymphocytes which were prestimulated with the mitogen, phytohemagglutinin (PHA) (Zagon et al., 2011). Most relevant to this review was their finding that OGF suppressed T-cell proliferation in prestimulated (but not unstimulated) cultures. It must be noted however that this study used relatively high levels of OGF (10^{-7} to 10^{-4} M) and thus considering the bimodal effects of opioids reported by others, this result is not altogether surprising. Further, the influence of OGF on the various subpopulations of T cells was not considered in this study (Zagon et al., 2011). Using a similar model however, Shahabi and Sharp using a molecule similar to OGF, the delta receptor agonist, [D-Ala^2]-met-enkephalinamide, reported an antiproliferative effect on highly purified murine on both helper ($CD4^+$) and cytotoxic ($CD8^+$) T cells (Shahabi & Sharp, 1995).

In contrast to these findings, the results of others studies report a positive effect of OGF on lymphocyte proliferation (Hucklebridge et al., 1990; Ye et al., 1989; Kowalski, 1998; Gabrilovac & Marotti, 2000: Sorensen & Claesson, 1998; Heagy et al., 1990). Hucklebridge et al. found that stimulation involving OGF acting through the delta receptor stimulated in vitro proliferation of T cell in human peripheral blood (Hucklebridge et al., 1990). Of particular significance was the fact that proliferation did not require prestimulation by a lymphocyte mitogen. Further, while the mu-selective agonist DAGON had no effect, OGF stimulation was blocked by the delta-selective antagonist ICI-174864, suggesting the involvement of the classical delta opioid receptors in OGF-induced T-cell proliferation. Likewise, the results of in vivo animal work support the notion of a positive effect on proliferation (Kowalski, 1998; Gabrilovac & Marotti, 2000). Gabrilovac and Marotti investigated the proliferative effects of OGF on splenic lymphocytes that were stimulated by the mitogen concanavalin-A (Con-A) (Gabrilovac & Marotti, 2000). These investigators found that 14 h after a single intraperitoneal (IP) injection of OGF at 2.5 mg/kg body weight, mitogen-induced proliferation of T cells was significantly enhanced in male mice. This OGF-induced stimulatory effect on T-cell proliferation was reversed by naloxone (10 mg/kg body weight) injection prior to administration to OGF treatment, suggesting an involvement of opioid receptors. It was further found however that, similar to the bimodal effect reported by Zagon's group, a higher concentration of OGF (10 mg/kg body weight) had on effect on T-cell proliferation. Likewise, Kowalski found that systemic administration of OGF via a single IP injection increased mitogen-induced (Con-A) proliferation of T cells in mice and the effect was by naloxone pretreatment (Kowalski, 1998). In addition to these studies, Bajpai and coworkers found that in addition to naturally occurring OGF, two of its more stable analogs, Tyr-D-Ala-Gly-MePhe-Met-NHC3H gamma-iso and Tyr-D-Ala-Gly-MePhe-Gly-NHC3H gamma-iso, were also able to significantly enhanced human T-cell proliferation in vitro after 5 days of incubation (Bajpai et al., 1995, 1997). This enhancement in proliferation occurred even in the absence of mitogen and was completely inhibited by naloxone. Additionally in this study, treatment with OGF or either of these analogs led to significant enhancement of active T-cell rosettes and the production of IL-2 production. Finally, regarding specific subpopulations of T cells, Hua and associates found that in vitro OGF treatment (10^{-12} M) was associated with expansion of several populations of cells including both $CD4^+$ helper T cells and $CD8^+$ cytotoxic T cells (Hua et al., 2012).

While the studies reporting a positive immunomodulatory effect of OGF on various T-cell subpopulations are numerous, it must be pointed that it has suggested that at least in certain cancers, OGF may have an overall negative influence on cancer immunity. Ohmori et al. have suggested that the OGF that is secreted by human colorectal cancer cells leads to suppression of T lymphocyte numbers (Ohmori et al., 2009). In this instance, it was suggested that the effect of OGF may be to help the cancer cells escape the host's anticancer immunity through inhibition of the expression of $CD4^{+}$ helper T cells. Thus, despite the fact that the direct effect of OGF on T cells seems to be a positive immunomodulatory response, this effect on the T-cell response to cancer may in at least some situations be tempered by interaction with tumor cells.

Decrease in the Frequency and Activity of Regulatory T Cells

Regulatory T cells (Tregs) have been identified as a subpopulation of T lymphocytes that work to suppress the immune system, helping to insure a balanced immunity for the body. By suppressing activation of the immune system, they function to maintain immune system homeostasis and to assure tolerance of the system to self-antigens. Treg cells function to inactivate cytotoxic T cells, and this T-cell-mediated immunosuppression is accomplished through secretion of IL-10 and adenosine. Since these cells express CD4, CD25, and Foxp3 markers, they are referred to by several designations [e.g., CD4+CD25+; CD4+ Foxp3+ CD25 (high)]. Foxp3 functions as a master regulator in the development and function of Treg cells. It is a multifaceted factor of cancer biology, and while being associated with suppression of carcinogenesis as a potent repressor of several oncogenes, in regard to cancer immunity, it has a tumor enhancing role through Tregs and the effect of these cells on tumor tolerance.

OGF has been clearly shown to have a immunomodulatory effect on Treg cells, with marked reduction cell population being universally reported (Hua et al., 2012; Li et al., 2015; Wang et al., 2014). The work of Hua and associates demonstrated that a 35% decrease in the number of $CD4^{+}CD25^{+}$ Treg cells was induced by treatment of cultures of human peripheral blood with 10^{-12} M OGF (Hua et al., 2012). As indicated earlier, this decline in the Treg population was associated with expansion of several populations of cells including both helper T cells and cytotoxic T cells. As discussed in detail below, this same group examined the proliferation of subpopulations of nucleated cells in the peripheral blood of cancer patients before and after treatment with OGF. Treatment with 10^{-12} M OGF induced a strong inhibitory effect on Treg cells and led to a population decrease of approximately 38%. This decline in Treg numbers was associated with a significant decrease in proliferation of both helper T cells and cytotoxic T cells. It is thus concluded that OGF can inhibit the activity of Treg cells and that this inhibition is associated with both an expansion of several major types of immune cells including subpopulations of lymphocyte and a slowing of tumor growth (Hua et al., 2012; Li et al., 2015; Wang et al., 2014).

Role of T Cells in Antitumor Immunity

Historically, cancer immunotherapy has focused on eliciting and strengthening the cytotoxic T-cell response to kill tumor cells. The importance of this as well as enhancing this response as well as the supportive role played by helper T cells is apparent. Thus, a major effect of OGF that relate to involvement of T cells in anti-cancer immunity seems to involve an increased activity of both cytotoxic and helper T cells that may result from a decrease in activity of Treg cells. It is well established from in vivo studies of animals and humans that cancer is associated with an increased number of Treg cells. During an immune response to cancer, it appears that Tregs are preferentially trafficked to the tumor microenvironment. Thus, while Tregs normally make up approximately 4% of $CD4^+$ T cells, they may account for as much as 20–30% of the total $CD4^+$ population within and around a tumor. Evidence continues to emerge which suggest that suppression of immunity by Tregs contributes to tumor growth and progression. Following recent clinical trials in cancer patients, the researchers concluded that the beneficial effect of OGF on expansion of T-cell subpopulations could be mainly due to inhibition of Tregs (Wang et al., 2014). Inhibition of Treg proliferation is thus an important mechanism by which OGF may exert positive effects on anticancer activities of the immune system.

While cancer immunotherapy has focused largely activity of cytotoxic T-cell response to kill tumor cells, more recently increasing attention has been given to efforts to stimulate T_H cells. T_H cells exist predominantly as one of two subtypes which are referred to as T_H1 and T_H2. Differentiation of T_H cells is dependent on the local cytokine environment, with commitment to T_H1 development depending on local production of IL-12. As discussed below, OGF's effects on macrophages results in a local environment that leads to proliferation of T_H1 cells, which in turn leads to enhanced function of both cytotoxic T cells and macrophages. As is the case with autoimmune diseases, modulation of the T_H1 cell response against a tumor antigen could be part of an effective immune-based anticancer therapy. T_H1 cells may promote anticancer immunity by release of cytokines that lead to activation of death receptors on the surface of tumor cells.

OGF Enhances the Immunological Function of Dendritic Cells

Dendritic cells (DCs) are found in tissues that are in direct contact with the external environment of the body, such as the skin and the digestive and respiratory tracts. In these locations, they serve as immunological sentinels for these potentially vulnerable peripheral tissues where they continuously sample the environment for pathogens. The primary function of a DC is to serve as an antigen presenting cell (APC). In this role, they capture and process the antigens that they encounter and subsequently present the processed antigen material to T cells, thereby initiating an immune response. In its peripheral tissue, a DC resides in a resting or immature state which is capable of capturing and processing antigens. This activity results in a differentiation to a cell with less antigen processing capabilities and an enhanced expression of MHC protein and necessary costimulatory molecules. The active DC migrates to a secondary lymphoid organ where it presents processed antigen to a naïve T cells. In addition, the activated DC functions to induce appropriate polarization of helper T cells to a T_H1 or T_H2 phenotype, a process that depends on the pathological context. DCs thus serve the vital function of instructing the adaptive arm of the immunity and directing it to respond to challenges from exposure to environmental pathogens.

Induction of DC Differentiation and Maturation and Enhancement of Antigen Processing

The work of Shan's laboratory has focused on the effects of OGF on differentiation, maturation, and function of DCs (Liu et al., 2012; Li et al., 2012; Meng et al., 2013, 2017; Shan et al., 2011). Using murine bone marrow derived dendritic cells (BMDCs), this group demonstrated that OGF can induce DCs to polarize predominantly to the myeloid or "conventional" subtype (cDC) (Liu et al., 2012). It is the cDC phenotype that secretes the key immune cytokines, IL-6, IL-12, and tumor necrosis factor α (TNFα) and is a major stimulator of T-cell function. Further, it was shown that OGF led to an increase in expressions of MHC-II and expression of the maturation markers CD40, CD80, CD83, and CD86, which are key costimulatory molecules on the DC cell surface (Liu et al., 2012; Li et al., 2012; Meng et al., 2013; Shan et al., 2011). These changes were accompanied by an increase in secretion of proinflammatory cytokines of IL-12 and TNFα (Liu et al., 2012; Li et al., 2012; Meng et al., 2013; Shan et al., 2011).

In addition to these molecular changes, OGF was found to enhance DCs' ability to induce T-cell proliferation and to intensify the immune response mounted by cytotoxic T cells (Liu et al., 2012; Li et al., 2012; Meng et al., 2017). It was further shown that OGF treatment augmented the DCs' capacity to induce apoptosis of tumor cells in vitro (Liu et al., 2012), and to inhibit the growth of tumors in vivo (Liu et al., 2012; Meng et al., 2017). Another key finding of this group was an effect of OGF on subpopulations of the T cells. OGF alone or in combination with IL-2 or IFN-γ, markedly increased the proliferation of helper T cells, including increased expression the CD4 molecule (Shan et al., 2011). This was accompanied by an increased production of IL-2 by helper T cells. Finally, it was shown by Meng et al. that OGF treatment of DCs led to increased activation of both helper T cells and cytotoxic T cells and decreased capacity of DCs to induce differentiation of $Foxp3^+$ Treg cells (Meng et al., 2017). It is thus concluded from this work that OGF is able to induced phenotypic and functional maturation of DCs, as evidenced by increased expression of key surface markers and an increase in production of cytokines which in turn alters the function of other immune cells.

Role of Dendritic Cells in Antitumor Immunity

DCs are central regulators of adaptive immunity, and as such are required for a proper T-cell-mediated immune response against cancer. Several studies have reported changes in DCs with OGF treatment that are consistent with enhancement of immunity, including polarization of DCs and an enhancement of helper and cytotoxic T-cell function. OGF appears to induce DCs to polarize to the "conventional" (cDC) subtype. It is the cDC that is particularly adept at initiating a T-cell response and presenting antigen on either MHC-I or MHC-II. By phagocytizing dead cancer cells or cellular debris, tumor-associated cDCs are able to transport cancer-associated antigens to a local lymph node where T-cell priming and activation occur.

The two major functions of cDCs, antigen presentation and T-cell sensitization, are thus vital for induction of a de novo T-cell response and consequently for development of antitumor immunity. The vital role of DC function in anticancer immunity is apparent, as all cancer vaccines are dependent upon the ability of DCs to act as antigen-presenting cells to T cells. The future translational potential of DC-targeted therapies will not doubt rely on molecules that can drive the antigen presenting and stimulatory capacity of DCs. Additionally, OGF seems to exert a positive modulation on DC-CD4^{+}helper T-cell interaction by increasing expression and activity of delta receptors on the surface of the CD 4^{+} helper cells (Shan et al., 2011). Shan's group has recently suggested that this effect of OGF could potentially help to overcome the bottleneck involved in clinical application of IL-2 and IFN-γ for T-cell-mediated immune responses (Shan et al., 2011). Thus, in sum, a positive immunomodulation of DC function by OGF could significantly enhance the efficiency of such anticancer immunotherapy.

Phagocytosis and Cytokine Secretion by Macrophages Is Enhanced by Exposure to OGF

A macrophage is highly specialized phagocyte that engulfs and digests foreign material, cellular debris, microbes, and cancer cells. Found in essentially all tissues of the body and having a variety of forms with specific names, a macrophage uses an amoebic-like movement to patrol its resident tissue for potential pathogens. The function of these versatile cells extends well beyond that of phagocytosis. Like DCs, macrophages play an important role in initiating an immune response by serving as antigen presenting cells. In addition to these functions, macrophages play a vital role in regulation of the immune response through secretion of a number of immunomodulatory cytokines that help to regulate the immune response through communication with other cell and by induction of an inflammatory response. Depending on the cytokine microenvironment, macrophages adapt their phenotype and as such they can be classified as an M1 ("classically activated") macrophage which is proinflammatory and is adapted for phagocytosis or the M2 ("alternatively activated") phenotype which is better adapted for tissue repair.

Stimulation of Phagocytosis and Cytokine Secretion and Upregulation of Surface Markers

Foris and coworkers found that antibody-dependent cell-mediated cytotoxicity (ADCC) in rat peritoneal macrophages was stimulated by OGF at a relatively low concentration range of 10^{-9} to 10^{-7} M (Foris et al., 1986). Within this range (10^{-8}), the effects of OGF appeared to result from specific binding to delta opioid receptors with a corresponding activation of guanylate cyclase and an elevation of cGMP. In contrast, a higher concentration of OGF (10^{-6} to 10^{-5}) did not stimulate ADCC, but instead was associated with an increase in Fcγ receptor (FcγR)-mediated phagocytosis in a manner that was not mediated by the delta receptor. Likewise, the work of Petty et al. examined the effect of OGF on antibody-dependent effector activity of RAW264 cells (Petty & Berg, 1988; Petty & Martin, 1989). It was shown that OGF enhanced the ability of these murine cells to phagocytize red blood cells from sheep. The optimal dose that stimulated ADCC in this study was 10^{-8} M (Petty & Berg, 1988). In addition, their results suggest an additional positive effect on macrophage function as OGF induced a dose-dependent increase in cell spreading, again with a maximal response seen at 10^{-8} M (Petty & Martin, 1989).

Likewise, Vujic's group demonstrated a positive modulatory effect of OGF on macrophage function (Vujic et al., 2004). These researchers treated rat peritoneal macrophages with OGF in the presence of inhibitors for specific subtypes of opioid receptors. Results indicated that OGF treatment led to increased H_2O_2 production in activated macrophages that was mediated mainly through delta-1 opioid receptors. Further, OGF enhanced NO production in activated cells through delta-1 and mu receptors. A facilitation of these effects through blockade of kappa receptors illustrates the complex nature of the opioid system. A subsequent study by this group involving different combinations of receptor blockers showed an even greater level of complexity and in that study OGF-induced enhancement of H_2O_2 release was mediated by delta-1 or delta-2 receptors or by combined activation of mu and kappa receptors (Stanojevic et al., 2008).

The effects of OGF on IL-6 production by activated peritoneal murine macrophages were studied by Kowalski and associates (Kowalski et al., 2000). Macrophages were activated with IL-1β or INF-γ in the presence of OGF and the results indicated an increase in IL-6 release. Yang and Li examined the influence of OGF on IL-1 production from murine peritoneal macrophages that were induced by lipopolysaccharide (Kowalski et al., 1995). It was shown that over a wide range of concentrations in cell culture (1 pmol–0.1 μmol/L), OGF significantly increased both the production and release of IL-1 by macrophages. This effect was not affected by the presence of Naloxone. In addition, similar results were obtained from experiments involving IP injection of OGF into mice. The results thus suggest that OGF mediates IL-1 synthesis and release through a mechanism that does not require binding to classical opioid receptors (Kowalski et al., 1995).

More recent investigations by Shan's group have examined the influence of OGF on macrophage polarization (Chen et al., 2012; Wang et al., 2018a). In the study by Chen et al., murine cells of the Sarcoma 180 (S180) line were injected subcutaneously into C57BL/6 mice. Mice in the treatment group were injected with OGF (20 mg/kg) once a day for 21 days. The results showed that OGF resulted in decreased expression of CD206 and a decrease in production of arginase-1 in tumor associated macrophages, both of which are markers for M2 macrophages. Further, OGF treatment led to a marked increase in expression of CD64 and MHC-II and an increase in nitric oxide synthase activity, changes which are markers for M1 macrophages. In another investigation, Wang and associates found an increase in tumor associated macrophages expressing CD206 and increase in TNF-α and decrease in IL-10 in human gastric cancer xenografts in mice treated with OGF (Wang et al., 2018a). Consistent with these results, in vitro studies showed that OGF treatment (10^{-12} M) of bone marrow-derived macrophages resulted in a marked increase in tumoricidal activity (Chen et al., 2012). Also, OGF treatment has been shown to enhance the release of reactive oxidant species (ROS) and the production of TNFα and IL-12, but caused a decrease in the production of IL-10. In a similar in vivo study, Xu et al. found similar results working with microglial cells, which are phagocytic cells of the central nervous system (Xu et al., 2016). OGF in this study was associated with increases in the M1 markers, CD86, CD40, IL-12, and TNF-α and markedly greater phagocytic capacity, including increases in ROS and NO production. Thus, like their counterparts in the periphery, microglia was induced to attain an M1 phenotype (Xu et al., 2016). Similarly to OGF, low-dose naltrexone treatment has recently been shown to increase proportion of macrophages expressing M1 markers in colorectal cancer xenografts while increasing TNF-α and decreasing IL-10 (Ma et al., 2020).

Role of Macrophages in Antitumor Immunity

It is thus well demonstrated that OGF modulates various functions of macrophages that relate to both immune and inflammatory reactions. Because of the multifunctionality of macrophages, the effects of OGF are far-reaching. An increase in the functional capacity to phagocytize cells is apparent and is accompanied by enhancement of cytotoxic activity through an increased synthesis of H_2O_2 and NO. The significance of such augmentation of macrophage function in an immune system that has the ability to effective recognize cancer cells is apparent.

OGF-induced secretion of IL-6 by macrophages has special clinical significance, as secretion of this cytokine has been associated with tumor progression. Although chronic IL-6 signaling within in a tumor microenvironment supports cancer cell proliferation, it must be remembered that IL-6 has a more recently discovered beneficial effect, namely its involvement in proliferation and activation of lymphocytes. During an immune response, IL-6 seems to play an important role in enhancing

T-cell trafficking to lymph nodes for activation and then to tumors where they execute their cytotoxic functions. Thus, an enhanced secretion of IL-6 by macrophages would support this anticancer function of T cells.

An important feature of macrophages is the capacity to adapt their phenotype to the cytokine microenvironment and as such they can be classified as M1 or M2 macrophages. The M2 phenotype secretes high levels of IL-10, the immunosuppressive, protumor cytokine. By contributing to an environment of immunosuppression and angiogenesis, the M2 phenotype supports cancer cell invasive and proliferation. Tumor associated macrophages are thus primarily of the M2 phenotype. The M1 macrophage on the other hand stimulates inflammation and has the unique ability to metabolize arginine to the cytotoxic molecule NO. OGF thus polarizes the macrophage population toward a phenotype that it is more cytotoxic to cancer cells.

The overall effect of OGF on macrophages is far more extensive than polarization toward a "killer" phenotype. OGF causes to a marked increase in expression of MHC-II, one of the markers for M1 macrophages. Presentation of antigen in the context of MHC-II on the surface of a macrophage results in secretion of IL-12, which is the cytokine triggers proliferation of the T_H1 subtype of helper T cells. Therefore, enhancement of M1 macrophages by OGF results in a cytokine environment that leads to preferential proliferation of T_H1 cells. In turn, the main effector cells of T_H1-induced immunity are cytotoxic T cells and macrophages. Indeed, T_H1-derived IFN-γ is now known to render these cells more cytotoxic to cancer cells.

It must be pointed out that the role of inflammation in cancer is still somewhat controversial and may be potentially confusing as both and tumor-suppressive and tumor-promoting aspects have been reported. This and other characteristics of cancer highlight the complex nature of the tumor microenvironment which includes stromal, immune and inflammatory cells, all of which produce adhesion molecules, cytokines, and growth factors that promote tumor growth. Thus, context is required in order to understand how the proinflammatory cytokines IL-1, IL-6, and TNFα are involved in suppression of cancer.

Chronic nonresolving inflammation involving these cytokines is clearly associated with tumor-promoting processes. Cancer cells are known to not only produce IL-1, but to induce cells within the tumor microenvironment of many types of cancers to do so as well. Patients with IL-1 producing tumors generally have bad prognoses. The mechanisms by which IL-1 promotes tumor growth remain unclear but they are generally believed to primarily act by inducing expression of metastatic genes such as matrix metalloproteinases (MMP) and by stimulating stimulates nearby cells to produce angiogenic proteins and growth factors such as VEGF, IL-8, and transforming growth factor beta (TGFβ).

In clear contrast, however, and as indicated above, that the ability of the immune system to target and destroy cancer cells is dependent upon the collaborative efforts of tumor-infiltrating, antigen-presenting macrophages and tumor-specific T_H1 helper T cells. Further, this interaction is dependent upon proper cell–cell communication and through T_H1 activity, and enhancement of IL-1 and IL-6 secretion by macrophages is known to occur. Thus, perhaps the most important role of OGF in stimulating the immune to eradicate cancer cells is stimulation of release of the canonical proinflammatory cytokine IL-1 by macrophages which in turn enhances T_H1-mediated immunity against cancer.

OGF Stimulates the Natural Killer Cell Response

A natural killer (NK) cell is a cytotoxic lymphocyte that is unique in its ability to recognize and kill virally-infected cells and certain cancer cells without the stimulation of antigens or MHC. These cells (not to be confused with a natural killer T cell [NKT cell]) are thus able to respond very rapidly to viral infection. In addition, the distinctive abilities of NK cells permit a response to the "altered-self" properties of cancerous cells. The chronic downregulation of class I MHC protein on the surface of damaged or cancer cells makes them undetectable to T cells and thus allows them to evade T-cell-mediated immunity. The response of NK cells thus fills a vital niche for the immune system. NK cells can thus be thought of as an innate immune system cell that is analogous to the cytotoxic T cell of the adaptive arm of the system. They are especially noted for providing what is considered a type of immune memory for the innate arm of the system. The cytotoxic function of NK cells is especially important because altered-self cells are missing MHC I markers on their surface and cannot be detected and destroyed by other immune cells, such as cytotoxic T cells. Since they do not require normal activation to kill altered cells such as those found with cancer, they can be thought of as "natural killers." In addition to their ability to kill aberrant cells which involves their special capacity to recognize these stressed cells in the absence of MHC and antibodies, NK cells are vital components the innate immune response by virtue of their capacity to secrete a wide variety of chemokines and cytokines.

Increased Target Killing Activity of NK Cells and Induction of Cytokine Production

Using a model of experimental metastasis of the murine B-16 melanoma, Faith and Murgo found that OGF was able to significantly enhance splenic NK cell activity. Of particular significance it was demonstrated that this marked increase in cytotoxic potential of NK cells was associated with a substantial inhibition of pulmonary metastasis (Faith & Murgo, 1988). Since that time, several other studies support a role for OGF in enhancement of NK cell activity (Kowalski, 1998; Bajpai et al., 1997; Kowalski et al., 2000; Ghanta et al., 1991; Puente et al., 1992; Mozzanica et al., 1991; Zalys et al., 2000; Hsueh et al., 1992; Martin-Kleiner & Gabrilovac, 1996; Sutlu & Alici, 2009). The results to these studies indicate that OGF's effect on NK cells induces the production of the important cytokines, IL-2, IL-12, and TNFα.

The studies of Kowalski reported earlier, for example, showed that a single IP injection of OGF in mice increased NK cell proliferation and cytotoxic activity and intracerebroventricular injection of OGF (100 μg) had a marked effect on splenic

NK cell cytotoxic activity with an increase observed 96 h after injection (Kowalski, 1998; Kowalski et al., 2000). These studies are worthy of special note since enhancement of activity was inhibited by pretreatment with naloxone (20 μg) suggesting that the effect of OGF on NK activity involved binding to opioid receptors. In other murine work, the central effect of OGF on NK cell activity in BALB/c mice was determined (Hsueh et al., 1992). Injection of 0.02 μg of OGF directly into the cisterna magna of the brain resulted in a significant enhancement of NK cell activity, an effect that was blocked by opiate antagonist, naltrexone. This same small dose of OGF was found to have no effect on NK cell activity when injected intraperitoneally or intravenously. The results thus indicate that activation of an OGF-mediated pathway in the central nervous system is capable of stimulating NK cell activity in the periphery (Hsueh et al., 1992).

Simulation of NK cell activity by OGF has also been observed in human peripheral blood obtained from a group of healthy human subjects (Puente et al., 1992). Treatment of samples with either 10^{-6} or 10^{-8} OGF resulted in a significant increase in NK cell activity of at least 20% over base value in 7 out of 15 subjects, with changes in the rest of the subject being considered not significant. In a similar human study, a relatively low concentration of OGF [10^{-12} to 10^{-8}] did not produce significant changes in NK-activity in peripheral blood from 30 normal healthy subjects when the whole group was considered (Martin-Kleiner & Gabrilovac, 1996). While OGF had no significant on the group as a whole, at the individual level, 60% of the samples responded to OGF with a slight enhancement of basal NK activity and 12% of the samples showed a mild attenuation. The response at this relatively low level of OGF was not significantly altered when samples were prestimulated with IL-2. Thus, while the results of human studies are more variable, they indicate that OGF has the potential to stimulate human NK cell activity.

Role of NK Cells in Antitumor Immunity

NK cells play an important role in antitumor immunity by direct cytolytic activity that is possible even in the absence of surface adhesion molecules or antigenic peptides. In addition, tumor cell detection results in activation of NK cells leading to subsequent cytokine production and release. The importance of NK cell function is obvious in certain situations such as prostate cancer, where immune evasion is common. These tumor cells are often able to evade cytotoxic T-cell recognition due to their ability to downregulate expression of MHC I. However, owing to the continuous tumor immunosurveillance and response of NK cells, cytotoxic T cells can consequently only act on tumor cells through a response that is made possible by NK cell initiated cytokine production. The results of several in vitro and in vivo studies report that OGF can stimulate the proliferation and cytotoxic activity of NK cells and the secretion of cytokines that are essential to response of the innate immune system.

Summary of the Effects of OGF on Immunity

This review presents evidence that the endogenous opioid peptide OGF is part of a vital regulatory loop between the neuroendocrine and immune systems. Moreover, since OGF is also produced locally in the periphery by immune cells such as T cells and macrophages and because of its apparent influence on the proliferation and functions of immunocompetent cells, it is suggested that OGF may function as an important signaling molecule for the immune system.

Much of the research results to date seem to indicate that certain concentrations of OGF can enhance immune function in a way that may lead to inhibition of cancerous cell growth. Given the fact that (1) OGF arises from a variety of sources including not only the central nervous system, but also chromaffin cells of the adrenal medulla and immune cells such as lymphocytes, (2) a variety of immune system cell types are effected, and (3) the effects on cell function are concentration-dependent, there are many uncertainties that still exist. Thus, while there is much work yet to be done before an overall mechanism can be proposed, the following beneficial effects on of OGF on immune function are supported by the literature:

- Enhanced functional responses cytotoxic T cells and helper T cells;
- Inhibition of regulatory T-cell activity
- Enhanced T-cell response brought about by an increased maturation, differentiation and activity of dendritic cells
- Increased phagocytic activity of macrophages;
- Enhanced activity of NK cells
- Enhanced interactions between immune cells through increased secretion of positive immunomodulatory cytokines including IL-1, IL-6, TNFα, and INFγ

Most of the reports reviewed thus suggest a positive immunomodulatory role for OGF which leads to increased responses from a variety of immune cells such as lymphocytes, macrophages, NK cells and dendritic cells. Therefore, in addition to a direct effect on cancer cells, OGF may provide a beneficial effect on immune-related diseases such as neoplasia by augmenting the activity of cells involved in both the innate and acquired immunity.

In Vivo Effects of OGF in Anticancer Immunity

A considerable amount of work over the past few decades has focused on the use of a variety of cytokines and other molecules to direct the host immune response toward recognition and eradication of cancerous cells. For example, among the endogenous substances investigated, considerable advances have been made with IL-2 and IFNα and the successful use of these cytokines in treatment of cancers such as renal cell carcinoma and melanoma is apparent. The results of the numerous reports presented here clearly suggest that OGF has the capacity to boost anticancer immunity.

There are a number of in vivo animal and human studies support the idea that OGF immunoreactivity may play a positive role in the body's response to neoplasia. Murgo demonstrated an antitumor effect of OGF in C57BL/6J mice which had been inoculated with B16-BL6 melanoma cells (Murgo, 1985). In these animals, subcutaneous tumor growth was inhibited with a 50 μg daily dose of OGF for 1 to 2 weeks. This antitumor effect was inhibited by the administration naloxone suggesting that the effect may have been mediated through delta receptors. Likewise, the work of Kita and coworkers showed a positive effect of OGF on tumoricidal activity of spleen cells from nude mice (Kita et al., 1992). When mouse splenic cells were incubated with 10 μM OGF, there was a 3.7-fold increase in the ability of the cells to lyse human ovarian cancer cells. These animal studies thus suggest that OGF may play a critical role in cellular immunity associated with cancer.

As noted previously, much of the benefit of OGF may lie in its ability to alter cytokine secretion. Among these cytokines, the role of IL-6 is perhaps the most complex and potentially confusing. Within the microenvironment of a tumor, chronic IL-6 signaling is generally believed to act intrinsically on cancer cells to promote tumor progression by supporting cancer cell proliferation, metastasis, and angiogenesis. However, a perhaps lesser known and beneficial role of IL-6 that has recently emerged is its involvement in proliferation, activation, and survival of lymphocytes during an active immune response. IL-6 seems to play an important role in enhancing T-cell trafficking to lymph nodes for activation and subsequently to tumors where they execute their cytotoxic effector functions. Zhong et al. found that OGF stimulates IL-6 expression in mice and in humans (Zhong et al., 1998). In their study, six days of intraperitoneal exposure of mice to 0.1 or 1.0 mg/kg OGF was associated with increases in IL-6 mRNA expression and plasma IL-6 levels. Further, in seven coronary bypass patients, plasma OGF levels were found to be markedly elevated after surgery (8.9 ± 2.9 to 135 ± 18 pg/ml). This increase in OGF was associated with a substantial increase in plasma levels of IL-6 (2.1 ± 1.5 to 237 ± 50 pg/ml). These in vivo data support the notion of OGF's role as a component of cytokine signaling in immune cells.

More recently, Hua and associates examined the effects of OGF on subpopulations of lymphocytes in the peripheral blood of healthy human subjects (Hua et al., 2012). Cultures of cells collected from these individuals were treated with 10^{-12} M OGF. The results showed OGF treatment was associated with expansion of several populations of cells including $CD8^+$ cytotoxic T cells, $CD4^+$ helper T cells, dendritic cells (DCs), NK cells, and other subpopulations of lymphocytes. The authors concluded that OGF stimulated expansion of lymphocyte subpopulations probably occurred due to an inhibition of $CD4^+CD25^+$ Treg cells (Hua et al., 2012). Subsequent to this work, the same group studied the effect of OGF treatment on changes in lymphocyte subpopulations in the peripheral blood of 50 patients with various types of terminal cancer with broad metastases and whose immune systems had been exposed to the effects of chemotherapy (Wang et al., 2014). Again, proliferation of subpopulations of nucleated blood cells was determined before and after treatment with 10^{-12} M OGF, and similar results were found. OGF stimulation in these cancer patients resulted in a marked proliferation of lymphocyte subpopulations including increases of 40%, 72%, and 264% being reported for cytotoxic T cells, helper T cells, and NK cells, respectively. As in the previous study, OGF induced a strong inhibitory effect on Treg cells, with a population decrease of 38% (Wang et al., 2014). It is thus clear from these in vivo animal and human studies that OGF has a positive immunomodulatory effect and it thus has the potential to be part of a therapy that mobilizes the body's immune defenses to fight cancer.

Conclusion

The activation of OGFr activity by OGF has been well established as a key control of cellular proliferation in normal development, homeostasis and tumorigenesis. Upon entry into the nucleus the OGF-OGFr complex acts to increase expression of the cyclin-dependent kinase inhibitors $p16^{INK4A}$ and $p21^{CIP1}$, although whether acts directly in a transcriptional complex for these genes needs to be firmly established. These cyclin-dependent kinase inhibitors act by preventing activation of cyclin-dependent kinases necessary for sequential phosphorylation of the retinoblastoma tumor suppressor and passage through the G1 restriction point in the cell cycle. This is the mostly firmly established mechanism of OGF-OGFr activity on tumor cells to date with most studies indicating that it does not function in differentiation, migration, adhesion or apoptosis of cancer cells (Zagon & McLaughlin, 2005; Zagon et al., 2007). However, more recent investigations have indicated functions in cell cancer cell migration and apoptosis. OGF induction of apoptosis has been reported in human melanoma cells through significantly reduced expression of the inhibitor of apoptosis protein, survivin; however, this may have involved binding of OGF to a classical opioid receptor as expressions of the delta, mu, and kappa receptors were all upregulated in response to OGF (Wang et al., 2016). Recently, more convincing evidence has been presented for OGF-OGFr signaling leading to apoptosis in at least some cancer cells (Wang et al., 2018a, b). It should also be noted that, at low doses, naltrexone binding to OGFr can also lead to apoptosis (Ma et al., 2020). Clinical studies to date are limited, with the only positive reported results coming from patients with advanced pancreatic cancer. From preclinical studies, combination of OGF with other therapies look promising and could be one way of effecting apoptosis as well as decreasing cell proliferation. OGFr activation and decreased cancer cell proliferation have also been reported after binding to morphine and synthetic peptide OGF analogues (Kim et al., 2016; Horvat et al., 2006). In the later study, treatment with one of the synthetic peptides resulted in both reduction in cell proliferation and increased apoptosis in SW-620 metastatic colon carcinoma cell while treatment with OGF only affected proliferation. In addition to a direct effect on cancer cells, OGF may provide a beneficial effect on immune-related diseases such as neoplasia by augmenting the activity of cells involved in both innate and acquired immunity. Used as an adjunct to chemotherapy agents, OGF may help to sustain and maximize the efficiency of the immune system of patients during the process of the treatment. In addition, OGF could be used to help restore a patient's immune system that has been compromised by traditional chemotherapy or radiation treatment. Taken together, the evidence supports consideration of opioid peptides such OGF as a new strategy for cancer immunotherapy. Therefore, development of analogues to specifically target OGF and one or two classical opioid receptors could also be an approach to target more than cell proliferation.

References

Avella, D. M., Kimchi, E. T., Donahue, R. N., Tagaram, H. R. S., McLaughlin, P. J., Zagon, I. S., & Staveley-O'Carroll, K. F. (2010). The opioid growth factor-opioid growth factor receptor axis regulates cell proliferation of human hepatocellular cancer. *American Journal of Physiology Regulative, Intergrative and Comparitive Physiology, 298*(2), 459–466. https://doi.org/10.1152/ajpregu.00646.2009

Bajpai, K., Singh, V. K., Agarwal, S. S., Dhawan, V. C., Naqvi, T., Haq, W., & Mathur, K. B. (1995). Immumodulatory activity of met-enkephelin and its two potent analogs. *International Journal of Immunopharmacolcoy, 17*(3), 207–212. https://doi.org/10.1016/0192-0561(94)00080-8

Bajpai, K., Singh, V. K., Dhawan, V. C., Haq, W., Mathur, K. B., & Agarwal, S. S. (1997). Immunomodulation by two potent analogs of met-enkephalin. *Immunopharmacology, 35*, 213–220. https://doi.org/10.1016/s0162-3109(96)00147-6

Barron, B. A., Oakford, L. X., Gaugl, J. F., & Caffrey, J. L. (1995). Methionine-enkephalin-Arg-Phe immunoreactivity in heart tissue. *Peptides, 16*(7), 1221–1227. https://doi.org/10.1016/0196-9781(95)02012-l

Bisignani, G. J., McLaughlin, P. J., Ordille, S. D., Beltz, M. S., Jarowenko, M. V., & Zagon, I. S. (1999). Human renal cell cancer proliferation in tissue culture is tonically inhibited by opioid growth factor. *The Journal of Urology, 162*(2), 2186–2191. https://doi.org/10.1016/s0022-5347(05)68157-8

Blebea, J., Mazo, J. E., Kihara, T. K., Jonathan-Hien, V., McLaughlin, P. J., Atnip, R. G., & Zagon, I. S. (2000). Opioid growth modulates angiogenesis. *Journal of Vascular Surgery, 32*(2), 364–373. https://doi.org/10.1067/mva.2000.107763b

Boluyt, M. O., Younes, A., Caffrey, J. L., O'Neill, L., Barron, B. A., Crow, M. T., & Lakatta, E. G. (1993). Age-associated increase in rat cardiac opioid production. *American Journal of Physiology, 265*(1 Pt 2), H212–H218. https://doi.org/10.1152/ajpheart.1993.265.1.H212

Chen, Y. L., Law, P. Y., & Loh, H. H. (2008). The other side of the opioid story: Modulation of cell growth and survival signaling. *Current Medicinal Chemistry, 15*(8), 772–778. https://doi.org/10.2174/092986708783955518

Chen, W., Liu, J., Meng, J., Lu, C., Li, X., Wang, E., & Shan, F. (2012). Macrophage polarization induced by neuropeptide methionine enkephalin (MENK) promotes tumoricidal responses. *Cancer Immunology, Immunotherapy, 61*(10), 1755–1768. https://doi.org/10.1007/s00262-012-1240-6

Cheng, F., Zagon, I. S., Verderame, M. F., & McLaughlin, P. J. (2007). The opioid growth factor (OGF)-OGF receptor axis uses the p16 pathway to inhibit head and neck cancer. *Cancer Research, 67*(21), 10511–10518. https://doi.org/10.1158/0008-5472.CAN-07-1922

Cheng, F., McLaughlin, P. J., Verderame, M. F., & Zagon, I. S. (2008). The OGF-OGFr axis utilizes the p21 pathway to restrict progression of human pancreatic cancer. *Molecular Cancer, 7*, 5. https://doi.org/10.1186/1476-4598-7-5

Cheng, F., McLaughlin, P. J., Banks, W. A., & Zagon, I. S. (2009a). Passive diffusion of naltrexone into human and animal cells and upregulation of cell proliferation. *American Journal of Physiology, Regulatory, Integrative and Comparative Physiology, 297*(3), R844–R852. https://doi.org/10.1152/ajpregu.00311.2009

Cheng, F., McLaughlin, P. J., Verderame, M. F., & Zagon, I. S. (2009b). Dependence on nuclear localization signals of the opioid growth factor receptor in the regulation of cell proliferation. *Experimental Biology and Medicine, 234*(5), 532–541. https://doi.org/10.3181/0901-RM-16

Cheng, F., McLaughlin, P. J., Verderame, M. F., & Zagon, I. S. (2009c). The OGF-OGFr axis utilizes the p16INK4a and p21WAF1/CIP1 pathways to restrict normal cell proliferation. *Molecular Biology of the Cell, 20*(1), 319–327. https://doi.org/10.1091/mbc.e08-07-0681

Cheng, F., McLaughlin, P. J., Banks, W. A., & Zagon, I. S. (2010a). Internalization of the opioid growth factor [Met[5]]-enkephalin is dependent on clathrin-mediated endocytosis for downregulation of cell proliferation. *American Journal of Physiology, Regulatory, Integrative and Comparative Physiology, 299*(3), R744–R785. https://doi.org/10.1152/ajpregu.00318.2010

Cheng, F., McLaughlin, P. J., & Zagon, I. S. (2010b). Regulation of cell proliferation by the opioid growth factor is dependent on karyopherin β and ran for nucleocytoplasmic trafficking. *Experimental Biology and Medicine, 235*(9), 1093–1101. https://doi.org/10.1258/ebm.2010.010139

Donahue, R. N., McLaughlin, P. J., & Zagon, I. S. (2009). Cell proliferation of human ovarian cancer is regulated by the opioid growth factor-opioid growth factor receptor axis. *American Journal of Physiology, Regulatuve Intergrative and Comparative Physiology, 296*(6), R1716–R1725. https://doi.org/10.1152/ajpregu.00075.2009

Donahue, R. N., McLaughlin, P. J., & Zagon, I. S. (2011a). The opioid growth factor (OGF) and low dose naltrexone (LDN) suppress human ovarian cancer in mice. *Gynecologic Oncology, 122*(2), 382–388. https://doi.org/10.1016/j.ygyno.2011.04.009

Donahue, R. N., McLaughlin, P. J., & Zagon, I. S. (2011b). Low dose naltrexone suppresses ovarian cancer and exhibits enhanced inhibition in combination with cisplatin. *Experimental Biology and Medicine, 236*(7), 883–895. https://doi.org/10.1258/ebm.2011.011096

Faith, R. E., & Murgo, A. J. (1988). Inhibition of pulmonary metastases and enhancement of natural killer cell activity by methionine-enkephalin. *Brain, Behavior and Immunity, 2*(2), 114–122. https://doi.org/10.1016/0889-1591(88)90012-8

Fanning, J., Hossler, C. A., Kesterson, J. P., Donahue, R. N., McLaughlin, P. J., & Zagon, I. S. (2012). Expression of the opioid growth factor-opioid growth factor receptor axis in human ovarian cancer. *Gynecologic Oncology, 124*(2), 319–324. https://doi.org/10.1016/j.ygyno.2011.10.024

Fichna, J., & Janecka, A. (2004). Opioid peptides in cancer. *Cancer and Metastasis Reviews, 239*(3–4), 351–366. https://doi.org/10.1023/B:CANC.0000031773.46458.63

Foris, G., Medgyesi, G. A., & Hauck, M. (1986). Bidirectional effect of Met-enkephalin on macrophage effector functions. *Molecular and Cellular Biochemistry, 69*(2), 127–137. https://doi.org/10.1007/BF00224759

Gabrilovac, J., & Marotti, T. (2000). Gender-related differences in murine T- and B-lymphocyte proliferative ability in response to *in vivo* [Met(5)]enkephalin administration. *European Journal of Pharmacology, 392*(1–2), 101–108. https://doi.org/10.1016/s0014-2999(00)00118-7

Ghanta, V. K., Rogers, C. F., Hsueh, C. M., Hiramoto, N. S., Soong, S. J., & Hiramoto, R. N. (1991). *In vivo* enhancement of NK cell activity with met-enkaphalin and glycyl-glutamine: Their possible role in the conditioned response. *International Journal of Neuroscience, 61*(1–2), 135–143. https://doi.org/10.3109/00207459108986281

Goldenberg, D., Zagon, I. S., Fedok, F., Crist, H. S., & McLaughlin, P. J. (2008). Expression of opioid growth factor (OGF)-OGF receptor (OGFr) axis in human nonmedullary thyroid cancer. *Thyroid, 18*(11), 1165–1170. https://doi.org/10.1089/thy.2008.0112

Harlan, R. E., Shivers, B. D., Romano, G. J., Howells, R. D., & Pfaff, D. W. (1987). Localization of preproenkephalin mRNA in the rat brain and spinal cord by in situ hybridization. *Journal of Comparative Neurology, 258*(2), 159–184. https://doi.org/10.1002/cne.902580202

Heagy, W., Laurance, M., Cohen, E., & Finberg, R. (1990). Neurohormones regulate T cell function. *Journal of Experimental Medicine, 171*(5), 1625–1633. https://doi.org/10.1084/jem.171.5.1625

Horvat, S., Mlinaric-Majerski, K., Glavas-Obrovac, L., Jakas, A., Veljkovic, J., Marezi, S., Kragol, G., Roscic, M., Matkovic, M., & Milostic-Srb, A. (2006). Tumor cell-targeted methionine-enkephalin analogues containing unnatural amino acids: Design, synthesis, and *in vitro* anti-tumor activity. *Journal of Medicinal Chemistry, 49*(11), 3136–3142. https://doi.org/10.1021/jm051026+

Howells, R. D., Kilpatrick, D. L., Bailey, L. C., Noe, M., & Udenfriend, S. (1986). Proenkephalin mRNA in rat heart. *Proceedings of the National Academy of Sciences USA, 83*(6), 1960–1963. https://doi.org/10.1073/pnas.83.6.1960

Hsueh, C. M., Hiramoto, R. N., & Ghanta, V. K. (1992). The central effect of methionine enkephalin on NK cell activity. *Brain Research, 578*(1–2), 142–148. https://doi.org/10.1016/0006-8993(92)90241-z

Hua, H., Lu, C., Li, W., Meng, J., Wang, D., Plotnikoff, N. P., Wang, E., & Shan, F. (2012). Comparison of stimulating effect on subpopulations of lymphocytes in human peripheral blood by methionine enkephalin with IL-2 and IFN-γ. *Human Vaccines & Immunotherapeutics, 8*(8), 1082–1089. https://doi.org/10.4161/hv.20759

Hucklebridge, F. H., Hudspith, B. N., Lydyard, P. M., & Brostoff, J. (1990). Stimulation of human peripheral lymphocytes by methionine enkephalin and δ-selective opioid analogues. *Immunopharmacology, 19*(2), 87–91. https://doi.org/10.1016/0162-3109(90)90043-e

Immonen, J. A., Zagon, I. S., & McLaughlin, P. J. (2014). Topical naltrexone as treatment for type 2 diabetic cutaneous wounds. *Advances in Wound Care, 3*(6), 419–427. https://doi.org/10.1089/wound.2014.0543

Innovive Pharmaceuticals. (2005). *INNO-105 in patients with solid tumors*. ClinicalTrials.gov. Last update May 14, 2007. ClinicalTrials.gov identifier NCT00255333.

Isayama, T., & Zagon, I. S. (1991). Localization of preproenkephalan A mRNA in the neonatal rat retina. *Brain Research Bulletin, 27*(6), 805–808. https://doi.org/10.1016/0361-9230(91)90212-3

Isayama, T., Hurst, W. J., McLaughlin, P. J., & Zagon, I. S. (1995). Ontogeny of the opioid growth factor, [Met^5]enkephalin and its binding activity in the rat retina. *Visual Neuroscience, 12*(5), 939–950. https://doi.org/10.1017/s0952523800009494

Isayama, T., McLaughlin, P. J., & Zagon, I. S. (1996). Ontogeny of preproenkephalin mRNA expression in the rat retina. *Visual Neuroscience, 13*(4), 695–704. https://doi.org/10.1017/s0952523800008580

Jaglowski, J. R., Zagon, I. S., Stack, B. C., Jr., Verderame, M. F., Leure-duPree, A. E., Manning, J. D., & McLaughlin, P. J. (2005). Opioid growth factor enhances tumor growth inhibition and increases the survival of paclitaxel-treated mice with squamous cell carcinoma of the head and neck. *Cancer Chemotherapy and Pharmacology, 56*(1), 97–104. https://doi.org/10.1007/s00280-004-0929-4

Keshet, E., Polakiewicz, R. D., Itin, A., Ornoy, A., & Rosen, H. (1989). Proenkephalin A is expressed in mesodermal lineages during organogenesis. *The EMBO Journal, 8*(10), 2919–2923. https://doi.org/10.1002/j.1460-2075.1989.tb08441.x

Kew, D., & Kilpatrick, D. L. (1990). Widespred organ expression of the rat proenkekphalin gene during early postnatal development. *Molecular Endocrinology, 4*(2), 337–340. https://doi.org/10.1210/mend-4-2-337

Kim, J. Y., Ahn, H. J., Kim, J. K., Kim, J., Lee, S. H., & Chae, H. B. (2016). Morphine suppresses lung cancer cell proliferation through the interaction with opioid growth factor receptor: An *in vitro* and human lung tissue study. *Anesthesia & Analgesia, 123*(6), 1429–1436. https://doi.org/10.1213/ANE.0000000000001293

Kita, T., Kikuchi, Y., Oomori, K., & Nagata, I. (1992). Effects of opioid peptides on the tumoricidal activity of spleen cells from nude mice with or without tumors. *Cancer Detection and Prevention, 16*(4), 211–214.

Klocek, M. S., Sassani, J. W., Donahue, R. N., McLaughlin, P. J., & Zagon, I. S. (2010). Regulation of Tenon's capsule fibroblast cell proliferation by the opioid growth factor and the opioid growth factor receptor axis. *Investigative Ophthalmology and Visual Science, 51*(10), 5054–5061. https://doi.org/10.1167/iovs.09-4949

Környei, J. L., Vértes, Z., Kovàcs, K. A., Göcze, P. M., & Vértes, M. (2003). Developmental changes in the inhibition of cultured rat uterine cell proliferation by opioid peptides. *Cell Proliferation, 36*(3), 151–163. https://doi.org/10.1046/j.1365-2184.2003.00263.x

Kowalski, J. (1998). Immunologic action of [Met5]enkephalin fragments. *European Journal of Pharmacology, 347*(1), 95–99. https://doi.org/10.1016/s0014-2999(98)00079-x

Kowalski, J., Belowski, D., & Wieigus, J. (1995). Bidirectional modulation of mouse natural killer cell and macrophage cytotoxic activities by enkephalins. *Polish Journal of Pharmacology, 47*(4), 327–331.

Kowalski, J., Makowiecka, K., Belowski, D., & Herman, Z. S. (2000). Augmenting effect of methionine-enkephalin on interleukin-6 production by cytokine-stimulated murine macrophages. *Neuropeptides, 34*(3–4), 187–192. https://doi.org/10.1054/npep.2000.0812

Krajnik, M., Schäfer, M., Sobanski, P., Kowalewski, J., Bloch-Boguslawska, E., Zylicz, Z., & Mousa, S. A. (2010). Enkephalin its precursor, processing enzymes, and receptor as part of a local opioid network throughout the respiratory system of lung cancer patients. *Human Pathology, 41*(5), 632–642. https://doi.org/10.1016/j.humpath.2009.08.025

Kren, N. P., Zagon, I. S., & McLaughlin, P. J. (2016). Nuclear export of opioid growth receptor is CRM1 dependent. *Experimental Biology and Medicine, 241*(3), 273–281. https://doi.org/10.1177/1535370215605585

Levin, R. J., Wu, Y., McLaughlin, P. J., & Zagon, I. S. (1997). Expression of the opioid growth factor, [Met5]-enkephalin, and the zeta opioid receptor in head and neck squamous cell carcinoma. *Laryngoscope, 107*(3), 335–339. https://doi.org/10.1097/00005537-199703000-00011

Li, W., Meng, J., Li, X., Hua, H., Meng, Y., Wang, Q., Wang, E., & Shan, F. (2012). Methionine enkephalin (MENK) improved the functions of bone marrow-derived dendritic cells (BMDCs) loaded with antigen. *Human Vaccines & Immunotherapeurics, 8*(9), 1236–1242. https://doi.org/10.4161/hv.21128

Li, X., Meng, Y., Plotnikoff, N. P., Youkilis, G., Griffin, N., Wang, E., Lu, C., & Shan, F. (2015). Methionine Enkephalin (MENK) Inhibits tumor growth through regulating CD4+Foxp3+ Regulatory T cells (Tregs) in mice. *Cancer Biology and Therapy, 16*(3), 450–459. https://doi.org/10.1080/15384047.2014.1003006

Li, Z., You, Y., Griffin, N., Feng, Y., & Shan, F. (2018). Low-dose naltrexone (LDN): A promosing treatment in immune-related diseases and cancer therapy. *International Immunopharmacology, 61*(2018), 178–184. https://doi.org/10.1016/j.intimp.2018.05.020

Liang, X., Lui, R., Chen, C., Ji, F., & Li, T. (2016). Opioid system modulates immune function: A review. *Translational Perioperative and Pain Medicine, 1*(1), 5–13. https://doi.org/10.1007/s10753-011-9329-1

Liu, J., Chen, W., Meng, J., Lu, C., Wang, E., & Shan, F. (2012). Induction on differentiation and modulation of bone marrow progenitor of dendritic cell by methionine enkephalin (MENK). *Cancer Immunology, Immunotherapy, 61*(10), 1699–1711. https://doi.org/10.1007/s00262-012-1221-9

Lu, W., Xie, H., Tie, X., Wang, R., Wu, A., & Shan, F. (2018). NFAT-1 hyper-activation my methionine enkephalin (MENK) significantly induces cell apoptosis of rats C6 glioma *in vivo* and *in vitro*. *International Immunopharmacology, 56*, 1–4. https://doi.org/10.1016/j.intimp.2018.01.005

Ma, M., Wang, X., Liu, N., Shan, F., & Feng, Y. (2020). Low-dose naltrexone inhibits colorectal cancer progression and promotes apoptosis by increasing M1-type macrophages and activating the Bax/Bcl-2/caspase-3/PARP pathway. *International Immunopharmacology, 83*(2020), 106388. https://doi.org/10.1016/j.intimp.2020.106388

Martin-Kleiner, I., & Gabrilovac, J. (1996). Pretreatment of human peripheral blood lymphocytes with interleukin-2 or dexamethasone does not alter their response to Met-enkephalin in a NK-cytotoxicity assay. *Immunopharmacology and Immunotoxicology, 18*(1), 37–57. https://doi.org/10.3109/08923979609007109

McLaughlin, P. J. (1994). Opioid antagonist modulation of rat heart development. *Life Sciences, 54*(19), 1423–1431. https://doi.org/10.1016/0024-3205(94)00597-4

McLaughlin, P. J., & Allar, M. A. (1998). Preproenkephalin gene expression and [Met^3]-enkephalin levels in the developing rat heart. *Molecular Brain Research, 60*(2), 160–167. https://doi.org/10.1016/s0169-328x(98)00178-8

McLaughlin, P. J., & Wu, Y. (1998). Opioid gene expression in the developing and adult rat heart. *Developmental Dynamics, 211*(2), 153–163. https://doi.org/10.1002/(SICI)1097-0177(199802)211:2<153::AID-AJA4>3.0.CO;2-G

McLaughlin, P. J., & Zagon, I. S. (2006). Progression of squamous cell carcinoma of the head and neck is associated with down-regulation of the opioid growth factor receptor. *International Journal of Oncology, 28*(6), 1577–1583. https://doi.org/10.3892/ijo.28.6.1577

McLaughlin, P. J., & Zagon, I. S. (2012). The opioid factor-opioid growth factor receptor axis: Homeostatic regulator of cell proliferation and its implications for health and disease. *Biochemical Pharmacology, 84*(6), 746–755. https://doi.org/10.1016/j.bcp.2012.05.018

McLaughlin, P. J., & Zagon, I. S. (2015). Duration of opioid blockage determines biotherapeutic response. *Biochemical Pharmacology, 97*(3), 236–246. https://doi.org/10.1016/j.bcp.2015.06.016

McLaughlin, P. J., Levin, R. J., & Zagon, I. S. (1999). Regulation of human head and neck squamous cell carcinoma growth in tissue culture by opioid growth factor. *International Journal of Oncology, 14*(5), 991–998. https://doi.org/10.3892/ijo.14.5.991

McLaughlin, P. J., Levin, R. J., & Zagon, I. S. (2000). The opioid growth factor receptor in human head and neck squamous cell carcinoma. *International Journal of Molecular Medicine, 5*(2), 191–196. https://doi.org/10.3892/ijmm.5.2.191

McLaughlin, P. J., Levin, R. J., & Zagon, I. S. (2003a). Opioid growth factor (OGF) inhibits the progression of human squamous cell crcinoma of the head and neck transpanted into nude mice. *Cancer Letters, 199*(2), 209–217. https://doi.org/10.1016/s0304-3835(03)00341-0

McLaughlin, P. J., Stack, B. C., Jr., Levin, R. J., Fedok, F., & Zagon, I. S. (2003b). Defects in the opioid growth receptor in human squamous cell carcinoma of the head and neck. *Cancer, 97*(7), 1701–1710. https://doi.org/10.1002/cncr.11237

McLaughlin, P. J., Verderame, M. F., Hankins, J. L., & Zagon, I. S. (2007). Overexpression of the opioid growth factor receptor downregulates cell proliferation of human squamous carcinoma cells of the head and neck. *International Journal of Molecular Medicine, 19*(3), 421–428. https://doi.org/10.3892/ijmm.19.3.421

McLaughlin, P. J., Zagon, I. S., Park, S. S., Conway, A., Donahue, R. N., & Goldenberg, D. (2009). Growth inhibition of thyroid follicular cell-derived cancers by the opioic growth factor (OGF)-opioid growth receptor (OGRr) axis. *BMC Cancer, 9*, 369. https://doi.org/10.1186/1471-2407-9-369

McLaughlin, P. J., Keiper, C. L., Verderame, M. F., & Zagon, I. S. (2012). Targeted overexpression of OGFr in epithelium of transgenic mice suppresses cell proliferation and impairs full-thickness wound closure. *American Journal of Physiology, Integrative and Comparative Physiology, 302*(9), R1084–R1090. https://doi.org/10.1152/ajpregu.00670.2011

Meng, Y., Wang, Q., Zhang, Z., Wang, E., Plotnikoff, N. P., & Shan, F. (2013). Synergistic effect of methionine encephalin (MENK) combined with pidotimod(PTD) on the maturation of murine dendritic cells (DCs). *Human Vaccines & Immunotherapeutics, 9*(4), 773–783. https://doi.org/10.4161/hv.23137

Meng, Y., Gao, X., Chen, W., Plotnikoff, N. P., Griffin, M., Zhang, G., & Shan, F. (2017). Methionine enkephalin (MENK) mounts antitumor effect via regulating dendritic cells (DCs). *International Immunopharmacology, 44*, 61–71. https://doi.org/10.1016/j.intimp.2017.01.004

Miller, G. C., Murgo, A. J., & Plotnikoff, N. P. (1983). Enkephalins-enhancement of active T-cell rosettes from lymphoma patients. *Clinical Immunology and Immunopathology, 26*(3), 446–451. https://doi.org/10.1016/0090-1229(83)90131-9

Miller, G. C., Murgo, A. J., & Plotnikoff, N. P. (1984). Enkephalins-enhancement of active T-cell rosettes from normal volunteers. *Clinical Immunology and Immunopathology, 31*(1), 132–137. https://doi.org/10.1016/0090-1229(84)90198-3

Milton S. Hershey Medical Center. (2008). *Opioid growth factor (OGF) and gemcitabine: Novel treatment for pancreatic cancer.* ClinicalTrials.gov. Last update September 6, 2018. ClinicalTrials.gov identifier NCT00783172.

Milton S. Hershey Medical Center. (2009). *Treatment of advanced head and neck cancer.* ClinicalTrials.gov. Last update September 30, 2014. ClinicalTrials.gov identifier NCT00982696.

Mozzanica, N., Finzi, A. F., Foppa, S., Vignati, G., & Villa, M. L. (1991). Association between circadian rhythms of endogenous hyopthalamic opioid peptides and of natural killer cell activity. *International Journal of Immunopharmacology, 13*(2–3), 317–321. https://doi.org/10.1016/0192-0561(91)90113-l

Murgo, A. J. (1985). Inhibition of B16-BL6 melanoma growth in mice by methionine-enkephalin. *Journal of the National Cancer Institute, 75*(2), 341–344. https://doi.org/10.1093/jnci/75.2.341

Murgo, A. J., Plotnikoff, N. P., & Faith, R. E. (1985). Effect of methionine-enkephalin plus $ZnCl_2$ on active T cell rosettes. *Neuropeptides, 5*(4–6), 367–370. https://doi.org/10.1016/0143-4179(85)90030-7

Nissen, J. B., & Kragballe, K. (1997). Enkephalins modulate differentiation of normal human keratinocytes *in vitro*. *Experimental Dermatology, 6*(5), 341–344. https://doi.org/10.1111/j.1600-0625.1997.tb00166.x

Ohmori, H., Fujii, K., Sasahira, T., Luo, Y., Isobe, M., Tatsumoto, N., & Kuniyasu, H. (2009). Methionine-enkephalin secreted by human colorectal cancer cells suppresses T lymphocytes. *Cancer Science, 100*(3), 497–502. https://doi.org/10.1111/j.1349-7006.2008.01073.x

Osborne, J. G., Kindy, M. S., & Hauser, K. F. (1991). Expression of proenkephalin mRNA in developing cerebellar cortex of the rat: Expression levels coincide with maturational gradients in Purkinje cells. *Developmental Brain Research, 63*(1–2), 63–69. https://doi.org/10.1016/0165-3806(91)90067-s

Palmer, M. R., Miller, R. G., Olson, L., & Seiger, A. (1982). Prenatal ontogeny of neurons with enkephalin-like immunoreactivity in the rat central nervous system: An immunological mapping investigation. *Medical Biology, 60*(2), 61–88.

Petty, H. R., & Berg, K. A. (1988). Combinative ligand-receptor interactions: Epinephrine depresses RAW264 macrophage antibody-dependent phagocytosis in the absence and presence of met-enkephalin. *Journal of Cellular Physiology, 134*(2), 281–286. https://doi.org/10.1002/jcp.1041340215

Petty, H. R., & Martin, S. M. (1989). Combinative ligand-receptor interactions: Effects of cAMP, epinephrine, and met-enkephalin on RAW264 macrophage morphology, spreading, adherence, and microfilaments. *Journal of Cellular Physiology, 138*(2), 247–256. https://doi.org/10.1002/jcp.1041380205

Plotnikoff, N. P., Miller, G. C., Nimeh, N., Faith, R. E., Murgo, A. J., & Wybran, J. (1987). Enkephalins and T-cell enhancement in normal volunteers and cancer patients. *Annals of the New York Academy of Sciences, 496*, 608–619. https://doi.org/10.1111/j.1749-6632.1987.tb35821.x

Plotnikoff, N. P., Faith, R. F., Murgo, A. J., Herberman, R. B., & Good, R. A. (1997). Methionine enkephalin: A new cytokine- human studies. *Clinical Immunology and Immunopathology, 82*(2), 93–101. https://doi.org/10.1006/clin.1996.4287

Polakiewicz, R. D., & Rosen, H. (1990). Regulated expression of proenkephalin A during ontogenic development of mesenchymal derivative tissues. *Molecular and Cellular Biology, 10*(2), 736–742. https://doi.org/10.1128/mcb.10.2.736

Puente, J., Maturana, P., Miranda, D., Navarro, C., Wolf, M. E., & Mosnaim, A. D. (1992). Enhancement of human natural killer cell activity by opioid peptides: Similar response to methionine-enkephalin and beta-endorphin. *Brain, Behavior, and Immunity, 6*(1), 32–39. https://doi.org/10.1016/0889-1591(92)90057-u

Rosen, H., Polakieweicz, R. D., Benzakine, S., & Bar-Shavit, Z. (1991). Proenkephalin A in bone-derived cells. *Proceedings of the National Acadamy of Sciences USA, 88*(9), 3705–3709. https://doi.org/10.1073/pnas.88.9.3705

Rosen, H., Krichevsky, A., Polakiewicz, R. D., Benzakine, S. S., & Bar-Shavit, Z. (1995). Developmental regulation of proenkephalin gene expression in osteoblasts. *Molecular Endocrinology, 9*(11), 1621–1631. https://doi.org/10.1210/mend.9.11.8584038

Shahabi, N. A., & Sharp, B. M. (1995). Delta opioid agonists inhibit proliferation of highly purified murine CD4+ and CD8+ T-cells. *Advances in Experimental Medicine and Biology, 373*, 29–36. https://doi.org/10.1007/978-1-4615-1951-5_5

Shan, F., Xia, Y., Wang, N., Meng, J., Lu, C., Meng, Y., & Plotnikoff, N. P. (2011). Functional modulation of the pathway between dendritic cells (DCs) and CD4+T cells by the neuropeptide: Methionine enkephalin (MENK). *Peptides, 32*(5), 929–937. https://doi.org/10.1016/j.peptides.2011.01.033

Sibinga, N. E. S., & Goldstein, A. (1988). Opioid peptides and opioid receptors in cells of the immune system. *Annual Reviews in Immunology, 6*, 219–249. https://doi.org/10.1146/annurev.iy.06.040188.001251

Sikong, Y., Wang, Q., Cai, M., Zhang, A., Pang, F., & Cui, X. (2019). Exogenous OGF enhances the anti-tumor activity of cisplatin on hepatocellular carcinoma. *International Journal of Clinical & Experimental Pathology, 12*(2), 590–598.

Smith, E. M., Harbour-McMenamin, D., & Blalock, J. E. (1985). Lymphocyte production of endorphins and endorphin-mediated immunoregulatory activity. *Journal of Immunology, 135*, 779S–782S.

Smith, J. P., Conter, R. L., Demers, L. M., McLaughlin, P. J., & Zagon, I. S. (2000). Elevated levels of opioid growth factor in the plasma of patients with pancreatic cancer. *Pancreas, 21*(2), 158–164. https://doi.org/10.1097/00006676-200008000-00009

Smith, J. P., Conter, R. L., Bingaman, S. I., Harvey, H. A., Mauger, D. T., Ahmad, M., Demers, L. M., Stanley, W. B., McLaughlin, P. J., & Zagon, I. S. (2004). Treatment of advanced pancreatic cancer with opioid growth factor: Phase I. *Anti-Cancer Drugs, 15*(3), 203–209. https://doi.org/10.1097/00001813-200403000-00003

Smith, J. P., Bingaman, S. I., Mauger, D. T., Harvey, H. H., Demers, L. M., & Zagon, I. S. (2010). Opioid growth factor improves clinical benefit and survival in patients with advanced pancreatic cancer. *Open Access Journal of Clinical Trials, 2010*(2), 37–48. https://doi.org/10.2147/oajct.s8270

Sorensen, A. N., & Claësson, M. H. (1998). Effect of the opioid methionine enkephalinamide on signal transduction in human T-lymphocytes. *Life Sciences, 62*(14), 1251–1259. https://doi.org/10.1016/s0024-3205(98)00055-1

Springhorn, J. P., & Claycomb, W. C. (1989). Preproenkephalin mRNA expression in developing rat heart and in cultured ventricular cardiac muscle cells. *Biochemical Journal, 258*(1), 73–78. https://doi.org/10.1042/bj2580073

Springhorn, J. P., & Claycomb, W. C. (1992). Translation of heart preproenkephalin mRNA and secretion of enkephalin peptides from cultured cardia myocytes. *American Journal of Physiology, 263*(5 Pt 2), H1560–H1566. https://doi.org/10.1152/ajpheart.1992.263.5.H1560

Srisuchart, B., Fuchs, B. A., Sikorski, E. E., Munson, A. E., & Loveless, S. E. (1989). Antitumor activity of enkephalin analogues in inhibiting PYB6 tumor growth in mice and immunological effects of methionine enkephalinamide. *International Journal of Immunopharmacology, 11*(5), 487–500. https://doi.org/10.1016/0192-0561(89)90178-1

Stanojevic, S., Vujic, V., Mitic, K., Kustrimovic, N., Kovacevic-Jovanovic, V., Miletic, T., & Dimitrijevic, M. (2008). Methionine-enkephalin modulation of hydrogen peroxide (H_2O_2) release by rat peritoneal macrophages involves different types of opioid receptors. *Neuropeptides, 42*(2), 147–158. https://doi.org/10.1016/j.npep.2007.12.004

Sutlu, T., & Alici, E. (2009). Natural killer cell-based immunotherapy in cancer: Current insights and future prospects. *Journal of Internal Medicine, 266*(2), 154–181. https://doi.org/10.1111/j.1365-2796.2009.02121.x

Suzuki, M., Chiwaki, F., Sawada, Y., Ashikawa, M., Aoyagi, K., Fujita, T., Yanagihara, K., Komatsu, M., Narita, M., Suzuki, T., Nagase, H., Kushima, R., Sakamoto, H., Fukagawa, T., Katai, H., Nakagama, H., Yoshida, T., Uezono, Y., & Sasaki, H. (2015). Peripheral antagonist enhances the effect of anti-tumor drug by blocking a cell growth-suppressive pathway *in vivo. PLoS One, 10*(4), e0123407. https://doi.org/10.1371/journal.pone.0123407

Tischler, A. S., Delellis, R. A., Slayton, V. W., Blount, M. W., & Wolfe, H. J. (1983). Enkephalin-like immunoreactivity in human adrenal medullary cultures. *Laboratory Investigation, 48*(1), 13–18.

University of Missouri-Columbia. (2008). *Treatment of unresectable hepatocellular cancer with opioid growth factor*. ClinicalTrials.gov. Last update August 2, 2017. ClinicalTrials.gov identifier NCT00706576.

Vértes, Z., Környei, J. L., Kovàcs, S., & Vértes, M. (1996). Opioids regulate cell proliferation in the developing uterus: Effects during the period of sexual maturation. *The Journal of Steroid Biochemistry and Molecular Biology, 59*(2), 173–178. https://doi.org/10.1016/s0960-0760(96)00101-x

Vujic, V., Stanojevic, S., & Dimitrijevic, M. (2004). Methionine-enkephalin stimulates hydrogen peroxide and nitric oxide production in rat peritoneal macrophages: Interaction of mu, delta and kappa opioid receptors. *Neuroimmunomodulation, 11*(6), 392–403. https://doi.org/10.1159/000080150

Wang, Q., Gao, X., Yuan, Z., Wang, Z., Meng, Y., Cao, Y., Plotnikoff, N. P., Griffin, N., & Shan, F. (2014). Methionine enkephalin (MENK) improves lymphocyte subpopulations in human peripheral blood of 50 cancer patients by inhibiting regulatory T cells (Tregs). *Human Vaccines & Immunotherapeutics, 10*(7), 1836–1840. https://doi.org/10.4161/hv.28804

Wang, D.-M., Wang, G.-C., Yang, J., Plotnikoff, N. P., Griffen, N., Han, Y.-M., Qi, R.-Q., Gao, X.-H., & Shan, F.-P. (2016). Inhibition of the growth of human melanoma cells by methionine enkephalin. *Molecular Medicine Reports, 14*(6), 5521–5527. https://doi.org/10.3892/mmr.2016.5941

Wang, X., Jiao, X., Meng, Y., Chen, H., Liu, N., Gao, X., Griffin, N., Gao, Y., & Shan, F. (2018a). Methionine enkephalin (MENK) inhibits human gastric cancer through regulating tumor associated macrophages (TAMS) and PI3K/AKT/mTor signaling pathway inside cancer cells. *International Immunopharmacology, 65*(2018), 312–322. https://doi.org/10.1016/j.intimp.2018.10.023

Wang, X., Tian, J., Jiao, X., Geng, J., Wang, R., Griffin, N., Gao, X., & Shan, F. (2018b). The novel mechanism of anticancer effect on gastric cancer through inducing G0/G1 cell cycle arrest and caspase-dependent apoptosis *in vitro* and *in vivo* by methionine enkephalin. *Cancer Management and Research, 10*, 4773–4787. https://doi.org/10.2147/CMAR.S178343

Wang, R., Zhang, Y., & Shan, F. (2019). Interaction of opioid growth factor (OGF) and opioid antagonist and their significance in cancer therapy. *International Immunopharmacology, 75*, 105785. https://doi.org/10.1016/j.intimp.2019.105785

Weinberg, R. A. (2014). *The biology of cancer* (2nd ed.). Garland Science, Taylor and Francis Group. Chapter 8.

Worley, B., Zagon, I., & McLaughlin, P. (2015). Opioid growth factor receptor (OGFR) expression is downregulated with progression of triple negative breast cancer. *International Journal of Cancer Therapy and Oncology, 3*(4), 347.

Wu, Y., McLaughlin, P. J., & Zagon, I. S. (1998). Ontogeny of the opioid growth factor, [Met5]enkephalin, preproenkephalin gene expression, and the ξ receptor in the developing and adult aorta of rat. *Developmental Dynamics, 211*(4), 327–337. https://doi.org/10.1002/(SICI)1097-0177(199804)211:4<327::AID-AJA4>3.0.CO;2-J

Wu, Y., Xiong, Q., Li, S., Yang, X., & Ge, F. (2018). Integrated proteomic and transcriptomic analysis reveals long noncoding RNA HOX transcript antisense intergenic RNA (HOTAIR) promotes hepatocellular carcinoma cell proloferation by regulating opioid growth factor (OGFr). *Molecular & Cellular Proteomics, 17*(1), 146–159. https://doi.org/10.1074/mcp.RA117.000277

Wybran, J., Appelboom, T., Famaey, J. P., & Govaerts, A. (1979). Suggestive evidence for receptors for morphine and methionine-enkephalin on normal human blood T lymphocytes. *Journal of Immunology, 123*(3), 1068–1070.

Xu, X., Gao, Y., Wen, L., Zhai, Z., Zhang, S., Shan, F., & Feng, J. (2016). Methionine enkephalin regulates microglia polarization and function. *International Immunopharmacology, 40*, 90–97. https://doi.org/10.1016/j.intimp.2016.08.037

Ye, S., Applegren, R. R., Davis, J. M., & Cheung, H. T. (1989). Modulation of lymphocyte motility by beta-endorphin and met-enkephalin. *Immunopharmacology, 17*(2), 81–89. https://doi.org/10.1016/0162-3109(89)90053-2

Zagon, I. S., & McLaughlan, P. J. (2014). Opioid growth factor and the treatment of human pancreatic cancer: A review. *World Journal of Gastroenterology, 20*(9), 2218–2223. https://doi.org/10.3748/wjg.v20.i9.2218

Zagon, I. S., & McLaughlin, P. J. (1983a). Increased brain size and cellular content in infant rats treated with an opiate antagonist. *Science, 21*(4616), 1179–1180. https://doi.org/10.1126/science.6612331

Zagon, I. S., & McLaughlin, P. J. (1983b). Naltrexone modulates growth in infant rats. *Life Sciences, 33*(24), 2449–2454. https://doi.org/10.1016/0024-3205(83)90639-2

Zagon, I. S., & McLaughlin, P. J. (1984). Naltrexone modulates body and brain development in rats: A role for endogenous opioids in growth. *Life Sciences, 35*(20), 2057–2064. https://doi.org/10.1016/0024-3205(84)90563-0

Zagon, I. S., & McLaughlin, P. J. (1987). Endogenous opioid systems regulate cell proliferation in the developing rat brain. *Brain Research, 412*(1), 68–72. https://doi.org/10.1016/0006-8993(87)91440-5

Zagon, I. S., & McLaughlin, P. J. (1991). Identification of opioid peptides regulating proliferation of neurons and glia in the developing nervous system. *Brain Research, 542*(2), 318–323. https://doi.org/10.1016/0006-8993(91)91585-o

Zagon, I. S., & McLaughlin, P. J. (1993). Production and characterization of polyclonal and monoclonal antibodies to the zeta (ζ)-opioid receptor. *Brain Research, 630*(1–2), 295–302. https://doi.org/10.1016/0006-8993(93)90669-e

Zagon, I. S., & McLaughlin, P. J. (2003). Opioids and the apoptotic pathway in human cancer cells. *Neuropeptides, 37*(2), 79–88. https://doi.org/10.1016/s0143-4179(03)00007-6

Zagon, I. S., & McLaughlin, P. J. (2004). Gene expression of OGFr in the developing and adult rat brain and cerebellum. *Brain Research Bulletin, 63*(1), 57–63. https://doi.org/10.1016/j.brainresbull.2003.12.002

Zagon, I. S., & McLaughlin, P. J. (2005). Opioids and differentiation in human cancer cells. *Neuropeptides, 39*(5), 495–505. https://doi.org/10.1016/j.npep.2005.07.001

Zagon, I. S., Goodman, S. R., & McLaughlin, P. J. (1989). Characterization of zeta (ζ): A new opioid receptor involved in growth. *Brain Research, 482*(2), 297–305. https://doi.org/10.1016/0006-8993(89)91192-x

Zagon, I. S., Goodman, S. R., & McLaughlin, P. J. (1990). Demonstration and characterization of zeta (ζ), a growth-related opioid receptor, in a neuroblastoma cell line. *Brain Research, 511*(2), 181–186. https://doi.org/10.1016/0006-8993(90)90159-9

Zagon, I. S., Gibo, D. M., & McLaughlin, P. J. (1991). Zeta (ζ), a growth-related opioid receptor in developing rat cerebellum: Identification and characterization. *Brain Research, 551*(1–2), 28–35. https://doi.org/10.1016/0006-8993(91)90909-f

Zagon, I. S., Goodman, S. R., & McLaughlin, P. J. (1993). Zeta (ζ), the opioid growth factor receptor, identification and characterization of binding subunits. *Brain Research, 605*(1), 50–56. https://doi.org/10.1016/0006-8993(93)91355-v

Zagon, I. S., Isayama, T., & McLaughlin, P. J. (1994a). Preproenkephalin mRNA expression in the developing and adult rat brain. *Molecular Brain Research, 21*(1–2), 85–98. https://doi.org/10.1016/0169-328x(94)90381-6

Zagon, I. S., Wu, Y., & McLaughlin, P. J. (1994b). Opioid growth factor inhibits DNA synthesis in mouse tongue epithelium in a circidian rhythm-dependent manner. *American Journal of Physiology, 267*(3 Pt 2), R645–R652. https://doi.org/10.1152/ajpregu.1994.267.3.R645

Zagon, I. S., Sassani, J. W., Allison, G. M., & McLaughlin, P. J. (1995). Conserved expression of the opioid growth factor, [Met[3]]-enkephalin, and the zeta opioid receptor in vertebrate cornea. *Brain Research, 671*(1), 105–111. https://doi.org/10.1016/0006-8993(94)01314-8

Zagon, I. S., Hytrek, S. D., & McLaughlin, P. J. (1996a). Opioid growth factor (OGF) tonically inhibits human colon cancer cell proliferation in tissue culture. *American Journal of Physiology Integrative and Comparative Physiology, 271*(3 Pt 2), R511–R518. https://doi.org/10.1152/ajpregu.1996.271.3.R511

Zagon, I. S., Hytrek, S. D., Lang, C. M., Smith, J. P., McGarrity, T. J., Wu, Y., & McLaughlin, P. J. (1996b). Opioid growth factor ([Met[5]]enkephalin) prevents the incidence and retards the growth of human colon cancer. *American Journal of Physiology, Integrative and Comparative Physiology, 271*(3 Pt 2), R780–R786. https://doi.org/10.1152/ajpregu.1996.271.3.R780

Zagon, I. S., Hytrek, S. D., Smith, J. P., & McLaughlin, P. J. (1997a). Opioid growth factor (OGF) inhibits human pancreatic cancer implanted into nude mice. *Cancer Letters, 112*(2), 167–175. https://doi.org/10.1016/s0304-3835(96)04566-1

Zagon, I. S., Sassani, J. W., Kane, E. R., & McLaughlin, P. J. (1997b). Homeostasis of ocular surface epithelium in the rat is regulated by opioid growth factor. *Brain Research, 759*(1), 92–102. https://doi.org/10.1016/s0006-8993(97)00238-2

Zagon, I. S., Sassani, J. W., Wu, Y., & McLaughlin, P. J. (1998). The autocrine derivation of the opioid growth factor, [Met5]-enkephalin, in ocular surface epithelium. *Brain Research, 792*(1), 72–78. https://doi.org/10.1016/s0006-8993(98)00123-1

Zagon, I. S., Smith, J. P., & McLaughlin, P. J. (1999a). Human pancreatic cell proliferation in tissue culture is tonically inhibited by opioid growth factor (OGF). *International Journal of Oncology, 14*(3), 577–584. https://doi.org/10.3892/ijo.14.3.577

Zagon, I. S., Verderame, M. F., Allen, S. S., & McLaughlin, P. J. (1999b). Cloning, sequencing, expression and function of a cDNA encoding the opioid growth factor receptor, [Met[5]]enkephalin. *Brain Research, 849*(1–2), 147–154. https://doi.org/10.1016/s0006-8993(99)02046-6

Zagon, I. S., Wu, Y., & McLaughlin, P. J. (1999c). Opioid growth factor and organ development in rat and human embryos. *Brain Research, 839*(2), 313–322. https://doi.org/10.1016/s0006-8993(99)01753-9

Zagon, I. S., Roesener, C. D., Verderame, M. F., Ohlsson-Wilhelm, B. M., Levin, R. J., & McLaughlin, P. J. (2000a). Opioid growth factor regulates the cell cycle of human neoplasias. *International Journal of Oncology, 17*(5), 1053–1061. https://doi.org/10.3892/ijo.17.5.1053

Zagon, I. S., Sassani, J. W., & McLaughlin, P. J. (2000b). Reepithelialization of the human cornea is regulated by endogenous opioids. *Investigative Ophthalmology and Visual Science, 41*(1), 73–81.

Zagon, I. S., Verderame, M. F., Allen, S. S., & McLaughlin, P. J. (2000c). Cloning, sequencing, chromosomal location and function of cDNAs encoding the opioid growth factor receptor (OGFr) in humans. *Brain Research, 856*(1–2), 75–83. https://doi.org/10.1016/s0006-8993(99)02330-6

Zagon, I. S., Verderame, M. F., Zimmer, W. E., & McLaughlin, P. J. (2000d). Molecular characterization and distribution of the opioid growth factor receptor (OGFr) in mouse. *Molecular Brain Research, 84*(1–2), 106–114. https://doi.org/10.1016/s0169-328x(00)00232-1

Zagon, I. S., Verderame, M. F., & McLaughlin, P. J. (2002). The biology of the opioic growth factor receptor. *Brain Research Reviews, 38*(2002), 351–376. https://doi.org/10.1016/s0165-0173(01)00160-6

Zagon, I. S., Essis, F. M., Jr., Verderame, M. F., Healy, D. A., Atnip, R. G., & McLaughlin, P. J. (2003a). Opioid growth factor inhibits intimal hyperplasia in balloon-injured rat carotid artery. *Journal of Vascular Surgery, 37*(3), 636–643. https://doi.org/10.1067/mva.2003.165

Zagon, I. S., Ruth, T. B., Leure-duPree, A. E., Sassani, J. W., & McLaughlin, P. J. (2003b). Immunoelectron microscopic localization of the opioid growth receptor (OGFr) and OGF in the cornea. *Brain Research, 967*(1–2), 37–47. https://doi.org/10.1016/s0006-8993(02)04172-0

Zagon, I. S., Verderame, M. F., & McLaughlin, P. J. (2003c). The expression and function of the OGF-OGFr axis – a tonically active negative regulator of growth – in COS cells. *Neuropeptides, 37*(5), 290–297. https://doi.org/10.1016/j.npep.2003.07.001

Zagon, I. S., Jaglowski, J. R., Vederame, M. F., Smith, J. P., Leure-duPree, A. E., & McLaughlin, P. J. (2005a). Combination chemotherapy with gemcitabine and biotherapy with opioid growth factor (OGF) enhances the growth inhibition of pancreatic adenocarcinoma. *Cancer Chemotherapy and Pharmacology, 56*(5), 541–520. https://doi.org/10.1007/s00280-005-1028-x

Zagon, I. S., Ruth, T. B., & McLaughlin, P. J. (2005b). Nucleocytoplasmic distribution of opioid growth and its receptor in tongue epithelium. *The Anatomical Record Part A, 282*(1), 24–37. https://doi.org/10.1002/ar.a.20161

Zagon, I. S., Sassani, J. W., Verderame, M. F., & McLaughlin, P. J. (2005c). Particle-mediated gene transfer of OGFr cDNA regulates cell proliferation of the corneal epithelium. *Cornea, 24*(5), 614–619. https://doi.org/10.1097/01.ico.0000153561.89902.57

Zagon, I. S., Rahn, K. A., & McLaughlin, P. J. (2007). Opioids and migration, chemotaxis, invasion, and adhesion of human cancer cells. *Neuropeptides, 41*(2), 441–452. https://doi.org/10.1016/j.npep.2007.08.002

Zagon, I. S., Donahue, R. N., & McLaughlin, P. J. (2009). Opioid growth factor-opioid growth factor receptor axis is a physiological determinant of cell proliferation in diverse human cancers. *American Journal of Physiology, Integrative and Comparative Physiology, 297*(4), R1154–R1161. https://doi.org/10.1152/ajpregu.00414.2009

Zagon, I. S., Donahue, R. N., Bonneau, R. H., & McLaughlin, P. J. (2011). T lymphocyte proliferation is suppressed by the opioid growth factor ([Met^5]-enkephalin)-opioid growth factor receptor axis: Implication for the treatment of autoimmune diseases. *Immunobiology, 216*, 579–590. https://doi.org/10.1016/j.imbio.2010.09.014

Zalys, R., Zagon, I. S., Bonneau, R. H., Lang, C. M., & McLaughlin, P. J. (2000). *In vivo* effects of chronic treatment with [MET5]-enkephalin on hematological values and natural killer cell activity in athymic mice. *Life Sciences, 66*(9), 829–834. https://doi.org/10.1016/s0024-3205(99)00655-4

Zhao, D., Ploknikoff, N., Griffin, N., Song, T., & Shan, F. (2016). Methionine enkephalin, its role in immunoregulation and cancer therapy. *International Immunopharmocology, 37*, 59–64. https://doi.org/10.1016/j.intimp.2016.02.015

Zheng, T., Qiu, J., Li, C., Lin, X., Tang, X., & Hua, K. (2019). Long noncoding RNA LINC00673 promotes the proliferation and metastasis of epithelia ovarian cancer by associating with opioid growth factor receptor. *Oncotargets and Therapy, 12*, 6145–6152. https://doi.org/10.2147/OTT.S209784

Zhong, F., Li, X. Y., Yang, S., Stefano, G. B., Fimiani, C., & Bilfinger, T. V. (1998). Methionine-enkephalin stimulates interleukin-6 mRNA expression: Human plasma levels in coronary artery bypass grafting. *International Journal of Cardiology, 64*(1), S53–S59. https://doi.org/10.1016/s0167-5273(98)00036-9

Role of Endorphins in Breast Cancer Pathogenesis and Recovery

David H. Nguyen

Abstract Understanding the relationship between stress and breast cancer development is essential to preventing and alleviating the cancer. Recent research has shed light on the cognitive, physiological, cellular, and molecular underpinnings of how the endorphin pathway and stress pathway affect breast cancer. This chapter consists of two parts. Part 1 will discuss the role of endorphins in breast cancer development. This includes a discussion of three topics: (1) the neurophysiological effect of endorphins on breast tumor growth in vivo, along with further experiments that will deepen our knowledge of how β-endorphin affects breast cancer; (2) how both the opioid receptor and somatostatin receptor classes alter intracellular signaling in breast cancer cells; and (3) genetic alleles in the opioid signaling pathway that are correlated with increased breast cancer risk. Part 2 will discuss the role of endorphins in recovery from breast cancer. This includes a discussion of three topics: (1) the relationship between breast cancer diagnosis and depression; (2) the effectiveness of cognitive behavioral therapy in reducing stress in breast cancer patients; and (3) the effect of psychotherapy and exercise on preserving telomere length in breast cancer patients.

Keywords Endorphin · Breast cancer · Telomeres · Stress

Beta-Endorphin Neuron Regulates Tumor Development

β-endorphin (βE) is a polypeptide secreted from neurons located in the arcuate nucleus of the hypothalamus. βE is produced as a part of the proopiomelanocortin (POMC) protein from which adrenocorticotropic hormone (ACTH) also comes.

D. H. Nguyen (✉)
BrainScanology, Inc, Concord, CA, USA

Molecular Biophysics and Integrated Bioimaging Division, Lawrence Berkeley National Laboratory, Berkeley, CA, USA
e-mail: david.hh.n@berkeley.edu

P. L. Kerr et al. (eds.), *Endogenous Opioids*, Advances in Neurobiology 35,
https://doi.org/10.1007/978-3-031-45493-6_5

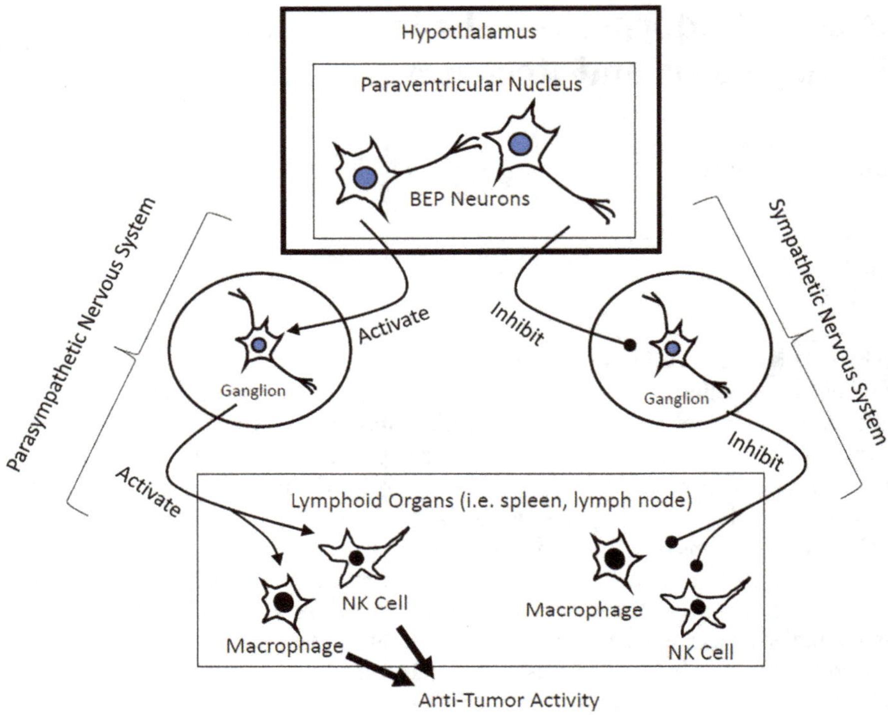

Fig. 1 Schematic of how βE augmentation activates the parasympathetic nervous system to activate macrophages and NK cells. βE neurons in the paraventricular nucleus of the hypothalamus release βE, which activates the parasympathetic nervous system while inhibiting the sympathetic nervous system. Activation of the parasympathetic pathway leads to activation of macrophages and NK cells against tumor cells

Sarkar and colleagues performed fascinating experiments in which prostate (Sarkar et al., 2008) or mammary (Sarkar et al., 2011) tumors in rats were inhibited by augmenting the pathway that produces βE (Fig. 1 summarizes the mechanism). The researchers differentiated stem cells from fetal rat brains into mature cells that produced the β-endorphin polypeptide (BEP). Fifty-day-old female virgin Sprague-Dawley rats were treated with the chemical carcinogen N-Nitroso-N-methylurea (NMU), which effectively induces mammary cancers that start to appear 12–15 weeks later. Six weeks after injection with NMU, the rats were transplanted with either cortical neurons (as controls) or the differentiated BEP-producing neurons.

Notably, Sarkar et al. (2011) showed that transplantation of βE neurons reduced the number of rats that formed tumors by half (~60% vs. ~35%). In the control group, 50% of the rats developed a tumor by 12 weeks after transplantation of the neurons (60% formed tumors by 16 weeks), while only ~35% of the βE group developed a tumor even at 16 weeks. Tumors that formed in the βE group were nearly 3 times smaller than in the control rats. More than that, 75% of the tumors in the control group were of a malignant type, while only 25% of tumors in the βE group were malignant. Consistent with a previous study by the same investigators (Boyadjieva et al., 2009), the researchers found that transplanting βE neurons enhanced innate immune function. βE transplantation resulted in NK cells having higher cytolytic activity within the peripheral blood monocyte population (PBMC) and within splenocytes. Also, macrophages were more migratory in βE-transplanted rats. In order to test the effect of βE transplantation on breast cancer metastasis, the authors injected MADB106 rat mammary cancer cells through the rat's jugular vein. Four weeks later, 70–80% of the control rats had lung tumors, while none of the βE-transplanted rats had visible lung metastasis.

Opioid Receptor and Somatostatin Receptor Signaling in Breast Cancer

Effect of Opioids on Breast Cancer Cells

A family of three opioid receptor classes serves as the cell-surface receptors of endorphins: mu (μ; MOR), delta (δ; DOR), and kappa (κ; KOR). Each class has several isoforms: μ1 and μ2; δ1 and δ2; and κ1, 2, and 3. Human breast cancer cell lines express opioid receptors and have been shown to respond differently to morphine treatment depending on experimental context. When discussing the intracellular signaling of opioid receptors, it is important to also consider the activity of somatostatin receptors because they also bind opioids.

In vitro experiments performed in two-dimensional petri dish cultures show that morphine inhibits breast cancer growth. This was observed for both ER-positive human breast cancer cell lines and ER-negative ones (Hatzoglou et al., 1996; Tegeder et al., 2003). Furthermore, treating ER-positive breast cancer cells with morphine in addition to the chemotherapeutic agent 5-Fluoroacil showed synergistic effects blocking cell growth growth (Ge et al., 2014).

Interestingly, the effects of morphine on breast cancer cells may not be due to direct binding of morphine to the aforementioned opioid receptors. Morphine inhibited the growth of T47D, MCF7, and MDA-MB-231 breast cancer cell lines, but this effect was not reversed by treatment with morphine antagonists, such as diprenorphine and naloxone (Hatzoglou et al., 1996; Tegeder et al., 2003). There is evidence suggesting that morphine may interact with the estrogen receptor, not just the opioid receptors when acting upon the MCF7 and T47D breast cancer cell lines (Panagiotou, 1998).

In the discussion of opioid receptor signaling, it is worth discussing the somatostatin receptor (SSTR) class members, SSTR1–5. Somatostatin receptors are expressed in breast cancer cells and can bind various opioids (see Hatzoglou, 2005). Morphine has been shown to bind SSTR2 when inhibiting the growth of T47D breast cancer cells (Hatzoglou, 1995). All five SSTR receptors are expressed in primary human breast cancers. The protein form of the receptors are detected in both the tumor epithelia and the tumor stroma via immunohistochemistry. The expression of SSTR1, 2, and 4 showed correlation with the expression of ER-alpha or progesterone receptor (PR), but these correlations varied depending on the patients' age and stage of cancer (Kumar et al., 2005).

Cross-Talk Between Somatostatin Receptor or Opioid Receptor with Epidermal Growth Factor Receptors

The epidermal growth factor receptor (EGFR) family—also called the ErbB or HER family—consists of four members: EGFR (ErbB1, HER1), ErbB2 (HER2, *neu* in rodents), ErbB3 (HER3), and ErbB4 (HER4). These receptors play important roles in the progression of breast cancer. The ErbB receptors homo- and heterodimerize with each other. Of interest here is the fact that ErbB receptors also heterodimerize with somatostatin receptors.

Somatostatin receptors have been shown to heterodimerize with ErbB receptors and even disrupt the heterodimerization of two ErbB family receptors (Watt et al., 2009; Kharmate, 2011a, b). Specifically, somatostatin receptors interact with ErbB receptors in human breast cancer cells (Watt & Kumar, 2006; Watt et al., 2009). Among the opioid receptors, the μ opioid receptor has been shown to activate the extracellular regulated kinase (ERK) signaling pathway in a way that is dependent upon the ErbB pathway (Al-Hasani & Bruchas, 2011).

Mouse Models in Which Morphine Promotes Breast Cancer

Two studies examining the role of morphine on breast tumors grown in immunocompromised mice showed that it promoted the cancer. Bimonte et al. (2015) found that while morphine inhibited the growth of ERα-negative breast cancer cells in vitro, it promoted the growth of these cells when grown as a tumor. The authors employed a xenograft approach, meaning they injected human breast cancer cells into a mouse. In order for this to work, the mouse's immune system must be genetically compromised. The mouse model they used is the Foxn1 mutant mouse, which does not have T lymphocytes and thus an impaired adaptive immune defense. Under this condition, morphine treatment caused the xenografted tumors to grow faster than tumors that received a control treatment. The faster growing tumors were better at neoangiogenesis, the formation of new blood vessels, which explains their faster growth rate because more blood flow through a tumor means more oxygen and nutrients to feed that tumor (Bimonte et al., 2015). A separate study by Niu et al. (2015) also revealed that morphine promotes human breast cancer growth in an immunocompromised mouse (Niu et al., 2015). In this study, the authors injected human breast cancer cells into a NOD/Scid mouse, which lacks B lymphocytes and T lymphocytes and so has no adaptive immune defense. They found that giving the mouse morphine caused the tumors to grow faster. Interestingly, they also found that giving the mouse morphine along with nalmefene, an antagonist of morphine, they were able to reverse the tumor-promoting effects of morphine.

A third study, this time in a mouse model with a complete immune system, also suggests that morphine promotes breast cancer. Using the C3Tag mouse that was engineered to produce a viral protein that promotes breast cancer, Nguyen et al. (2014) showed that giving a normal C3Tag mouse morphine did not cause cancer. However, if the mouse already had developed breast cancer, then giving morphine enhanced the cancer (Nguyen et al., 2014).

Mouse Models in Which Morphine Has No Effect on Breast Cancer

The in vivo data regarding the role of morphine in regulating breast cancer progression in mice is conflicting. Countering the studies discussed in the previous section suggesting morphine's tumor promoting activity are two mouse models from a study that suggests morphine has no effect on breast cancer.

Doonerbal and colleagues (2015) tested the effect of morphine on two mouse models of breast cancer. One model transplanted a fragment of a mammary tumor from a mutant mouse into a normal mouse. The fragments were from a tumor that had two mutated genes (*Cdh1*$^{-/-}$ and *Trp53*$^{-/-}$) and thus was aggressive. The second model was the MMTV-Neu mouse, which contained a mutated growth signal receptor and also forms very aggressive tumors. Both models exhibited metastasis of the mammary tumor to other organs, particularly the lungs. The authors reported that treatment with morphine had no effect on how long the mice lived, when the tumors appeared, how severe the lung metastasis was, nor how many blood vessels were in the tumors. In support of these this mouse study, a prospective study of Danish human patients recovering from breast cancer showed no effect of opioid prescriptions on the recurrence of breast cancer (Cronin-Fenton et al., 2015).

Resolving the Conflicting In Vivo Data About Morphine and Breast Cancer

The previous two sections presented conflicting evidence as to whether morphine promotes breast cancer in mouse models or has no effect. This section proposes a hypothesis as to why this is the case. Morphine is an appetite suppressant, so one of the effects of giving morphine is that rodents are inadvertently put on a calorie restricted (CR) diet. Though caloric restriction is known to prolong the lifespan of many organisms, it does have immunological disadvantages. Macrophages, and thus the innate immune system, are negatively affected by caloric restriction. Peritoneal macrophages from CR mice produce lower amounts of cytokines, TNF-α and IL-6, in response to endotoxin treatment compared to normally fed mice (McCarter et al., 1998). CR also inhibits macrophage activity in an intestinal puncture model of infection, resulting in the production of less cytokines and lowered phagocytic action (Sun et al., 2001). Putting mice on a high-fat diet and then reducing caloric intake by 30% caused a four-fold reduction in the number of macrophages in adipose tissue (Wasinkski et al., 2013).

The immunocompromised mouse models in which morphine promoted breast cancer (Bimonte et al., 2015; Niu et al., 2015) already had an impaired adaptive immune system (T cells and B cells). Though they still had an innate immune system (macrophages and natural killer cells) that could fight off a developing tumor, treatment with morphine suppresses macrophage function. Therefore, morphine made their incomplete immune system even less effective, allowing tumors to grow faster.

As to why the C3Tag mouse, which had both innate and adaptive immune systems functional, also exhibited an increased growth of breast cancer after morphine treatment (Nguyen et al., 2014), the CR hypothesis is also applicable. CR affects both the innate and adaptive immune systems. CR results in T cells that have problems processing antigens, maturing into effective immune regulators, and dividing to produce more copies of themselves. CR also makes antigen presenting cells, which includes macrophages, less effective in activating T cells into attack mode (Christadoss et al., 1984; Shi et al., 1998).

This hypothesis that morphine promotes breast cancer in mouse models because CR impairs the immune system requires experimental validation. Aside from suppressing appetite, morphine has many other effects. In addition to the effect of CR induced by morphine's appetite suppressant effects, morphine has been reported to block the ability of 4T1 mouse mammary cancer cells from activating macrophages (Khabbazi et al., 2015). Factors such as the dose of morphine, the length of treatment, the age of the mouse, the stage of the cancer at the time of morphine treatment, and the genetic background of the mouse are important in reconciling discrepant data. These same factors should be considered when considering the role of morphine on human cancer patients or on human tumors xenografted into mice. Humans differ in their pain tolerance and responsiveness to the pain-relieving effects of opioids, so the dosage needs to be normalized according to patient/specimen weight or other relevant factors. The complicated nature of mouse models and validated methods to minimize some of the complexity has been addressed extensively elsewhere, and is beyond the scope of this chapter (see Nguyen, 2016 for a thorough treatment of this topic).

The implications of the hypothesis that morphine treatment mimics caloric restriction, which inhibits the innate immune system's ability to fight the development of breast cancer, suggests that augmenting the βE pathway in humans to fight breast cancer may require modulating the immune system in a manner that does not introduce an opioid into the circulatory system. One way to do this is to isolate patient-specific innate or adaptive immune cells, prime them to become what they would be if βE neurons would have been transplanted into the patient, and then reintroduce the cells into the patient.

Opioid Receptor Pathway Allelic Variants and Breast Cancer Risk

Several polymorphisms in the opioid signaling pathway have been found to be associated with an increased risk for breast cancer. An allelic variant in the μ opioid receptor, the OPRM1 gene, itself has been identified as a significant risk factor for breast cancer in Northeastern Polish women (Cieślińska et al., 2015). The replacement of adenine (A) by guanine (G) at sequence position 118 changes the glycosylation pattern of the μ opioid receptor. However, a study of Korean women did not find any association between the A118G polymorphism with breast cancer (Oh et al., 2016). Adding to the complex nature of OPRM1 polymorphisms in breast cancer, another study showed that the 118G allele was associated with fewer deaths due to breast cancer in women 10 years after diagnosis, suggesting a beneficial effect. However, this association was only present in women who had invasive breast cancer (Bortsov et al., 2012).

It is noteworthy that SSTR5, which exhibits intracellular cross-talk with the opioid receptors, has polymorphisms that are associated with high circulating levels of IGF1 (Gu et al., 2010). This is relevant because high circulating levels of IGF1 are associated with an increased risk of breast cancer and pancreatic cancer (Gu et al., 2010; Li et al., 2011). Thus, there is likely a genetic background that predisposes a person to hyper- or hypo-actively respond to opioid signaling such that breast cancer is indirectly promoted or inhibited via the secretion of IGF1.

Role of Endorphins in Recovery from Breast Cancer

There is a strong correlation between the onset of depression after diagnosis of breast cancer. Compared to noncancer patients, breast cancer patients are much more likely to be on antidepressants, have a higher body mass index (BMI), and to have the metabolic syndrome (Serra et al., 2016). Various forms of therapy have been shown to reduce symptoms of depression while enhancing feels of well-being in breast cancer survivors. These include the expression of emotions through verbal or written communication (Marroquin et al., 2016); and several forms of cognitive behavioral therapy (Gudenkauf et al., 2015; Stagl et al., 2015a, b, c). In light of this evidence, it is important for basic science to uncover the underpinning mechanisms that make these therapies effective against depression caused by breast cancer diagnosis and treatment.

Telomere Length, Endogenous Opioids, and Recovery from Breast Cancer

Telomeres are sections of DNA that reside at the end of the human chromosomes. They play crucial roles in maintain genome stability and integrity. Telomeres shorten with every cell division and thus shorten as a person ages. Restoring or maintaining telomere length by activating the telomerase enzyme in aging cells is purported to be the secret to the fountain of youth. However, the dark side of long telomeres is that cancer cells, in part, are resistant to therapy because they can maintain their telomeres. Thus, telomere length is an indication of aging, health, and disease (Blackburn et al., 2015).

Zhu et al. (2016) performed a meta-analysis of 56 studies associating telomere length and cancer risk, which included eight studies on breast cancer. Four of the eight breast cancer studies suggested that longer telomere lengths associate with increased risk for breast cancer (Gramatges et al., 2010; Pellatt et al., 2013; Shen et al., 2009, 2007); one of the eight studies found the opposite relationship (Qu et al., 2013); and three of the eight studies found no relationship between telomere length and breast cancer risk (De Vivo et al., 2009; Kim et al., 2011; Zheng et al., 2010). Given the conflicting results of the eight studies, it is no surprise that Zhu et al. found no relationship between telomere length and the risk for breast cancer in a subsequent study (Zhu et al., 2016). However, there may be a discrepancy in methodology between the original papers and Zhu's meta-analysis. Zhu et al. (2016) define the first quartile or quintile as the shortest telomere lengths, but Shen et al. (2007, 2009) and Kim et al. (2011) define the first quartile/quintile (Q1) as the longest telomere lengths. An inversion of definitions would certainly nullify any correlations that are present in the meta-data.

People who release high amounts of stress hormone, cortisol, in response to psychological stress have been reported to have shorter telomeres (Epel et al., 2006; Gotlib et al., 2015; Kroenke et al., 2011; Tomiyama et al., 2012). Lengacher et al. (2014) examined the effects of stress-reducing activities on the lengths of telomeres in women with breast cancer. Women with Stage I-III cancer were enrolled in a 6-week mindfulness-based stress reduction (MBSR) program, which included gentle Hatha yoga and meditation. The control group received the usual health program. Measures were taken at the start of the program, 6-weeks, and 12-weeks after. Telomere length did not increase between the MBSR or control groups, but the activity of telomerase, the enzyme that maintains telomere length, steadily increased in MBSR patients by 17% (Lengacher et al., 2014).

Similarly, Carlson et al. (2015) compared the effect of mindfulness-based cancer recovery (MBCR) and supportive-expressive group therapy (SET) against a 1-day stress management seminar. Breast cancer patients who underwent either program had maintained their telomere length compared to patients who received the one-day seminar (Carlson et al., 2015). Garland et al. (2014) examined the levels of physical activity in 293 postmenopausal women with Stage I-III breast cancer. Breast cancer patients who participated in no physical activity had shorter telomeres than those who were active (Garland et al., 2014).

It is important to note that the studies discussed in this section measure telomere length in immune cells that were obtained from a patient's blood and not from a patient's cancer. Understanding that circulating immune cells are only surrogate markers that link telomere length with a cancer patient's health status will help clarify the confusion between why longer telomeres are associated with an increased risk of developing breast cancer, which is a bad thing, while also being associated with breast cancer patients who have benefited from behavioral therapy and physical activity, which is a good thing. The meaning of telomere length for well-ness or illness depends on the contexts in which someone develops breast cancer as well as who already has breast cancer. The further clarify potential confounding factors that affect the interpretation of what telomere length means, chronic social stress factors have been shown to shorten telomere length in both children and adults; an effect that occurs independent of breast cancer diagnoses and breast cancer treatment. Chronic social stress factors include poverty, exposure to violence, or caregiving for family members (reviewed in Oliveira et al., 2016).

In summary, telomere length is a surrogate marker for the amount of chronic stress that an individual has undergone. This chapter presents experimental evidence in rodent models that the βE pathway, when augmented, is able to inhibit mammary cancer development. Since the βE pathway is responsible to countering the stress response in mammals, one would hypothesize that: (1) stress reduction via the βE pathway would correlate with circulating immune cells that have longer telomeres than those found in identical cells from individuals who did not receive stress reduction treatment; (2) stress reduction via the βE pathway would also correlate with tumor infiltrating immune cells that have shorter telomeres compared to those found in identical cells from individuals who did not receive stress reduction treatment.

Future Experiments That Will Further Our Knowledge and Bring Us Closer to Human Treatments

This section proposes a number of experiments that will clarify the cellular and physiological mechanisms by which βE augmentation inhibits mammary cancer. Knowledge from these experiments will bring us closer to translating research findings from animals to treatments for humans. The seminal studies in this area were done using rats, which is not a species that is easily genetically engineered. This section outlines experiments in mice because they are the most common preclinical animal model in biomedical research. However, certain proposed experiments can also be implemented in rats. Transplantation of differentiated neurons into the human hypothalamus is an invasive and potentially dangerous medical procedure, so alternative methods of mimicking the βE neuron transplantation experiments are highly desirable. The following experiments were designed to dissect cellular and molecular mechanisms that are activated downstream of the augmented βE pathway. The results from these experiments will reveal which of the downstream mechanisms are best candidates to modify in hopes of mimicking the anticancer effects of transplanting βE neurons.

Test the Effect of BEP Neuron Transplantation on a Mouse Model of De Novo Metastasis That Originates in the Mammary Gland

Metastasis is a multiple-step process that involves invasion of the cancer cells into the stroma, intravasation of the cancer cell into a blood vessel within the stroma, traversal in the blood stream to a distant organ, extravasation out of a blood vessel and into the new organ, finding a niche in the new organ, and then dividing to produce a new tumor (DeNardo et al., 2008; Fidler, 2003; Liotta et al., 1991). The published studies thus far employ metastatic mammary cancer cells that are injected in the jugular vein. Vein injection models bypass many steps in the metastatic process, thus limiting the data regarding the metastatic process that can be derived. Furthermore, vein injection models of breast cancer metastasis often result in cancer cells being lodged in the lung, which is why the lung is a common organ that exhibits metastasis in these models. Vein injection models are useful as a first test of whether a treatment has antimetastatic potential, but other models of metastasis are needed to dissect which steps that come before entry into the blood stream is inhibited by a treatment.

A mouse or rat model that produces metastasis from a mammary tumor that forms de novo in the mammary gland will allow investigators to better understand how βE augmentation inhibits metastasis. Metastasis of de novo human breast cancer is a multiple step process requiring a breast tumor cell evolve the ability to escape its natural environment, survive traversal in the blood stream to a distant organ, and then be able to invade the normal environment of that organ. The microenvironment of each organ is highly specialized in terms of what types of cells can grow there, so a breast cancer cell has to be able to overcome many obstacles before it can metastasize to the lung. βE augmentation may or may not inhibit early steps in the progression of a breast cancer from being only locally invasive as opposed to metastatic. Augmenting βE in genetically engineered models of mice will reveal clues as to how what steps of metastasis in humans might also be inhibited through the βE pathway.

Time-Course Studies of the Effect of BEP Augmentation by Dibutyryl-cAMP Injection

It will be useful to determine if there is a window of opportunity during which augmentation of the βE neuron pathway is most effective. In rodents, chemical carcinogens produce the most severe effects when the exposure occurs before puberty is complete (reviewed in Alvarado et al., 2017); in humans, ionizing radiation (i.e., X-rays, gamma rays) exhibits this peripubertal window of maximum potency (Land et al., 2003; Tokunaga et al., 1987, 1994). There is a window of susceptibility that creates potential for strategic preventive measures and treatments. Puberty is believed to be a window of susceptibility because of stem cells that need to be active in order to produce a rapidly branching ductal system (Brisken & Duss, 2007). The expansion of the ductal system is dramatic, governed by the hormones and their receptors: estrogen, progesterone, prolactin, and insulin-like growth factor-1 (IGF1) (reviewed in Brisken & O'Malley, 2010). In humans, this window of susceptibility is also true for breast cancers caused by exposure to ionizing radiation (McPherson et al., 2000).

In studies by both Sarkar et al. (2011) and Zhang and colleagues (2016), βE neuron transplantation or injection of dibutyryl-cAMP occurred after the rats formed tumors that had a maximum diameter of 0.5 cm. It would be informative to know if augmenting βE levels by transplantation or injection before the 0.5 cm mark is more effective against the tumor. Also, identification of the extent to which βE augmentation remains effective in tumors that exceed 0.5 cm will be important for clarifying the limits of this therapy. In humans, breast cancers are diagnosed at different sizes and stages, so knowing windows of opportunity for βE augmentation therapy to be effective is essential for stage-specific targeted therapies.

Testing Mouse Models That Lack Macrophages, Natural Killer Cells, or Lymphocytes

βE augmentation resulted in macrophages and natural killer (NK) cells that were more phagocytic, suggesting that they were key components of the mechanism by which βE augmentation inhibited mammary cancer (Fig. 1 summarizes how βE affects macrophages and NK cells). Thus, this immune mechanism should be clarified by augmenting βE levels in rats or mice that have macrophages or NK cells ablated.

Clodronate is a commonly used chemical that can be injected into rodents, which destroys macrophages (van Rooijen & Sanders, 1994; van Rooijen et al., 1996). Clodronate is encapsulated in liposomes that macrophages and other phagocytic cells like to engulf. Inside the cells, clodronate is released and turned into a nonhydrolyzable form of the energy molecule adenosine triphosphate (ATP). This "fake ATP" causes tears in the mitochondria of phagocytes, which releases apoptotic signals that cause cell death. Though simple to administer, clodronate has drawbacks that include off-target effects that kill other types of phagocytes aside from macrophages. The widespread death of macrophages through multiple organs in the body may cause an inflammatory milieu that confounds the interpretation of how effective βE augmentation is against mammary cancer. It will be essential to identify the extent to which βE augmentation is reliant on macrophages.

It is also essential to deplete NK cells after βE augmentation in order to determine the contribution of these types of phagocytes in the anticancer effect. NK cells can be inhibited by treatment with anti-CD81 antibodies or the major envelope protein of Hepatitis C Virus (HCV-E2) (Crotta et al., 2002). Another method of NK cell depletion is to inject antibodies against the membrane receptor Nk1.1 (Cook & Whitmire, 2013), which is involved in NK cell activation, cytotoxic granule release, and secretion of interferon gamma-1 (IFNg1). A fourth method is injecting the anti-asialo GM1 antibody (Christaki et al., 2015).

NK cells are of particular relevance to understanding the mechanisms by which βE augmentation affects breast cancer. Boyadjieva et al. (2009) showed that βE augmentation resulted in NK cells having higher cytolytic activity. This was observed in NK cells that were in circulation as part of the PBMC, and for NK cells located in the spleen. In humans, there are five subsets of NK cells that are associated with breast cancer. The proportion of these five subsets of NK cells in a tumor changes according to how advanced the breast cancer has become (Mamessier et al., 2013). NK subsets are defined by their expression of two cell surface markers, CD56 and CD16, as measured by flow cytometry. The three subsets that can exist apart from a breast tumor are CD56(dim)/CD16(+), CD56(bright)/CD16(−), and CD56(−)/CD16(+). A breast tumor can induce two other NK subsets, CD56(bright)/CD16(+) and CD56(dim)/CD16(−) [Description of flow cytometry notation: "dim" means that the protein is expressed at low levels; "bright" means that the protein is expressed at high levels; "+" means that the protein is expressed; "–" means that the protein is not expressed.]. Thus, it is important to determine if mammary tumors that are inhibited by βE augmentation exhibit different NK subsets than those in tumors in control rodents. This knowledge would allow researchers to determine ways of inducing NK cells into behaving as if they were from βE-augmented hosts, but without having to transplant βE neurons into the hosts brain. Patient-specific immunotherapy using NK cells is a promising route through which the βE pathway can be augmented in humans against breast cancer.

NK cells in mice do not express CD56, but they have been found to exist as subsets based on the expression of CD27 and CXCR3 (Marquardt et al., 2010). The subset of CXCR3(−) NK cells exhibited higher cytolytic activity than the CXCR3(+) ones, while CXCR3(+) ones secreted more inflammatory cytokines than the CXCR3(−) ones (IFN-gamma, TNF-alpha, and MIP1-alpha); this pattern correlates with human NK cell subsets defined as CD56(dim) and CD56(bright). The authors reported that upon activation with exogenous chemicals that mimic infection, CXCR3 expression levels changed. This means that researchers who characterize tumors associated NK cells in mice should be aware that the tumor will likely alter the expression level of CXCR3 on these NK cells compared to those that are in circulation.

There is extant biotechnology that can be applied to determining the contribution of other immune cells types to the anticancer effects of βE augmentation. Indeed, multiple models of genetically engineered mice have been developed. *Foxn1* mutant mice have no T cells. NOD/Scid mice have no mature B and T cells. *Rag1* mutant mice have no mature B and T cells. *Prf1* mutant mice lack the lytic pathway by which NK cells and CD8 T cells kill their targets.

Identify the Estrogen Receptor-α Status of Tumors Inhibited by BEP Neuron Transplantation Compared to Controls

Estrogen Receptor-α (ERα) is a key clinical marker of the clinical subtype of a breast cancer. The presence of ERα protein is also predictive of whether a patient is likely responsive to hormone therapy. Tamoxifen is one of the most common forms of hormone therapy for ERα + breast cancers (Eggemann et al., 2018; Tremont et al., 2017). The Luminal-A and Luminal-B subtypes of breast cancer both express ERα protein, while the HER2, Triple Negative, and Basal-like subtypes do not express it (Coleman & Anders, 2017).

Knowing if βE augmentation affects the ERα status of breast cancer can reveal the potential for more effective therapy. If βE augmentation selects for tumors that are ERα positive, then this suggests that behavioral and chemical interventions that reduce stress in humans may also influence the ERα status of their future breast cancers. Because ERα positive breast cancers have a better clinical outcome than ERα negative ones, these interventions can significantly improve the rate of survival. Behavioral interventions that reduce stress include exercise, cognitive behavioral therapy, and meditation.

It may also be informative to determine if βE augmentation influences the expression of estrogen receptor beta (ERβ), though it is not one of the markers that are routinely tested in classifying breast cancer molecular subtypes (Luminal A, HER2, etc.). Evidence has been accumulating that the expression of ERβ in breast cancers is predictive of favorable response to tamoxifen. Patients who had a high level of expression of ERβ showed response to tamoxifen after breast cancer surgery (Hopp et al., 2004). The expression of ERβ is also predictive of reduced Ki67 levels—an indicator of how many cells are actively dividing—after treatment with tamoxifen or the aromatase inhibitor anastrazole (Madeira et al., 2013).

Identify the Genomic Breast Cancer Subtypes of the Tumors Inhibited by BEP Neuron Transplantation Compared to Controls

The six breast cancer subtypes that are recognized in clinical settings (Luminal A, Luminal B, HER2, Basal-like, Triple-Negative, and Normal-like) each have distinct genomic signatures in the form of whole-genome signatures of: mRNA levels, miRNA levels, DNA copy number, or DNA mutations (Dai et al., 2015). It will be informative to know if augmenting the βE pathway works well for certain genomic/transcriptomic subtypes and not for others. This will reveal mechanisms for potentially sensitizing nonresponsive breast cancer subtypes by another treatment in addition to βE augmentation.

Test the Malignant Potential of Tumors Inhibited by βE Augmentation by Reseeding into New Hosts

Tumors contain cancer stem cells and cells-of-origin that can repopulate an entirely new tumor though they are only one cell (reviewed in Nguyen, 2016). Additionally, when fragments of a tumor are transplanted into a new host animal, the fragment grows into a new tumor that resembles the features of the original tumor—though this is not the case for every single tumor. Whether or not a transplanted fragment, or injected solution of isolated cancer stem cells, recreates a tumor that is similar to the original tumor (i.e., histopathology, growth rate, latency, etc.) provides critical information about the mechanisms by which the original tumor developed. This knowledge will clarify what compartment of the tumor and/or host should be targeted by therapies. Tumor development is a complex interplay between the tumor cells, nontumor cells within a tumor (fibroblasts, blood vessels, etc.), and the physiology of the host (hormonal milieu, growth factor milieu, etc.). The tumor may be dependent on the host for a supply of growth factors, exosomes, and inflammatory cells; many of which are not produced by the tumor cells themselves. It is important to understand whether primary tumors in βE-augmented hosts can regrow in new hosts and, if so, why. Since the eventual goal is to augment the βE pathway in humans to treat breast cancer, it is important to understand how much βE-augmentation reduces the malignant potential of breast cancers, which includes a tumor's ability to metastasize (a sign of its "aggressiveness").

Knowing the mechanisms of a tumor's development is not the same as understanding why a tumor is resistant against therapies, because the therapies modify the tumor and the host in a way that selects for the most resilient tumor cells to remain. The tumor that reoccurs after therapeutic intervention may or may not exhibit some of the same weaknesses as the tumor from which it came. Testing the degree of therapeutic resistance of tumors arising from βE-augmentation is also a future direction of research.

Test the Ability of βE Augmentation to Inhibit Mammary Cancer Caused by a Common Human Carcinogen

Many carcinogens cause or promote breast cancer in humans, including ionizing radiation (e.g., X-rays, gamma rays) and chemicals that leach from plastic products (e.g., bisphenol A; BPA). It is important to know if augmenting the βE pathway can stop tumors that were caused by, or promoted by, such human carcinogens given to a mouse or rat model as a first step toward identifying the potential for inhibiting tumorigenesis in humans. These carcinogen exposure studies should be done on genetically engineered mice that carry mutations that produce mammary cancer, on mice that are normal but that received a transplanted tumor, and on normal mice that naturally develop breast cancer over time. These experiments will reveal molecular and cellular mechanisms by which human breast cancers caused, or promoted, by environmental carcinogen can be inhibited through augmenting the βE pathway.

Conclusions

This chapter is among the first in the field of breast cancer research to blend a synthesis of animal model research on endogenous opioids with breast cancer research in humans. The chapter was written with a specific intent to propose experiments for moving the lines of research on rodent models forward into the human arena. The fact that augmenting the βE pathway in rodents can inhibit the growth of mammary cancer has tremendous implications for how psychosocial interventions have the potential as preventive measures against and treatments for breast cancer in humans. Repeating the experiments in mouse models that mimic various aspects of human breast cancer will reveal further insights into how the βE pathway can be modulated in human patients. Further work on the relationships between telomere length and chronic social stress, breast cancer diagnosis, or breast cancer treatment is the next step toward translating rodent data into clinical practice. The minute analysis of research studies in this area that conflict with each other will hopefully guide future work toward eradicating breast cancer.

References

Al-Hasani, R., & Bruchas, M. R. (2011). Molecular mechanisms of opioid receptor-dependent signaling and behavior. *The Journal of the American Society of Anesthesiologists, 115*(6), 1363–1381.

Alvarado, A., Faustino-Rocha, A. I., Colaco, B., & Oliveira, P. A. (2017). Experimental mammary carcinogenesis-Rat models. *Life Sciences, 173*, 116–134.

Bimonte, S., Barbieri, A., Rea, D., Palma, G., Luciano, A., Cuomo, A., et al. (2015). Morphine promotes tumor angiogenesis and increases breast cancer progression. *BioMed Research International*, 1–8. https://doi.org/10.1155/2015/161508

Blackburn, E. H., Epel, E. S., & Lin, J. (2015). Human telomere biology: A contributory and interactive factor in aging, disease risks, and protection. *Science, 350*(6265), 1193–1198.

Bortsov, A. V., Millikan, R. C., Belfer, I., Boortz-Marx, R. L., Arora, H., & McLean, S. A. (2012). μ-Opioid receptor gene A118G polymorphism predicts survival in patients with breast cancer. *Anesthesiology: The Journal of the American Society of Anesthesiologists, 116*(4), 896–902.

Boyadjieva, N. I., Ortigüela, M., Arjona, A., Cheng, X., & Sarkar, D. K. (2009). β-Endorphin neuronal cell transplant reduces corticotropin releasing hormone hyperresponse to lipopolysaccharide and eliminates natural killer cell functional deficiencies in fetal alcohol exposed rats. *Alcoholism: Clinical and Experimental Research, 33*(5), 931–937.

Brisken, C., & Duss, S. (2007). Stem cells and the stem cell niche in the breast: An integrated hormonal and developmental perspective. *Stem Cell Reviews, 3*(2), 147–156.

Brisken, C., & O'Malley, B. (2010). Hormone action in the mammary gland. *Cold Spring Harbor Perspectives in Biology, 2*, a003178.

Carlson, L. E., Beattie, T. L., Giese-Davis, J., Faris, P., Tamagawa, R., Fick, L. J., et al. (2015). Mindfulness-based cancer recovery and supportive-expressive therapy maintain telomere length relative to controls in distressed breast cancer survivors. *Cancer, 121*(3), 476–484.

Christadoss, P., Talal, N., Lindstrom, J., & Fernandes, G. (1984). Suppression of cellular and humoral immunity to T-dependent antigens by calorie restriction. *Cellular Immunology, 88*(1), 1–8.

Christaki, E., Diza, E., Giamarellos-Bourboulis, E. J., Papadopoulou, N., Pistiki, A., Droggiti, D. I., et al. (2015). NK and NKT cell depletion alters the outcome of experimental pneumococcal pneumonia: Relationship with regulation of interferon-γ production. *Journal of Immunology Research, 532717*, 1–10.

Cieślińska, A., Sienkiewicz-Szłapka, E., Kostyra, E., Fiedorowicz, E., Snarska, J., Wroński, K., et al. (2015). μ-Opioid receptor gene (OPRM1) polymorphism in patients with breast cancer. *Tumor Biology, 36*(6), 4655–4660.

Coleman, W. B., & Anders, C. K. (2017). Discerning clinical responses in breast cancer based on molecular signatures. *The American Journal of Pathology, 187*(10), 2199–2207.

Cook, K. D., & Whitmire, J. K. (2013). The depletion of NK cells prevents T cell exhaustion to efficiently control disseminating virus infection. *The Journal of Immunology, 190*(2), 641–649.

Cronin-Fenton, D. P., Heide-Jørgensen, U., Ahern, T. P., Lash, T. L., Christiansen, P. M., Ejlertsen, B., et al. (2015). Opioids and breast cancer recurrence: A Danish population-based cohort study. *Cancer, 121*(19), 3507–3514.

Crotta, S., Stilla, A., Wack, A., D'Andrea, A., Nuti, S., D'Oro, U., et al. (2002). Inhibition of natural killer cells through engagement of CD81 by the major hepatitis C virus envelope protein. *Journal of Experimental Medicine, 195*(1), 35–42.

Dai, X., Li, T., Bai, Z., Yang, Y., Liu, X., Zhan, J., Shi, B. (2015). Breast cancer intrinsic subtype classification, clinical use and future trends. *American Journal of Cancer Research, 5*(10), 2929–2943.

De Vivo, I., Prescott, J., Wong, J. Y., Kraft, P., Hankinson, S. E., & Hunter, D. J. (2009). A prospective study of relative telomere length and postmenopausal breast cancer risk. *Cancer Epidemiology and Prevention Biomarkers, 18*(4), 1152–1156.

DeNardo, D. G., Johansson, M., & Coussens, L. M. (2008). Immune cells as mediators of solid tumor metastasis. *Cancer and Metastasis Reviews, 27*(1), 11–18.

Doornebal, C. W., Vrijland, K., Hau, C. S., Coffelt, S. B., Ciampricotti, M., Jonkers, J., et al. (2015). Morphine does not facilitate breast cancer progression in two preclinical mouse models for human invasive lobular and HER2+ breast cancer. *Pain, 156*(8), 1424–1432.

Eggemann, H., Altmann, U., Costa, S. D., & Ignatov, A. (2018). Survival benefit of tamoxifen and aromatase inhibitor in male and female breast cancer. *Journal of Cancer Research and Clinical Oncology, 144*(2), 337–341.

Epel, E. S., Lin, J., Wilhelm, F. H., Wolkowitz, O. M., Cawthon, R., Adler, N. E., et al. (2006). Cell aging in relation to stress arousal and cardiovascular disease risk factors. *Psychoneuroendocrinology, 31*(3), 277–287.

Fidler, I. J. (2003). The pathogenesis of cancer metastasis: The'seed and soil'hypothesis revisited. *Nature Reviews Cancer, 3*(6), 453.

Garland, S. N., Johnson, B., Palmer, C., Speck, R. M., Donelson, M., Xie, S. X., et al. (2014). Physical activity and telomere length in early stage breast cancer survivors. *Breast Cancer Research, 16*(4), 413–421.

Ge, Z. H., Wang, Z. X., Yu, T. L., Yang, N., Sun, Y., Hao, C. L., & Sun, L. X. (2014). Morphine improved the antitumor effects on MCF-7 cells in combination with 5-Fluorouracil. *Biomedicine & Pharmacotherapy, 68*(3), 299–305.

Gotlib, I. H., LeMoult, J., Colich, N. L., Foland-Ross, L. C., Hallmayer, J., Joormann, J., et al. (2015). Telomere length and cortisol reactivity in children of depressed mothers. *Molecular Psychiatry, 20*(5), 615–620.

Gramatges, M. M., Telli, M. L., Balise, R., & Ford, J. M. (2010). Longer relative telomere length in blood from women with sporadic and familial breast cancer compared with healthy controls. *Cancer Epidemiology and Prevention Biomarkers, 19*(2), 605–613.

Gu, F., Schumacher, F. R., Canzian, F., Allen, N. E., Albanes, D., Berg, C. D., et al. (2010). Eighteen insulin-like growth factor pathway genes, circulating levels of IGF-I and its binding protein, and risk of prostate and breast cancer. *Cancer Epidemiology and Prevention Biomarkers, 19*(11), 2877–2887.

Gudenkauf, L. M., Antoni, M. H., Stagl, J. M., Lechner, S. C., Jutagir, D. R., Bouchard, L. C., et al. (2015). Brief cognitive–behavioral and relaxation training interventions for breast cancer: A randomized controlled trial. *Journal of Consulting and Clinical Psychology, 83*(4), 677–688.

Hatzoglou, A., Ouafik, L., Bakogeorgou, E., Thermos, K., Castanas, E. (1995). Morphine Cross-Reacts with Somatostatin Receptor SSTR2 in the T47D Human Breast Cancer Cell Line and Decreases Cell Growth. Cancer Research, 55(23), 5632–5636.

Hatzoglou, A., Bakogeorgou, E., & Castanas, E. (1996). The antiproliferative effect of opioid receptor agonists on the T47D human breast cancer cell line, is partially mediated through opioid receptors. *European Journal of Pharmacology, 296*(2), 199–207.

Hatzoglou, A., Kampa, M., Castanas, E. (2005). Opioid-somatostatin interactions in regulating cancer cell growth. *Frontiers in Bioscience, 10*, 244–56.

Hopp, T. A., Weiss, H. L., Parra, I. S., Cui, Y., Osborne, C. K., & Fuqua, S. A. (2004). Low levels of estrogen receptor β protein predict resistance to tamoxifen therapy in breast cancer. *Clinical Cancer Research, 10*(22), 7490–7499.

Khabbazi, S., Goumon, Y., & Parat, M. O. (2015). Morphine modulates interleukin-4-or breast cancer cell-induced pro-metastatic activation of macrophages. *Scientific Reports, 5*, 11389.

Kharmate, G., Rajput, P. S., Watt, H. L., Somvanshi, R. K., Chaudhari, N., Qiu, X., & Kumar, U. (2011a). Dissociation of epidermal growth factor receptor and ErbB2 heterodimers in the presence of somatostatin receptor 5 modulate signaling pathways. *Endocrinology, 152*(3), 931–945.

Kharmate, G., Rajput, P. S., Watt, H. L., Somvanshi, R. K., Chaudhari, N., Qiu, X., & Kumar, U. (2011b). Role of somatostatin receptor 1 and 5 on epidermal growth factor receptor mediated signaling. *Biochimica et Biophysica Acta (BBA)-Molecular Cell Research, 1813*(6), 1172–1189.

Kim, S., Sandler, D. P., Carswell, G., De Roo, L. A., Parks, C. G., Cawthon, R., et al. (2011). Telomere length in peripheral blood and breast cancer risk in a prospective case-cohort analysis: Results from the Sister Study. *Cancer Causes & Control, 22*(7), 1061–1066.

Kroenke, C. H., Epel, E., Adler, N., Bush, N. R., Obradović, J., Lin, J., et al. (2011). Autonomic and adrenocortical reactivity and buccal cell telomere length in kindergarten children. *Psychosomatic Medicine, 73*(7), 533–540.

Kumar, U., Grigorakis, S. I., Watt, H. L., Sasi, R., Snell, L., Watson, P., & Chaudhari, S. (2005). Somatostatin receptors in primary human breast cancer: Quantitative analysis of mRNA for subtypes 1–5 and correlation with receptor protein expression and tumor pathology. *Breast Cancer Research and Treatment, 92*(2), 175–186.

Land, C. E., Tokunaga, M., Koyama, K., Soda, M., Preston, D. L., Nishimori, I., & Tokuoka, S. (2003). Incidence of female breast cancer among atomic bomb survivors, Hiroshima and Nagasaki, 1950–1990. *Radiation Research, 160*(6), 707–717.

Lengacher, C. A., Reich, R. R., Kip, K. E., Barta, M., Ramesar, S., Paterson, C. L., et al. (2014). Influence of mindfulness-based stress reduction (MBSR) on telomerase activity in women with breast cancer (BC). *Biological Research for Nursing, 16*(4), 438–447.

Li, D., Tanaka, M., Brunicardi, F. C., Fisher, W. E., Gibbs, R. A., & Gingras, M. C. (2011). Association between somatostatin receptor 5 gene polymorphisms and pancreatic cancer risk and survival. *Cancer, 117*(13), 2863–2872.

Liotta, L. A., Steeg, P. S., & Stetler-Stevenson, W. G. (1991). Cancer metastasis and angiogenesis: An imbalance of positive and negative regulation. *Cell, 64*(2), 327–336.

Madeira, M., Mattar, A., Logullo, Â. F., Soares, F. A., & Gebrim, L. H. (2013). Estrogen receptor alpha/beta ratio and estrogen receptor beta as predictors of endocrine therapy responsiveness–a randomized neoadjuvant trial comparison between anastrozole and tamoxifen for the treatment of postmenopausal breast cancer. *BMC Cancer, 13*(1), 425–436.

Mamessier, E., Pradel, L. C., Thibult, M. L., Drevet, C., Zouine, A., Jacquemier, J., et al. (2013). Peripheral blood NK cells from breast cancer patients are tumor-induced composite subsets. *The Journal of Immunology, 190*, 2424–2436.

Marquardt, N., Wilk, E., Pokoyski, C., Schmidt, R. E., & Jacobs, R. (2010). Murine CXCR3+ CD27bright NK cells resemble the human CD56bright NK-cell population. *European Journal of Immunology, 40*(5), 1428–1439.

Marroquín, B., Czamanski-Cohen, J., Weihs, K. L., & Stanton, A. L. (2016). Implicit loneliness, emotion regulation, and depressive symptoms in breast cancer survivors. *Journal of Behavioral Medicine, 39*(5), 832–844.

McCarter, M. D., Naama, H. A., Shou, J., Kwi, L. X., Evoy, D. A., Calvano, S. E., & Daly, J. M. (1998). Altered macrophage intracellular signaling induced by protein–calorie malnutrition. *Cellular Immunology, 183*(2), 131–136.

McPherson, K., Steel, C., & Dixon, J. M. (2000). Breast cancer--epidemiology, risk factors, and genetics. *British Medical Journal, 321*(7261), 624–624.

Nguyen, D. H. (2016). Using mouse models and making sense of them. In *Systems biology of tumor physiology* (pp. 33–54). Springer Science and Business Media.

Nguyen, J., Luk, K., Vang, D., Soto, W., Vincent, L., Robiner, S., et al. (2014). Morphine stimulates cancer progression and mast cell activation and impairs survival in transgenic mice with breast cancer. *British Journal of Anaesthesia, 113*(suppl_1), i4–i13.

Niu, D. G., Peng, F., Zhang, W., Guan, Z., Zhao, H. D., Li, J. L., et al. (2015). Morphine promotes cancer stem cell properties, contributing to chemoresistance in breast cancer. *Oncotarget, 6*(6), 3963–3976.

Oh, C. S., Lee, S. H., Yoo, Y. B., Yang, J. H., & Kim, S. H. (2016). Mu-opioid receptor polymorphisms and breast cancer in a Korean female adult population: A retrospective study. *Journal of Breast Cancer, 19*(2), 206–209.

Oliveira, B. S., Zunzunegui, M. V., Quinlan, J., Fahmi, H., Tu, M. T., & Guerra, R. O. (2016). Systematic review of the association between chronic social stress and telomere length: A life course perspective. *Ageing Research Reviews, 26*, 37–52.

Panagiotou, S., Hatzoglou, A., Calvo, F., Martin, P.M., Castanas, E. (1998). Modulation of the estrogen-regulated proteins cathepsin D and pS2 by opioid agonists in hormone-sensitive breast cancer cell lines (MCF7 and T47D): evidence for an interaction between the two systems. *Journal of Cellular Biochemistry 71*(3):416–28.

Pellatt, A. J., Wolff, R. K., Torres-Mejia, G., John, E. M., Herrick, J. S., Lundgreen, A., et al. (2013). Telomere length, telomere-related genes, and breast cancer risk: The breast cancer health disparities study. *Genes, Chromosomes and Cancer, 52*(7), 595–609.

Qu, S., Wen, W., Shu, X. O., Chow, W. H., Xiang, Y. B., Wu, J., et al. (2013). Association of leukocyte telomere length with breast cancer risk: Nested case-control findings from the Shanghai Women's Health Study. *American Journal of Epidemiology, 177*(7), 617–624.

Sarkar, D. K., Boyadjieva, N. I., Chen, C. P., Ortigüela, M., Reuhl, K., Clement, E. M., et al. (2008). Cyclic adenosine monophosphate differentiated β-endorphin neurons promote immune function and prevent prostate cancer growth. *Proceedings of the National Academy of Sciences, 105*, 9105–9110.

Sarkar, D. K., Zhang, C., Murugan, S., Dokur, M., Boyadjieva, N. I., Ortiguela, M., et al. (2011). Transplantation of β-endorphin neurons into the hypothalamus promotes immune function and restricts the growth and metastasis of mammary carcinoma. *Cancer Research, 71*(19), 6282–6291.

Serra, M. C., Goldberg, A. P., & Ryan, A. S. (2016). Increased depression and metabolic risk in postmenopausal breast cancer survivors. *Diabetology & Metabolic Syndrome, 8*(1), 44–49.

Shen, J., Terry, M. B., Gurvich, I., Liao, Y., Senie, R. T., & Santella, R. M. (2007). Short telomere length and breast cancer risk: A study in sister sets. *Cancer Research, 67*(11), 5538–5544.

Shen, J., Gammon, M. D., Terry, M. B., Wang, Q., Bradshaw, P., Teitelbaum, S. L., et al. (2009). Telomere length, oxidative damage, antioxidants and breast cancer risk. *International Journal of Cancer, 124*(7), 1637–1643.

Shi, H. N., Scott, M. E., Stevenson, M. M., & Koski, K. G. (1998). Energy restriction and zinc deficiency impair the functions of murine T cells and antigen-presenting cells during gastrointestinal nematode infection. *The Journal of Nutrition, 128*(1), 20–27.

Stagl, J. M., Lechner, S. C., Carver, C. S., Bouchard, L. C., Gudenkauf, L. M., Jutagir, D. R., et al. (2015a). A randomized controlled trial of cognitive-behavioral stress management in breast cancer: Survival and recurrence at 11-year follow-up. *Breast Cancer Research and Treatment, 154*(2), 319–328.

Stagl, J. M., Bouchard, L. C., Lechner, S. C., Blomberg, B. B., Gudenkauf, L. M., Jutagir, D. R., et al. (2015b). Long-term psychological benefits of cognitive-behavioral stress management for women with breast cancer: 11-year follow-up of a randomized controlled trial. *Cancer, 121*(11), 1873–1881.

Stagl, J. M., Antoni, M. H., Lechner, S. C., Bouchard, L. C., Blomberg, B. B., Glück, S., et al. (2015c). Randomized controlled trial of cognitive behavioral stress management in breast cancer: A brief report of effects on 5-year depressive symptoms. *Health Psychology, 34*(2), 176–180.

Sun, D., Muthukumar, A. R., Lawrence, R. A., & Fernandes, G. (2001). Effects of calorie restriction on polymicrobial peritonitis induced by cecum ligation and puncture in young C57BL/6 mice. *Clinical and Diagnostic Laboratory Immunology, 8*(5), 1003–1011.

Tegeder, I., Grösch, S., Schmidtko, A., Häussler, A., Schmidt, H., Niederberger, E., et al. (2003). G protein-independent G1 cell cycle block and apoptosis with morphine in adenocarcinoma cells: Involvement of p53 phosphorylation. *Cancer Research, 63*(8), 1846–1852.

Tokunaga, M., Land, C. E., Yamamoto, T., Asano, M., Tokuoka, S., Ezaki, H., & Nishimori, I. (1987). Incidence of female breast cancer among atomic bomb survivors, Hiroshima and Nagasaki, 1950-1980. *Radiation Research, 112*(2), 243–272.

Tokunaga, M., Land, C. E., Tokuoka, S., Nishimori, I., Soda, M., & Akiba, S. (1994). Incidence of female breast cancer among atomic bomb survivors, 1950-1985. *Radiation Research, 138*(2), 209–223.

Tomiyama, A. J., O'Donovan, A., Lin, J., Puterman, E., Lazaro, A., Chan, J., et al. (2012). Does cellular aging relate to patterns of allostasis?: An examination of basal and stress reactive HPA axis activity and telomere length. *Physiology & Behavior, 106*(1), 40–45.

Tremont, A., Lu, J., & Cole, J. T. (2017). Endocrine therapy for early breast cancer: Updated review. *The Ochsner Journal, 17*(4), 405–411.

Van Rooijen, N., & Sanders, A. (1994). Liposome mediated depletion of macrophages: Mechanism of action, preparation of liposomes and applications. *Journal of Immunological Methods, 174*(1–2), 83–93.

Van Rooijen, N., Sanders, A., & van den Berg, T. K. (1996). Apoptosis of macrophages induced by liposome-mediated intracellular delivery of clodronate and propamidine. *Journal of Immunological Methods, 193*(1), 93–99.

Wasinski, F., Bacurau, R. F., Moraes, M. R., Haro, A. S., Moraes-Vieira, P. M., Estrela, G. R., et al. (2013). Exercise and caloric restriction alter the immune system of mice submitted to a high-fat diet. *Mediators of Inflammation, 395672*, 1–8.

Watt, H. L., & Kumar, U. (2006). Colocalization of somatostatin receptors and epidermal growth factor receptors in breast cancer cells. *Cancer Cell International, 6*(1), 5–23.

Watt, H. L., Kharmate, G. D., & Kumar, U. (2009). Somatostatin receptors 1 and 5 heterodimerize with epidermal growth factor receptor: Agonist-dependent modulation of the downstream MAPK signalling pathway in breast cancer cells. *Cellular Signaling, 21*(3), 428–439.

Zhang, C., Franklin, T., Sarkar, D.K. (2016). Inhibition of Mammary Cancer Progression in Fetal Alcohol Exposed Rats by β-Endorphin Neurons. *Alcohol Clinical & Experimental Research, 40*(1), 134–140.

Zheng, Y. L., Ambrosone, C., Byrne, C., Davis, W., Nesline, M., & McCann, S. E. (2010). Telomere length in blood cells and breast cancer risk: Investigations in two case–control studies. *Breast Cancer Research and Treatment, 120*(3), 769–775.

Zhu, X., Han, W., Xue, W., Zou, Y., Xie, C., Du, J., & Jin, G. (2016). The association between telomere length and cancer risk in population studies. *Scientific Reports, 6*, 22243.

Modulatory Processes in Craniofacial Pain States

Barry J. Sessle

Abstract Pain is a common symptom associated with many disorders affecting the craniofacial tissues that include the teeth and their supporting structures, the jaw, face and tongue muscles, and the temporomandibular joint. Most acute craniofacial pain states are easily recognized and readily treated, but chronic craniofacial pain states (e.g., temporomandibular disorders [TMD], trigeminal neuropathies, and some headaches) may be especially challenging to manage successfully. This chapter provides an overview of the processes that underlie craniofacial pain, with a focus on the pain-modulatory mechanisms operating in craniofacial tissues and in the central nervous system (CNS), including the role of endogenous chemical processes such as those involving opioids. The chapter outlines in particular findings from preclinical studies that have provided substantial information about the neural as well as nonneural (e.g., glial) processes involved in the initiation, transmission, and modulation of nociceptive signals in the trigeminal system, and also draws attention to their clinical correlates. The increased understanding gained from these preclinical studies of how nociceptive signals can be modulated will contribute to improvements in presently available therapeutic approaches to manage craniofacial pain as well as to the development of novel analgesic approaches.

Keywords Acute pain · Craniofacial pain · Central sensitization · Chronic pain · Descending modulation · Peripheral sensitization · Trigeminal system

B. J. Sessle (✉)
Department of Oral Physiology Faculty of Dentistry, Department of Physiology Faculty of Medicine, Centre for the Study of Pain, University of Toronto, Toronto, ON, Canada
e-mail: barry.sessle@utoronto.ca

P. L. Kerr et al. (eds.), *Endogenous Opioids*, Advances in Neurobiology 35,
https://doi.org/10.1007/978-3-031-45493-6_6

Introduction

Some of the most common pains in the body occur in the craniofacial region and carry a considerable socioeconomic burden (LeResche & Drangsholt, 2008; Institute of Medicine, 2011; Macfarlane, 2014; Sessle, 2014). Research findings of the past five decades have provided a substantive knowledge base of the structural and functional features of trigeminal somatosensory mechanisms including those contributing to acute and chronic craniofacial pain states (for detailed reviews, see Darian-Smith, 1966; Dubner et al., 1978; Hargreaves, 2011; Hargreaves & Ruparel, 2016; Sessle, 2000, 2014, 2021). This chapter provides an overview of these findings, with a particular focus on the physiological and endogenous chemical processes involved in the initiation, transmission and modulation of nociceptive signals in the trigeminal system that may result in the accentuation or attenuation of craniofacial pain.

Trigeminal Primary Afferent Mechanisms

The craniofacial tissues (e.g., oral mucosa, periodontium, facial skin, temporomandibular joint [TMJ], masticatory muscles, meninges) are innervated by large-, medium-, and small-diameter primary afferent nerve fibers, most of which occur in the trigeminal nerve and have their primary afferent cell bodies in the trigeminal ganglion (Cairns et al., 2014; Dubner et al., 1978, 2013, 2014; Iwata et al., 2011; Sessle, 2000). They include large- and medium-diameter afferents, which are myelinated and referred to respectively as A-beta and A-delta afferents, and unmyelinated afferents termed *C-fibers*. The endings of many of the A-beta afferents and of some of the A-delta afferents in the tissues terminate as specialized sense organs (receptors) and function as mechanoreceptors which are activated by tactile or proprioceptive stimuli applied to the tissues. Some of the smaller afferents (A-delta and C-fiber afferents) terminate as receptors sensitive to cooling or warming stimuli, but the majority terminate as so-called free nerve endings, most of which function as nociceptors since they are activated by noxious stimulation of the tissues.

Activation of the afferent endings may evoke action potentials in their associated afferent fibers, and then these action potentials are conducted along the afferents to provide the central nervous system (CNS) with sensory-discriminative information about the quality, location, intensity, and duration of the stimulus which activated the endings. An additional feature of the nociceptive afferent endings and their cell bodies in the trigeminal ganglion is that following tissue injury or inflammation, they may also develop a prolonged increase in excitability that is termed *peripheral (or nociceptor) sensitization*. The sensitized nociceptive afferents may develop spontaneous, unprovoked activity and also become more responsive to noxious stimulation or even respond to stimuli that under normal conditions are innocuous and do not activate them (Cairns et al., 2014; Iwata et al., 2011). Each of these functional changes in the properties of the nociceptive afferents contributes respectively to the spontaneous pain, hyperalgesia (increased sensitivity and/or excessive response to a stimulus that is normally painful) and allodynia (pain resulting from a stimulus that does not normally evoke pain) that occur in many pain conditions. Examples in clinical dentistry are the increased sensitivity of an inflamed tooth, the sensitivity to TMJ or muscle palpation that is typical of temporomandibular disorders (TMD), and the severe pain that is characteristic of several neuropathic pain conditions that can arise from injury to trigeminal nerve branches.

Several different types of chemical mediators are involved in the activation or peripheral sensitization of the nociceptive endings, and these include some that also have actions in the CNS (e.g., glutamate, serotonin [5-HT], and noradrenaline; Cairns, 2009; Cairns et al., 2014; Dubner et al., 2013, 2014; Iwata et al., 2011; Ringkamp et al., 2013; Sessle, 2009). The endings of the nociceptive afferents manifest ion channels and membrane receptors, and these include opioid (e.g., enkephalinergic) receptors as well as serotonergic, purinergic, cholinergic, bradykinin, prostaglandin, histamine, cannabinoid, excitatory amino acid and acid-sensitive receptors, adrenoreceptors, and vanilloid receptors (Fig. 1). Some ion channels and membrane receptors are activated or their afferent endings sensitized relatively directly by several types of noxious mechanical, chemical and thermal stimuli; for example, some vanilloid receptors (transient receptor potential [TRPV1]) respond to heat, protons (H^+) and some algesic chemicals (e.g., capsaicin) and thus are important peripheral elements in our sensitivity to heat, acid, and spicy foods. Other receptors or ion channels instead are acted upon by intermediary chemical mediators released in the peripheral tissues or from afferent nerve endings (see below) as a result of the noxious stimulus (Fig. 1). Some of these mediators are soluble factors such as neurotrophins, or act on the nociceptive afferent endings by integrin binding to extracellular matrix molecules and regulate the expression or trafficking of neuronal proteins including ion channels and receptors, or second messenger signaling pathways (see Cairns, 2009; Hargreaves, 2011; Hargreaves & Ruparel, 2016).

It is noteworthy that noxious stimuli producing tissue damage may also cause the release of chemicals that are synthesized in the trigeminal ganglion cell bodies of the nociceptive afferents themselves and that are released from their afferent endings in the peripheral tissues. These neurochemicals include calcitonin gene-related peptide (CGRP), substance P, somatostatin, glutamate, and nerve growth factors. Some of these neurochemicals can act on platelets, mast cells, macrophages, and other cells of the immune system to cause them to release inflammatory mediators such as 5-HT, histamine, bradykinins, and cytokines (Fig. 1). The redness, edema, and local temperature increase that result from the actions of these inflammatory mediators reflect a so-called neurogenic inflammation since the inflammation may be initiated by the chemical mediators that are released from the nerve fibers themselves. The chemical mediators may also spread through the tissues and act on the ion channels and membrane receptors of adjacent nociceptive afferent endings and thus contribute to the production of peripheral sensitization in these endings.

Space limitations and the vast number of these neurochemicals and other chemical mediators and associated processes do not allow a detailed review of each of

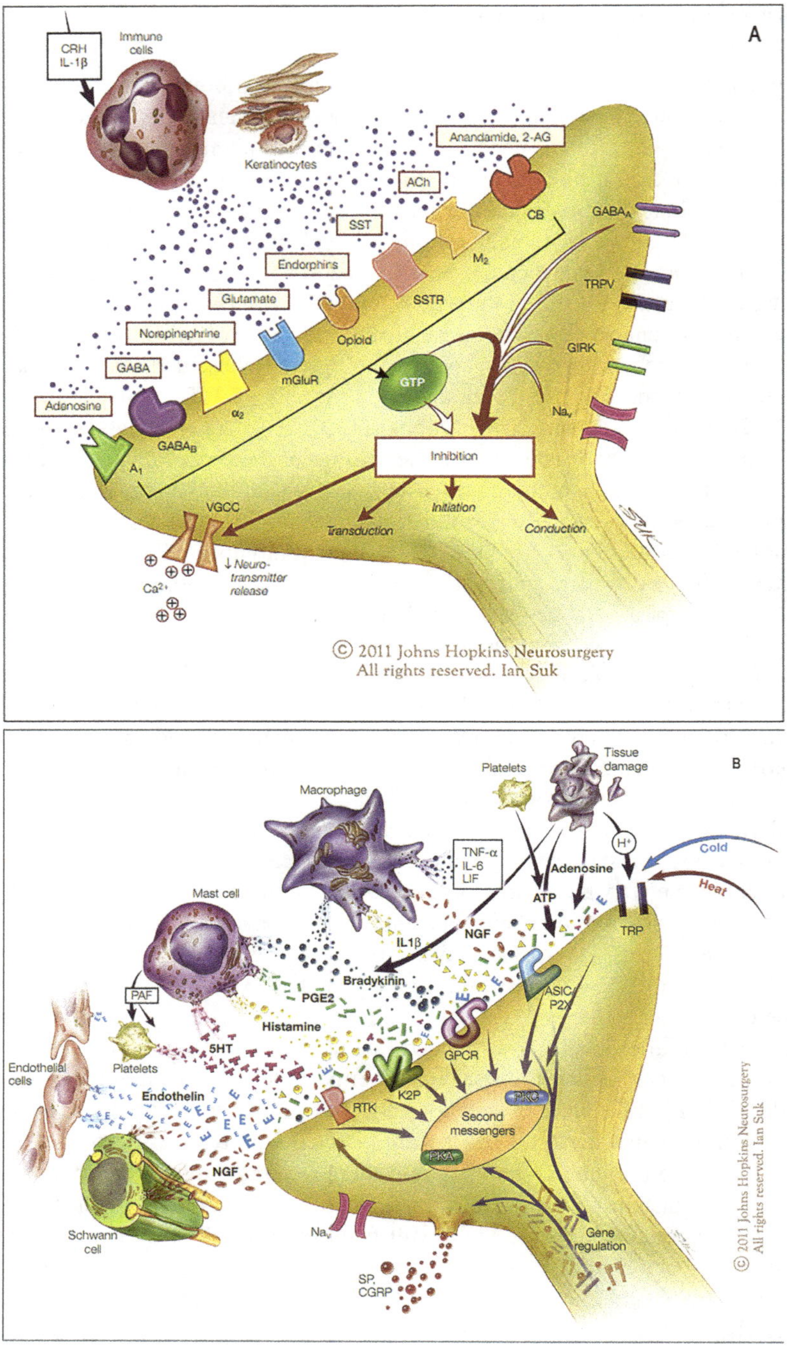

Fig. 1 Mediators involved in peripheral sensitization following inflammation of craniofacial tissues. As part of the inflammatory process as shown in A, numerous chemicals are released from macrophages, mast cells, immune cells, and injured cells and act on ion channels or membrane receptors on peripheral nociceptive afferent nerve endings and thereby may alter the sensitivity of the endings; only a few of the mediators are shown in A and B. Some of these mediators produce

them, so the following gives examples of three (TRP, glutamate, opioids); more detailed and comprehensive reviews are available (Cairns, 2009; Cairns et al., 2014; Hargreaves, 2011; Hargreaves & Ruparel, 2016). In the case of the TRP vanilloid receptor system, the application of TRP agonists such as capsaicin or heat to craniofacial tissues can have excitatory actions on the nociceptive afferents supplying the tissues and on trigeminal ganglion neurons in vivo or in vitro (Lam et al., 2009; Cairns, 2009; Cairns et al., 2014; Dostrovsky et al., 2014; Hargreaves, 2011; Hargreaves & Ruparel, 2016). Noxious heat for example causes the release of oxidized linoleic acid metabolites (OLAMs), some of which selectively activate TRPV1 receptors, leading to inward currents, increased intracellular calcium levels, and exocytosis of neuropeptides from trigeminal afferent neurons, and resulting in hypersensitivity to heat; these effects can be antagonized by TRPV1 or OLAM antagonists (see Hargreaves, 2011; Hargreaves & Ruparel, 2016). Findings such as these have led to the view that the TRPV1 and OLAM systems might especially come into play in conditions of tissue injury and inflammation and contribute to acute heat sensitivity as well as to more persistent inflammatory states such as those characteristic of dental inflammatory pain conditions.

An example of a neurochemical that has excitatory actions in craniofacial tissues is glutamate. This excitatory amino acid is well known for its role as the major excitatory neurotransmitter in the CNS. But glutamate is also synthesized in the primary afferent cell bodies in the trigeminal ganglion and is released from not only the central endings of the primary afferents in the CNS (i.e., brainstem; see below) but also from the endings of the afferents in the craniofacial tissues. Glutamate once released may then excite or sensitize the nociceptive afferent endings by acting on glutamatergic receptors (N-methyl-D-aspartate [NMDA] and non-NMDA receptors) occurring on the endings (Fig. 1). This excitatory effect has been well documented in TMJ and jaw muscle nociceptive afferents, and correlated human studies have shown that glutamate applied to these tissues has pain-producing effects in humans (see Cairns, 2009; Cairns et al., 2014; Sessle, 2021). Many of the nociceptive endings may also have ion channels and membrane receptors which once activated can instead produce a decrease in excitability of the nociceptive afferents. These include opioid, cannabinoid, and δ-amino butyric acid [GABA]) receptors (Fig. 1). For example, in animal models of craniofacial pain, the peripheral

Fig. 1 (continued) an increase in the excitability of the nociceptive afferent endings (i.e., peripheral sensitization), and other mediators such as opioids may exert inhibitory effects. ASIC acid-sensing ion channel, CRH corticotrophin-releasing hormone, GIRK G-protein-coupled inward rectifying potassium channel, 5-HT serotonin, iGluR ionotropic glutamate receptor, IL-1b interleukin-1-beta, IL-6 interleukin-6, LIF leukemia inhibitory factor, μ-opioid receptor, M2 muscarinic receptor, mGluR metabotropic glutamate receptor, NGF nerve growth factor, PAF platelet-activating factor, PGE2 prostaglandin E2, PKA protein kinase A, PKC protein kinase C, SSTR2A somatostatin receptor 2A, TNF-α tumor necrosis factor alpha, TrkA tyrosine kinase receptor A, TRPV1 transient receptor potential vanilloid 1, TTXr tetrodotoxin-resistant sodium channel. (Reproduced with permission from Ringkamp et al., 2013)

application of morphine or other opioid agonists to craniofacial tissues can induce inhibitory effects on nociceptive afferent excitability and decreased pain behavior. This is particularly evident when afferent excitability and pain behavior are induced by inflammatory irritants such as mustard oil, formalin or Complete Freund's Adjuvant, although it is also evident when these excitatory events are induced by glutamate which appears not to induce overt inflammation (Cairns, 2009; Cairns et al., 2014; Castrillon et al., 2008). The findings of opioid actions in peripheral tissues are supported by evidence of opioid receptors expressed on trigeminal sensory neurons, although under basal experimental conditions they do not appear to be coupled to inhibitory signaling pathways; however, they may rapidly gain functional competence following application to tissues of inflammatory mediators or in the presence of tissue inflammation (see Cairns, 2009; Hargreaves, 2011; Hargreaves & Ruparel, 2016).

It is of interest that there are sex differences in the peripheral actions of glutamate and the opioid agonist morphine. For example, there is a sex difference in the pain-inducing effects of glutamate when applied to the masseter muscle of humans that is consistent with findings in animal models (Cairns, 2009; Cairns et al., 2014; Castrillon et al., 2008). Nociceptive behavioral responses and TMJ or jaw muscle nociceptive afferents in these models show a greater sensitivity in female rats than in male rats to the application to the tissues of glutamate, and the excitatory effects of glutamate are modulated by estrogen levels. These findings are consistent with evidence of estrogen receptors being expressed on trigeminal afferent neurons and that estradiol selectively alters gene transcription in the neurons, with increased expression of neuropeptides (e.g., prolactin) that can sensitize neuronal responses to the TRP agonist capsaicin or to noxious heat. In the case of the sex difference in effects of the peripheral application of opioid agonists such as morphine, it has been shown that the craniofacial nociceptive afferents and nociceptive behavioral responses in females are less sensitive than in males to the peripheral application of morphine or other opioid agonists. Again, sex hormones may modulate these effects. For example, inflammation of craniofacial tissues such as the jaw muscles and tooth pulp upregulates μ, δ or κ opioid receptors in the rat trigeminal ganglion, and testosterone can modulate this opioid receptor expression via transcriptional activities (e.g., Huang et al., 2015; Hargreaves, 2011; Hargreaves & Ruparel, 2016; Lee et al., 2016; Zhang et al., 2014). Such physiologically based sex differences in the sensitivity of TMJ and jaw muscle nociceptive afferents to glutamate and opiate-related substances may contribute to the sex differences that have been documented clinically in many craniofacial pain conditions involving these tissues, such as TMD (see LeResche & Drangsholt, 2008; Macfarlane, 2014). Sex differences documented in environmental and psychosocial influences and sex differences in CNS nociceptive mechanisms may also be important contributory factors (see below).

The excitability changes in the nociceptive afferents that result from craniofacial tissue injury or inflammation may produce an altered afferent input into the brainstem that influences the function of neurons in the trigeminal nociceptive pathways in the CNS and induce altered pain behavior, such as is seen with the enhanced sensitivity of inflamed tissue. Another example is the abnormal sensory experiences that may accompany an injury to afferent nerve fibers that also produces an abnormal afferent barrage into the CNS. It is also noteworthy that these effects of injury or inflammation, and the actions of the excitatory or inhibitory mediators, are not limited to the peripheral endings of the nociceptive afferents. Craniofacial tissue inflammation or injury can also cause changes in the properties of the trigeminal ganglion cell bodies of the nociceptive afferents; many of these changes involve the action of chemical mediators released in the ganglion from the cell bodies themselves or from the satellite glial cells that encapsulate the cell bodies (see Dubner et al., 2014; Iwata et al., 2011). Thus, the abnormal sensory input into the brainstem may also result from excitability changes in the afferent cell bodies produced by glial-neuronal interactions. Furthermore, some of these effects in the trigeminal ganglion are not limited to the cell bodies of the afferents directly involved in the injury or inflammation of a particular part of the craniofacial region, but can also spread to ganglion neurons innervating another part of the craniofacial region. Thus, peripheral changes in the ganglion and the resulting ectopic afferent input into the brainstem may contribute to the extra-territorial spread of sensitivity that characterizes some craniofacial pain states.

Trigeminal Brainstem Mechanisms

The trigeminal primary afferents project via the trigeminal ganglion to the brainstem where the vast majority of them terminate in the trigeminal brainstem sensory nuclear complex (VBSNC; see Fig. 2). This structure is conventionally subdivided into the main (or principal) sensory nucleus and the trigeminal spinal tract nucleus, which comprises three subnuclei (oralis, interpolaris, and caudalis; Darian-Smith, 1966; Dubner et al., 1978, 2014; Sessle, 2000). The larger-diameter primary afferents that conduct craniofacial tactile or proprioceptive information terminate throughout the VBSNC and activate low-threshold mechanosensitive (LTM) neurons in subnucleus caudalis and the more rostral components of the VBSNC. In contrast, those small-diameter trigeminal primary afferents activated by innocuous thermal or noxious craniofacial stimuli predominantly terminate in subnucleus caudalis which merges caudally with the upper cervical dorsal horn of the spinal somatosensory system. Although there are some notable differences, subnucleus caudalis is often referred to as the medullary dorsal horn since it has several morphological and physiological features that are analogous to those of the spinal dorsal horn which processes nociceptive and thermosensory signals initiated by innocuous thermal or noxious stimuli applied to tissues supplied by spinal afferents (see Bereiter et al., 2000; Dubner et al., 2014; Sessle, 2000, 2021; Woda, 2003). For example, subnucleus caudalis, like the spinal dorsal horn, is a laminated structure that includes a substantia gelatinosa. And like their spinal afferent counterparts, the central terminals of the trigeminal nociceptive afferents release excitatory neurochemicals such as glutamate and substance P that can excite the second-order nociceptive neurons on which the afferents synapse (Dubner et al., 1978; Iwata et al.,

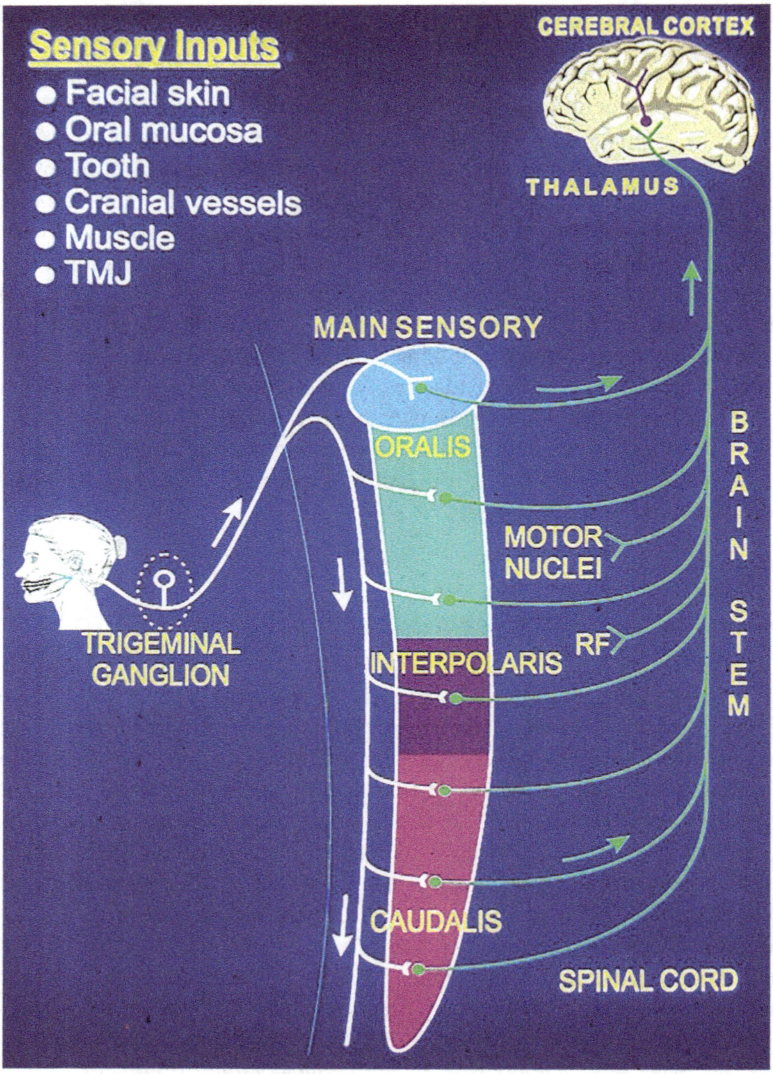

Fig. 2 Diagram of major somatosensory pathway from the orofacial region. The cell bodies of most trigeminal primary afferents are in the trigeminal ganglion and project to second-order neurons in the trigeminal brainstem sensory nuclear complex (VBSNC) which comprises the trigeminal main sensory nucleus and also the trigeminal spinal tract nucleus that has three subnuclei: oralis, interpolaris, and caudalis. These neurons may project to neurons at higher levels of the brain (e.g., in the somatosensory thalamus) or to brainstem regions such as the reticular formation or the cranial nerve motor nuclei. Some afferents (not shown) in cranial nerves VII, IX, X, and XII and in cervical nerves project to the VBSN and many V, VII, IX, and X afferents project to the solitary tract nucleus. (Reproduced with permission from Sessle, 2000)

2011; Sessle, 2000, 2021). These neurochemical processes are considered further below (section "Modulation of Trigeminal Nociceptive Processes in the CNS").

It is noteworthy that subnucleus caudalis not only receives trigeminal afferent inputs but also afferent inputs from some other cranial nerves and even cervical nerves; furthermore, the upper cervical spinal dorsal horn adjoining the caudal end of subnucleus caudalis also receives trigeminal as well as cervical nociceptive afferent inputs. These convergent input patterns are discussed further below in terms of mechanisms underlying the referral of pain. Also of note is the transitional zone between the rostral end of subnucleus caudalis and the caudal end of subnucleus interpolaris. It has special features in processing afferent inputs from deep as well as superficial tissues in the craniofacial region and contributes to autonomic and endocrine functions related to craniofacial pain, and there may be sex differences in the properties of the caudalis and transitional zone nociceptive neurons (for review, see Dubner et al., 2013, 2014; Ren & Dubner, 2011; Tashiro et al., 2008). Furthermore, some nociceptive neurons have also been described in more rostral components of the VBSNC (oralis, main sensory nucleus), and it has been suggested that these neurons, which especially are activated by oral noxious stimuli, may play a particular role in intraoral pain mechanisms (Sessle, 2000; Woda, 2003).

This brings us to consider the properties of the nociceptive neurons in subnucleus caudalis and its associated structures. Like the nociceptive neurons existing in the spinal dorsal horn, these nociceptive neurons predominate in the superficial and deep laminae and are of two main types: nociceptive specific (NS) neurons which normally respond only to noxious stimuli, and wide dynamic range (WDR) neurons which respond to nonnoxious stimuli (e.g., tactile) as well as to noxious stimuli. Also, like their spinal dorsal horn counterparts, some of these NS and WDR neurons receive nociceptive afferent inputs only from "cutaneous" tissues (facial skin, oral mucosa) and have properties indicative of a role in encoding the detection and discrimination of superficial pain. In contrast, nociceptive information from "deep" tissues (e.g., TMJ, muscle, tooth pulp, meninges) supplied by trigeminal nerve branches and from other tissues supplied by the upper cervical nerves and other cranial nerves is processed predominantly by subsets of these "cutaneous" nociceptive neurons that receive extensive convergent afferent inputs from these deep and other tissues as well as from facial skin and/or oral mucosa. These patterns of convergent afferent inputs appear to underlie the CNS mechanisms contributing to deep craniofacial and cervical pain, and in addition may explain the poor localization, spread, and referral of pain that are typical of craniofacial pain states involving deep tissues (e.g., TMD) and some cervical pain states (Dubner et al., 2014; Sessle, 2000, 2021).

The projection sites of VBSNC nociceptive and nonnociceptive neurons are numerous. Some neurons in the VBSNC give rise to axons that ramify within the VBSNC itself and modulate the activity of other neurons in the VBSNC. Many neurons project to other areas outside the VBSNC; these include the reticular formation, raphe nuclei, parabrachial nucleus and cranial nerve motor nuclei in the brainstem, and ventral horns in the spinal cord. These projections provide the central circuitry underlying autonomic and muscle reflex responses elicited by stimulation of craniofacial tissues (Dubner et al., 2014; Iwata et al., 2011; Sessle, 2000). Some of these brainstem areas also may contribute to so-called descending modulatory systems that can modify nociceptive transmission (see below) and other functions (e.g., sleep and awake states) that may be influenced by nociceptive afferent inputs into the CNS (see Lavigne & Sessle, 2016). Many neurons in the VBSNC also (or instead) contribute to pathways ascending through the brainstem to the ipsilateral and especially contralateral thalamus (Fig. 2).

Thalamocortical Mechanisms

Craniofacial somatosensory information that is relayed from the brainstem to the thalamus accesses mainly thalamic sites that are known in animals as the ventrobasal complex (this is analogous to the ventroposterior nucleus in humans), the posterior nuclear group and the medial thalamic nuclei including nucleus submedius (see Dostrovsky & Craig, 2013; Dubner et al., 2013, 2014; Sessle, 2000; Tang et al., 2009). These thalamic areas contain nonnociceptive (LTM and thermoreceptive) neurons which project to analogous neurons in the overlying somatosensory cerebral cortex. Here, their relayed signals are processed so as to provide for the detection and localization of tactile and warm or cool stimuli. These thalamic areas also contain NS and WDR neurons, most of which project to the somatosensory cortex which also has NS and WDR neurons. These thalamic and somatosensory cortical nociceptive neurons have spatiotemporal coding properties that indicate they have a role in encoding the spatiotemporal features of noxious stimuli, and thus in the sensory-discriminative dimension of pain. By contrast, nociceptive neurons in the medial thalamic nuclei and posterior nuclear group project mainly to other cortical areas such as the anterior cingulate and insula cortices. The spatiotemporal coding properties of these thalamic nociceptive neurons and analogous neurons in these particular cortical areas suggest a role for them more in the motivational or affective dimensions of pain, consistent with findings from brain imaging studies in humans (see Davis & Stohler, 2014).

Modulation of Trigeminal Nociceptive Processes in the CNS

The transmission of craniofacial somatosensory information can be modified at brainstem and thalamocortical levels, and be operational to varying degrees during different behavioral states, e.g., states of alertness, attention, emotion, and the hypotonia of rapid eye movement (REM) sleep (see Dubner et al., 1978, 2014; Lavigne & Sessle, 2016; Sessle, 2000, 2021). These modulatory processes involve neural circuits within the VBSNC and adjacent brainstem regions and the thalamic and cortical areas noted above that process craniofacial somatosensory information.

These circuits utilize inputs from primary afferents and from descending projections originating in several CNS areas including the reticular formation, raphe nuclei, locus coeruleus, hypothalamus, and cerebral cortex (for review, see Chichorro et al., 2017; Dubner et al., 2013; Ossipov et al., 2014; Sessle, 2000; Tang et al., 2009).

In the case of craniofacial nociceptive transmission, the variety of inputs and interconnections that are especially evident in trigeminal subnucleus caudalis and its associated regions (see above) provide the basis for considerable interaction between the various afferent inputs derived from peripheral tissues and the intrinsic brain areas noted above. Opioids (e.g., enkephalins and endophins) play a key role in modulating trigeminal nociceptive transmission, but several other chemical mediators, including GABA, glycine, 5-HT, noradrenaline, dopamine, and hypocretin (also known as orexin) also contribute to the rich endogenous chemical substrate by which many of the primary afferent and descending inputs can exert their modulatory effects on trigeminal nociceptive transmission. Some of these descending influences exert predominantly inhibitory effects on nociceptive transmission, but some exert facilitatory effects (see Chichorro et al., 2017; Dubner et al., 2013; Ossipov et al., 2014).

Other chapters in this book detail the processes underlying the actions of the opioids (see chapter by Felicione et al., chapter "Pain, Fear, Anxiety, and Stress: Relations to the Endogenous Opioid System", this volume), and the following focuses on their role in the modulation of trigeminal nociceptive processes in the CNS and associated craniofacial nociceptive behavior in preclinical craniofacial pain models. The narcotic analgesic drug morphine has been shown to exert powerful modulatory effects on trigeminal nociceptive transmission and nociceptive behavior by mimicking the action of endogenous opioid chemicals such as enkephalins. These are peptides that are pharmacologically similar to opiate drugs and that act on μ- and non-μ opioid receptors present in numerous CNS sites. These include neurons or afferent nerve terminals in the VBSNC such as subnucleus caudalis as well as neural elements in some of the other CNS sites noted above that process craniofacial nociceptive information or that exert descending modulatory influences upon nociceptive transmission (e.g., Bornhof et al., 2011; Coffield & Miletic, 1993; DosSantos et al., 2012; Hirata et al., 2000; Macedo et al., 2016; Mitchell et al., 2004; Tashiro et al., 2008; Wang et al., 1999; for review, see Chichorro et al., 2017; Dubner et al., 2013; Sessle, 2000; Tang et al., 2009). The CNS sites exerting modulatory influences include cerebral cortical areas such as sensorimotor, anterior cingulate, insula, and ventrolateral orbital areas, as well as the amygdala, hypothalamus, basal ganglia, raphe nuclei, and locus coeruleus.

Some of these effects involve interactions between opioids and other chemical mediators such as GABA and sex hormones. Indeed, as noted above, sex differences may occur in the nociceptive transmission process, but it also has been found that opioid modulatory influences on this process may themselves be influenced by sex hormone levels. For example, morphine effects on craniofacial nociceptive behavior and the response properties of nociceptive neurons in the superficial laminae of subnucleus caudalis to TMJ stimulation in female rats depend on estrogen status, and testosterone can modulate nociceptive responses to noxious TMJ stimulation in male rats via the activation of central μ- and κ-opioid receptors (Tashiro et al., 2008; Macedo et al., 2016). It is also noteworthy that opioid processes operating in the raphe nuclei and the associated raphe-induced modulation of nociceptive neurons in trigeminal subnucleus caudalis are involved in the clinical condition of morphine withdrawal-induced headache (Hitomi et al., 2017), and this brings us to consider the clinical significance of these opioid and related modulatory processes.

There are indeed several clinically relevant points about these modulatory influences. The sex differences in opioid processes modulating TMJ nociceptive processes in subnucleus caudalis, together with the sex differences noted earlier in peripheral nociceptive mechanisms and sex differences in processes in higher centers of the CNS, are physiologically based mechanisms that undoubtedly contribute to the sex differences that occur in several craniofacial pain states (e.g., TMD, burning mouth syndrome). In addition, the influences on the nociceptive neurons of state of alertness, sleep, distraction and attention, emotion, anxiety, and cognitive behavioral therapy are examples whereby descending influences emanating from CNS regions involved in these behavioral functions and operating at the VBSNC and at higher brain levels may affect craniofacial pain (see Chichorro et al., 2017; Dubner et al., 2013; Lavigne & Sessle, 2016; Ossipov et al., 2014; Sessle, 2000; Tang et al., 2009). Placebo analgesia, which contributes to the effect of most pain-relieving procedures, also involves some of these CNS regions utilizing chemical mediators such as opioids (Carlino et al., 2014; Colloca et al., 2013; Dubner et al., 2013; see also Kerr & Gregg, chapter "The Roles of Endogenous Opioids in Placebo and Nocebo Effects: From Pain to Performance to Prozac", this volume). Descending influences also have been implicated as intrinsic mechanisms contributing to the analgesic effects of several other procedures used to control pain. The analgesic effects of deep brain stimulation and some physical procedures (e.g., acupuncture; transcutaneous electrical nerve stimulation) appear to involve some of these endogenous neurochemical processes and intrinsic modulatory circuits that inhibit nociceptive transmission (Dubner et al., 2013; Sessle, 2000). Likewise, the actions of several pain-relieving drugs appear to act, at least in part, through these processes and circuits. Morphine, for example, suppresses the activity of the nociceptive neurons by its action of endogenous opiate receptors, which, as noted earlier, exist in the VBSNC and higher levels of the trigeminal system and in some of the intrinsic modulatory pathways.

Several of the chemical mediators contributing to the pain-modulatory processes also may be involved in other related CNS-based functions and dysfunctions such as conditions that may be comorbid with craniofacial chronic pain states (e.g., TMD and burning mouth syndrome). These comorbid conditions include depression, stress, anxiety and fear, and may involve opioid and serotonergic processes (see Pettrey et al., chapter "Physical Exercise as an Intervention for Depression: Evidence for Efficacy and Mu-Opioid Receptors as a Mechanism of Action", this volume; Felicione et al., chapter "Pain, Fear, Anxiety, and Stress: Relations to the Endogenous Opioid System", this volume). This feature explains, for example, the analgesic effectiveness of some antidepressant drugs that act through 5-HT receptor processes. Gabapentin and pregabalin, shown to be effective in many neuropathic pain states, act on voltage-gated calcium channels (see Cao et al., 2013; Dharmshaktu et al., 2012; Dooley et al., 2007) and carbamazepine, which is the drug of choice to manage the excruciatingly painful attacks of trigeminal neuralgia, depress VBSNC neuronal excitability (although its exact mechanism of action is still unclear; see Fromm & Sessle, 1991; Sessle & Watson, 2017).

Most of the examples noted above relate to suppression of nociceptive behavior and nociceptive transmission in the CNS. However, some processes acting on nociceptive neurons instead have a role in facilitating nociceptive transmission. Central sensitization is a good example of how craniofacial nociceptive transmission can be modulated at brainstem and thalamocortical levels, but in this case resulting in facilitation of transmission. Trigeminal central sensitization can be produced by nociceptive afferent inputs evoked by tissue injury or inflammation and is manifested in nociceptive neurons in subnucleus caudalis and its associated areas (e.g., C1–2 dorsal horn; the transitional zone) as well as in subnucleus oralis and the higher levels of the CNS (e.g., somatosensory thalamus, somatosensory cortex) (see Burstein et al., 2010; Dubner et al., 2013, 2014; Sessle, 2000, 2021). Trigeminal central sensitization in subnucleus caudalis has been the most studied in the trigeminal system, and involves some of the convergent nociceptive afferent inputs to the NS and WDR neurons mentioned earlier. These nociceptive afferent inputs release excitatory neurotransmitters such as the excitatory amino acid glutamate, and the neuropeptides substance P and CGRP, which respectively act through glutamatergic and neuropeptide receptors on the nociceptive neurons to induce neuroplastic changes in the nociceptive neurons that are reflected in an increase in excitability (central sensitization) of the neurons (see Dubner et al., 2014; Iwata et al., 2011; Sessle, 2000, 2021). The increased excitability may also involve chemical mediators released from other sources including interneurons and even nonneural (e.g., glial) cells (see below). The central sensitization of the nociceptive neurons has been shown to involve also changes in some of the descending modulatory influences (e.g., from the raphe nuclei) that operate through opioid or other chemical mediators, producing a loss in their inhibitory effect and/or an increase in their facilitatory effect. The resulting hyperexcitability of the nociceptive neurons that is the hallmark of central sensitization has features that include increases in neuronal spontaneous activity, receptive field size, and responses to stimulation of craniofacial tissues, plus a decrease in activation threshold, including phenotypic switching of NS neurons that become activated by tactile inputs and thus take on properties typical of WDR neurons. Central sensitization is reflected behaviorally in spontaneous pain behavior and hypersensitivity to mechanical, chemical, or thermal stimulation of the injured or inflamed tissue or of even body areas remote from this tissue.

These neuronal and behavioral features that represent a centrally sensitized state have clinical parallels in humans. Trigeminal central sensitization is evident in acute as well as chronic pain states, but of particular clinical relevance are findings indicating that the maintenance of a central sensitization state in the CNS nociceptive pathways is a major factor in the development and persistence of chronic pain conditions. In addition, along with peripheral sensitization (see above), its features can explain the spontaneous (i.e., unprovoked) pain, hyperalgesia, allodynia and pain spread and referral that are characteristic of many craniofacial pain states such as persistent dental pulpitis pain, TMD, headache, and neuropathic pain conditions that may develop following nerve injury (see Burstein et al., 2010; Cheng et al., 2015; Drummond & Finch, 2006; Dubner et al., 2014; Iwata et al., 2011; Konopka et al., 2012; Sessle, 2000, 2021; Sharav & Benoliel, 2015).

The occurrence of neuroplasticity reflected in central sensitization has particular clinical significance in underscoring the point that the nociceptive circuits in the CNS, including those in the brainstem, thalamus and cerebral cortex, are not "hard-wired." Rather, they are "plastic" and modifiable by afferent inputs evoked by events in peripheral tissues related to injury or inflammation as well as by descending modulatory inputs to the nociceptive circuits from CNS areas that, as noted above, utilize opioids and several other chemical mediators. Moreover, as mentioned earlier, these CNS areas are involved in a variety of functions such as memory, anxiety, mood, motivation, reward and attention/distraction, all of which can modify pain behavior. These features have clinical significance, firstly in emphasizing the importance of timely approaches to reduce nociceptive inputs that are primarily responsible in the first place for the induction of the central sensitization. A good example of such an approach is the standard of clinical practice in dentistry and other medical fields in utilizing local anesthesia and perioperative analgesics and anti-inflammatory drugs when a surgical or other procedure producing tissue injury and inflammation is part of a clinical management scenario. Secondly, the clinical significance of the descending influences is their contribution to pain modulation by a patient's mood, emotional and motivational state. Furthermore, as briefly noted above, several management approaches may have an action on the descending inputs and thereby their modulatory influences on nociceptive neurons and the appearance of central sensitization; these include among others the use of narcotic analgesics and antidepressants, cognitive behavioral therapy, placebo, and distraction.

Also of clinical relevance is the documentation that central sensitization, including that shown to occur in nociceptive neurons of trigeminal subnucleus caudalis and its associated areas (e.g., upper cervical dorsal horn; the transitional zone), involves nonneuronal (i.e., glial) as well as neural processes (see Chiang et al., 2011; Dubner et al., 2014; Iwata et al., 2011; Sessle, 2000, 2021). It has been shown that in association with the development of central sensitization, both astroglia and microglia in these regions become "activated" following craniofacial tissue injury or inflammation and that the administration of glial inhibitors can attenuate the trigeminal central sensitization and the nociceptive behavior associated with it. Furthermore, the pain states that may be associated with morphine withdrawal (see above) may involve glial cells and changes in opioid receptor function, resulting in a central sensitization. The role of glia is particularly noteworthy in terms of its clinical significance since glial mechanisms and associated processes may provide novel targets for developing new or improved analgesic approaches to control pain.

Summary and Conclusions

This chapter has focused on findings from preclinical pain models that bear on peripheral and central mechanisms underlying acute and chronic craniofacial pain states. Opioid-related processes operating within craniofacial tissues as well as in components of the trigeminal system in the CNS are one of several processes utilizing chemical mediators that exert modulatory effects on craniofacial nociceptive mechanisms and thereby accentuate or attenuate acute and chronic craniofacial pain. These processes are clinically relevant to the development and chronification of craniofacial pain and to the analgesic efficacy of a wide range of therapeutic approaches to manage craniofacial pain states. Nonneural as well as neural elements are involved and utilize several chemical mediators including the opioids. Further research advances in our understanding of these processes hold out the promise of new or improved therapeutic procedures to diagnose and treat craniofacial pain states.

References

Bereiter, D. A., Hirata, H., & Hu, J. W. (2000). Trigeminal subnucleus caudalis: Beyond homologies with the spinal dorsal horn. *Pain, 88*(3), 221–224.

Bornhof, M., Ihmsen, H., Schwilden, H., Yeomans, D. C., & Tzabazis, A. (2011). The orofacial formalin test in mice revisited—effects of formalin concentration, age, morphine and analysis method. *The Journal of Pain, 12*(6), 633–639.

Burstein, R., Jakubowski, M., Garcia-Nicas, E., Kainz, V., Bajwa, Z., Hargreaves, R., et al. (2010). Thalamic sensitization transforms localized pain into widespread allodynia. *Annals of Neurology, 68*(1), 81–91.

Cairns, B. E. (2009). *Peripheral receptor targets for analgesia: Novel approaches to pain management.* Wiley.

Cairns, B. E., Ren, K., & Tambeli, C. H. (2014). Musculoskeletal orofacial pain mechanisms: Insights from animal models. In B. J. Sessle (Ed.), *Orofacial pain* (pp. 351–372). International Association for the Study of Pain/IASP Press.

Cao, Y., Wang, H., Chiang, C. Y., Dostrovsky, J. O., & Sessle, B. J. (2013). Pregabalin suppresses nociceptive behavior and central sensitization in a rat trigeminal neuropathic pain model. *The Journal of Pain, 14*(2), 193–204.

Carlino, E., Frisaldi, E., & Benedetti, F. (2014). Pain and the context. *Nature Reviews Rheumatology, 10*(6), 348–355.

Castrillon, E. E., Cairns, B. E., Ernberg, M., Wang, K., Sessle, B., Arendt-Nielsen, L., & Svensson, P. (2008). Glutamate-evoked jaw muscle pain as a model of persistent myofascial TMD pain? *Archives of Oral Biology, 53*(7), 666–676.

Cheng, C. F., Cheng, J. K., Chen, C. Y., Rau, R. H., Chang, Y. C., & Tsaur, M. L. (2015). Nerve growth factor–induced synapse-like structures in contralateral sensory ganglia contribute to chronic mirror-image pain. *Pain, 156*(11), 2295–2309.

Chiang, C. Y., Dostrovsky, J. O., Iwata, K., & Sessle, B. J. (2011). Role of glia in orofacial pain. *The Neuroscientist, 17*(3), 303–320.

Chichorro, J. G., Porreca, F., & Sessle, B. (2017). Mechanisms of craniofacial pain. *Cephalalgia, 37*(7), 613–626.

Coffiel, J. A., & Miletic, V. (1993). Responses of rat nucleus submedius neurons to enkephalins applied with micropressure. *Brain Research, 630*(1–2), 252–261.

Colloca, L., Klinger, R., Flor, H., & Bingel, U. (2013). Placebo analgesia: Psychological and neurobiological mechanisms. *Pain, 154*(4), 511–514.

Darian-Smith, I. (1966). Neural mechanisms of facial sensation. *International Review of Neurobiology, 9*, 301–395.

Davis, K. D., & Stohler, C. S. (2014). Neuroimaging and orofacial pain. In B. J. Sessle (Ed.), *Orofacial pain: Recent advances in assessment, management and understanding of mechanisms* (pp. 165–184). IASP Press.

Dharmshaktu, P., Tayal, V., & Kalra, B. S. (2012). Efficacy of antidepressants as analgesics: A review. *The Journal of Clinical Pharmacology, 52*(1), 6–17.

Dooley, D. J., Taylor, C. P., Donevan, S., & Feltner, D. (2007). Ca2+ channel α2δ ligands: Novel modulators of neurotransmission. *Trends in Pharmacological Sciences, 28*(2), 75–82.

DosSantos, M. F., Martikainen, I. K., Nascimento, T. D., Love, T. M., Deboer, M. D., Maslowski, E. C., et al. (2012). Reduced basal ganglia μ-opioid receptor availability in trigeminal neuropathic pain: A pilot study. *Molecular Pain, 8*(1), 74–79.

Dostrovsky, J. O., & Craig, A. D. (2013). Ascending projection systems. In S. B. McMahon, M. Koltzenburg, I. Tracey, & D. C. Turk (Eds.), *Wall and Melzack's textbook of pain* (6th ed., pp. 182–197). Elsevier.

Dostrovsky, J. O., Sessle, B. J., & Lam, D. K. (2014). Inflammatory and cancer-related orofacial pain mechanisms: Insights from animal models. In B. J. Sessle (Ed.), *Orofacial pain: Recent advances in assessment, management and understanding of mechanisms* (pp. 305–328). IASP Press.

Drummond, P. D., & Finch, P. M. (2006). Sensory changes in the forehead of patients with complex regional pain syndrome. *Pain, 123*(1–2), 83–89.

Dubner, R., Sessle, B. J., & Storey, A. T. (1978). *The neural basis of oral and facial function*. Plenum.

Dubner, R., Ren, K., & Sessle, B. J. (2013). Sensory mechanisms of orofacial pain. In C. Greene & D. Laskin (Eds.), *Treatment of TMDs: Bridging the gap between advances in research and clinical patient management* (pp. 3–16). Quintessence.

Dubner, R., Iwata, K., & Wei, F. (2014). Neurophatic orofacial pain mechanisms: Insights from animal models. In B. J. Sessle (Ed.), *Orofacial pain: Recent advances in assessment, management and understanding of mechanisms* (pp. 331–349). IASP Press.

Fromm, G., & Sessle, B. J. (Eds.). (1991). *Trigeminal neuralgia: Current concepts regarding pathogenesis and treatment*. Butterworth-Heinemann.

Hargreaves, K. M. (2011). Congress orofacial pain. *Pain, 152*(3 Suppl), S25–S32.

Hargreaves, K. M., & Ruparel, S. (2016). Role of oxidized lipids and TRP channels in orofacial pain and inflammation. *Journal of Dental Research, 95*(10), 1117–1123.

Hirata, H., Takeshita, S., Hu, J. W., & Bereiter, D. A. (2000). Cornea-responsive medullary dorsal horn neurons: Modulation by local opioids and projections to thalamus and brain stem. *Journal of Neurophysiology, 84*(2), 1050–1061.

Hitomi, S., Kross, K., Kurose, M., Porreca, F., & Meng, I. D. (2017). Activation of dura-sensitive trigeminal neurons and increased c-Fos protein induced by morphine withdrawal in the rostral ventromedial medulla. *Cephalalgia, 37*(5), 407–417.

Huang, J., Lv, Y., Fu, Y., Ren, L., Wang, P., Liu, B., et al. (2015). Dynamic regulation of delta-opioid receptor in rat trigeminal ganglion neurons by lipopolysaccharide-induced acute pulpitis. *Journal of Endodontics, 41*(12), 2014–2020.

Institute of Medicine, Committee on Advancing Pain Research, Care, and Education. (2011). *Relieving pain in America: A blueprint for transforming prevention, care, education, and research. Report brief 1-4*. The National Academies Collection.

Iwata, K., Imamura, Y., Honda, K., & Shinoda, M. (2011). Physiological mechanisms of neuropathic pain: The orofacial region. *International Review of Neurobiology, 97*, 227–250.

Konopka, K. H., Harbers, M., Houghton, A., Kortekaas, R., van Vliet, A., Timmerman, W., et al. (2012). Bilateral sensory abnormalities in patients with unilateral neuropathic pain; a quantitative sensory testing (QST) study. *PLoS One, 7*(5), e37524.

Lam, D. K., Sessle, B. J., & Hu, J. W. (2009). Glutamate and capsaicin effects on trigeminal nociception I: Activation and peripheral sensitization of deep craniofacial nociceptive afferents. *Brain Research, 1251*, 130–139.

Lavigne, G. J., & Sessle, B. J. (2016). The neurobiology of orofacial pain and sleep and their interactions. *Journal of Dental Research, 95*(10), 1109–1116.

Lee, K. S., Zhang, Y., Asgar, J., Auh, Q. S., Chung, M. K., & Ro, J. Y. (2016). Androgen receptor transcriptionally regulates μ-opioid receptor expression in rat trigeminal ganglia. *Neuroscience, 331*, 52–61.

LeResche, L., & Drangsholt, M. (2008). Epidemiology of orofacial pain: Prevalence, incidence, and risk factors. In B. J. Sessle, G. L. Lavigne, J. P. Lund, & R. Dubner (Eds.), *Orofacial pain: From basic science to clinical management* (pp. 13–18). Quintessence.

Macedo, C. G., Fanton, L. E., Fischer, L., & Tambeli, C. H. (2016). Coactivation of μ- and κ-opioid receptors may mediate the protective effect of testosterone on the development of temporomandibular joint nociception in male rats. *Journal of Oral & Facial Pain & Headache, 30*(1), 61–67.

Macfarlane, T. V. (2014). Epidemiology of orofacial pain. In B. J. Sessle (Ed.), *Orofacial pain: Recent advances in assessment, management and understanding of mechanisms* (pp. 33–51). IASP Press.

Mitchell, J. L., Silverman, M. B., & Aicher, S. A. (2004). Rat trigeminal lamina I neurons that project to thalamic or parabrachial nuclei contain the μ-opioid receptor. *Neuroscience, 128*(3), 571–582.

Ossipov, M. H., Morimura, K., & Porreca, F. (2014). Descending pain modulation and chronification of pain. *Current Opinion in Supportive and Palliative Care, 8*(2), 143–151.

Ren, K., & Dubner, R. (2011). The role of trigeminal interpolaris-caudalis transition zone in persistent orofacial pain. *International Review of Neurobiology, 97*, 207–225.

Ringkamp, M., Raja, S. N., Campbell, J. N., & Meyer, R. A. (2013). Peripheral mechanisms of cutaneous nociception. In S. B. McMahon, M. Koltzenburg, M., I. Tracey, & D. C. Turk (Eds.), Wall & Melzack's Textbook of Pain (6th ed., pp. 1–30). Saunders.

Sessle, B. J. (2000). Acute and chronic craniofacial pain: Brainstem mechanisms of nociceptive transmission and neuroplasticity, and their clinical correlates. *Critical Reviews in Oral Biology & Medicine, 11*(1), 57–91.

Sessle, B. J. (2009). Role of peripheral mechanisms in craniofacial pain conditions. In B. E. Cairns (Ed.), *Peripheral receptor targets for analgesia* (pp. 3–20). Wiley.

Sessle, B. J. (2014). The societal, political, educational, scientific, and clinical context of orofacial pain. In B. J. Sessle (Ed.), *Orofacial pain: Recent advances in assessment, management and understanding of mechanisms* (pp. 1–15). IASP Press.

Sessle B. J. (2021). Chronic orofacial pain: Models, mechanisms, and genetic and related environmental influences. *International Journal of Molecular Sciences, 22*(13), 7112.

Sessle, B. J., & Watson, P. (2017). A comparison of clinical features, and mechanisms of trigeminal postherpetic neuralgia and trigeminal neuralgia. In C. P. N. Watson, A. A. Gershon, & M. N. Oxman (Eds.), *The varicella zoster virus: Varicella, herpes zoster, postherpetic neuralgia and persistent disease activity: A focus on treatment and prevention* (pp. 197–226). Springer.

Sharav, Y., & Benoliel, R. (2015). *Orofacial pain and headache* (2nd ed.). Quintessence.

Tang, J. S., Qu, C. L., & Huo, F. Q. (2009). The thalamic nucleus submedius and ventrolateral orbital cortex are involved in nociceptive modulation: A novel pain modulation pathway. *Progress in Neurobiology, 89*(4), 383–389.

Tashiro, A., Okamoto, K., & Bereiter, D. A. (2008). Morphine modulation of temporomandibular joint-responsive units in superficial laminae at the spinomedullary junction in female rats depends on estrogen status. *European Journal of Neuroscience, 28*(10), 2065–2074.

Wang, X. M., Zhang, K. M., Long, L. O., & Mokha, S. S. (1999). Orphanin FQ (nociceptin) modulates responses of trigeminal neurons evoked by excitatory amino acids and somatosensory stimuli, and blocks the substance P-induced facilitation of N-methyl-D-aspartate-evoked responses. *Neuroscience, 93*(2), 703–712.

Woda, A. (2003). Pain in the trigeminal system: From orofacial nociception to neural network modeling. *Journal of Dental Research, 82*(10), 764–768.

Zhang, X., Zhang, Y., Asgar, J., Niu, K. Y., Lee, J., Lee, K. S., et al. (2014). Sex differences in μ-opioid receptor expression in trigeminal ganglia under a myositis condition in rats. *European Journal of Pain, 18*(2), 151–161.

Enkephalin Rescues Temporomandibular Joint Pain-Related Behavior in Rats

Karin N. Westlund and A. Caitlynn Iddings

Abstract Temporomandibular joint disorders include a variety of clinical syndromes that are difficult to manage if associated with debilitating severe jaw pain. Thus, seeking additional experimental therapies for temporomandibular joint pain reduction is warranted. Targeted enkephalin gene therapy approaches provide clear promise for pain control. The studies detailed here indicate significant analgesia and protection of joint tissue are provided after injection of an overexpression viral vector gene therapy near the joint. The viral vector gene therapy described provides overexpression of naturally occurring opioid peptides after its uptake by trigeminal nerve endings. The viral vectors act as independent "minipump" sources for the opioid peptide synthesis in the neuronal cytoplasm producing the intended biological function, reduction of pain, and tissue repair. The antinociceptive effects provided with this delivery method of opioid expression persist for over 4 weeks. This is coincident with the expected time frame for the duration of the transgene overproduction of the endogenous opioid peptide before its diminution due to dormancy of the virus. These experimental studies establish a basis for the use of replication-defective herpes simplex type 1-based gene therapy for severe chronic inflammatory temporomandibular joint destruction and pain. As innovative means of significantly reducing joint inflammation and preserving tissue architecture, gene therapies may extend their clinical usefulness for patients with temporomandibular joint disorders.

Keywords Met-enkephalin · Gene therapy · Arthritis · HSV viral vector · Hypersensitivity

---

K. N. Westlund (✉)
Department of Anesthesiology, University of New Mexico, Albuquerque, NM, USA
e-mail: khigh@salud.unm.edu

A. C. Iddings
Pediatric and Adolescent Associates, Lexington, KY, USA

P. L. Kerr et al. (eds.), *Endogenous Opioids*, Advances in Neurobiology 35,
https://doi.org/10.1007/978-3-031-45493-6_7

Introduction

Inflammatory temporomandibular joint disorders (TMD) are a public health problem worldwide (List & Jensen, 2017). Persistent and chronic inflammatory pain in the temporomandibular joint (TMJ) and hypersensitivity in the overlying orofacial region are encountered in day-to-day dental practice and are the overwhelming reason these patients seek professional care with the hope of successful treatment. The regional orofacial pain, myofascial pain, headache, joint noise, and limitation in jaw mobility associated with TMD affect up to 15% of adults and 7% of adolescents (de Leeuw & Klasser, 2013; Drangsholt, 1999; Nilsson et al., 2005). Chronic orofacial pain also affects cognitive, psychological, and emotional responses (Carlson, 2008). A conservative estimate is that over $100 billion a year is spent in diagnosis and treatment of orofacial pain and TMD in the United States (Sessle, 2014).

Like other types of orofacial pain, chronic TMD pain is considered to be a disorder characterized by trigeminal nerve sensitization and central hyperexcitability of the trigeminal system (Dubner & Ruda, 1992). The first line of clinical treatment for TMD inflammatory pain is nonsteroidal anti-inflammatory drugs (NSAIDs) which act as nonselective inhibitors of cyclooxygenases. In chronic conditions, corticosteroids, opioids, antidepressants, and anticonvulsants may be somewhat effective. Standard therapies relying on higher and higher levels of circulating exogenous opiates, on the other hand, have been shown to result in decreased spinal cord met-enkephalin, intolerable side effects, as well contribute to the development of opiate tolerance and dependence. However, evidence supporting the effectiveness of drugs for the management of pain due to TMD remains equivocal and inconclusive (Mujakperuo et al., 2010). Additional experimental study is necessary to provide evidence in support of improved therapeutics.

Transgene Enkephalin Delivery

Provided below is experimental data in support of the use of overexpression vectors to deliver opioids as gene therapy for TMD. Trigeminal nerve nociceptors innervating the TMJ and adjacent region send information about the joint pain to the brainstem for relay to higher brain centers. Introducing gene therapies by uptake and retrograde delivery of viral vectors along trigeminal sensory and other peripheral nerves has been proposed as an ideal therapeutic route for effecting positive central nervous changes including reduction of pain (Antunes Bras et al., 1998; Davidson & Breakefield, 2003; Frampton et al., 2005; Goss et al., 2001, 2002; Wilson et al., 1999). We used a defective, nonreplicating herpes simplex type 1 (HSV-1) viral vector with a transgene cassette encoding the human preproenkephalin gene provided by Dr. Steven Wilson (Wilson et al., 1999; Yeomans & Wilson, 2009). Previous studies have shown the vector produces profound analgesic and antihyperalgesic effects in acute and chronic pain models in both rodents and nonhuman primates

Met-ENKEPHALIN

Morphine-like Analgesic Small Protein

Nervous System

- Has anti-nociceptive (pain relieving) properties
- Modulates development and growth
- Expressed by peripheral neurons with terminations in trigeminal and spinal cord dorsal horn
- Expressed by some central neurons and cells of adrenal medulla

Periphery

- Protective for heart ischemia
- Anti-inflammatory properties
- Decreases release of RANTES and other inflammatory mediators

Fig. 1 Localization and function of met-enkephalin

(Braz et al., 2001; Yeomans et al., 2006). The gene therapy employed successfully utilizes a replication-defective herpes simplex-1 viral (HSV-1) vector encoding preproenkephalin (HSV-ENK) in a rat CFA orofacial hyperalgesia model. The overexpressed endogenous opioids, met-enkephalin (ENK), β-endorphin, and other related peptides, are morphine-like small proteins with analgesic properties (Fig. 1).

In a site-directed manner, physiologically relevant levels of naturally occurring opioid peptides such as enkephalins can be overexpressed in the sensory afferent nerve fibers that transmit information about pain since the HSV-1 gene product is selectively expressed in nociceptive neurons that have internalized the vector at the injection site. This vector has been shown to induce overexpression of met-enkephalin in inflamed knee joint tissue and selectively in the spinal cord level innervating the knee joint in rats (Lu et al., 2008). Local injection of the HSV-1 viral constructs with a human proenkephalin gene insert results in effective, targeted tissue expression of opioid protein (Yeomans et al., 2006), whereas systemic administration of opiate compounds often have undesirable or intolerable side effects, including reduction of preproenkephalin gene expression (Uhl et al., 1988; Gudehithlu et al., 1991).

Methods

Orofacial Inflammation Model

Our basic study investigated the effectiveness of the opioid overexpression in the trigeminal nerve in an experimental inflammatory TMD arthritis model in rats induced with complete Freund's adjuvant (CFA) (Fig. 2). The CFA (100 µl) was

Methods

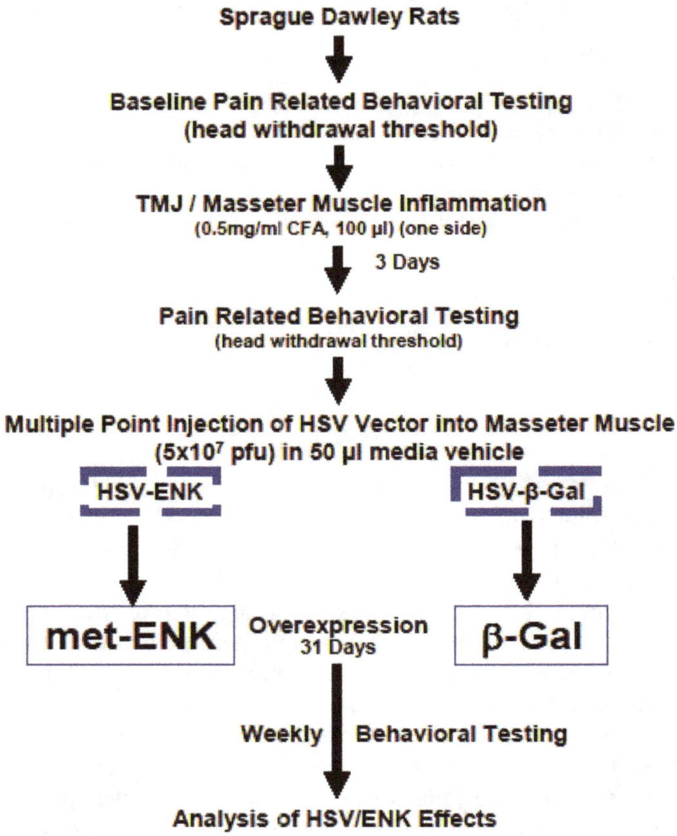

Fig. 2 Experimental flow chart

injected directly into the masseter muscle insertion near the TMJ on one side in male Sprague-Dawley rats to produce a localized inflammation (Fig. 3). After CFA injection, local tissue swelling developing immediately in the affected orofacial TMJ region persisted for 3 to 4 days, as has been shown previously by others associated with inflammation (Imbe & Ren, 1999; Takeda et al., 2005; Xu et al., 2010). Injection of CFA into TMJ has been shown to be associated with hypertrophy of the synovial lining and fibrin-like deposition (Xu et al., 2010). We have shown similar synovial hypertrophy for injection of CFA into the rat knee joint (Fig. 4b).

Three days later, after the CFA injection had produced maximal inflammation, the HSV-1 vector encoding preproenkephalin was injected in one group of rats. An HSV-1 vector with a neutral β-Galactosidase cDNA insert (HSVβ-Gal) was used as a control. A control HSVβ-Gal vector was injected in the same manner in another

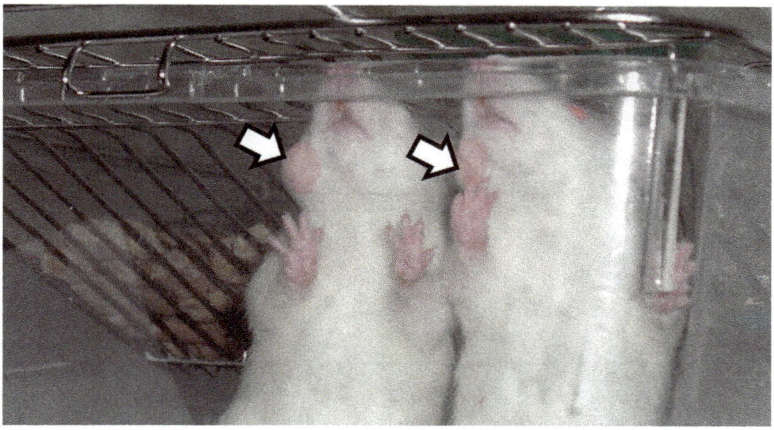

Fig. 3 Swelling (arrows) induced by CFA injection into the TMJ region in rats

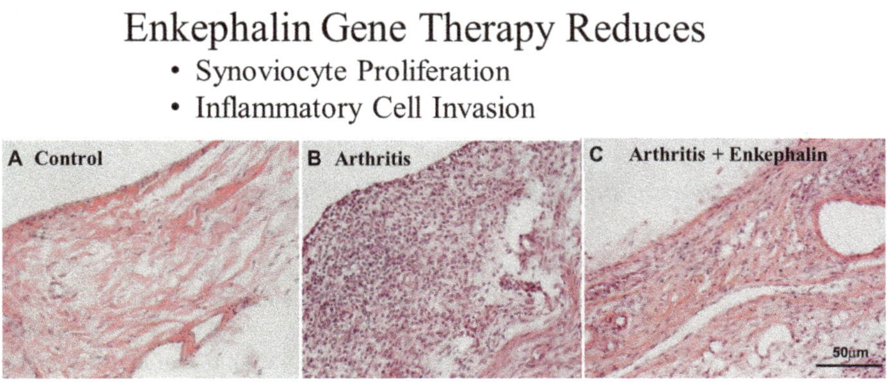

Fig. 4 Histological images of knee joint synovial lining from (**a**) naive control rat, (**b**) 2 weeks after CFA injection, and (**c**) after both CFA and HSV-ENK injections (H&E staining)

group of rats with CFA orofacial inflammation. Each vector was directed into the masseter muscle insertion near the TMJ joint at multiple injections points (5×10^7 pfu/50 μl total). The mechanical threshold was assessed at baseline, 3 days after HSV injection, and weekly thereafter.

Clear hypersensitivity to mechanical stimulation was detected within 3 days in the orofacial region ipsilateral to the CFA injected TMJ and masseter muscle region compared to baseline. The inflammatory hyperalgesia associated with the inflamed masseter muscle was assessed by measuring the decrease in response threshold of mechanical von Frey fiber stimuli on the skin over the masseter/TMJ region as an index of hypersensitivity (red ■, Fig. 5). The mechanical hypersensitivity persisted for over 4 weeks in untreated rats.

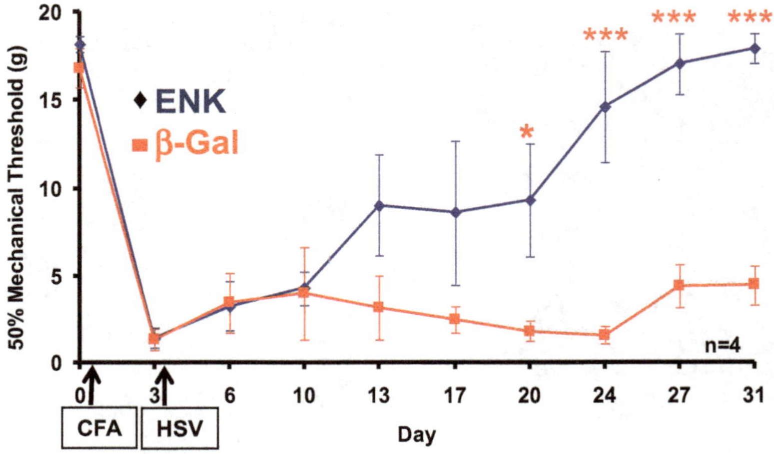

Fig. 5 Decrease in the mechanical threshold evident within 3 days after TMJ injection of CFA is attenuated in rats receiving the HSV-ENK overexpression vector, while the rats receiving the control HSV-β-galactosidase expression vector remain hypersensitive to the von Frey fiber stimulation applied on the skin over the joint. ($*p \leq 0.05$, $***p \leq 0.001$)

Results

HSV-ENK Attenuation of Mechanical Hypersensitivity

Treatment with HSV-1 vector encoding preproenkephalin significantly attenuated the mechanical hypersensitivity of the CFA induced orofacial inflammation compared to control vector treatment (dark blue ◆, Fig. 5) ($*\ p < 0.05$; $***\ p < 0.001$; two way ANOVA, HSV-ENK vs. HSVβ-Gal). The attenuation began in week 2 and persisted through the subsequent 3 weeks with return to the baseline threshold. This experiment clearly showed the HSV-ENK applied to deep tissues near the TMJ can have a profoundly positive effect for relief of orofacial hypersensitivity.

HSV-ENK Attenuation of Synovial Hypertrophy and Inflammatory Cell Invasion

Herpes vector-mediated gene transfer of met-enkephalin into CFA induced TMD rat model has been shown to reduce hypertrophy of the synovial lining and fibrin-like deposition (Xu et al., 2010). Histological evidence of reduced synovial hypertrophy and decreased inflammatory cell invasion was also evident in the knee joint after HSV-ENK protection (Fig. 4b) as in our previous study (Lu et al., 2008).

In this study, we found enkephalin peptide expression (green) in trigeminal ganglion neurons and nerve fibers on the same side where the injured masseter muscle and joint were injected with the HSV-ENK vector (Fig. 6a). Immunohistochemical dual labeling of the HSV was evident in neuronal nuclei shown in red (Fig. 6a). In the trigeminal ganglion contralateral to the inflamed masseter muscle with reverse color localization, there was only positive staining for endogenous ENK (red) in the nerve bundles typical of trigeminal nerve expression with little or no HSV (green) staining detected in trigeminal ganglia (Fig. 6b). The trigeminal ganglia were taken from rats one week after receiving an HSV-ENK injection.

The histogram shows the fluorescent immunostaining intensities for met-enkephalin or β-galactosidase immunoreactivity in rat trigeminal ganglia taken from rats with DMEM vehicle and HSV viral vector treatment groups (Fig. 7). Immunoreactive met-enkephalin was significantly greater on the injured side injected with the HSV-Enk viral vector indicating the vector induced met-enkephalin overexpression than is naturally expressed in the sensory nerves and ganglia. Immunohistochemical detection of leu-enkephalin in monkey dorsal root ganglia has also been reported (Yeomans et al., 2006).

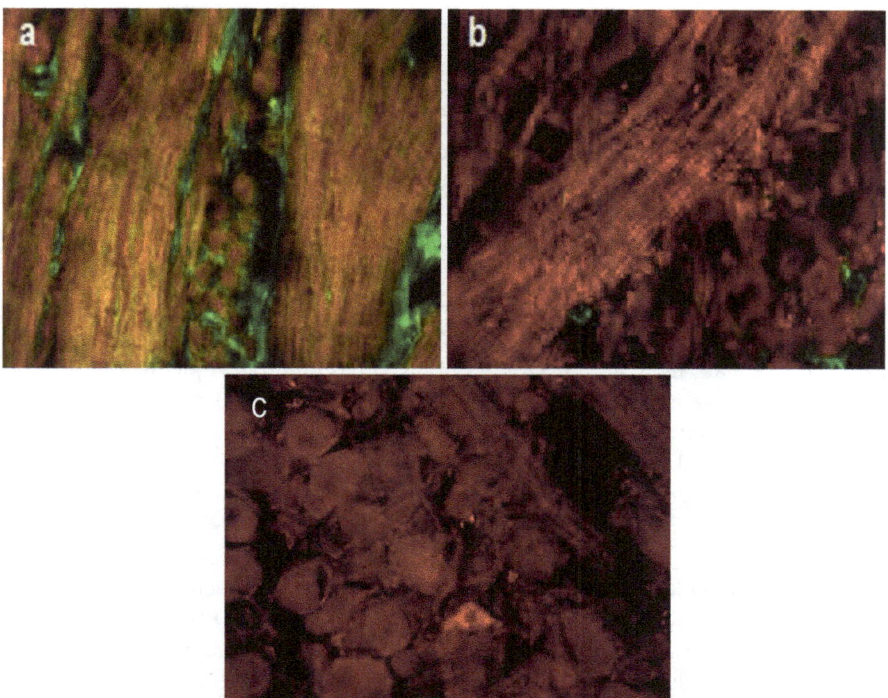

Fig. 6 HSV and met-ENK double labeled trigeminal ganglion neurons and axons in rat were stained for (**a**) met-ENK (green), HSV (red); (**b**) met-ENK (red), HSV (green); (**c**) Overexpressed met-Enk in a small trigeminal ganglion neuron (red), potentially a pain nociceptor (20X). Yellow (**a**) indicates dual labeling in axons

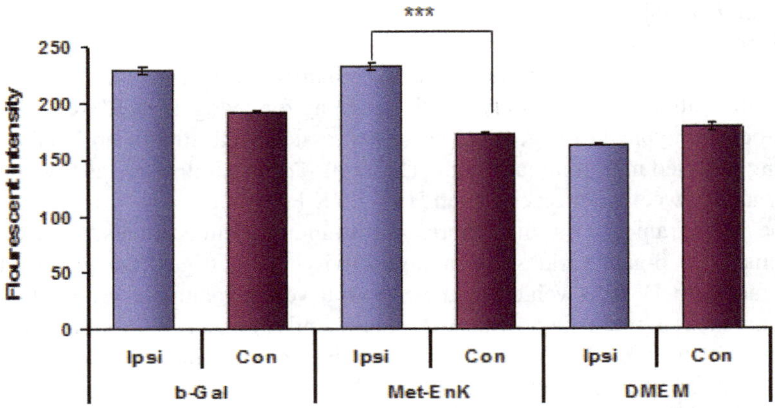

Fig. 7 Fluorescent intensity histogram for β-glactosidase, met-enkephalin, and DMEM vehicle control in the trigeminal ganglia after injections of each HSV overexpression vector into the ipsilateral and contralateral TMJ. (***$p < 0.001$)

Advantages and Safety Features of HSV-1 Gene Therapy

Limited histological examination indicated the HSV vector did not enter the brainstem spinal trigeminal nucleus, the site of nociceptor terminations in the brainstem. However, the vector was found in the mesencephalic nucleus in the pons, the internally located ganglion cell bodies of proprioceptive sensory neurons. No HSV was found in any other brainstem site in our cursory examination, indicating the replication defective HSV remained captured in sensory neurons and had no central nervous system spread.

One limitation of our study was insufficient information about the specific site of enkephalin release, centrally and peripherally, that is providing the attenuation of the pain-related behavior. Nonetheless, the data presented here and our previous experimental data indicate retrogradely transported preproenkephalin vectors can provide innovative means for significantly reducing joint inflammation and preserving tissue architecture coincident with spinal cord and joint met-enkephalin vector expression (Lu et al., 2008).

Thus, these experimental studies establish a basis for the use of HSV-based gene therapy for chronic inflammatory joint pain. Significant analgesia and protection of joint tissue are provided for the duration of the transgene expression of the endogenous opioid peptide (~4–6 weeks). The important issue of desensitization and tolerance appears to be a nonissue with enkephalin generated by the HSV-Enk viral vector. In studies with trigeminal neuropathic pain models, the preproenkephalin vector not only reduced heat stress responses in animals with nerve ligations but also independently prevents morphine tolerance effects (Hao et al., 2003).

Many gene therapy clinical trials have relied on retroviruses or adenoviruses to deliver the desired gene. The major advantages of using herpes simplex virus-1 vectors are its capacity to encode, insert, and promote cellular production of large protein molecules selectively in sensory nerves for uptake to the ganglia without invading central nervous system neurons. In contrast, lentiviral vector spread via the trigeminal nerve is reported centrally (Kyrkanides et al., 2007). The expression of the inserted gene product is provided at a specific locale, in the absence of productive HSV-1 infection and without integration into the host genome (Wilson et al., 1999). The recombinant viral vectors designed for gene therapy are rendered replication defective so they do not produce productive viral infection in vivo, but the gene product can persist for weeks to months despite negligible viral protein synthesis. These viral constructs establish a quiescent state similar to natural viral latency but do not reactivate to cause active infection in neuronal cells in vivo (Wilson et al., 1999; Goss et al., 2001; Fink et al., 2011; Goins et al., 2011). The limited potential to generate a host response also improves the potential for using repeated doses of HSV-based viral constructs. HSV-1 viral constructs offer unique advantages in treatment of peripheral inflammation and pain, as they selectively infect primary sensory neurons but do not integrate into the host genome. These are very important properties of HSV-1-based viral vectors. Adenoviruses cause productive infection, induce host inflammation above the inflammation already generated in target tissues and increase the potential for extended injury and host innate immune response to the virus making repeated doses problematic (Minter et al., 2001). These adverse reactions are avoided with HSV viral vectors. Safety with the HSV-1 vector inoculation is not a consideration in comparison to other viral vectors which are not peripherally restricted. An estimated 70–90% of the human adult population has evidence of prior HSV-1 infection known to cause small oral fever blisters.

Orofacial pain affects cognitive, psychological, and emotional responses, perhaps more substantially than with other types of somatic pain (Carlson, 2008). In a previous experimental study, we have shown that chronic orofacial pain reduces the ability to redirect behaviors in a paradigm designed to demonstrate cognitive extinction of an irrelevant lever-pressing task (Kniffin et al., 2015). Human trials involving injection of replication-competent HSV delivered into tumors of the brain (glioblastoma), breast, liver, or skin have shown the virus is well tolerated and safe with no serious adverse events even with intracerebral inoculation (Harrow et al., 2004). The viruses did not reactivate endogenous HSV, they did not cause encephalitis, nor did they spread appreciably from the inoculation site (Rampling et al., 2000; Markert et al., 2000; Papanastassiou et al., 2002). Subsequent human trials employed recombinant defective HSV gene therapy similar to the one reported here which can replicate in skin but is rendered incapable of replicating HSV genes in neurons with deletions of the required promoters (Fink et al., 2011). This feature forces the virus into a latent phase in nervous tissue. Improvement was reported for patients with intractable cancer pain assessed with a drop to 20–50 from ≥70 to 80 of 100 for over a month on the numeric visual analogue rating scale (NRS) and 70–80% improvement on the Short Form McGill Pain Questionnaire (SF-MPQ) for the two higher doses tested (10^8, 10^9 pfu) (Fink et al., 2011). Thus, the HSV-ENK treatment for intractable pain was effective in this small clinical trial, demonstrating the potential for clinical use.

Our previous study using quantitative real-time PCR reported 8- to 50-fold greater viral transduction in the trigeminal ganglia after topical application than subcutaneous injection and ≥ 100-fold greater efficiency for replication-conditional than replication-defective vectors (Ma et al., 2016). Another study found a nine-fold increase in proenkephalin A mRNA levels in trigeminal ganglion ipsilateral to HSV-ENK infected vibrissal pad territory in the neuropathic orofacial pain model induced by chronic constriction injury to the infraorbital nerve branch of the trigeminal nerve (Meunier et al., 2005). Targeted enkephalin gene therapy approaches are providing clear promise for pain control. As innovative means of significantly reducing joint inflammation and preserving tissue architecture, gene therapies may extend their clinical usefulness for patients with TMD pain. Although the precise mechanisms allowing the protracted effectiveness of the endogenous opioids are not known at this time, we propose that it is the available, continuous release of overexpressed met-enkephalin directly onto receptors at nerve terminals both centrally and peripherally that provides additive effectiveness for reducing hyperalgesia and tissue protection from inflammatory responses without tolerance. In this site-directed manner, physiologically relevant levels of enkephalin can affect the very same afferent nerve fibers that receive information from the inflamed TMJ. The studies provided here and other previous studies indicate the use of gene therapies for overexpression of opioids might be extended for patients with TMD pain.

Conclusion

The experimental data clearly demonstrate enkephalin-encoding HSV-1 viral vector can be applied to deep tissues and have a profoundly positive effect when used for relief of orofacial hypersensitivity. This immunohistochemical study indicates HSV-ENK is capable of expressing met-enkephalin in primary trigeminal neurons in rats in vivo.

Acknowledgments This study was supported by University Professor funding from the University of Kentucky President's Research Fund (KWH).

References

Antunes Bras, J. M., Epstein, A. L., Bourgoin, S., Hamon, M., Cesselin, F., & Pohl, M. (1998). Herpes simplex virus 1-mediated transfer of preproenkephalin A in rat dorsal root ganglia. *Journal of Neurochemistry, 70*(3), 1299–1303.

Braz, J., Beaufour, C., Coutaux, A., Epstein, A. L., Cesselin, F., Hamon, M., & Pohl, M. (2001). Therapeutic efficacy in experimental polyarthritis of viral-driven enkephalin overproduction in sensory neurons. *Journal of Neuroscience, 21*(20), 7881–7888.

Carlson, C. R. (2008). Psychological considerations for chronic orofacial pain. *Oral and Maxillofacial Surgery Clinics, 20*(2), 185–195.

Davidson, B. L., & Breakefield, X. O. (2003). Viral vectors for gene delivery to the nervous system. *Nature Reviews Neuroscience, 4*(5), 353–364.

de Leeuw, R., & Klasser, G. D. (Eds.). (2013). *Orofacial pain: Guidelines for assessment, diagnosis, and management* (5th ed.). Quintessence Publishing.

Drangsholt, M. (1999). Temporomandibular pain. In I. K. Crombie, P. R. Croft, S. J. Linton, L. LeResche, & M. Von Korff (Eds.), *Epidemiology of pain*. IASP Press.

Dubner, R., & Ruda, M. A. (1992). Activity-dependent neuronal plasticity following tissue injury and inflammation. *Trends in Neurosciences, 15*(3), 96–103.

Fink, D. J., Wechuck, J., Mata, M., Glorioso, J. C., Goss, J., Krisky, D., & Wolfe, D. (2011). Gene therapy for pain: Results of a phase I clinical trial. *Annals of Neurology, 70*(2), 207–212.

Frampton, A. R., Jr., Goins, W. F., Nakano, K., Burton, E. A., & Glorioso, J. C. (2005). HSV trafficking and development of gene therapy vectors with applications in the nervous system. *Gene Therapy, 12*(11), 891–901.

Goins, W. F., Krisky, D. M., Wechuck, J. B., Wolfe, D., Huang, S., & Glorioso, J. C. (2011). Generation of replication-competent and-defective HSV vectors. *Cold Spring Harbor Protocols, 2011*(5), pdb-prot5615.

Goss, J. R., Mata, M., Goins, W. F., Wu, H. H., Glorioso, J. C., & Fink, D. J. (2001). Antinociceptive effect of a genomic herpes simplex virus-based vector expressing human proenkephalin in rat dorsal root ganglion. *Gene Therapy, 8*(7), 551–556.

Goss, J. R., Harley, C. F., Mata, M., O'Malley, M. E., Goins, W. F., Hu, X., et al. (2002). Herpes vector–mediated expression of proenkephalin reduces bone cancer pain. *Annals of Neurology, 52*(5), 662–665.

Gudehithlu, K. P., Tehwani, G. A., & Bhargava, H. N. (1991). β-Endorphin and methionine-enkephalin levels in discrete brain regions, spinal cord, pituitary gland and plasma of morphine tolerant-dependent and abstinent rats. *Brain Research, 553*(2), 284–290.

Hao, S., Mata, M., Goins, W., Glorioso, J. C., & Fink, D. J. (2003). Transgene-mediated enkephalin release enhances the effect of morphine and evades tolerance to produce a sustained antiallodynic effect in neuropathic pain. *Pain, 102*(1–2), 135–142.

Harrow, S., Papanastassiou, V., Harland, J., Mabbs, R., Petty, R., Fraser, M., et al. (2004). HSV1716 injection into the brain adjacent to tumour following surgical resection of high-grade glioma: Safety data and long-term survival. *Gene Therapy, 11*(22), 1648–1658.

Imbe, H., & Ren, K. (1999). Orofacial deep and cutaneous tissue inflammation differentially upregulates preprodynorphin mRNA in the trigeminal and paratrigeminal nuclei of the rat. *Molecular Brain Research, 67*(1), 87–97.

Kniffin, T. C., Danaher, R. J., Westlund, K. N., Ma, F., Miller, C. S., & Carlson, C. R. (2015). Persistent neuropathic pain influences persistence behavior in rats. *Journal of Oral & Facial Pain and Headache, 29*(2), 183–192.

Kyrkanides, S., Fiorentino, P. M., Miller, J. N. H., Gan, Y., Lai, Y. C., Shaftel, S. S., et al. (2007). Amelioration of pain and histopathologic joint abnormalities in the Col1-IL-1βXAT mouse model of arthritis by intraarticular induction of μ-opioid receptor into the temporomandibular joint. *Arthritis & Rheumatology, 56*(6), 2038–2048.

List, T., & Jensen, R. H. (2017). Temporomandibular disorders: Old ideas and new concepts. *Cephalalgia, 37*(7), 692–704.

Lu, Y., McNearney, T. A., Wilson, S., Yeomans, D. C., & Westlund, K. N. (2008). Joint capsule treatment with enkephalin-encoding HSV-1 recombinant vector reduces inflammatory damage and behavioural sequelae in rat CFA monoarthritis. *European Journal of Neuroscience, 27*(5), 1153–1165.

Ma, F., Wang, C., Yoder, W. E., Westlund, K. N., Carlson, C. R., Miller, C. S., & Danaher, R. J. (2016). Efficacy of Herpes simplex virus vector encoding the human preproenkephalin gene for treatment of facial pain in mice. *Journal of Oral & Facial Pain and Headache, 30*(1), 42–50.

Markert, J. M., Medlock, M. D., Rabkin, S. D., Gillespie, G. Y., Todo, T., Hunter, W. D., et al. (2000). Conditionally replicating herpes simplex virus mutant, G207 for the treatment of malignant glioma: Results of a phase I trial. *Gene Therapy, 7*(10), 867–874.

Meunier, A., Latrémolière, A., Mauborgne, A., Bourgoin, S., Kayser, V., Cesselin, F., et al. (2005). Attenuation of pain-related behavior in a rat model of trigeminal neuropathic pain by viral-driven enkephalin overproduction in trigeminal ganglion neurons. *Molecular Therapy, 11*(4), 608–616.

Minter, R. M., Ferry, M. A., Murday, M. E., Tannahill, C. L., Bahjat, F. R., Oberholzer, C., et al. (2001). Adenoviral delivery of human and viral IL-10 in murine sepsis. *The Journal of Immunology, 167*(2), 1053–1059.

Mujakperuo, H. R., Watson, M., Morrison, R., & Macfarlane, T. V. (2010). Pharmacological interventions for pain in patients with temporomandibular disorders. *Cochrane Database of Systematic Reviews*. CD004715. The Cochrane Library.

Nilsson, I. M., List, T., & Drangsholt, M. (2005). Prevalence of temporomandibular pain and subsequent dental treatment in Swedish adolescents. *Journal of Orofacial Pain, 19*(2), 144–150.

Papanastassiou, V., Rampling, R., Fraser, M., Petty, R., Hadley, D., Nicoll, J., et al. (2002). The potential for efficacy of the modified (ICP 34.5−) herpes simplex virus HSV1716 following intratumoural injection into human malignant glioma: A proof of principle study. *Gene Therapy, 9*(6), 398–406.

Rampling, R., Cruickshank, G., Papanastassiou, V., Nicoll, J., Hadley, D., Brennan, D. A., et al. (2000). Toxicity evaluation of replication-competent herpes simplex virus (ICP 34.5 null mutant 1716) in patients with recurrent malignant glioma. *Gene Therapy, 7*(10), 859–866.

Sessle, B. J. (2014). The societal, political, educational, scientific and clinical context of orofacial pain. In B. J. Sessle (Ed.), *Orofacial pain. Recent advances in assessment, management, and understanding of mechanisms* (pp. 1–15). IASP Press.

Takeda, M., Tanimoto, T., Ikeda, M., Nasu, M., Kadoi, J., Shima, Y., et al. (2005). Temporomandibular joint inflammation potentiates the excitability of trigeminal root ganglion neurons innervating the facial skin in rats. *Journal of Neurophysiology, 93*(5), 2723–2738.

Uhl, G. R., Ryan, J. P., & Schwartz, J. P. (1988). Morphine alters preproenkephalin gene expression. *Brain Research, 459*(2), 391–397.

Wilson, S. P., Yeomans, D. C., Bender, M. A., Lu, Y., Goins, W. F., & Glorioso, J. C. (1999). Antihyperalgesic effects of infection with a preproenkephalin-encoding herpes virus. *Proceedings of the National Academy of Sciences, 96*(6), 3211–3216.

Xu, W., Wu, Y., Bi, Y., Tan, L., Gan, Y., & Wang, K. (2010). Activation of voltage-gated KCNQ/Kv7 channels by anticonvulsant retigabine attenuates mechanical allodynia of inflammatory temporomandibular joint in rats. *Molecular Pain, 6*(1), 49.

Yeomans, D. C., & Wilson, S. P. (2009). Herpes virus-based recombinant herpes vectors: Gene therapy for pain and molecular tool for pain science. *Gene Therapy, 16*(4), 502–508.

Yeomans, D. C., Lu, Y., Laurito, C. E., Peters, M. C., Vota-Vellis, G., Wilson, S. P., & Pappas, G. D. (2006). Recombinant herpes vector-mediated analgesia in a primate model of hyperalgesia. *Molecular Therapy, 13*(3), 589–597.

Endogenous Opioids and Exercise-Related Hypoalgesia: Modern Models, Measurement, and Mechanisms of Action

Allan H. Goldfarb, Robert R. Kraemer, and Brandon A. Baiamonte

Abstract This chapter will focus on the role exercise appears to have on activation and modulating factors within the central nervous system related to endogenous like opioids and its possible contribution to exercise-induced hypoalgesia. The implications for the exercise-mediated alterations of CNS activation factors related to opioids, specifically endorphins and enkephalins, will be presented. In this update, we discuss utilization of new technology and methods to monitor mechanisms of opioid involvement to suggest their contribution with exercise mediated hypoalgesia as well as their relationships to alterations of perceptions of pain and mood. Several special populations were included to suggest that not all individuals will respond to the exercise by mediating hypoalgesia. Factors that may confound the current understanding and suggestions from the recent literature will be presented as well as suggestions for future investigations.

Keywords Endorphins · Enkephalins · Endocannabinoid · Pain · Mood

Introduction: The Neurophysiology of Exercise

It has been postulated for many years that exercise of sufficient intensity and duration has positive implications for modulating central nervous system (CNS) activity through alterations in neurotransmitters, including those that activate opioid-like agents known broadly as endogenous opioids. There are many circuits that can

A. H. Goldfarb (✉)
University of North Carolina Greensboro, Department of Kinesiology, Greensboro, NC, USA
e-mail: ahgoldfa@uncg.edu

R. R. Kraemer
Southeastern Louisiana University, Department of Kinesiology and Health Studies, Hammond, LA, USA

B. A. Baiamonte
Southeastern Louisiana University, Department of Psychology, Hammond, LA, USA

P. L. Kerr et al. (eds.), *Endogenous Opioids*, Advances in Neurobiology 35,
https://doi.org/10.1007/978-3-031-45493-6_8

influence pain or its attenuation. Endogenous opioids bind to specific receptors both within and outside the CNS and include dynorphins (κ-receptors), endorphins (μ-receptors), endomorphins (μ-receptors), enkephalins (δ receptors), and nociceptin (NOP-receptors). These substances are typically thought to be too large to cross the blood brain barrier. Therefore, the actions of these molecules within the CNS are most likely attributable to their activation and release within the CNS. However, it is also possible for peripheral sensory receptors to activate central mediated opioid pathways (Koltyn et al., 2014; Thoren et al., 1990; Woolf & Mannion, 1999).

In addition, there is evidence that serotonin release, which modifies behavior, is modified by opioid-activated receptors which influences GABA involvement (Ferreira & Mensecal-de-Oliveria, 2012), as well as gene-to-gene interactions whereby the μ-opioid receptor interacts with serotonergic structures involved with both reuptake and release of serotonin (Tour et al., 2017). Furthermore, numerous nonopioid analgesics may influence both acute and chronic pain stress (Hebbes, 2016) and some are related to cannabinoid action (Crombie et al., 2017; Guindon & Beaulieu, 2009; Hohmann & Suplita, 2006; Tsou et al., 1998). Therefore, caution should be taken when considering the interpretation of changes that only measure opioid-like agents with or without blockers or activators to their specific receptors and not assessing alternative pain influencing agents.

One common index of endogenous opioid response is the plasma level of beta-endorphin (βE), a μ-opioid that is secreted in response to physical exertion of various types and intensities (Goldfarb & Jamurtas, 1997). An increase in circulating βE arises primarily from the pituitary gland, which releases βE from the larger molecule pro-opiomelanocortin (POMC), but which does not necessarily reflect actions within the CNS (Goldfarb, 2013; Goldfarb & Jamurtas, 1997). Reports on the association of plasma βE and pain initially were few and equivocal (Droste et al., 1991; Hartmann et al., 2005; Janal et al., 1984). However, recent research suggests a more robust and contextual relationship between plasma βE and pain (e.g., Bruehl et al., 2012; see also Bodnar, 2017, 2018 for summaries of recent findings). As described in this chapter, the neurophysiological substrates that are activated by and during exercise vary based on type, duration, intensity, and other factors.

Exercise-Induced Hypoalgesia

Physical activity is known to be critical for health, longevity, and high quality of life (O'Donovan et al., 2010) and is effective for treatment as well as prevention of diseases (Lukacs & Barkai, 2015; Clark, 2015; Johnson et al., 2015; Sahni et al., 2015). Regular exercise has been shown to increase coordination and aerobic fitness, decrease risk of various cardiovascular diseases (Shephard, 1986; Lyu et al., 2016), and enhance body image, self-efficacy, and emotional stability while alleviating depression and reducing anxiety and stress (Gurevich et al., 1994; see also Pettrey et al., 2024, this volume). In addition, research has indicated health benefits associated with acute exercise. Some forms of acute exercise influence pain perception by decreasing pain sensitivity in healthy individuals following participation in exercise (Koltyn, 2000; O'Connor & Cook, 1999). This decrease in pain sensitivity, which is known as exercise-induced hypoalgesia (EIH), occurs during and after higher intensities and longer periods of aerobic exercise (Hoffman et al., 2004; Koltyn, 2002).

Recent evidence strongly suggests that signaling molecules carried in circulation are important for exercise-induced hypoalgesia. Jones et al. (2017) measured pressure pain thresholds in subjects following 5 min of high-intensity cycling with one arm occluded and the other with normal blood flow. These investigators reported that there was a reduced EIH effect in the occluded arm. This analgesic phenomenon is of great interest as exercise regimens are becoming the focal point of most pain management programs (Hayden et al., 2005; Nijs et al., 2015; Raithel, 1989). A review of the literature on EIH reveals that healthy individuals will demonstrate hypoalgesia following most modalities of exercise including aerobic, isometric, and dynamic resistance exercise (Naugle et al., 2012; Baiamonte et al., 2017). However, there are differences in the degree to which each modality alters pain perception. According to Naugle et al.'s (2012) meta-analysis, aerobic exercise produces EIH in response to both pressure and thermal pain stimuli and seems to be the strongest when performed at moderate-to-high intensity (Naugle et al., 2012). Isometric exercise produced the largest effect size of the modalities and this was consistent regardless of pain stimulus and intensity of exercise.

There is a paucity of findings (Baiamonte et al., 2017; Focht & Koltyn, 2009; Koltyn & Arobast, 1998) on dynamic resistance exercise; however, the effect sizes were large when pain was assessed immediately following the exercise. While these studies have provided insight into the effects of dynamic resistance exercise on EIH, there are a few key aspects of these studies that should be addressed to elucidate the effects of dynamic resistance exercise on pain perception. The first important discrepancy in research methodology is the inconsistency in time points in which pain was assessed after exercise. Koltyn and Arobast (1998) measured pain perception at 5 and 15 min postexercise, whereas Focht and Koltyn (2009) assessed pain at 1- and 15-min time points, and Baiamonte et al. (2017) utilized all three time points (1, 5 and 15 min postexercise). In addition, the exercise protocol implemented in each study varied in terms of sets, repetitions, intensity, and duration. Baiamonte et al.'s (2017) procedure consisted of nine lifts where participants were required to perform 3 sets of 12 repetitions at 60% 1RM for 45 min with a 1:1 work to rest ratio. In contrast, Focht and Koltyn (2009) and Koltyn and Arobast (1998) consisted of only four movements of three sets of 10 repetitions at 75% 1RM for 45 min than unspecified work-to-rest ratio. The structure of the exercise protocol is of importance to the outcome of exercise on pain perception and the mechanisms responsible for EIH. In summary, all three modalities of exercise appear to produce moderate to large effects in healthy individuals, but these effects also appear to be transient, with the optimal dose along with the mechanisms responsible for this phenomenon still yet to be identified.

Aerobic exercise and resistance exercise both are known to elicit EIH for a brief period (Naugle et al., 2012; Baiamonte et al., 2017). Because aerobic exercise (Goldfarb & Jamurtas, 1997; Goldfarb et al., 1998) and resistance exercise (Kraemer et al., 1996) of high enough intensity (Goldfarb et al., 1990; Kraemer et al., 1989, 1996) are known to release endogenous opioids, some researchers have speculated that EIH is due to pain modulating substances such as βE (Koltyn & Arobast, 1998, 2000; Nijs et al., 2012); however, this has not been fully supported by empirical data.

While the EIH findings have been consistent for these exercise modalities at greater intensities in healthy participants, the evidence supporting the effectiveness of exercise on people being treated for chronic pain has been limited. The EIH findings in healthy individuals are much more consistent when compared to EIH in chronic pain populations, which have produced more variable outcomes, ranging from small to large effects in individuals with regional chronic pain conditions (Naugle et al., 2012). This variability could be explained by the intensity of the exercise, location of chronic pain condition in relation to experimental pain induction site, and severity of chronic pain condition. Interestingly, there is no evidence of EIH in patients with widespread chronic pain conditions and at times, exercise at moderate to high intensity exacerbated the pain. Moreover, it has been suggested that greater sensitivity to pain (i.e., hyperalgesia) in response to pressure in muscles after static contractions in patients with fibromyalgia (Kosek & Ekhom, 1995, 1996a, b; Nijs et al., 2012) indicates that patients with fibromyalgia may experience a dysfunction of endogenous analgesia during exercise compared with reduced pain sensitivity in healthy patients during exercise (Kosek & Lundberg, 2003). However, an acute exercise session by women with chronic neck pain was shown to reduce pain intensity and sensitivity which was associated with greater circulating βE and cortisol concentrations (Karlsson et al., 2015). This suggests that different pain conditions (e.g., chronic pain versus fibromyalgia) may have different underpinning pathways leading to differential responses to exercise (i.e., hyperalgesia versus hypoalgesia).

Electro-acupuncture has also been used to treat chronic pain. Research has found that electro-acupuncture at 2 Hz will enhance release of βE and enkephalin (Han, 2004). There is evidence that patients with chronic low back pain have greater activity in pain-related areas of the brain, whereas there is reduced activity in analgesic regions of the brain (Kregel et al., 2015). Recent evidence for patients with fibromyalgia suggests that after 10 treatments of transcranial direct current stimulation, pain was reduced, mood was improved, and these changes were related to circulating concentrations of βE (Khedr et al., 2017).

Chronic pain disorders such as lower back pain (van Middelkoop et al., 2010) and fibromyalgia (Mannerkorpi & Henriksson, 2007; Brosseau et al., 2008) are often treated with exercise therapy (Nijs et al., 2012). Patients with low back pain have significant pain reduction following treatments of aquatic exercise (Shi et al., 2017) and unsupervised, low-volume trunk exercises have also been shown to reduce pain in these individuals (Haufe et al., 2017); however, patients with long-term whiplash disorder do not show reductions in long-term pain after exercise treatments (Griffin et al., 2017).

As previously stated, sufficient exercise intensity and duration produces activation of the endogenous opioid system to produce EIH. Research has indicated that stimulation of afferent A-delta and C fibers via muscle contractions during exercise will activate spinal and supraspinal inhibitory signaling to dampen pain perception (Thoren et al., 1990; Koltyn et al., 2014). Both animal and human studies have been conducted to verify this mechanism, but the findings have been controversial. Most studies have utilized administration of opioid antagonists (naltrexone or naloxone) prior to exercise, which bind to the μ-opioid receptors and theoretically prevent or reduce EIH. In both human and animal studies, the results have been mixed, with attenuation of EIH following opioid antagonist administration prior to exercise in some studies and insensitivity to opioid antagonists in others (Koltyn, 2000). Researchers have suggested that the equivocal findings are due in part to methodological differences, which have included different exercise intensities, durations, and variations in opioid antagonist administration (Koltyn et al., 2014). In fact, previous research has indicated that manipulation of the exercise protocol in animal research can produce differences in EIH following opioid antagonist administration (Cook & Koltyn, 2000). Therefore, animal research has indicated that there may be multiple mechanisms (nonopioid systems) involved in EIH, other than the endogenous opioid system (Hohmann & Suplita, 2006, Koltyn et al., 2014). Researchers have suggested involvement of the endocannabinoid system in EIH due to the presence of CB1 receptors in pain processing areas (Guindon & Beaulieu, 2009; Herkenham et al., 1991; Tsou et al., 1998) and evidence of increased endocannabinoid concentrations following exercise (Crombie et al., 2017; Galdino et al., 2014a; Koltyn et al., 2014). A recent study conducted by Crombie et al. (2017) demonstrated an interaction between the opioid system and endocannabinoid system. When participants were administered naltrexone, the endocannabinoid 2-arachidonoylglycerol (2-AG) increased significantly following exercise. Even more interesting, the endocannabinoid N-arachidonylethanolamine (AEA) did not increase following administration of naltrexone and exercise. Therefore, increases in AEA typically observed following exercise were blocked by administration of an opioid antagonist, which suggests an interaction between the two systems. Recent findings that μ-opioid receptor inhibition decreases voluntary wheel running in rats, which involved signaling in a dopamine-dependent manner, supports complex regulation of pain at multiple levels within the brain (Ruegsegger et al., 2016). This particular study suggested that this overlap may at least partially explain why some individuals sense pleasure with exercise and others do not. Such findings may have implications and be instructive for interventions aimed at increasing physical exercise as a form of pain management, mood regulation, or health promotion in general.

Future research should focus on the complex interplay between the opioid and nonopioid systems on EIH rather than concentrating on each system independently. Investigation into the interaction between these two systems as well as other pathways should provide further evidence of the multiple mechanisms involved in EIH. This could provide insights into more appropriate treatments for prescribing dose and intensity of exercise needed to take advantage of all the physical and mental benefits that exercise has to offer besides just EIH.

Aerobic Exercise

Many studies have examined the influence of aerobic exercise in the circulation of βE and noted there was a critical intensity level to increase circulating βE or opioid-like peptides. The intensity needed to be at least 60% VO_2 max to result in elevations in circulating βE in most studies, and βE increased to a greater extent with higher intensities and longer duration (Goldfarb & Jamurtas, 1997). More recent studies with humans have reported alterations within the CNS with some of the newer techniques. One of the earlier studies showed that ^{133}Xe clearance was affected by cycling in the frontal region (Herholz et al., 1987). Another early study examined brain regions in eight males who ran for 35 min prior to [^{18}F] FDG injection or 20 min after running and compared to nine controls using MRI brain images and reported enhanced metabolic rate in many regions for the runners but some (temporal cortex, prefrontal cortex, and brain stem) showed reduced activity (Tashiro et al., 2001).

Often, the molecules infused are not metabolized and can thus suggest uptake over the time measured. This uptake is rapid in the first 10 min after injection with [^{18}F] FDG and then gradually decreases over time within the brain. Therefore, this method is only valid for a short duration and is not valid for long-duration exercise that would show minimal [^{18}F] FDG concentration. In addition, the three-dimensional mode of data acquisition has increased the sensitivity compared to conventional two-dimensional mode systems (Fujiwara et al., 1997). Healthy males (n = 8) scanned twice at least 2 h apart using [^{11}C] carfentanil PET to assess receptor binding in tissues and arterial plasma combined with MRI to confirm anatomical placement reported reproducible results with high binding in basal ganglia, thalamus and cortical regions (Hirvonen et al., 2009). One of the first exercise studies in this line of research used [^{18}F]FDPN PET to determine altered opioid binding in 10 males (Boecker et al., 2008). Boeker and colleagues (2008) performed two scans, one at rest and one 30 min postexercise (running for an average of 115 min), with an average heart rate of 144 beats per minute. Decreased binding to receptors in the frontal, temporal and parietal areas and several subcortical areas were reported. Ratings of euphoria from a visual analog scale significantly and negatively correlated with receptor binding. Central μ-opioid activation with PET analysis and imaging was also compared in seven young males before and after moderate- and high-intensity exercise (Hiura et al., 2017). Untrained males cycled for 20 min at 60% VO_2 max and then 80% VO_2 max on separate days with μ-opioid binding decreased in certain brain regions (limbic system, insula cortex and rostral anterior cingulate cortex) with both exercises but only decreased in pituitary gland at 80% intensity. Significant positive correlations were reported in mood ratings for vigor and depression for the moderate exercise and endogenous opioid binding.

Another recent study utilized PET analysis to study moderate exercise of 60 min duration versus high-intensity interval training (HIIT; Saanijoki et al., 2017). HIIT resulted in decreased μ-receptor binding in the frontolimbic region and was related to negative emotion. In contrast, moderate-intensity exercise μ-receptor binding was unchanged, but subjects reported euphoria. Therefore, the researchers concluded different intensities may influence μ-receptor binding differently. Increased blood flow during exercise has also been reported using PET (Hiura et al., 2014), single photon emission tomography (Fukuyama et al., 1997; Hanakawa et al., 1999), and near-infrared spectroscopy (Miyai et al., 2001).

In a recent review of research on brain activity during gait (real walking vs. imagery), methods used in the literature included MRI/fMRI or fNIRS and PET; however, only 17 studies with apparently healthy individuals were included comparing standing to gait walking (Hamacher et al., 2015). Synthesized results from several studies indicated enhanced blood flow in multiple brain regions, with most reporting enhanced activation in cortical and subcortical structures. However, certain regions also showed decreased blood flow. A key limitation is that most of the studies did not quantify the intensity of the walking nor indicate the duration of activity, making interpretation cautious and replication difficult. It should be noted that the enhanced blood flow with walking probably did not elevate opioid levels in the brain but likely alters metabolic activity within the brain.

Exercise training effects on brain blood flow during walking were recently studied in older adults who walked prior to and after a three-month intervention. The intervention included biweekly 90-min sessions of aerobic, strength training, and physical therapy activity (Shimada et al., 2017). Participants ($n = 2\ 4$) were 75–83 years old and reported 100% compliance with intervention instructions. No differences were found in PET measurements or [18F]fluorodeoxyglucose (FDG) uptake; however, specific regions of the brain demonstrated enhanced activity with walking. This suggests that even older individuals can alter their blood flow to certain areas of the brain with modest activity.

Animal studies have reported research findings both supportive and unsupportive of opioid system involvement in EIH. Acute aerobic exercise was reported to increase the expression of μ-opioid receptors in the hippocampal formation of rats that ran on a treadmill (forced) or voluntary (wheel running; de Oliveria et al., 2010). In contrast, both forced and voluntary training rats did not show any significant changes in μ-opioid receptors in the hippocampus region compared to control rats. This study was limited by the use of a small sample size for each group, and the investigators may not have endurance-trained their rats sufficiently to modify μ-opioid receptor expression. It is important to note that hippocampal progenitor cell proliferation has been suggested to be influenced by endogenous opioids and running (Persson et al., 2004) and has previously been shown to be promoted by βE in hippocampal cell proliferation in rats (Koehl et al., 2008).

In a recent study, voluntary wheel running in rats for 9 weeks was shown to influence opioid signaling in the nucleus accumbens (Ruegsegger et al., 2015). Ruegsegger et al. (2015) noted that high-volume running modulated μ-opioid receptors to express an increased motivational drive through opioid signaling compared to rats with low voluntary running. In a follow-up study, μ-opioid receptor inhibition was reported to decrease voluntary wheel running in rats (Ruegsegger et al., 2016). The researchers found that the inhibition of the μ-opioid receptor in the nucleus accumbens worked through a dopamine-dependent mechanism. They suggested that the increased dopamine in the high-volume runners but not in the low-volume runners may induce motivational signals through the opioidergic and dopaminergic systems to enhance activity.

The role of endogenous opioids in physiological response to aerobic exercise is complex. For example, Galdino et al. (2014a) found no evidence that endogenous opioids were associated with changes in pain thresholds in response to aerobic exercise. However, this research group did find that the endocannabinoid system was involved in mediating aerobic exercise-induced antinociception in rats. This same research group previously reported that release of nitric oxide may be involved with aerobic EIH (Galdino et al., 2010a, b). Clearly more research is needed to clarify the involvement and the mechanisms of opioid inputs to EIH with aerobic exercise.

Resistance Exercise

There are only a few studies that have examined the influence of resistance exercise and opioid-mediated responses in either the circulation or the CNS and related to EIH and have utilized animal models. Exercise rats trained to resistance exercise were reported to have elevated nitrite levels in plasma and cerebrospinal fluid increased (Galdino et al., 2014b). These data suggested antinociception was induced by resistance exercise, but it was related to the nitric oxide pathway. In an earlier study, nociceptive thresholds were elevated immediately after weight lifting exercise after 12 weeks of training in rats (Galdino et al., 2010a, b). It was also noted that, when cannabinoid receptor blockers were given 10 min before exercise, the EIH protection was reversed. Therefore, resistance exercise appears to induce an EIH response, but may not be directly related to the opioid pathways based on these data. Research with humans and resistance training has suggested that exercise load was important to induce circulating changes in beta-endorphins (Kraemer et al., 1993). However, the researchers in this study did not measure EIH measures in this study. More research on the relationship between EIH and resistance exercise in both animals and humans will be essential to further elucidate the role of endogenous opioids and EIH in resistance exercise.

Role of Enkephalins

It is possible that enkephalin peptides play a role in altering pain sensation. Exercise or ischemia induced enkephalin release from various nonneuronal tissues was examined in rat, mouse, pig and human cells (Denning et al., 2008). Using real time

PCR, Western blot analysis, ELISA, and immunofluorescence microscopy they reported extensive expression of preproenkephalin mRNA as well as enkephalin precursor protein proenkephalin. Isolated ex vivo tissue studies revealed that skeletal muscle, heart muscle, and intestinal tissue also released enkephalins. The investigators concluded that nonneuronal tissues could aid in inducing local and systemic enkephalin effects.

From a physiological standpoint, most researchers fail to agree on the exact mechanism responsible for pain reduction following bouts of exercise. The most commonly proposed mechanism includes the activation of the endogenous opioid system, in particular the release of βE with the CNS (Hoffman et al., 1990) and from muscle contractions during intense exercise stimulating pain receptors in skeletal muscle which can activate the endogenous opioid system (Umeda et al., 2016). However, previous research on humans and animals has been unclear regarding the involvement of the endogenous opioid system in EIH following the administration of an opioid antagonist (Koltyn et al., 2014). In animal studies, the opioid antagonists attenuated the hypoalgesia effect of exercise whereas human studies have produced conflicting results (Koltyn et al., 2014). Therefore, further investigation into the role of endorphins in EIH is warranted to address the inconsistent findings in the literature.

Methods of Measurement

Monitoring or evaluating the binding of endogenous opioid molecules to targeted receptors in vivo within the CNS has been a challenge. Methodologies utilized to determine ligand binding have used labeled isotopes and thermal imaging, the latter only indicating greater metabolic activity via altered blood flow in these areas. However, the measurement of these ligands typically only captures a single time point of analysis. Therefore, the results may not reflect the receptor activation over time for activities like exercise. The results from these studies analyzing the binding characteristics provides a picture of receptor occupancy and number of receptors, but does not necessarily translate directly to downstream actions.

Another methodology to assess neural changes is microdialysis measurements in animals and comparing changes at rest versus exercise (Meeusen et al., 2001). The microdialysis process can measure the amount of a substance collected over a selected time frame. In addition, the material collected may not reflect what is happening in the synaptic cleft or the receptor and may reflect spillover. Microdialysis often measures the accumulation of substances within the extracellular fluid, which is spillover or leakage and may not reflect ligand interaction at the receptor level. More recent advances in technology have utilized positron emission tomography (PET) to monitor ligand binding within the CNS and combined this with neuroimaging (Tashiro et al., 2001, 2008). An early study to examine PET in humans was with ^{133}Xe clearance and brain activity during an exercise task and noted the brain frontal region had increased blood flow compared to rest (Herholz et al., 1987). This

PET procedure evaluated the ligand [^{18}F] fluorodeoxyglucose ([^{18}F] FDG) to assess perfusion changes within brain regions as an index of cerebral metabolic activation. This [^{18}F] FDG tracer has a half-life of about 110 min and thus is suitable for observations ranging from 30- to 60-min duration. Because brain activated areas demonstrated enhanced oxygen usage and increased glucose uptake, tracers can be used to monitor these processes. The addition of radio-labeled water ([^{15}O]H_2O) enables measurement of brain regions with enhanced perfusion (Hirura et al., 2014). However, the use of PET together with [^{15}O]H_2O is cumbersome and has limitations and restrictions. For example, the subject has to be in the supine position during PET, thus limiting movement activities. Furthermore, the half-life for [^{15}O] H_2O is short (~ 2 min) which makes this technique a risk for radiation exposure if repeated frequently. Fortunately, this is not the case with functional MRI (fMRI), and near-infrared spectroscopy (NIRS), both of which measure regional cerebral blood flow changes. Recent advances with fMRI have demonstrated higher spatial and temporal resolution (1–2 mm level) that is currently greater than near-infrared spectroscopy (NIRS) spatial resolution. In addition, NIRS has some limitation in the ability of the subject to move as they are connected to the data acquisition system with cables.

Results from studies using PET have indicated regional shifts in brain blood flow, with some areas showing decreased flow and other areas increased regional flow with total brain glucose metabolism and total blood flow with exercise not significantly different to rest (Herholz et al., 1987; Ide et al., 1999; Tashiro et al., 2001). Decreased global brain glucose uptake in humans associated with high-intensity exercise has also been reported (Kemppainen et al., 2005). A more recent review noted that a few studies did not show increased cerebral blood flow with exercise but that the overwhelming evidence suggests that cerebral blood flow does increase with exercise (Shimada et al., 2017). It should be noted that other substrates have been suggested to compensate for glucose, such as lactate for the increased energy demands of neuronal activity during high-intensity exercise (Ide et al., 1999).

Recent research using PET within the central nervous system (CNS) and comparing these CNS changes to those of opioid alterations within the circulation are limited (Boecker et al., 2010). These studies and others need to evaluate both the opioid changes within the brain and spinal cord as well as monitor changes in pain perception. In addition, studies that have utilized opioid receptor blockers need to confirm the role of these opioids in altering EIH. Many factors can influence pain; accordingly, studies that have reported a decrease in pain with exercise need to confirm opioid actions were involved and possibly compare these changes with other pain modifiers.

Advanced Techniques to Investigate βE and Pain

Thermal heat pain challenges were used before and after running and walking trials to determine the effects of exercise intensity on pain (Stagg et al., 2011). The pattern of pain-related activity in response to heat/pain treatment using fMRI analysis before and after both sessions was compared. The mesial and lateral pain systems and periaqueductal gray (PAG) were the key areas of the descending antinociceptive pathway that were investigated. Running reduced affective pain ratings whereas walking did not. fMRI revealed that there was a reduction in pain activation in the PAG with decreases after running but pain activation was elevated after walking. For the pregenual anterior cingulate cortex and middle insular cortex, there were similar trends of activation for running vs. walking. Importantly, the authors concluded that increased circulating βE levels that were found in running, but not walking suggested involvement of the opioidergic system. Another study using fMRI examined pain modulation in 20 athletes both before and after running or walking (Scheef et al., 2012). They examined the PAG as well as pain ratings and reported that enhanced antinociceptive mechanisms was attenuated by the running (23 km, HR = 148) but no walking (10 km, HR = 84) activity. Elevated plasma βE were also reported only in the running treatment. These results supported previous studies suggesting that intensity and duration of exercise would influence βE levels in the blood. A recent fMRI study reported that resistance exercise training two times a week for 15 weeks in fibromyalgia patients did not significantly alter distraction-induced analgesia nor influence brain activity (Martinsen et al., 2017). This group previously published data on the effects of the exercise training in a larger cohort of subjects (Larsson et al., 2015).

β-Endorphin and Pain in Clinical Populations

There are a number of studies that investigated the effects of exercise on βE in different clinical populations affected by pain. A recent study determined whether circulating concentrations of several neuropeptides, steroid hormones and metabolites after exercise were different in women with chronic neck/shoulder pain than healthy women (Karlsson et al., 2015). The investigators used microdialysis to analyze substance P, βE, cortisol, glutamate, lactate, and pyruvate before and after an exercise training regimen. They also measured pain intensity and pain threshold. Before the training regimen, women with neck/shoulder pain had higher circulating levels of glutamate and βE and lower concentrations of cortisol than healthy women. Following the exercise regimen, women with shoulder/neck pain had less circulating substance P (and possibly glutamate) and greater circulating concentrations on βE and cortisol as well as reduced pain intensity and higher pain pressure thresholds. These investigators concluded there were greater algesic and analgesic changes in the affected trapezius muscle at baseline. Moreover, they suggested that exercise training could alter pain intensity and sensitivity as well as peripheral substances related to pain. The study provides more suggestive effects of opioid mediated pain modification following exercise training.

A study conducted in coronary artery bypass graft surgery patients, transcutaneous electrical nerve stimulation (TENS) or sham TENS was applied over the posterior cervical region (C7-T4) to access the stellate ganglion region, five days after surgery (Cipriano et al., 2014). The treatment was conducted four times per day for 30 min per session. Patients who had TENS treatment had less postoperative pain and less opiate requirements as well as greater circulating βE. They also had greater limb blood flow during a sympathetic stimulation (cold pressor) procedure. Thus, TENS, which elicits muscle contractions, appears to increase circulating βE which could lead to pain reduction.

Contemporary Pain Models

In a recent review, Nijs et al. (2012) determined that increased pain threshold following exercise were attributed to the release of endogenous opioids. More specifically, EIH was demonstrated in healthy participants due to the activation of μ-opioid receptors peripherally and centrally. However, evidence of EIH in individuals with chronic pain has been equivocal. Research has indicated that exercise of many modalities can decrease pain symptoms, resulting in improved daily function for individuals who suffer from chronic pain (Chatzitheodorou et al., 2007; Ellingson et al., 2014; Gowans et al., 2004; Hayden et al., 2005; Malmros et al., 1998; 27, McCain et al., 1988; Scheef et al., 2012). In contrast, exercise may not produce pain facilitation in certain chronic pain groups (Nijs et al., 2012). For example, patients with fibromyalgia, whiplash, and chronic fatigue all demonstrate pain sensitivity following exercise (Lannersten & Kosek, 2010; Meeus et al., 2010; Van Oosterwijck et al., 2010, 2012). Nijs et al. (2012) suggested that these patients had "dysfunctional endogenous analgesia" in response to exercise resulting in abnormalities in the central pain modulation system, which includes βE (Nijs et al., 2012). While this topic has been extensively investigated over the last 20 years, research has mainly focused on the hypoalgesic effects of exercise following a single episode of exercise as a result of increases in endogenous opioids (Koltyn, 2000).

It has been hypothesized that regular aerobic exercise leads to sustained reversal of neuropathic pain by activating endogenous opioid-mediated pain modulatory systems (Tashiro et al., 2008). Following nerve ligation, rats displayed thermal and mechanical sensitivities that were attenuated within 3 weeks of exercise training (Stagg et al., 2011). However, hyperalgesia returned 5 days after cessation of exercise. These authors provided evidence of βE and met-enkephalin involvement and injected naltrexone into the intracerebroventricular region which reversed EIH. Recent studies have indicated increased βE and met-enkephalin in the medulla and periaqueductal gray area, regions of the brain that are also involved in the descending pain pathway.

Immobilization

One possible cause of chronic pain can be long durations of limb immobilization as a result of cast. Research has indicated that cast immobilization induces chronic pain associated with central sensitization in both animal and human models (Hamaue et al., 2013; Thoren et al., 1990). Immobilization of limbs is a common and important method of medical treatment for tissue damage, and exercise is typically prescribed for the nonimmobilized limb but not the immobilized limb which can exacerbate hyperalgesia in the immobilized limb. Therefore, a better understanding of the hyperalgesia associated with cast immobilization and treatments for this phenomenon are imperative for the proper healing and return to normal daily functioning. A recent study demonstrated cast immobilization hyperalgesia in a rat model using Von Frey filaments to test for mechanical nociception (Chuganji et al., 2015). Interestingly, the data demonstrated a significant reduction in the hyperalgesia in the immobilized limb following a daily exercise regimen in the nonimmobilized limb. Furthermore, βE levels were increased in the hypothalamus and periaqueductal gray of the immobilized/exercise group. This finding suggests a role of βE upregulation in the attenuation of cast immobilization hyperalgesia following exercise.

Conclusion

Clearly, exercise of sufficient intensity and duration can induce transient pain modification. How and why this occurs is not entirely clear as of this writing. Nonetheless, more evidence is needed to substantiate the role of endogenous opioids in EIH, and clarify the underpinning mechanisms of action. βE can bind to opioid receptors within the CNS and can modify pain. However, there are some important unanswered questions that future research must aim to resolve. First, it is unclear if the actions of βE binding function solely to induce EIH, versus serving other simultaneous functions. Second, it is not clear whether these binding actions occur exclusively within the CNS or are also triggered by molecules arising from exercising muscles to help alleviate pain. Finally, not all individuals respond in a uniform manner to exercise. Thus, elucidating the mechanisms that explain differential responses to exercise is essential. Identifying why exercise may work to alleviate pain in some, while yielding little benefit or added discomfort in others, is important for understanding the full range of exercise applications in clinical practice with both clinical and healthy populations.

References

Baiamonte, B. A., Kraemer, R. R., Chabreck, C. N., Reynolds, M. L., McCaleb, K. M., Shaheen, G. L., et al. (2017). Exercise-induced hypoalgesia: Pain tolerance, preference and tolerance for exercise intensity, and physiological correlates following dynamic circuit resistance exercise. *Journal of Sports Sciences, 35*(18), 1–7.

Bodnar, R. J. (2017). Endogenous opiates and behavior. *Peptides 88*, 126–158.

Bodnar, R. J. (2018). Endogenous opiates and behavior. *Peptides 101*, 167–212.

Boecker, H., Sprenger, T., Spilker, M. E., Henriksen, G., Koppenhoefer, M., Wagner, K. J., et al. (2008). The runner's high: Opioidergic mechanisms in the human brain. *Cerebral Cortex, 18*(11), 2523–2531.

Boecker, H., Othman, A., Mueckter, S., Scheef, L., Pensel, M., Daamen, M., et al. (2010). Advocating neuroimaging studies of transmitter release in human physical exercise challenges studies. *Open Access Journal of Sports Medicine, 1*, 167–175.

Brosseau, L., Wells, G. A., Tugwell, P., Egan, M., Wilson, K. G., Dubouloz, C. J., et al. (2008). Ottawa Panel evidence-based clinical practice guidelines for aerobic fitness exercises in the management of fibromyalgia: Part 1. *Physical Therapy, 88*(7), 857–871.

Bruehl, S., Burns, J. W., Chung, O. Y., & Chont, M. (2012). What do plasma beta-endorphin levels reveal about endogenous opioid analgesic function? *European Journal of Pain, 16*(3), 370–380.

Chatzitheodorou, D., Kabitsis, C., Malliou, P., & Mougios, V. (2007). A pilot study of the effects of high-intensity aerobic exercise versus passive interventions on pain, disability, psychological strain, and serum cortisol concentrations in people with chronic low back pain. *Physical Therapy, 87*(3), 304–312.

Chuganji, S., Nakano, J., Sekino, Y., Hamaue, Y., Sakamoto, J., & Okita, M. (2015). Hyperalgesia in an immobilized rat hindlimb: Effect of treadmill exercise using non-immobilized limbs. *Neuroscience Letters, 584*, 66–70.

Cipriano, G., Neder, J. A., Umpierre, D., Arena, R., Vieira, P. J., Chiappa, A. M. G., et al. (2014). Sympathetic ganglion transcutaneous electrical nerve stimulation after coronary artery bypass graft surgery improves femoral blood flow and exercise tolerance. *Journal of Applied Physiology, 117*(6), 633–638.

Clark, J. E. (2015). Diet, exercise or diet with exercise: Comparing the effectiveness of treatment options for weight-loss and changes in fitness for adults (18–65 years old) who are overfat, or obese; systematic review and meta-analysis. *Journal of Diabetes & Metabolic Disorders, 14*(1), 31.

Cook, D. B., & Koltyn, K. F. (2000). Pain and exercise. *International Journal of Sport Psychology, 31*, 256–277.

Crombie, K. S., Brellenthin, A. G., Hillard, C. J., & Koltyn, K. F. (2017). Endocannabinoid and opioid system interactions in exercise-induced hypoalgesia. *Pain Medicine*. https://doi.org/10.1093/pm/pnx058

de Oliveira, M. S. R., da Silva Fernandes, M. J., Scorza, F. A., Persike, D. S., Scorza, C. A., da Ponte, J. B., et al. (2010). Acute and chronic exercise modulates the expression of MOR opioid receptors in the hippocampal formation of rats. *Brain Research Bulletin, 83*(5), 278–283.

Denning, G. M., Ackermann, L. W., Barna, T. J., Armstrong, J. G., Stoll, L. L., Weintraub, N. L., et al. (2008). Proenkephalin expression and enkephalin release are widely observed in non-neuronal tissues. *Peptides, 29*(1), 83–92.

Droste, C., Greenlee, M. W., Schreck, M., & Roskamm, H. (1991). Experimental pain thresholds and plasma beta-endorphin levels during exercise. *Medicine & Science in Sports Exercise, 23*, 334–342.

Ellingson, L. D., Koltyn, K. F., Kim, J. S., & Cook, D. B. (2014). Does exercise induce hypoalgesia through conditioned pain modulation? *Psychophysiology, 51*(3), 267–276.

Ferreira, M. D., & Menescal-de-Oliveira, L. (2012). Opioidergic, GABAergic and serotonergic neurotransmission in the dorsal raphe nucleus modulates tonic immobility in guinea pigs. *Physiology & Behavior, 106*(2), 109–116.

Focht, B. C., & Koltyn, K. F. (2009). Alterations in pain perception after resistance exercise performed in the morning and evening. *Journal of Strength Conditioning Research, 23*(3), 891–897.

Fujiwara, T., Watanuki, S., Yamamoto, S., Miyake, M., Seo, S., Itoh, M., et al. (1997). Performance evaluation of a large axial field-of-view PET scanner: SET-2400W. *Annals of Nuclear Medicine, 11*(4), 307–313.

Fukuyama, H., Ouchi, Y., Matsuzaki, S., Nagahama, Y., Yamauchi, H., Ogawa, M., et al. (1997). Brain functional activity during gait in normal subjects: A SPECT study. *Neuroscience Letters, 228*(3), 183–186.

Galdino, G. S., Cortes, S. F., Duarte, I. D., & Perez, A. C. (2010a). Involvement of the nitric oxide/(C)GMP/K(ATP) pathway in antinociception induced by exercise in rats. *Life Sci, 86*(13-14), 505–509.

Galdino, G. S., Duarte, I. D. G., & Perez, A. C. (2010b). Participation of endogenous opioids in the antinociception induced by resistance exercise in rats. *Brazilian Journal of Medical and Biological Research, 43*(9), 906–909.

Galdino, G., Romero, T. R., Silva, J. F. P., Aguiar, D. C., de Paula, A. M., Cruz, J. S., et al. (2014a). The endocannabinoid system mediates aerobic exercise-induced antinociception in rats. *Neuropharmacology, 77*, 313–324.

Galdino, G., Romero, T., da Silva, J. F. P., Aguiar, D., de Paula, A. M., Cruz, J., et al. (2014b). Acute resistance exercise induces antinociception by activation of the endocannabinoid system in rats. *Anesthesia and Analgesia, 119*(3), 702.

Goldfarb, A. H. (2013). Exercise and endogenous opiates. In N. Constantini & A. C. Hackney (Eds.), *Endocrinology of physical activity and sport* (pp. 21–36). Humana Press/Springer Science.

Goldfarb, A. H., & Jamurtas, A. Z. (1997). Beta-endorphin response to exercise: An update. *Sports Medicine, 24*(1), 8–16.

Goldfarb, A. H., Hatfield, B. D., Armstrong, D., & Potts, J. (1990). Plasma beta-endorphin concentration: Response to intensity and duration of exercise. *Medicine and Science in Sports and Exercise, 22*(2), 241–244.

Goldfarb, A. H., Jamurtas, A. Z., Kamimori, G. H., Hegde, S., Otterstetter, R., & Brown, D. A. (1998). Gender effect on beta-endorphin response to exercise. *Medicine and Science in Sports and Exercise, 30*(12), 1672–1676.

Gowans, S. E., Dehueck, A., Voss, S., Silaj, A., & Abbey, S. E. (2004). Six-month and one-year followup of 23 weeks of aerobic exercise for individuals with fibromyalgia. *Arthritis Care & Research, 51*(6), 890–898.

Griffin, A., Leaver, A., & Moloney, N. (2017). General exercise does not improve long-term pain and disability in individuals with whiplash-associated disorders: A systematic review. *Journal of Orthopaedic & Sports Physical Therapy, 47*(7), 472–480.

Guindon, J., & Beaulieu, P. (2009). The role of the endogenous cannabinoid system in peripheral analgesia. *Current Molecular Pharmacology, 2*(1), 134–139.

Gurevich, M., Kohn, P. M., & Davis, C. (1994). Exercise-induced analgesia and the role of reactivity in pain sensitivity. *Journal of Sports Science, 6*, 549–559.

Hamacher, D., Herold, F., Wiegel, P., Hamacher, D., & Schega, L. (2015). Brain activity during walking: A systematic review. *Neuroscience & Biobehavioral Reviews, 57*, 310–327.

Hamaue, Y., Nakano, J., Sekino, Y., Chuganji, S., Sakamoto, J., Yoshimura, T., et al. (2013). Immobilization-induced hypersensitivity associated with spinal cord sensitization during cast immobilization and after cast removal in rats. *The Journal of Physiological Sciences, 63*(6), 401–408.

Han, J. S. (2004). Acupuncture and endorphins. *Neuroscience Letters, 361*(1), 258–261.

Hanakawa, T., Katsumi, Y., Fukuyama, H., Honda, M., Hayashi, T., Kimura, J., & Shibasaki, H. (1999). Mechanisms underlying gait disturbance in Parkinson's disease: A single photon emission computed tomography study. *Brain, 122*(7), 1271–1282.

Hartmann, S., Bung, P., Schlebusch, H., & Hollmann, W. (2005). The analgesic effect of exercise during labor. *Zeitschrift fur Geburtshilfe und Neonatologie, 209*(4), 144–150.

Haufe, S., Wiechmann, K., Stein, L., Kück, M., Smith, A., Meineke, S., et al. (2017). Low-dose, non-supervised, health insurance initiated exercise for the treatment and prevention of chronic low back pain in employees. Results from a randomized controlled trial. *PloS One, 12*(6), e0178585.

Hayden, J. A., Van Tulder, M. W., & Tomlinson, G. (2005). Systematic review: Strategies for using exercise therapy to improve outcomes in chronic low back pain. *Annals of Internal Medicine, 142*(9), 776–785.

Hebbes, C. (2016). Non-opioid analgesics. *Anaesthesia and Intensive Care Medicine, 17*(9), 469–472.

Herholz, K., Buskies, W., Rist, M., Pawlik, G., Hollmann, W., & Heiss, W. D. (1987). Regional cerebral blood flow in man at rest and during exercise. *Journal of Neurology, 234*(1), 9–13.

Herkenham, M., Lynn, A. B., Johnson, M. R., Melvin, L. S., de Costa, B. R., & Rice, K. C. (1991). Characterization and localization of cannabinoid receptors in rat brain: A quantitative in vitro autoradiographic study. *Journal of Neuroscience, 11*(2), 563–583.

Hirvonen, J., Aalto, S., Hagelberg, N., Maksimow, A., Ingman, K., Oikonen, V., et al. (2009). Measurement of central μ-opioid receptor binding in vivo with PET and [11C] carfentanil: A test–retest study in healthy subjects. *European Journal of Nuclear Medicine and Molecular Imaging, 36*(2), 275–286.

Hiura, M., Nariai, T., Ishii, K., Sakata, M., Oda, K., Toyohara, J., & Ishiwata, K. (2014). Changes in cerebral blood flow during steady-state cycling exercise: A study using oxygen-15-labeled water with PET. *Journal of Cerebral Blood Flow & Metabolism, 34*(3), 389–396.

Hiura, M., Sakata, M., Ishii, K., Toyohara, J., Oda, K., Nariai, T., & Ishiwata, K. (2017). Central μ-opioidergic system activation evoked by heavy and severe-intensity cycling exercise in humans: A pilot study using positron emission tomography with 11C-carfentanil. *International Journal of Sports Medicine, 38*(01), 19–26.

Hoffman, P., Terenius, L., & Thoren, P. (1990). Cerebrospinal fluid immunoreactive beta-endorphin concentration is increased by long-lasting voluntary exercise in the spontaneously hypertensive rat. *Regulatory Peptides, 28*, 233–239.

Hoffman, M. D., Shepanski, M. A., Ruble, S. B., Valic, Z., Buckwalter, J. B., & Clifford, P. S. (2004). Intensity and duration threshold for aerobic exercise-induced analgesia to pressure pain. *Archives of Physical Medicine and Rehabilitation, 85*(7), 1183–1187.

Hohmann, A. G., & Suplita, R. L. (2006). Endocannabinoid mechanisms of pain modulation. *The AAPS Journal, 8*(4), E693–E708.

Ide, K., Horn, A., & Secher, N. H. (1999). Cerebral metabolic response to submaximal exercise. *Journal of Applied Physiology, 87*(5), 1604–1608.

Janal, M. N., Col, E. W., Clark, W. C., & Glusman, M. (1984). Pain sensitivity, mood and plasma endocrine levels in man following long-distance running: Effects of naloxone. *Pain, 19*, 13–25.

Johnson, E. J., Dieter, B. P., & Marsh, S. A. (2015). Evidence for distinct effects of exercise in different cardiac hypertrophic disorders. *Life Sciences, 123*, 100–106.

Jones, M. D., Taylor, J. L., & Barry, B. K. (2017). Occlusion of blood flow attenuates exercise-induced hypoalgesia in the occluded limb of healthy adults. *Journal of Applied Physiology, 122*, 1284–1291.

Karlsson, L., Gerdle, B., Ghafouri, B., Bäckryd, E., Olausson, P., Ghafouri, N., & Larsson, B. (2015). Intramuscular pain modulatory substances before and after exercise in women with chronic neck pain. *European Journal of Pain, 19*(8), 1075–1085.

Kemppainen, J., Aalto, S., Fujimoto, T., Kalliokoski, K. K., Långsjö, J., Oikonen, V., et al. (2005). High intensity exercise decreases global brain glucose uptake in humans. *The Journal of Physiology, 568*(1), 323–332.

Khedr, E. M., Omran, E. A., Ismail, N. M., El-Hammady, D. H., Goma, S. H., Kotb, H., et al. (2017). Effects of transcranial direct current stimulation on pain, mood and serum endorphin level in the treatment of fibromyalgia: A double blinded, randomized clinical trial. *Brain Stimulation, 10*(5), 893–901.

Koehl, M., Meerlo, P., Gonzales, D., Rontal, A., Turek, F. W., & Abrous, D. N. (2008). Exercise-induced promotion of hippocampal cell proliferation requires β-endorphin. *The FASEB Journal, 22*(7), 2253–2262.

Koltyn, K. F. (2000). Analgesia following exercise. A review. *Sports Medicine, 29*(2), 85–98.

Koltyn, K. F. (2002). Exercise-induced hypoalgesia and intensity of exercise. *Sports Medicine, 32*(8), 477–487.

Koltyn, K. F., & Arobast, R. W. (1998). Perception of pain after resistance exercise. *British Journal of Sports Medicine, 32*(1), 20–24.

Koltyn, K. F., Brellenthin, A. G., Cook, D. B., et al. (2014). Mechanisms of exercise-induced hypoalgesia. *Journal of Pain, 15*(12), 1294–1304.

Kosek, E., & Ekhom, J. (1995). Modulation of pressure pain thresholds during and following isometric contraction. *Pain, 61*(3), 481–486.

Kosek, E., & Lundberg, L. (2003). Segmental and plurisegmental modulation of pressure thresholds during static muscle contractions in healthy individuals. *European Journal of Pain, 7*(3), 251–258.

Kosek, E., Ekhom, J., & Hansson, P. (1996a). Increased pressure pain sensibility in fibromyalgia patients is located deep to the skin but not restricted to muscle tissue. *Pain, 63*(3), 335–339.

Kosek, E., Ekhom, J., & Hansson, P. (1996b). Modulation of pressure pain thresholds during and following isometric contraction in patients with fibromyalgia and in healthy controls. *Pain, 64*(3), 415–423.

Kraemer, R. R., Blair, S., Kraemer, G. R., & Castracane, V. D. (1989). Effects of treadmill running on plasma beta-endorphin, corticotropin, and cortisol levels in male and female 10K runners. *European Journal of Applied Physiology and Occupational Physiology, 58*(88), 45–51.

Kraemer, W. J., Dziados, J. E., Marchitelli, L. J., Gordon, S. E., Harman, E. A., Mello, R., et al. (1993). Effects of different heavy-resistance exercise protocols on plasma beta-endorphin concentrations. *Journal of Applied Physiology, 74*(1), 450–459.

Kraemer, R. R., Acevedo, E. O., Dzewaltowski, D., et al. (1996). Effects of low-volume resistive exercise on beta-endorphin and cortisol concentrations. *International Journal of Sports Medicine, 17*(1), 12–16.

Kregel, J., Meeus, M., Malfiet, A., et al. (2015). Structural and functional brain abnormalities in chronic low back pain: A systematic review. *Seminars in Arthritis and Rheumatism, 45*(2), 229–237.

Lannersten, L., & Kosek, E. (2010). Dysfunction of endogenous pain inhibition during exercise with painful muscles in patients with shoulder myalgia and fibromyalgia. *Pain, 151*(1), 77–86.

Larsson, A., Palstam, A., Löfgren, M., Ernberg, M., Bjersing, J., Bileviciute-Ljungar, I., et al. (2015). Resistance exercise improves muscle strength, health status and pain intensity in fibromyalgia—a randomized controlled trial. *Arthritis Research & Therapy, 17*(1), 161. https://doi.org/10.1186/s13075-015-0679-1

Lukács, A., & Barkai, L. (2015). Effect of aerobic and anaerobic exercises on glycemic control in type 1 diabetic youths. *World Journal of Diabetes, 6*(3), 534.

Lyu, X., Li, S., Peng, S., Cai, H., Liu, G., & Ran, X. (2016). Intensive walking exercise for lower extremity peripheral arterial disease: A systematic review and meta-analysis. *Journal of Diabetes, 8*(3), 363–377.

Malmros, B., Mortensen, L., Jensen, M. B., & Charles, P. (1998). Positive effects of physiotherapy on chronic pain and performance in osteoporosis. *Osteoporosis International, 8*(3), 215–221.

Mannerkorpi, K., & Henriksson, C. (2007). Non-pharmacological treatment of chronic widespread musculoskeletal pain. *Best Practice & Research Clinical Rheumatology, 21*(3), 513–534.

Martinsen, S., Flodin, P., Berrebi, J., Löfgren, M., Bileviciute-Ljungar, I., Mannerkorpi, K., et al. (2017). The role of long-term physical exercise on performance and brain activation during the Stroop colour word task in fibromyalgia patients. *Clinical Physiology and Functional Imaging*. https://doi.org/10.1111/cpf.12449

Mccain, G. A., Bell, D. A., Mai, F. M., & Halliday, P. D. (1988). A controlled study of the effects of a supervised cardiovascular fitness training program on the manifestations of primary fibromyalgia. *Arthritis & Rheumatology, 31*(9), 1135–1141.

Meeus, M., Roussel, N. A., Truijen, S., & Nijs, J. (2010). Reduced pressure pain thresholds in response to exercise in chronic fatigue syndrome but not in chronic low back pain: An experimental study. *Journal of Rehabilitation Medicine, 42*(9), 884–890.

Meeusen, R., Piacentini, M. F., & De Meirleir, K. (2001). Brain microdialysis in exercise research. *Sports Medicine, 31*(14), 965–983.

Miyai, I., Tanabe, H. C., Sase, I., Eda, H., Oda, I., Konishi, I., et al. (2001). Cortical mapping of gait in humans: A near-infrared spectroscopic topography study. *NeuroImage, 14*(5), 1186–1192.

Naugle, K. M., Fillingim, R. B., & Riley, J. L. (2012). A meta-analytic review of the hypoalgesic effects of exercise. *The Journal of Pain, 13*(12), 1139–1150.

Nijs, J., Kosek, E., Van Oosterwijck, J., & Meeus, M. (2012). Dysfunctional endogenous analgesia during exercise in patients with chronic pain: To exercise or not to exercise? *Pain Physician, 3S*, ES205–ES213.

Nijs, J., Girbés, E. L., Lundberg, M., Malfliet, A., & Sterling, M. (2015). Exercise therapy for chronic musculoskeletal pain: Innovation by altering pain memories. *Manual Therapy, 20*(1), 216–220.

O'connor, P. J., & Cook, D. B. (1999). 5 exercise and pain: The neurobiology, measurement, and laboratory study of pain in relation to exercise in humans. *Exercise and Sport Sciences Reviews, 27*(1), 119–166.

O'Donovan, G., Blazevich, A. J., Boreham, C., Cooper, A. R., Crank, H., Ekelund, U., et al. (2010). The ABC of physical activity for health: A consensus statement from the British Association of Sport and Exercise Sciences. *Journal of Sports Sciences, 28*(6), 573–591.

Pettrey, C., Kerr, P. L., & Dickey, T. O. (2024). Physical exercise as an intervention for depression: Evidence for efficacy and mu-opioid receptors as a mechanism of action. In P. L. Kerr, C. Sirbu, & J. M. Gregg (Eds.), *Endogenous opioids*. Springer. this volume.

Persson, A. I., Naylor, A. S., Jonsdottir, I. H., Nyberg, F., Eriksson, P. S., & Thorlin, T. (2004). Differential regulation of hippocampal progenitor proliferation by opioid receptor antagonists in running and non-running spontaneously hypertensive rats. *European Journal of Neuroscience, 19*(7), 1847–1855.

Raithel, K. S. (1989). Chronic pain and exercise therapy. *The Physician and Sportsmedicine, 17*(3), 203–209.

Ruegsegger, G. N., Toedebusch, R. G., Will, M. J., & Booth, F. W. (2015). Mu opioid receptor modulation in the nucleus accumbens lowers voluntary wheel running in rats bred for high running motivation. *Neuropharmacology, 97*, 171–181.

Ruegsegger, G. N., Brown, J. D., Kovarik, M. C., Miller, D. K., & Booth, F. W. (2016). Mu-opioid receptor inhibition decreases voluntary wheel running in a dopamine-dependent manner in rats bred for high voluntary running. *Neuroscience, 339*, 525–537.

Saanijoki, T., Tuominen, L., Tuulari, J. J., Nummenmaa, L., Arponen, E., Kalliokoski, K., & Hirvonen, J. (2017). Opioid release after high-intensity interval training in healthy human subjects. *Neuropsychopharmacology, 43*(2), 1–9.

Sahni, S., Capozzi, B., Iftikhar, A., Sgouras, V., Ojrzanowski, M., & Talwar, A. (2015). Pulmonary rehabilitation and exercise in pulmonary arterial hypertension: An underutilized intervention. *Journal of Exercise Rehabilitation, 11*(2), 74–79.

Scheef, L., Jankowski, J., Daamen, M., Weyer, G., Klingenberg, M., Renner, J., et al. (2012). An fMRI study on the acute effects of exercise on pain processing in trained athletes. *Pain, 153*(8), 1702–1714.

Shephard, R. J. (1986). Exercise and malignancy. *Sports Medicine, 3*, 235–241.

Shi, Z., Zhou, H., Lu, L., Pan, B., Wei, Z., Yao, X., et al. (2017). Aquatic exercises in the treatment of low back pain: A systematic review of the literature and meta-analysis of eight studies. *American Journal of Physical Medicine & Rehabilitation*. https://doi.org/10.1097/PHM.0000000000000801

Shimada, H., Ishii, K., Makizako, H., Ishiwata, K., Oda, K., & Suzukawa, M. (2017). Effects of exercise on brain activity during walking in older adults: A randomized controlled trial. *Journal of Neuroengineering and Rehabilitation, 14*(1), 50. https://doi.org/10.1186/s12984-017-0263-9

Stagg, N. J., Mata, H. P., Ibrahim, M. M., et al. (2011). Regular exercise reverses sensory hypersensitivity in a rat neuropathic pain model: Role of endogenous opioids. *Anesthesiology, 114*(4), 940–948.

Tashiro, M., Itoh, M., Fujimoto, T., & Fujiwara, T. (2001). (18) F-FDG PET mapping of regional brain activity in runners. *Journal of Sports Medicine and Physical Fitness, 41*(1), 11–17.

Tashiro, M., Itoh, M., Fujimoto, T., Masud, M. M., Watanuki, S., & Yanai, K. (2008). Application of positron emission tomography to neuroimaging in sports sciences. *Methods, 45*(4), 300–306.

Thorén, P., Floras, J. S., Hoffmann, P., & Seals, D. R. (1990). Endorphins and exercise: Physiological mechanisms and clinical implications. *Medicine & Science in Sports & Exercise, 22*(4), 417–428.

Tour, J., Löfgren, M., Mannerkorpi, K., Gerdle, B., Larsson, A., Palstam, A., et al. (2017). Gene-to-gene interactions regulate endogenous pain modulation in fibromyalgia patients and healthy controls—antagonistic effects between opioid and serotonin-related genes. *Pain, 158*(7), 1194–1203.

Tsou, K., Brown, S., Sanudo-Pena, M. C., Mackie, K., & Walker, J. M. (1998). Immunohistochemical distribution of cannabinoid CB1 receptors in the rat central nervous system. *Neuroscience, 83*(2), 393–411.

Umeda, M., Lee, W., Marino, C. A., & Hilliard, S. C. (2016). Influence of moderate intensity physical activity levels and gender on conditioned pain modulation. *Journal of Sports Sciences, 34*(5), 467–476.

van Middelkoop, M., Rubinstein, S. M., Verhagen, A. P., Ostelo, R. W., Koes, B. W., & van Tulder, M. W. (2010). Exercise therapy for chronic nonspecific low-back pain. *Best Practice & Research Clinical Rheumatology, 24*(2), 193–204.

Van Oosterwijck, J., Nijs, J., Meeus, M., Lefever, I., Huybrechts, L., Lambrecht, L., & Paul, L. (2010). Pain inhibition and postexertional malaise in myalgic encephalomyelitis/chronic fatigue syndrome: An experimental study. *Journal of Internal Medicine, 268*(3), 265–278.

Van Oosterwijck, J., Nijs, J., Meeus, M., Van Loo, M., & Paul, L. (2012). Lack of endogenous pain inhibition during exercise in people with chronic whiplash associated disorders: An experimental study. *The Journal of Pain, 13*(3), 242–254.

Woolf, C. J., & Mannion, R. J. (1999). Neuropathic pain: Aetiology, symptoms, mechanisms, and management. *The Lancet, 353*(9168), 1959–1964.

Pain, Fear, Anxiety, and Stress: Relations to the Endogenous Opioid System

Nicholas J. Felicione, Melissa D. Blank, Casey D. Wright, and Daniel W. McNeil

Abstract Pain, fear, stress, and anxiety are separate yet interrelated phenomena. Each of these concepts has an extensive individual body of research, with some more recent work focusing on points of conceptual overlap. The role of the endogenous opioid system in each of these phenomena is only beginning to be examined and understood. Research examining the ways in which endogenous opioids (e.g., beta-endorphin; βE) may mediate the relations among pain, fear, stress, and anxiety is even more nascent. This chapter explores the extant evidence for endogenous opioid activity as an underpinning mechanism of these related constructs, with an emphasis on research examining βE.

Keywords Pain · Endorphins · Endogenous opioids · Anxiety

Pain, fear, anxiety, and stress all are ubiquitous experiences for higher-order animals, including humans. While typically considered negative affective and sensory states, they add a rich variegation to emotional life and are evolutionarily adaptive. Nevertheless, at intense levels acutely and chronically, these states are highly distressing, lead to avoidance behavior, and impair quality of life. Underlying biological substrates that might explain some variation in affective and sensory experiences involve endorphins, a class of endogenous opioid peptides. Endorphins include alpha (α), beta (β), and gamma (γ) subtypes, though most attention is paid to beta-endorphin (βE).

N. J. Felicione
University at Buffalo, Buffalo, NY, USA

M. D. Blank
West Virginia University, Morgantown, WV, USA

C. D. Wright
Marquette University, Milwaukee, WI, USA

D. W. McNeil (✉)
University of Florida, Gainesville, FL, USA
e-mail: dmcneil1@ufl.edu

P. L. Kerr et al. (eds.), *Endogenous Opioids*, Advances in Neurobiology 35,
https://doi.org/10.1007/978-3-031-45493-6_9

βE is synthesized primarily in the anterior pituitary gland (Guillemin et al., 1977). Specifically, the precursor protein proopiomelanocortin (POMC) is cleaved enzymatically to produce several peptides including βE. In the peripheral nervous system, βE binds to mu-opioid receptors (μORs) and inhibits Substance P, a protein that is released to signal pain (Steinhoff et al., 2014). In the central nervous system, βE acts as an agonist at μORs and acts through the periaqueductal gray. When released from this area, βE increases the levels of dopamine through the inhibition of *gamma*-aminobutyric acid. Interestingly, exogenous βE may be up to 33 times more potent than morphine in terms of producing analgesia in several preclinical models (Loh et al., 1976), which speaks to its importance biologically and its potential impact on behavior.

The relation between βE and pain is complex, in part because it is likely influenced by the close emotional relatives of pain – fear, anxiety, and stress. In fact, others have conceptualized the relation of opioid peptides and negative affect as incomplete without being described as "complex, contradictory, and inconsistent" (Drolet et al., 2001, p. 736; originally cited in Hayden-Hixon & Nemeroff, 1993). Thus, this chapter overviews what is known about βE and each of these psychological processes from an integrative viewpoint. Such an ecumenical, inclusive perspective may help further research in understanding the nature of these (typically) negative affective states and how βE may interact or be implicated. We begin by giving a brief overview on the literature of the psychological conceptualizations and definitions of pain, fear, anxiety, and stress and how they might be involved in βE processes directly. Then, we address these negative affect constructs from a biological perspective of the βE literature. Finally, we offer an integrative model followed by a discussion about future research needs and directions. As will be evident in the discussion of integration, we propose that there is considerable overlap among these states and common pathways of biological and behavioral response that may similarly involve βE.

Psychological Concepts of Pain, Fear, Anxiety, and Stress

In everyday life, and in many clinical settings, individuals often use similar language for different psychological experiences (e.g., *anxiety* and *fear*). Even in scientific inquiry, such language and operationalized definitions also are often used interchangeably. For obvious reasons, however, clinicians and researchers alike have attempted to conceptualize and operationalize a variety of negative affect experiences (e.g., pain, anxiety, fear, and stress) as independent constructs. Yet there remains a great deal of inconsistency across definitions and measurement approaches in the literature. Here we propose that there is considerable overlap among these states for both substrate biological processes and psychological response, which should be considered in the language chosen to describe them.

For the states of pain and stress, for example, there often is conceptual and definitional confusion between stimulus and response. That is, sometimes pain (or stress) can be considered a stimulus in that it elicits a response. Other times, pain (or stress) can be the response elicited by another stimulus. Even still, nociceptive stimulation (i.e., originating from stimuli which are noxious/painful) can lead to the experience of "pain" that is expressed in various modalities, such as verbal complaints of distress. Similarly, there is a distinction between "stressors" and "stress." The former term reflects the combination of environmental demands, problems, hassles, and aggravations that create negative, sometimes overwhelming, environmental circumstances. Stress itself, like pain, is the experience of the organism that results from the presence of these stressors, and stress, like pain, is expressed across modalities. This confusion necessitates a discussion about potential overlap or a reconceptualization of these terms to be more integrative. First, though, it is important to understand what is understood about each of the terms presently before moving on to a discussion of the relation among these concepts and also their relation to βE.

Pain The most current definition of pain comes from the International Association for the Study of Pain (Raja et al., 2020), which is "an unpleasant sensory and emotional experience associated with, or resembling that associated with, actual or potential tissue damage." There are six accompanying key notes that expand upon the definition along with a description of the etymology of *pain* as a word. In acute forms, pain is an adaptive warning signal (Grichnik & Ferrante, 1991). It is an evolutionarily selected response that is ubiquitous in animals and is a universal part of the human experience.

Importantly, pain as an experience is known largely by the expression of the organism, although external observers may assess the intensity of the noxious stimuli in assuming the presence of pain in others. Pain is a psychological, neurological, neuroendocrine, and physical state and/or response. Pain may be differentiated from noxious stimuli, environmental events, or internal processes that lead to a nociceptive response on the part of the organism. Pain exists along a continuum of intensity, ranging from mildly annoying to deadly.

Pain is a construct (Cleeland, 1986) that is both sensory and psychological (e.g., emotional) and is a response that is manifested across three response systems (Lang, 1968): (a) overt behavior, which can be motoric (e.g., facial grimacing) or reflexive (e.g., withdrawal of one's hand from a hot pan); (b) physiology (e.g., activation of certain brain regions, how one's body responds, such as via cardiovascular activity); and (c) verbal report (i.e., what one says about an event [e.g., "Having an injection in my lower jaw was terrible"] or how one describes one's internal state [e.g., "This headache is awful"]) (Lang, 1968). Cognitive processes also are involved in pain, but as "private events" (Skinner, 1969), typically can be measured only indirectly through one of the aforementioned responses (e.g., self-report, reaction time in cognitive processing tests), although neuroimaging may provide information as well.

Pain has been classified by the American College of Emergency Physicians (2009) in a variety of ways including acute, recurrent acute (or intermittent), chronic, and cancer pain, but the acute-chronic distinction is the most common. To be considered chronic, pain must have been experienced for at least 6 months, although more recently, a commonly adopted criterion has been 3 months (International Association for the Study of Pain, 1986; Merskey & Bogduk, 1994; Treede et al., 2015). There are numerous types of chronic pain, such as chronic primary, cancer, postsurgical and posttraumatic, neuropathic, headache, orofacial, visceral, and musculoskeletal ones (see Treede et al., 2015). Conversely, acute pain is less than 3–6 months in duration and is usually a result of immediate injury or disease (Grichnik & Ferrante, 1991). An important trigger for several physiological reactions (e.g., reflexes in the muscular, skeletal, and digestive systems), acute pain helps organisms avoid or escape from situations that are harmful (Grichnik & Ferrante, 1991). Of course, the body responses to both chronic and acute pain states, for instance, by having βE, serve an analgesic function.

Fear A psychological state, fear is most properly considered a response that is manifested in three systems (Lang, 1968), akin to pain, including motoric behaviors (e.g., retreating from a coiled, hissing snake), physiological activation (e.g., an increase in heart rate and blood pressure to being in an automobile accident), and verbal reports (e.g., what one says about an event or how one describes one's internal state, such as "I'm so scared"). Again, as in pain, cognitive processes also are involved in fear; perhaps like βE and other biological influences, cognitions may underlie the three directly measurable systems. Fear is considered to be relatively close in temporal proximity to a precipitating event (Craske, 2003) and typically is associated with a robust physiological response; upon sitting down in a dental chair to receive treatment, those who are fearful of dental care may experience an abrupt increase in blood pressure, heart rate, and/or cortisol.

Fear is a normal, natural emotion that is evolutionarily adaptive. It prompts us to be careful in certain situations (e.g., driving on an icy road in the winter) but, in controlled situations, can be stimulating for certain people in a very positive way (e.g., riding on a rollercoaster or going through a haunted house at Halloween). On the other hand, fear can be extremely distressing, chronic, and impairing of life and health functioning when it is out of proportion to evocative stimuli. Fear is often a result of conditioning, as organisms learn to fear stimuli that are associated with or may predict a painful stimulus. Fear typically is considered an acute response and, in more chronic forms, may be more closely akin to anxiety.

The relation between fear and βE is complex. Multiple theories regarding the relation of βE and fear tie into learning, whether it relates to classical conditioning or an evolutionary perspective. Generally, available research suggests that fearful stimuli may lead to an endorphinergic response, which leads to adaptive analgesia. Some studies, however, contrast these theories, and inter-study differences and unknown mechanisms make the relation between fear and βE more nuanced than it may initially seem.

Anxiety Topographically similar to fear, anxiety is expressed across three response systems as well (i.e., motoric behavior, physiology, and verbal reports), but is less physiological and more cognitive (i.e., worry) in nature (McNeil et al., 2017), meaning the measurement (at least in humans) is often through self-report. Given its more cognitive nature and reliance on self-report, anxiety becomes difficult to measure particularly in animals. Also in contrast to fear, it is more temporally distal from evocative environmental events (Craske, 2003) and more chronic in nature. For instance, one might be anxious about dental treatment generally, manifested when one is reminded about the need for regular professional dental care, and one thinks about the possible pain associated with prior dental appointments (e.g., friends and family members report attending dental visits) (see McNeil & Randall, 2014).

Similar to fear, the relation between anxiety and βE is complex and is more poorly characterized than the relationship between fear and βE. While it appears that anxiety can cause an endorphinergic response, the mechanisms are poorly understood. Much of the literature is focused on βE and pain, which often has fear-, anxiety-, and stress-inducing qualities itself. The relation between βE and anxiety needs to be more thoroughly researched, with a particular focus on measuring the βE response and ruling out related and perhaps confounding states such as fear.

Stress The concept of stress affecting humans and other animals emanated from the field of physics but was brought to popular attention by Hans Selye (1956) who defined it as "the non-specific response of the body to any demand." Interestingly, his first definition included the word *endocrine* (Szabo et al., 2012). A more contemporary definition of stress is "a negative emotional experience accompanied by predictable biochemical, physiological, cognitive, and behavioral changes that are directed either toward altering the stressful event or accommodating to its effects" (Taylor, 2018, p. 115). In essence, when an organism experiences a disruption in homeostasis, it experiences stress. The organism will either return to homeostasis or adapt and establish a new "normal." Stress, as an organismic response, must be differentiated from stressors, which are the environmental contexts, demands, and even the appraisal of stimuli as threatening (Lovallo, 2010; Taylor, 2018) that lead to the stress response. In some cases, as will be described later, researchers have used the term "stress" to describe the outcomes of pain-, fear-, and anxiety-provoking tasks, resulting in stress often being treated as an "umbrella term" of sorts. Various response, stimulus, and interactional/transactional models and theories of stress have been proposed, along with more comprehensive models that are inclusive of the role of individual differences in stress responses (Larkin, 2005).

The concept of stress has been conceptualized in the literature both as a stimulus (i.e., a stressor) and as a response to some constellation of environmental events (e.g., child care responsibilities along with demands at work) and/or internal states (e.g., illness, pain, anxiety, fear). That is, sometimes stress is spoken of in terms of what *leads* to a response, and other times it's spoken of in terms of *as being* the response. As a stimulus, stress has been defined as "… a physical or psychological *stimulus* that can produce mental tension or physiological reaction to produce illness" (Bali et al., 2015, p. 139, emphasis added). This definition arguably includes any event, task, activity, or phenomenon with the potential of producing a disruption in homeostasis, which can be manifested cognitively (e.g., disrupted or negative thinking patterns manifested via self-report), physiologically (e.g., increased cortisol production), or behaviorally (e.g., spending more time on a task with a closely approaching deadline, leaving other important tasks unaddressed). Indeed, sometimes researchers are specifically interested in what triggers a stress response, and sometimes they are interested in the response itself. More often, however, in the stress and βE literature, most work is done examining stress as a response, though the language may use the term stress to describe both the stimulus and/or the response.

In terms of the conceptualization of stress as a response, it has been defined as a "physical, emotional, cognitive, and behavioral *response* to events that are appraised as threatening or challenging" (Cartwright & Cooper, 1997, as cited in Bali et al., 2015, p. 139). A key here is the appraisal by the individual once something is deemed stressful, which then leads to a response. Importantly, part of this response may include a physiological reaction such as changes in βE levels.

Stress, like pain, fear, and anxiety, is positive in some ways. In fact, stress has more forthrightly been conceptualized as being either positive ("eustress") or negative ("distress") (Selye, 1974). Life events (Holmes & Rahe, 1967) are associated with stress, regardless of whether they are positive (e.g., parents' response to the birth of a healthy baby) or negative (e.g., financial shortfall in one's household), testimony to the fact that stressors can be positive or negative in valence.

Important distinctions have been made between acute and chronic stress. Broadly speaking, acute stress tends to be adaptive in helping deal with a threat, harm, or challenge (Taylor, 2018) through appropriate physiological and psychological arousal. Historically, this process has been explained through the "fight or flight" and the "freeze" response, also called tonic immobility (Gallup, 1977). When a stressor is appraised as being a threat, harmful, or challenging, the organism will work to reestablish homeostasis, as previously described. If successful (i.e., an acute stress response is alleviated), physiological resources are replenished. Once a stressor becomes more chronic (i.e., the organism has difficulty returning to a state of homeostasis), deleterious outcomes also become more common.

Stress is most often measured in the realm of verbal report and, in terms of biological markers, from the endocrine system (e.g., cortisol). The presence of stress in humans also is assumed from illnesses and negative life events, although the interpretation of these events has been found to be quite important (Taylor, 2018). Even just appraising events as harmful or threatening can create stress responses (Lovallo, 2010; Taylor, 2018).

Stress has been shown to affect βE (Cohen et al., 1982; Govoni et al., 1984) and the immune system (Hargreaves, 1984; Hartwig, 1991; Sprouse-Blum et al., 2010). The relation between βE and stress will be further addressed later, but suffice it to state here that stress as a broadly conceptualized state has been shown to increase and otherwise alter βE levels.

Conclusion The psychological conceptualizations and definitions of pain, fear, anxiety, and stress share many similarities; at a lower level of intensity, they can be important and functional warning signals or prompts for humans and other animals. Also similar are the methods of measurement for these concepts, particularly verbal report, but also including physiological response and overt behavior (Gross & Collins, 1981). As is covered in the next section, βE also is involved, in rather similar ways across these four states.

Beta-Endorphin and Relation to Pain, Fear, Anxiety, and Stress

The βE literature is extensive, and among the four related constructs of interest here, pain and stress are those most frequently studied. We will review selected literature on βE and the four states, with the caveat that there is an astounding lack of consistency in the use of terms, with frequent overlapping of concepts, and even misuse of labels. As a related issue, the measurement approaches and selected variables differ widely, so drawing conclusions often is difficult. In a later section, we attempt to begin a reconciliation of findings within an integrated model.

Pain Generally speaking, endogenous βE increases from resting level in response to painful stimuli such as that associated with surgery (Cork et al., 1985; Cohen et al., 1982; Dubois et al., 1982, Hargreaves et al., 1983; Matejec et al., 2003), in the later stages of pregnancy (Genazzani et al., 1981), and during labor (Facchinnetti et al., 1982; Thomas et al., 1982; see also Khajehei, chapter "Endorphins, Sexuality, and Reproduction", this volume). βE levels also increase in response to pain that is induced experimentally. For instance, both humans (al'Absi et al., 2004; Bruehl et al., 2007) and rodents (Cepeda et al., 1993; Rasmussen & Farr, 2003, 2009; Rossier et al., 1977; Zangen et al., 1998) demonstrate significant increases in βE after exposure to a painful thermal (e.g., heat and cold), chemical (e.g., formalin injection), or electrical (e.g., foot shock) stimulus. In each of these cases, βE is presumably serving an analgesic function.

Increases in βE can be altered by administration of an opioid agonist (often dispensed for pain alleviation) or antagonist, which target the same μOR as βE. Relative to placebo or other controls, fentanyl attenuates βE elevations observed in patients undergoing cardiac (Cork et al., 1985; see also Sirbu (chapter "The Role of Endogenous Opioids in Cardioprotection", this volume) for a more extensive review of the cardioprotective effects of endogenous opioid activity), laparotomic (DuBois et al., 1982), or oral surgery (Hargreaves, 1984; Hargreaves et al., 1986). The βE elevations observed during surgery can be increased by the opioid antagonist, naloxone. For instance, βE levels in patients having impacted molars extracted were compared as a function of administration of placebo, fentanyl, or naloxone (Hargreaves et al., 1986). βE levels were higher in patients who received naloxone than those who received placebo, which were higher than those who received fentanyl. Naloxone shows a similar pattern of results when administered in response to an experimentally based noxious stimulus. Using a cold pressor task (e.g., hand submerged in ice water), βE levels were significantly higher in human participants for naloxone relative to placebo (al'Absi et al., 2004).

In support of the idea that βE increases in response to painful stimuli, extant work shows that individuals with higher resting βE levels have greater analgesia. In patients undergoing prostate surgery, for example, those with higher resting βE levels self-reported lower levels of pain, experienced analgesia for a longer period of time, and used less opioid medication post-surgery than those patients with lower βE levels (Nader-Djalal et al., 1995). Similarly, higher resting levels of βE have been negatively associated with self-reported ratings of pain in adult patients undergoing oral surgery (Hargreaves et al., 1983) or pediatric patients undergoing dressing changes for burns (Szyfelbein et al., 1985). Additionally, following a transcranial direct current stimulation treatment for fibromyalgia, serum βE levels were positively correlated with pain threshold and negatively correlated with pain ratings (Khedr et al., 2017). In other work, levels of resting βE were negatively correlated with morphine (Cohen et al., 1982; Pickar et al., 1983) or pethidine (Tamsen et al., 1982) requirements post-surgery. Moreover, women in labor with higher βE levels have required less pain medication (Dabo et al., 2010), consistent with findings described by Khajehei (chapter "Endorphins, Sexuality, and Reproduction", this volume). Using an experimental model, levels of resting βE correlated positively with pain threshold evaluated during a pulpar test (i.e., electrical stimulation of healthy teeth) in otherwise healthy individuals (Guasti et al., 1996) and negatively correlated with pain ratings in response to cold pressor or heat thermode tests (al'Absi et al., 2004).

Nevertheless, the opposite pattern also has been observed, with higher resting βE levels associated with a weaker analgesic response (e.g., increased pain ratings, lower pain thresholds, higher medication needs, and/or lower medication response). In patients undergoing orthopedic surgery, postoperative pain ratings were correlated positively with resting βE levels (Matejec et al., 2003). This same finding has been shown with patients following spinal fusion surgery (Leonard et al., 1993), as well as among participants following the administration of an ischemic pain task (Bruehl et al., 2012).

Still other work demonstrates the complexity of these effects, with response to painful stimuli and opioid analgesia dependent on the measurement method used and the chronic pain status of the participant (Bruehl et al., 2017). Indeed, resting βE levels and their relation to pain response may be different in healthy individuals relative to those with a chronic pain condition. Relative to healthy controls, resting βE levels have been observed to be higher among individuals with temporomandibular disorder (Feldreich et al., 2012), but lower among individuals with chronic back pain (Rhodin et al., 2013), neuropathic pain (CSF; Bäckryd et al., 2014), migraine pain (Misra et al., 2017), or pain related to premenstrual dysphoric disorder (Straneva et al., 2002). In other work, however, resting βE levels between those in chronic pain (Dercum's disease; fibromyalgia) and controls were not significantly different (Hansson et al., 2012; Vaeroy et al., 1988). Other biological factors, such as age, may also impact βE levels and response. Baseline levels of βE may be higher in older adults compared to younger adults, and older adults may show larger increases in βE following several validated laboratory pain models (Riley et al., 2017). Cross-study differences also may be due to the method of measurement of βE derived from the peripheral versus the central nervous system, that is, whether βE is measured from serum or cerebrospinal fluid, respectively. As stated by Bruehl et al. (2012; p. 377): "This would not be surprising given their different sources: βE in circulation derives largely from the pituitary (Solomon, 1999), whereas βE in the CNS [central nervous system] derives from the hypothalamus, periaqueductal gray and other brain regions (Basbaum & Fields, 1984; Pilcher et al., 1988; Zubieta et al., 2001; Sprenger et al., 2006)." It has been proposed that βE derived from the PNS is a result of a stress rather than pain response, as outlined in more detail in the relevant subsection on stress.

Fear To our knowledge, no work exists that has examined directly the influence of fear on βE response and vice versa. That is, endogenous βE is not measured in this work. Rather, published studies have relied on a combination of measures of pain and administration of drugs that affect the opioid system to speculate about the relation between fear and βE. This same work describes these relations within the context of various theories, in particular those based on behavioral learning theory. We subsequently describe some of those approaches in detail and how they may explain associations between fear and βE.

One conceptualization is the perceptual-defensive-recuperative model (Bolles & Fanselow, 1980; Fanselow, 1986). In this evolutionary model, innate and learned stimuli that signal danger can lead to mobilization of an organism's defensive system (perceptual phase). Activating the defensive system typically leads to species-specific defensive reactions, such as freezing by a rat after seeing a predator (defensive phase). Separate but related to this model are the consequences of painful stimuli. Typically, nociceptive (i.e., painful) stimuli elicit reflexive behaviors, such as a withdrawal response. Stronger nociceptive stimuli that cause more serious damage may lead to complex recuperative behaviors directed toward the injury or to convalescence (recuperative phase). For example, injecting a rat's paw with formalin may lead initially to withdrawal of that paw (reflexive behavior) and later to paw licking (recuperative behavior). Ultimately, this model explains the adaptive and functional role of pain, with pain leading to behavior directed at recovering from a potential injury.

There may be situations, however, in which pain sensitivity must be reduced to increase the adaptability of an organism. Within the perceptual-recuperative-defensive model, for example, this situation involves the interaction of two independent systems: the nociceptive system and antipredator defensive system. Specifically, an initially neutral stimulus that reliably predicts a nociceptive stimulus will acquire the ability to activate the defensive system. Consequently, that neutral stimulus will take on the same properties as the nociceptive stimulus, thus becoming a conditioned stimulus and producing the same defensive response (i.e., the conditioned response, as posited by Pavlovian conditioning). Notably, activation of the defensive system may include activation of endogenous analgesic mechanisms (e.g., βE) to reduce the pain that the nociceptive stimulus produces in the organism.

To demonstrate the utility of this interaction, imagine an animal that has been bitten on the leg by a predator. Without endogenous-based analgesia, that animal would spend time tending to the wound, or attempting to escape while limping, neither of which are evolutionarily advantageous. On the other hand, if the endogenous analgesia system is able to attenuate the pain caused by that bite, the animal will have the greatest chance of survival by escaping, before attending to the wound. Thus, the leg bite situation provides an example of an endorphinergic response to pain. However, a similar response can be triggered in response to cues associated with that pain, which better resembles fear. More specifically, by learning to associate neutral stimuli with nociceptive stimuli, the animal can respond directly to the neutral stimuli before engagement with the nociceptive stimuli. For instance, the animal can escape an area upon seeing the stripes of a tiger or hearing the roar of a lion and thus before the tiger or lion actually approach that animal. This situation would also involve activation of the endogenous analgesia system to reduce pain sensitivity, in the event that the tiger or lion actually attacks the animal. This example may be akin to fear in humans, in which a stimulus indicating danger induces fear, thus leading the individual to avoid that stimulus altogether and for biological mechanisms (e.g., βE release) to be activated to prepare for the potential danger.

To test this theory, Fanselow (1986) focused on the consequences of the nociceptive system such as unconditioned reflexive behavior, conditioning of acquired danger stimuli, and recuperative behavior. Endogenous analgesia (e.g., βE release) should be able to suppress all of these reactions, by reducing the level of pain experienced by the organism. Fanselow (1986) presented a series of experiments to support the theory. One such experiment involved forward conditioning (i.e., the neutral stimulus is presented before the painful stimulus, such as tone followed by shock) and backward conditioning (i.e., the painful stimulus is presented before the neutral stimulus, such as shock followed by tone; Fanselow & Bolles, 1979). The forward conditioning condition resembles fear, as the animal may react to a stimulus that is associated with a later painful stimulus. Forward conditioning resulted in lower levels of the reflexive behavior (i.e., freezing response) than backward conditioning, perhaps due to activation of endogenous analgesic mechanisms in response to the neutral stimulus (i.e., tone). Notably, this effect was observed after multiple presentations of the fear-inducing stimulus, but not after a single instance, demonstrating the importance of the learned association between stimuli. It was theorized, therefore, that the release of endogenous opioids in response to the initial fearful stimulus would produce analgesia, thus reducing the aversiveness of subsequent presentations of that same stimulus. Moreover, blocking this analgesic effect would lead to more freezing, as the shock would be more painful or aversive. Indeed, this hypothesized result was observed when naloxone (opioid antagonist) was administrated to those rats that experienced forward conditioning, suggesting that the endogenous analgesia is a result of endogenous endorphin release, such as βE (Fanselow & Bolles, 1979). Additional studies involving human participants are consistent with the perceptual-defensive-recuperative model, demonstrating that fear can reduce pain sensitivity and induce analgesia to a painful stimulus (Rhudy et al., 2004; Rhudy & Meagher, 2000).

Some studies, however, show an opposite effect, in which fear can increase pain sensitivity (Meagher et al., 2001; Williams & Rhudy, 2007). For example, Williams and Rhudy (2007) exposed participants to happy and fearful facial expressions. Half of the participants received an aversive electrical stimulation paired with the fearful face and half with the happy face. To test pain sensitivity, the researchers assessed latency to finger withdrawal from a heat device, along with skin conductance response and self-reported emotion, when shown the facial expression without the electrical stimulation. Ultimately, pain threshold was influenced by fearful faces when they had been paired with the aversive stimulation. Specifically, these participants had a reduced pain threshold or increased pain sensitivity. In this study, it seemed that fear-relevant stimuli led to negative emotions and enhanced pain. A similar study demonstrated consistent conclusions, with fear cues reducing pain tolerance (Meagher et al., 2001). Together, these studies provide evidence for a motivational priming model, in which negative affective states can enhance pain and positive affective states can attenuate pain (Meagher et al., 2001; Williams & Rhudy, 2007).

Another approach posited to explain the relation between endogenous opioids like βE and fear is that of the Rescorla and Wagner (1972) model, which is also based on Pavlovian conditioning. At the start of fear conditioning, the painful stimulus (e.g., shock) is not predicted. That is, the neutral stimulus (e.g., tone) has not yet been associated with the painful stimulus and thus cannot predict impending danger (thus, no fear). As the number of pairings between the neutral and painful stimuli increase, however, endogenous opioids are more likely to be released in response to the neutral stimulus and before the painful stimulus is presented. Then, once the association is formed (i.e., fear learning has occurred), the neutral stimulus will be able to produce an endogenous opioid response regardless of whether the painful stimulus also is presented. In this situation, that neutral stimulus will have acquired fearful properties, as it is used by the organism to predict a later painful stimulus.

In other work based on behavioral learning theory (Eippert et al., 2008), the influence of the opioid antagonist naloxone on fear conditioning was examined in human participants. Simple geometric forms were used as neutral/conditioned stimuli, while the painful stimulus (i.e., unconditioned stimulus) was thermal stimulation of the forearm. Prior to presentations of these stimuli, participants were administered naloxone or a saline control. In this scenario, the pairing of the neutral and painful stimuli would lead to a reduction in pain sensitivity as a result of conditioning, likely due to the release of endogenous opioids like βE. Relative to those who received saline, naloxone administration resulted in stronger responses to the unconditioned painful stimulus in brain areas implicated in pain processing (e.g., the insula, dorsolateral prefrontal cortex, and basal ganglia), suggesting that naloxone administration led to increased pain sensitivity (i.e., hyperalgesia). The naloxone group also showed a stronger and more sustained conditioning effect than the saline group. Additionally, the saline group showed stronger habituation in pain ratings and skin conductance response to the unconditioned stimulus than the naloxone group. Ultimately, the results suggest that blocking endogenous opioids via naloxone inhibited the development of reduced pain sensitivity through conditioning and led to stronger responses to pain. Specifically, naloxone blocked processing in brain areas associated with anticipatory pain. The authors concluded that endogenous opioids have an inhibitory role in acquisition of conditioned fear in humans, which is consistent with rodent models assessing a similar phenomenon (Fanselow & Baackes, 1982; Helmsetter & Fanselow, 1987).

In conclusion, while a number of studies have assessed the influence of conditioned fear on human pain thresholds, they do not measure βE directly, disallowing any definitive conclusions here (Cole & McNally, 2007; Meagher et al., 2001; Rhudy et al., 2004; Rhudy & Meagher, 2000; Williams & Rhudy, 2007). A common paradigm used to study these relationships involves fear conditioning, in which a neutral stimulus precedes a painful stimulus and the neutral stimulus becomes a conditioned stimulus that signals upcoming pain. Some of these studies conclude that fear reduces pain sensitivity and can lead to analgesia to a later painful stimulus, consistent with the perceptual-defensive-recuperative model (Rhudy et al., 2004; Rhudy & Meagher, 2000). However, the opposite pattern has also been found, with fear causing a negative emotional state that leads to increased pain sensitivity (Meagher et al., 2001; Williams & Rhudy, 2007). While inconsistent findings may be related to inter-study differences involving varying intensities of stimuli (Vowles et al., 2006), it is difficult to understand the direct relationship between fear and pain reactivity. To add another layer of complexity, there are multiple possible mechanisms that may explain this relation, with some explanations involving factors other than βE alone (e.g., affective states, attributional factors, attention; Rhudy & Meagher, 2000; Meagher et al., 2001) . Thus, while the Bolles and Fanselow (1980) perceptual-defensive-recuperative model is a sensible evolutionary theory based in Pavlovian conditioning with data to support, it is clear that the relations are not as simple as initially described. Direct measurement of endogenous opioid activity may allow for more consistent findings; and, as noted by Kerr, Sirbu, and Gregg in the concluding chapter of this volume ("Emerging Horizons"), indirect measurement of endogenous opioids remains a limitation to be overcome in future research.

Anxiety An understanding of the relationship between anxiety and βE is limited by several factors. First, few studies exist. Second, work that does exist rarely measures βE directly. Finally, it is difficult to disentangle anxiety from the aforementioned concepts of fear and pain (Gross & Collins, 1981).

In the Hargreaves (1984) study already mentioned in the section on pain, patients undergoing oral surgery were administered diazepam (an anxiolytic), fentanyl (an opioid analgesic), or placebo. In addition to βE, patients' ratings of pain and anxiety were measured before, during, and after surgery. While βE levels were increased for those administered placebo, diazepam and fentanyl independently blocked this effect. Ratings of anxiety showed a similar pattern of results, but ratings of pain did not differ between groups. These results suggest that anxiety can be an independent stimulus from pain that can cause an endorphinergic response in humans. However, it is possible that anxiety and stress are entangled here, as the author refers to the paradigm as surgical stress.

Other work that has measured the influence of anxiety on βE has done so using models of phobia, focusing on this diagnosable anxiety disorder characterized by irrational fear and anxiety to certain stimuli or situations. For instance, in a single subject experiment, an individual with fear of snakes had changes in βE level that corresponded with presentation and removal of a snake stimulus (Thyer & Matthews, 1986). More specifically, βE concentrations peaked 8 min after the initiation of stimulus exposure and returned to normal levels once exposure had ended. Heart rate and ratings of anxiety followed a similar trend, peaking early during exposure and then declining. The authors hypothesized that βE release may be a potential mechanism of the therapeutic benefits of prolonged exposure therapy to treat anxiety. To place this finding in context of learning models, it is possible that these individuals may have anxiety due to fear of pain from snakes, and the reduced pain sensitivity associated with βE release can reduce anxiety. However, this is just one proposed explanation of this finding, as it is difficult to disentangle anxiety and fear.

Although not measuring βE directly, in other work by Janssen and Arntz (1996), individuals with an intense fear of spiders received mild painful shocks and were or were not treated with the opioid antagonist naltrexone. Additionally, the researchers manipulated anxiety levels (high versus low) and attention (focused on versus distracted from pain). For example, in the high-anxiety/attention condition, participants were exposed to a spider in a jar and asked to touch it with a pencil; they were also asked to concentrate on the painful stimuli when the shock occurred. In the low-anxiety/distraction condition, participants were asked to concentrate on a video and to attempt to ignore painful stimuli. Self-reported pain and physiological pain responses (measured via skin conductance response) were lower in the high-anxiety condition, suggesting that experiencing anxiety led to analgesia. This effect, however, was not maintained when controlling for attentional influences, suggesting that there may be mediating factors in the relation between anxiety and pain response. This effect of anxiety leading to pain reduction also was not reversed by naltrexone, suggesting that attention was a more important factor in analgesia than endorphin release. However, independent of anxiety condition (low versus high), a low dose of naltrexone increased pain, suggesting that there may still be an endorphinergic response to the pain. At the very least, this study suggests that any relations that may exist among anxiety, pain, and βE are complex, involve mediating factors, and are not fully understood. This study also may suggest the importance of attention to stimuli in the formation of an endorphinergic response to anxiety elicited by possible pain. As mentioned earlier, it is important to note that Thyer and Matthews (1986) and Janssen and Arntz (1996) both focused on individuals with a phobia who likely experience both anxiety and fear, which is a complication when trying to determine the impact of each of these states.

At least one study has attempted to assess the effect of a fearful stimulus separate from that of an anxiogenic stimulus (Rhudy & Meagher, 2000), although again not involving βE measurement. Pain response was measured via a cold pressor task in one of the three conditions: fear (exposure to brief shocks), anxiety (threat of shock), and neutral (no exposure or threat). Consistent with previous literature (Johnson & Helmstetter, 1994; Willer et al., 1981), the fear stimulus reduced pain reactivity. This finding might be explained by the perceptual-defensive-recuperative model (Bolles & Fanselow, 1980) in that the fearful stimulus may have led to endogenous opioid release, which would thereby decrease pain sensitivity, so the organism can respond to the situation rather than tend to pain. Also as expected (Chapman & Feather, 1983; Cornwall & Donderi, 1988; Dellemijn & Fields, 1994), the anxiogenic stimulus increased pain reactivity. This finding also may be explained by the perceptual-defensive-recuperative model (Bolles & Fanselow, 1980) in that anxiety would not be immediately competitive with pain for expression and, as a coexisting negative state, may amplify or be synergistic with the pain response. This finding of anxiety increasing pain sensitivity is consistent with the motivational priming theory discussed in the fear section, in which a negative affective state can lead to greater pain sensitivity.

In summary, while some studies suggest that anxiety reduces pain sensitivity (Hargreaves, 1984; Thyer & Matthews, 1986), others demonstrate the opposite pattern (Rhudy & Meagher, 2000; Cornwall & Donderi, 1988). As mentioned previously, different results are possibly a result of inter-study differences, the entanglement of pain and anxiety, or other mediating mechanisms in the relation (e.g., affect, attention). Of course, without measurement of βE levels in most studies, no definitive conclusions can be made regarding the role of the endorphin system in these responses.

Stress During a stress response, the hypothalamus stimulates the production of corticotrophin-releasing hormone (CRH) which in turn stimulates the production of POMC. As aforementioned, POMC is cleaved to produce peptides such as βE and also adrenocorticotropic hormone (ACTH). ACTH then activates the adrenal gland to initiate the stress response within the peripheral nervous system (Fink, 2017). βE, however, may inhibit subsequent CRH release, thereby attenuating the stress response (Calogero et al., 1988).

For example, in one study with patients undergoing surgery, βE levels prior to surgery were higher than those found one day following surgery, which the authors suggested was due to pre-surgical stress and may have been in preparation for or an innate effort to ameliorate the potential response (i.e., in this case tissue damage) to the stressor (i.e., the surgery itself) (Matejec et al., 2003). In a similar fashion, others have demonstrated how βE levels double in new parachute jumpers after jumping out of an airplane, in what seems to be a preparatory response to the possibility of pain or a dysregulation in homeostasis (Schedlowski et al., 1995). Interestingly, individuals with lower fatalistic attitude scores (e.g., more optimistic) did not experience as large of an increase in βE compared to those with higher fatalistic attitude scores, who had a more pronounced boost in βE after jumping (Schedlowski et al., 1995). This finding speaks to the importance of appraisal and other cognitive processes in the relation between stress and βE. Individuals also have shown elevated levels of βE in response to a public speaking task (Miller et al., 1993).

Whether stress is acute or chronic has relevance for βE responsivity. Depending on the region of the central nervous system studied, and the type of stress (i.e., acute versus chronic), the levels of βE are selectively increased (Amir et al., 1980). In one study with rats, with acute stress, greater levels of βE were produced by the anterior pituitary, whereas the opposite was true of chronically administered stress (Shiomi & Akil, 1982). In a different paradigm with rats, however, it was found that βE released from the pituitary, in general, involved a blunted response to acute stress and an enhanced response to chronic stress (Young & Akil, 1985). In another study with rats, βE levels in plasma in chronically stressed rats were not depleted relative to naïve rats, but neither were the levels increased (Przewlocki et al., 1987).

Stress as a response also is an established topic among researchers in psychoneuroimmunology, which has implications for endorphinergic processes. Especially when experienced chronically, the stress response has important implications for immune functioning (Coe, 2010). For example, it is well known that stress affects the HPA axis and, depending on the nature and chronicity of the stress, can lead to a compromised immune response, putting the organism in danger of disease or other malady (Segerstrom & Miller, 2004). Different explanations have been offered to explain how a stress response is associated with increased immune-related processes, which also may implicate βE. It has been posited that the stress response itself directly alters biological functioning (e.g., through sympathetic functioning or through the hypothalamic-pituitary-adrenal pathway, which consequently also is highly associated with endorphinergic processes; Mørch & Pedersen, 1995). For example, βE production has been shown to be a part of a physiological stress response (Cohen et al., 1982; Govoni et al., 1984). In fact, pain tasks have been shown to be a primary method for producing βE (Valentino & Bockstaele, 2008). Another explanation, which is not mutually exclusive, is that the demands of stress could alter individuals' lifestyles (e.g., diet, social functioning, exercise habit), which in turn alter biological functioning (Taylor, 2018). For example, perhaps a stressor (e.g., chronic pain) leads to a stress-induced behavioral response (e.g., less exercise) that then leads to a biological change (e.g., decreased βE), as physical exercise is known to increase the levels of βE (Thoren et al., 1990). However, recent research suggests that βE may not be responsible for exercise-induced euphoria and anxiolysis (Siebers et al., 2021). In this study, exercise was used to induce a "runner's high," but increased euphoria and reduced anxiety were still observed when participants were administered the opioid receptor antagonist naltrexone. The researchers concluded that endocannabinoids, and not endogenous opioids, were responsible for these exercise-induced changes.

In humans, βE is part of a defensive organismic response to higher-order physical and social actual and potential threats. Physiological pathways for the acute stress response typically lead to greater βE production. With chronic stress, there is less clarity about βE pathways, but it appears that the levels are tissue-specific and that there is βE responsivity to both acute and chronic stress.

Toward a Comprehensive and Integrated Model

Given some degree of commonality in the biological substrates for pain, fear, anxiety, and stress, it is not surprising that there is a lack of precision and in fact a morass of confusion in the scientific and lay literature regarding the distinctions and similarities in pain, fear, anxiety, and stress. At a phenomenological level of personally experienced internal conditions, humans perceive intense levels of all of these states as negative in valence and physiologically, cognitively, and behaviorally arousing. Consequently, humans may experience, describe, and respond physiologically, cognitively, and behaviorally to these states rather similarly. Moreover, the high intensity of any of these states may lead to a propensity to behavioral avoidance, if that is possible in the context in which the person finds himself or herself (e.g., avoidance or escape from a phobic situation versus experiencing acute internal pain of unknown origin). The use of words, numbers, and other symbols to describe these states likely is dependent upon the particular language being used by the respondent, in terms of its degree of variability and precision in terms that describe these states. Individual differences also are likely involved, as in the degree of emotional intelligence of the individual in recognizing and describing variations in his or her own internal states.

Also involved in the distinctions and similarities among these states is one's awareness of, and ability to accurately report on, one's private states (Leigland, 2014). As an example, a popular press comic strip depicts a dental patient who waits in the dental chair, and upon the dentist walking in, the patient exclaims, "Ouch!" (Shepherd, 1991). Could it be true that the dental patient in the comic strip is experiencing pain due to the mere presence of the dentist? A likely response to this inquiry is: "Of course it is not pain, the patient is afraid of the dentist." In reality, this person may not be experiencing pain in its most traditional sense, rather experiencing fear. Why is it that we sometimes use expressions of pain (e.g., "Ow!") when we feel fear?

Consider another example when an individual experiences the death of a loved one and they say: "My heart hurts." The literature on social exclusion and loss indicates that the same nervous system pathways are activated with external, physically induced pain as with loss of social contact and interaction (Eisenberger, 2011). In addition, it could be that losing a spouse results in an uncertainty about the future leading to a feeling of anxiety instead of pain. How can these various determinants be teased apart? Or is it necessary to do so? Perhaps should the focus be on the response of the organism and not on the stimuli or determinants?

In many instances, human beings likely experience two or more of these states simultaneously. A woman who is giving birth, for example, may experience acute pain with labor, fear about her ability to cope with possibly more intense pain later in the process, and anxiety about the health of the baby. Similarly, a person who is recovering from shoulder replacement surgery may experience acute postoperative pain and stress about being unable to go to work and temporary inability to drive a car or to type on a keyboard. The intermingling or comorbidity of these states is normal and natural, perhaps more typical than their not overlapping. Some theoretical development (e.g., Bolles & Fanselow, 1982) has been devoted to understanding how these states may interact with one another. The relation among these states has evolutionary significance, in that it may be more adaptive at times for fear to be the dominant response (as in threatening, crisis situations) or, on the other hand, for pain to be paramount, when convalescence and recovery are in the best interest (see Bolles & Fanselow, 1982). The duration and consistency of the environmental or physiological process that evokes a negative state or states are important in producing either an acute (e.g., transient or time-limited) response or a chronic one. It has been suggested that acute pain and fear may be more related, while chronic pain and anxiety may share more commonalities in response (Elman & Borsook, 2018). Trying to tease apart the relative contribution of these states is enormously complex. In fact, the interaction of the processes is varied and may be additive, competitive, or synergistic, or perhaps sometimes they even are unrelated to one another and act independently.

Clearly, these concepts are difficult to define as separate states and in fact likely are overlapping in terms of organismal responses and biological substrates. Nevertheless, it is necessary to understand both their similarities and their differences to more clearly elucidate the overall mechanisms involved in their expression, including the involvement of βE.

The ambiguity of the terms pain, fear, anxiety, and stress also is present in the extant literature on the endogenous opioid system, which leads to challenges in research. Bolles and Fanselow (1982) note that stress is a vague term and includes many stimuli ranging from electric shock (which could also be deemed pain) to food deprivation (which could be argued as anxiety-provoking). These definitions still leave room for confusion and overlap, as stimuli that induce pain, fear, or anxiety may all be categorized as stressors (Bolles & Fanselow, 1982). The nature of these stressors, however, may differ, along with the stress response of the organism.

Revisiting some of the literature reviewed here reveals numerous examples of confusion and well-intended but likely inaccurate assumptions on the part of researchers. These issues truly reveal problems with the literature, rather than the individual studies themselves.

Sometimes researchers use painful stimuli to invoke "stress" (Cohen et al., 1982). Additionally, researchers have used fear- or anxiety-provoking stimuli to invoke "stress" (Miller et al., 1993). Finally, these studies on "fear" actually may be studies of "anxiety." For example, in studies of individuals with phobias (Janssen & Arntz, 1996; Thyer & Matthews, 1986), it is difficult to determine to what degree the individuals are experiencing anxiety and to what degree they are experiencing fear. As a related issue, researchers often have used "stress" as an umbrella term to describe organisms' responses to stimuli and experiences that are or otherwise could be interpreted as painful or fearful.

A limited amount of research exists that attempts to tease apart different types of stressors or stress responses. One strategy to understand how anxiety and pain may relate to βE involves the administration of drugs. For example, Hargreaves (1984) assigned patients undergoing surgery for impacted molars to three different groups:

administration of fentanyl, diazepam, or placebo. This study design allows for separate analysis of endorphin levels through the different effects of these drugs, with fentanyl serving as a pain reliever through opiate agonism and diazepam serving as an anxiolytic through *gamma*-aminobutyric acid agonism. The levels of βE increased from pre- to intra-operation in the placebo group, but not in the fentanyl or diazepam groups. It is possible that the endorphin response in the placebo group was a function of surgical stress/anxiety and/or onset of acute pain. Thus, because of the ability of an anxiolytic or an analgesic to attenuate the βE response to surgery, the authors concluded that pain and anxiety constitute independent, and possible equipotent, stimuli for the release of beta-endorphin. Other studies have attempted to parse apart the influence of stress, fear, and anxiety on pain in animals (Lopez-Luna et al., 2017). Specifically, zebrafish were exposed to a stressor (air emersion), predatory fear cue (alarm substance), or anxiogenic (caffeine) before exposing them to a nociceptive stimulus (acetic acid) and tracking their swimming behavior. The painful stimulus alone reduced swimming behavior; however, exposure to the stressor and fear cue inhibited these changes, as evidenced by the lack of differences in swimming behavior pre- and post-nociceptive stimulation. Thus, the researchers concluded that stress or fear activated an antinociceptive mechanism in zebrafish, which inhibited the typical pain response. Despite not measuring a zebrafish analog of βE, one hypothesis may be that stress or fear in humans causes a release of βE, which may then attenuate pain. Other researchers may use different manipulations to tease apart concepts within stress, such as anxiety and fear. For example, fear may be induced through exposure to a painful stimulus (e.g., brief shocks), while anxiety may be induced through the threat of exposure to a painful stimulus (Rhudy & Meagher, 2000). Studies that attempt to independently manipulate different negative affective states, that are themselves stressful, may provide useful information about the similarities and differences experienced by individuals experiencing stress, anxiety, fear, and pain.

Given the very early stage of this literature, it may be helpful to advance theoretical models and theories that can provide guidance in developing methodologies for hypothesis testing. Considering already-existing models and theories, the perceptual-recuperative-defensive model (Bolles & Fanselow, 1980; Fanselow, 1986) provides a helpful conceptual scaffolding to promote the understanding of the relation of pain with fear and anxiety. The fear-avoidance model (Leeuw et al., 2007) elucidates a possible role for fear (including catastrophizing but even more likely anxiety) in the chronification of pain (also see chapters that discuss pain chronification in this volume by Kerr & Gregg, chapter "The Roles of Endogenous Opioids in Placebo and Nocebo Effects: From Pain to Performance to Prozac"; Sessle, chapter "Modulatory Processes in Craniofacial Pain States"; and Goldfarb and colleagues, chapter "Endogenous Opioids and Exercise-Related Hypoalgesia: Modern Models, Measurement, and Mechanisms of Action"). Finally, models of social exclusion and loss (MacDonald & Jensen-Campbell, 2011) may clarify how negative social events activate the pain responsivity and βE systems.

The development of a specific, comprehensive, and integrated model of the intersections among βE and the states of pain, fear, anxiety, and stress is beyond the

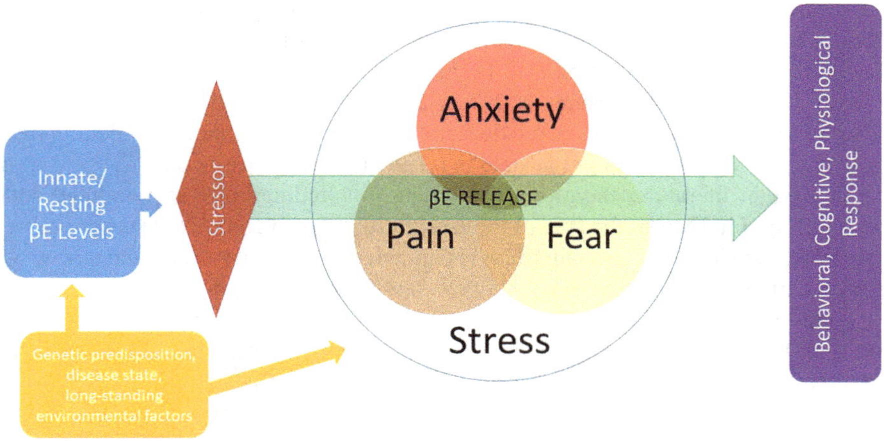

Fig. 1 Proposed interactions among pain, fear, anxiety, and stress

scope of this chapter. Nevertheless, the figure (Fig. 1) provides a graphical view of our proposed conceptualization of the interactions among pain, fear, anxiety, and stress. In light of this graphical representation, some considerations for a future model can be identified from this review and analysis. First, researchers should carefully define the state(s) being induced in experiments, with the knowledge that labeling a state in a certain way does not make it that state. Second, the likelihood of more than one response (e.g., pain and stress) being evoked should be considered, looking temporally at the progression of reactivity, rather than focusing solely on the first, immediate response. Third, research should carefully delineate the distinction between stimuli and the organism's response(s). Finally, consideration should be given to the measurement method used for ascertaining βE levels, such as plasma- versus CSF-derived βE.

Conclusions and Future Directions

The reviewed literature in this chapter allows us to conclude rather convincingly that βE is involved in the experience of various negative affective and sensory states, including pain, fear, anxiety, and stress. Other emotions that are known to be associated with these states, such as depression, should be included in future research, to fully consider the range of responses that may be mediated by βE. In particular, pain is known to be associated with depression (Bair et al., 2003) and fear and anxiety as well (McNeil et al., 2018).

This chapter also highlights critical conceptual and definitional issues that are hampering progress in this and related fields. Researchers are making assumptions about inducing particular states in experimental settings, without consideration of important definitional distinctions (and similarities) among these states. A related issue is observed in the clinical literature, in which self-report assessments are utilized with either "fear" or "anxiety" in their titles, or described in subscales, when clearly the instrument measures both fear and anxiety (McNeil & Vowles, 2004). Simply labeling a state with a name does not mean that one actually is inducing that state in a study or measuring that state in a verbal report instrument.

The extant literature, therefore, is a morass of findings that are muddled and bewildering. Findings in the area of the association of pain and βE are perhaps more extensive and clear-cut, but pain often is entangled with anxiety, fear, and stress, so even the findings in this area must be viewed with some caution.

Future research must more clearly distinguish among and define pain, fear, anxiety, and stress, with the acknowledgement that often more than one state is simultaneously (or subsequently) involved. In this way, we will be able to more clearly identify both common and unique pain- and emotion-related processes involved in βE pathways in the nervous system.

Clearly, βE is important in and perhaps integral to pain, anxiety, fear, and stress. Further elucidating the mechanisms involved may help to propel the field forward toward a more complete understanding of emotional and sensory responses that are innervated by βE.

References

al'Absi, M., Wittmers, L. E., Ellestad, D., Nordehn, G., Kim, S. W., Kirschbaum, C., & Grant, J. E. (2004). Sex differences in pain and hypothalamic-pituitary-adrenocortical responses to opioid blockade. *Psychosomatic Medicine, 66*, 198–206. https://doi.org/10.1097/01.psy.000016250.81254.5d

American College of Emergency Physicians. (2009). *Optimizing the treatment of pain in patients with acute presentations*. Retrieved from https://www.acep.org/Clinical-Practice-Management/Optimizing-the-Treatment-of-Pain-in-Patients-with-Acute-Presentations/

Amir, S., Brown, Z. W., & Amit, Z. (1980). The role of endorphins in stress: Evidence and speculations. *Neuroscience & Biobehavioral Reviews, 4*, 77–86.

Bäckryd, E., Ghafouri, B., Larsson, B., & Gerdle, B. (2014). Do low levels of beta-endorphin in the cerebrospinal fluid indicate defective top-down inhibition in patients with chronic neuropathic pain?: A cross-sectional, comparative study. *Pain Medicine, 15*, 111–119. https://doi.org/10.1111/pme.12248

Bair, M. J., Robinson, R. L., Katon, W., & Kroenke, K. (2003). Depression and pain comorbidity: A literature review. *Archives of Internal Medicine, 163*, 2433–2445. https://doi.org/10.1001/archinte.163.20.2433

Bali, A., Randhawa, P. K., & Jaggi, A. S. (2015). Stress and opioids: Role of opioids in modulating stress-related behavior and effect of stress on morphine conditioned place preference. *Neuroscience and Biobehavioral Reviews, 51*, 138–150. https://doi.org/10.1016/j.neubiorev.2014.12.018

Basbaum, A. I., & Fields, H. L. (1984). Endogenous pain control systems: Brainstem spinal pathways and endorphin circuitry. *Annual Review of Neuroscience, 7*, 309–338.

Bolles, R. C., & Fanselow, M. S. (1980). A perceptual-defensive-recuperative model of fear and pain. *Behavioral and Brain Sciences, 3*(2), 291–301. https://doi.org/10.1017/S0140525X0000491X

Bolles, R. C., & Fanselow, M. S. (1982). Endorphins and behavior. *Annual Review of Psychology, 33*, 87–101.

Bruehl, S., Chung, O. Y., Burns, J. W., & Diedrich, L. (2007). Trait anger expressiveness and pain-induced beta-endorphin release: Support for the opioid dysfunction hypothesis. *Pain, 130*(3), 208–215. https://doi.org/10.1016/j.pain.2006.11.013

Bruehl, S., Burns, J. W., Chung, O. Y., & Chont, M. (2012). What do plasma beta-endorphin levels reveal about endogenous opioid analgesic function? *European Journal of Pain, 16*(3), 370–380.

Bruehl, S., Burns, J. W., Gupta, R., Buvanendran, A., Chont, M., Orlowska, D., Schuster, E., & France, C. R. (2017). Do resting plasma beta-endorphin levels predict responses to opioid analgesics? *Clinical Journal of Pain, 33*(1), 12–20.

Calogero, A., Gallucci, W., Gold, P., & Chrousos, G. (1988). Multiple feedback regulatory loops upon rat hypothalmic corticotropin-releasing hormone secretion. *The Journal of Clinical Investigation, 82*, 767–774.

Cartwright, S., & Cooper, C. L. (1997). *Managing workplace stress*. Sage Publications.

Cepeda, M. S., Lipkowski, A. W., Langlade, A., Osgood, P. F., Ehrlich, H. P., Hargreaves, K., Szyfelbein, S. K., & Carr, D. B. (1993). Local increases of subcutaneous beta-endorphin immunoactivity at the site of thermal injury. *Immunopharmacology, 25*(3), 205–213.

Chapman, C. R., & Feather, B. W. (1983). Effects of diazepam on human pain tolerance and pain sensitivity. *Psychosomatic Medicine, 35*, 330–340.

Cleeland, C. S. (1986). Behavioral control of symptoms how to treat a construct. *Journal of Pain and Symptom Management, 1*(3), 161–162.

Coe, C. L. (2010). All roads lead to psychoneuroimmunology. In J. M. Suls, K. W. Davidson, & R. M. Kaplan (Eds.), *Handbook of health psychology and behavioral medicine* (pp. 182–199). The Guilford Press.

Cohen, M. R., Pickar, D., Dubois, M., & Bunney, W. E. (1982). Stress-induced plasma beta-endorphin immunoreactivity may predict postoperative morphine usage. *Psychiatry Research, 6*(1), 7–12. https://doi.org/10.1016/0165-1781(82)90032-4

Cole, S., & McNally, G. P. (2007). Opioid receptors mediate direct predictive fear learning: Evidence from one-trial blocking. *Learning & Memory, 14*, 229–235. https://doi.org/10.1101/lm.489507

Cork, R. C., Hameroff, S. R., & Weiss, J. L. (1985). Effects of halothane and fentanyl anesthesia on plasma β-endorphin immunoreactivity during cardiac surgery. *Anesthesia and Analgesia, 64*(7), 677–680.

Cornwall, A., & Donderi, D. C. (1988). Th effect of experimentally induced anxiety on the experience of pressure pain. *Pain, 35*, 105–113.

Craske, M. G. (2003). *Origins of phobias and anxiety disorders: Why more women than men?* Elsevier Ltd.

Dabo, F., Nyberg, F., Zhou, Q., Sundström-Poromaa, I., & Akerud, H. (2010). Plasma levels of beta-endorphin during pregnancy and use of labor analgesia. *Reproductive Science, 17*(8), 742–747. https://doi.org/10.1177/1933719110370059

Dellemijn, P. L. I., & Fields, H. L. (1994). Do benzodiazepines have a role in chronic pain management? *Pain, 57*, 137–152.

Drolet, G., Dumont, É. C., Gosselin, I., Kinkead, R., Laforest, S., & Trottier, J. F. (2001). Role of endogenous opioid system in the regulation of the stress response. *Progress in Neuro-Psychopharmacology & Biological Psychiatry, 25*(4), 729–741.

Dubois, M., Pickar, D., Cohen, M., Gadde, P., Macnamara, T. E., & Bunney, W. E. (1982). Effects of fentanyl on response of plasma beta-endorphin immunoreactivity to surgery. *Anesthesiology, 57*(6), 468–472.

Eippert, F., Bingel, U., Schoell, E., Yacubian, J., & Büchel, C. (2008). Blockade of endogenous opioid neurotransmission enhances acquisition of conditioned fear in humans. *Journal of Neuroscience, 28*(21), 5465–5472. https://doi.org/10.1523/JNEUROSCI.5536-07.2008

Eisenberger, N. I. (2011). The neural basis of social pain: Findings and implications. In G. MacDonald & L. A. Jensen-Campbell (Eds.), *Social pain: Neuropsychological and health implications of loss and exclusion* (pp. 53–78). American Psychological Association.

Elman, I., & Borsook, D. (2018). Threat response system: Parallel brain processes in pain vis-à-vis fear and anxiety. *Frontiers in Psychiatry, 9*, 29. https://doi.org/10.3389/fpsyt.2018.00029

Facchinnetti, F., Centini, G., Parrini, F., D'Antona, E., & Genazzani, A. (1982). Opioid plasma levels during labor. *Gynecologic and Obstetric Investigation, 13*, 155–163.

Fanselow, M. S. (1986). Conditioned fear-induced opiate analgesia: A competing motivational state theory of stress analgesia. *Annals of the New York Academy of Sciences, 467*, 40–54.

Fanselow, M. S., & Baackes, M. P. (1982). Conditioned fear-induced opiate analgesia on the formalin test: Evidence for two aversive motivational systems. *Learning and Motivation, 13*, 200–221.

Fanselow, M. S., & Bolles, R. C. (1979). Triggering of the endorphin analgesic reaction by a cue previously associated with shock: Reversal by naloxone. *Bulletin of the Psychonomic Society, 14*(2), 88–90.

Feldreich, A., Ernberg, M., Lund, B., & Rosén, A. (2012). Increased ß-endorphin levels and generalized decreased pain thresholds in patients with limited jaw opening and movement-evoked pain from the temporomandibular joint. *Journal of Oral and Maxillofacial Surgery, 70*(3), 547–556. https://doi.org/10.1016/j.joms.2011.09.013

Fink, G. (2017). *Stress: Neuroendocrinology and neurobiology*. Academic Press.

Gallup, G. G. (1977). Tonic immobility: The role of fear and predation. *Psychological Record, 27*, 41–61.

Genazzani, A. R., Facchinetti, F., & Parrini, D. (1981). ß-lipotrophin and ß-endorphin plasma levels during pregnancy. *Clinical Endocrinology, 14*(4), 409–418. https://doi.org/10.1111/j.1365-2265.1981.tb00628.x

Govoni, S., Pasinetti, G., Inzoli, M. R., Rozzini, R., & Trabucchi, M. (1984). Correlation between beta-endorphin/beta-lipotropin-immuno-reactivity and cortisol plasma concentrations. *Life Sciences, 35*, 2549–2552.

Grichnik, K. P., & Ferrante, F. M. (1991). The difference between acute and chronic pain. *Mount Sinai Journal of Medicine, 58*(3), 217–220.

Gross, R. T., & Collins, F. L. (1981). On the relationship between anxiety and pain: A methodological confounding. *Clinical Psychology Review, 1*, 375–386.

Guasti, L., Cattaneo, R., Daneri, A., Bianchi, L., Gaudio, G., Regazzi, M. B., et al. (1996). Endogenous beta-endorphins in hypertension: Correlation with 24-hour ambulatory blood pressure. *Journal of the American College of Cardiology, 28*(5), 1243–1248.

Guillemin, R., Vargo, T., Rossier, J., Minich, S., Ling, N., River, C., Vale, W., & Bloom, F. (1977). ß-endorphin and adrenocorticotropin are secreted concomitantly by the pituitary gland. *Science, 197*, 1367–1369.

Hansson, E., Manjer, J., Svensson, H., Åberg, M., Fagher, B., Ekman, R., & Brorson, H. (2012). Neuropeptide levels in Dercum's disease (adiposis dolorosa). *Reumatismo, 64*, 134–141. https://doi.org/10.4081/reumatismo.2012.134

Hargreaves, K. M. (1984). Plasma beta-endorphin levels in oral surgery patients following diazepam, fentanyl, or placebo. *Anesthesia Progress, 31*(3), 124–129.

Hargreaves, K. M., Dionne, R. A., & Mueller, G. P. (1983). Plasma beta-endorphin-like immunoreactivity, pain, and anxiety following administration of placebo in oral surgery patients. *Journal of Dental Research, 62*, 1170–1173.

Hargreaves, K. M., Dionne, R. A., Mueller, G., Goldstein, D. S., & Dubner, R. (1986). Naloxone, fentanyl, and diazepam modify plasma beta-endorphin levels during surgery. *Clinical Pharmacology & Therapeutics, 40*(2), 165–171.

Hartwig, A. C. (1991). Peripheral beta-endorphin and pain modulation. *Anesthesia Progress, 38*(3), 75–78.

Hayden-Hixon, D. M., & Nemeroff, C. B. (1993). Role(s) of neuropeptides in responding and adaptation to stress: A focus on corticotropin-releasing factor and opioid peptides. In S. C. Stanford & P. Salmon (Eds.), *Stress, from stress to syndrome* (pp. 356–391). Academic Press.

Helmsetter, F. J., & Fanselow, M. S. (1987). Effects of naltrexone on learning and performance of conditional fear-induced freezing and opioid analgesia. *Physiology & Behavior, 30*, 501–505.

Holmes, T. H., & Rahe, R. H. (1967). The social readjustment rating scale. *Journal of Psychosomatic Research, 11*, 213–218.

International Association for the Study of Pain. (1986). Classification of chronic pain. *Pain, 3*(Suppl), S1–S226.

Janssen, S. A., & Arntz, A. (1996). Anxiety and pain: Attentional and endorphinergic influences. *Pain, 66*, 145–150.

Johnson, N. A., & Helmstetter, F. J. (1994). Conditioned fear-induced hypoalgesia in humans using non-noxious UCS. *Society for Neuroscience Abstracts, 20*, 360.

Khedr, E. M., Omran, E. A. H., Ismail, N. M., El-Hammady, D. G., Goma, S. H., Kotb, H., et al. (2017). Effects of transcranial direct current stimulation on pain, mood and serum endorphin level in the treatment of fibromyalgia: A double blinded, randomized clinical trial. *Brain Stimulation, 10*, 893–901.

Lang, P. J. (1968). Fear reduction and fear behavior: Problems in treating a construct. *Research in Psychotherapy, 1*, 90–102. https://doi.org/10.1177/0192513X12437708

Larkin, K. E. (2005). *Stress & hypertension: Examining the relation between psychological stress and high blood pressure*. Yale University Press.

Leeuw, M., Goossens, M. E., Linton, S. J., Crombez, G., Boersma, K., & Vlaeyen, J. (2007). The fear-avoidance model of musculoskeletal pain: Current state of the scientific evidence. *Journal of Behavioral Medicine, 30*, 77–94.

Leigland, S. (2014). Contingency horizon: On private events and the analysis of behavior. *Behavior Analyst, 37*, 13–24. https://doi.org/10.1007/s40614-014-0002-5

Leonard, T. M., Klem, S. A., Asher, M. A., Rapoff, M. A., & Leff, R. D. (1993). Relationship between pain severity and serum beta-endorphin levels in postoperative patients. *Pharmacotherapy, 13*, 378–381.

Loh, H. H., Tseng, L. F., Wei, E., & Li, C. H. (1976). β-endorphin is a potent analgesic agent. *Proceedings of the National Academy of Sciences of the United States of America, 73*(8), 2895–2898.

Lopez-Luna, J., Al-Jubouri, Q., Al-Nuaimy, W., & Sneddon, L. U. (2017). Impact of stress, fear, and anxiety on the nociceptive responses of larval zebrafish. *PLoS One, 12*(8), e0181010. https://doi.org/10.1371/journal.pone.0181010

Lovallo, W. R. (2010). Emotions and stress. In J. M. Suls, K. W. Davidson, & R. M. Kaplan (Eds.), *Handbook of health psychology and behavioral medicine* (pp. 31–48). The Guilford Press.

Macdonald, G., & Jensen-Campbell, L. A. (2011). *Social pain: Neuropsychological and health implications of loss and exclusion*. American Psychological Association.

Matejec, R., Ruwoldt, R., Bödeker, R. H., Hempelmann, G., & Teschemacher, H. (2003). Release of the β-endorphin immunoreactive material under perioperative conditions into blood or cerebrospinal fluid: Significance for postoperative pain? *Anesthesia and Analgesia, 96*, 481–486. https://doi.org/10.1213/01.ANE.0000044420.02070.56

McNeil, D. W., & Randall, C. L. (2014). Dental fear and anxiety associated with oral health care: Conceptual and clinical issues. In D. I. Mostofsky & F. Fortune (Eds.), *Behavioral dentistry* (2nd ed.). Wiley.

McNeil, D. W., & Vowles, K. E. (2004). Assessment of fear and anxiety associated with pain: Conceptualization, methods, and measures. In G. J. G. Asmundson, J. W. S. Vlaeyen, & G. Crombez (Eds.), *Understanding and treating fear of pain* (pp. 189–211). Oxford University Press.

McNeil, D. W., Arias, M. C., & Randall, C. L. (2017). Anxiety versus fear. In A. E. Wenzel, D. Frideman-Wheeler, & E. Flannery-Schroeder (Eds.), *Encyclopedia of abnormal and clinical psychology* (Vol. 1, pp. 279–280). Sage.

McNeil, D. W., Kennedy, S. G., Randall, C. L., Addicks, S. H., Wright, C. D., Hursey, K. G., & Vaglienti, R. (2018). Fear of pain questionnaire-9: Brief assessment of pain-related fear and anxiety. *European Journal of Pain, 22*, 39–48. https://doi.org/10.1002/ejp.1074

Meagher, M. W., Arnau, R. C., & Rhudy, J. L. (2001). Pain and emotion: Effects of affective picture modulation. *Psychosomatic Medicine, 63*, 79–90.

Merskey, H., & Bogduk, N. (1994). Part III: Pain terms, a current list with definitions and notes on usage. In *Classification of chronic pain, 2nd Edition, IASP Task Force on Taxonomy* (pp. 209–214). IASP Press.

Miller, P. F., Light, K. C., Bragdon, E. E., Ballenger, M. N., Herbst, M. C., Maixner, W., et al. (1993). Beta-endorphin response to exercise and mental stress in patients with ischemic heart disease. *Journal of Psychosomatic Research, 37*(5), 455–465.

Misra, U. K., Kalita, J., Tripathi, G., & Bhoi, S. K. (2017). Role of β endorphin in pain relief following high rate repetitive transcranial magnetic stimulation in migraine. *Brain Stimuli, 10*, 618–623. https://doi.org/10.1016/j.brs.2017.02.006

Mørch, H., & Pedersen, B. K. (1995). Beta-endorophin and the immune system—possible role in autoimmune diseases. *Autoimmunity, 21*(3), 161–171.

Nader-Djalal, N., de Leon-Casasola, O. A., Peer, G. L., Vladutiu, A. O., & Lema, M. J. (1995). The influence of preoperative concentrations of beta-endorphin and met-enkephalin on the duration of analgesia after transurethral resection of prostate. *Anesthesia & Analgesia, 81*, 591–595.

Pickar, D., Cohen, M. R., & Dubois, M. (1983). The relationship of plasma cortisol and beta-endorphin immunoreactivity to surgical stress and postoperative analgesic requirement. *General Hospital Psychiatry, 5*, 93–98.

Pilcher, H. P., Joseph, S. A., & McDonald, J. V. (1988). Immunocytochemical localization of pro-opiomelanocortin neurons in human brain areas subserving stimulation analgesia. *Journal of Neurosurgery, 68*, 621–629.

Przewlocki, R., Lasoń, W., Höllt, V., Silberring, J., & Herz, A. (1987). The influence of chronic stress on multiple opioid peptide systems in the rat: Pronounced effects upon dynorhpin in spinal cord. *Brain Research, 413*(2), 213–219.

Raja, S. N., Carr, D. B., Cohen, M., Finnerup, N. B., Flor, H., Gibson, S., Keefe, F. J., Mogil, J. S., Ringkamp, M., Sluka, K. A., Song, X. J., Stevens, B., Sullivan, M. D., Tutelman, P. R., Ushida, T., & Vader, K. (2020). The revised international association for the study of pain definition of pain: Concepts, challenges, and compromises. *Pain, 161*(9), 1976–1982. https://doi.org/10.1097/j.pain.0000000000001939

Rasmussen, N. A., & Farr, L. A. (2003). Effects of morphine and time of day on pain amd beta-endorphin. *Biological Research for Nursing, 5*, 105–116.

Rasmussen, N. A., & Farr, L. A. (2009). Beta-endorphin response to an acute pain stimulus. *Journal of Neuroscience Methods, 177*, 285–288. https://doi.org/10.1016/j.jneumeth.2008.10.013

Rescorla, R. A., & Wagner, A. R. (1972). A theory of Pavlovian conditioning: Variations in the effectiveness of reinforcement and non reinforcement. In A. H. Black & W. F. Prokasy (Eds.), *Classical conditioning II: Current research and theory* (pp. 64–99). Appleton-Century-Crofts.

Rhodin, A., Grönbladh, A., Ginya, H., Nilsson, K. W., Rosenblad, A., Zhou, Q., et al. (2013). Combined analysis of circulating β-endorphin with gene polymorphisms in OPRM1, CACNAD2, and ABCB1 reveals correlation with pain, opioid sensitivity and opioid-related side effects. *Molecular Brain, 6*, 8. https://doi.org/10.1186/1756-6606-6-8

Rhudy, J. L., & Meagher, M. W. (2000). Fear and anxiety: Divergent effects on human pain thresholds. *Pain, 84*, 65–75.

Rhudy, J. L., Grimes, J. S., & Meagher, M. W. (2004). Fear-induced hypoalgesia in humans: Effects on low intensity thermal stimulation and finger temperature. *Journal of Pain, 5*(8), 458–468. https://doi.org/10.1016/j.pain.2004.08.001

Riley, J. L., III, Cruz-Almeida, Y., Dasilva Ribeiro, M. C., Simon, C. B., Eckert, N. R., Aguirre, M., Sorenson, H. L., Tighe, P. J., Edwards, R. R., & Wallet, S. M. (2017). Age differences in the time course and magnitude of changes in circulating neuropeptides after pain evocation in humans. *Journal of Pain, 18*(9), 1078–1086. https://doi.org/10.1016/j.jpain.2017.04.006

Rossier, J., French, E. D., Rivier, C., Ling, N., Guillemin, R., & Bloom, F. E. (1977). Foot-shock induced stress increases the β-endorphin levels in blood but not brain. *Nature, 270*, 618–620. https://doi.org/10.1038/270618a0

Schedlowski, M., Flüge, T., Richter, S., Tewes, U., Schmidt, R. E., & Wagner, T. O. F. (1995). B-endorphin, but not substance-P, is increased by acute stress in humans. *Psychoneuroendocrinology, 20*, 103–110.

Segerstrom, S. C., & Miller, G. E. (2004). Psychological stress and the human immune system: A meta-analytic study of 30 years of inquiry. *Psychological Bulletin, 130*(4), 601–630. https://doi.org/10.1037/0033-2909.130.4.601

Selye, H. (1956). *The stress of life*. McGraw-Hill.

Selye, H. (1974). *Stress without distress*. J.B. Lippincott Company.

Shepherd, S. (1991). *Francie* [cartoon]. United Feature Syndicate, Inc.

Shiomi, H., & Akil, H. (1982). Pulse-chase studies of the POMC/beta-endorphin system in the pituitary of acutely and chronically stressed rats. *Life Sciences, 31*, 2185–2188.

Siebers, M., Biedermann, S. V., Bindila, L., Lutz, B., & Fuss, K. (2021). Exercise-induced euphoria and anxiolysis do not depend on endogenous opioids in humans. *Psychoneuroendocrinology, 126*, 105172.

Skinner, B. F. (1969). Contingencies of reinforcement: A theoretical analysis. In *Behaviorism at fifty* (pp. 221–268). Appleton-Century-Crofts.

Solomon, S. (1999). POMC-derived peptides and their biological action. *Annals of the New York Academy of Science, 885*, 22–40.

Sprenger, T., Valet, M., Boecker, H., Henriksen, G., Spilker, M. E., Willoch, F., et al. (2006). Opioidergic activation in the medial pain system after heat pain. *Pain, 122*, 63–67. https://doi.org/10.1016/j.pain.2006.01.033

Sprouse-Blum, A. S., Smith, G., Sugai, D., & Parsa, F. D. (2010). Understanding endorphins and their importance in pain management. *Hawai'i Medical Journal, 69*, 70–71.

Steinhoff, M. S., von Mentzer, B., Geppetti, P., Pothoulakis, C., & Bunnett, N. W. (2014). Tachykinins and their receptors: Contributions to physiological control and the mechanisms of disease. *Physiological Reviews, 94*, 265–301.

Straneva, P. A., Maixner, W., Light, K. C., Pedersen, C. A., Costello, N. L., & Girdler, S. S. (2002). Menstrual cycle, beta-endorphins, and pain sensitivity in premenstrual dysphoric disorder. *Health Psychology, 21*(4), 358–367. https://doi.org/10.1037/0278-6133.21.4.358

Szabo, S., Tache, Y., & Somogyi, A. (2012). The legacy of Hans Selye and the origins of stress research: A retrospective 75 years after his landmark brief "letter" to the editor of nature. *Stress (Amsterdam, Netherlands), 15*(5), 472–478. https://doi.org/10.3109/10253890.2012.710919

Szyfelbein, S. K., Osgood, P. F., & Carr, D. B. (1985). The assessment of pain and plasma β-endorphin immunoactivity in burned children. *Pain, 22*(2), 173–182.

Tamsen, S. E., Sakurada, T., Wahlström, A., Terenius, L., & Hartvig, P. (1982). Postoperative demand for analgesics in relation to individual levels of endorphins and substance P in cerebrospinal fluid. *Pain, 13*, 171–183.

Taylor, S. E. (2018). *Health psychology*. McGraw Hill.

Thomas, T. A., Fletcher, J. E., & Hill, R. G. (1982). Influence of medication, pain and progress in labour on plasma beta-endorphin-like immunoreactivity. *British Journal of Anaesthesia, 54*(4), 401–408.

Thorén, P., Floras, J. S., Hoffmann, P., & Seals, D. R. (1990). Endorphins and exercise: Physiological mechanisms and clinical implications. *Medicine & Science in Sports & Exercise, 22*(4), 417–428.

Thyer, B. A., & Matthews, J. (1986). The effect of phobic anxiety on plasma beta-endorphin: A single-subject experiment. *Behaviour Research and Therapy, 24*(2), 237–241.

Treede, R., Rief, W., Barke, A., Aziz, Q., Bennett, M. I., Benoliel, R., Cohen, M., et al. (2015). A classification of chronic pain for ICD-11. *Pain, 156*(6), 1003–1007.

Vaeroy, H., Helle, R., Førre, O., Kåss, E., & Terenius, L. (1988). Cerebrospinal fluid levels of beta-endorphin in patients with fibromyalgia (fibrositis syndrome). *The Journal of Rheumatology, 15*(12), 1804–1806.

Valentino, R. J., & Bockstaele, E. V. (2008). Convergent regulation of locus coeruleus activity as an adaptive response to stress. *European Journal of Pharmacology, 583*, 194–203.

Vowles, K. E., McNeil, D. W., Sorrell, J. T., & Lawrence, S. M. (2006). Fear and pain: Investigating the interaction between two aversive states. *Journal of Abnormal Psychology, 115*, 821–833.
Willer, J. C., Dehen, H., & Cambier, J. (1981). Stress-induced analgesia in humans: Endogenous opioids and naloxone-reversible depression of pain reflexes. *Science, 212*, 689–691.
Williams, A. E., & Rhudy, J. L. (2007). The influence of conditioned fear on human pain thresholds: Does preparedness play a role? *Journal of Pain, 8*(7), 596–606. https://doi.org/10.1016/j.pain.2007.03.004
Young, E. A., & Akil, H. (1985). Corticotropin-releasing factor stimulation of adrenocorticotropin and *β*-endorphin release: Effects of acute and chronic stress. *Endocrinology, 117*, 23–30. https://doi.org/10.1210/endo-117-1-23
Zangen, A., Herzberg, U., Vogel, Z., & Yadid, G. (1998). Nociceptive stimulus induces release of endogenous the β-endorphin in the rat brain. *Neuroscience, 85*(3), 659–662.
Zubieta, J. K., Smith, Y. R., Bueller, J. A., Xu, Y., Kilbourn, M. R., Jewett, D. M., et al. (2001). Regional mu opioid receptor regulation of sensory and affective dimensions of pain. *Science, 293*(5528), 311–315. https://doi.org/10.1126/science.1060952

The Roles of Endogenous Opioids in Placebo and Nocebo Effects: From Pain to Performance to Prozac

Patrick L. Kerr and John M. Gregg

Abstract Placebo and nocebo effects have been well documented for nearly two centuries. However, research has only relatively recently begun to explicate the neurobiological underpinnings of these phenomena. Similarly, research on the broader social implications of placebo/nocebo effects, especially within healthcare delivery settings, is in a nascent stage. Biological and psychosocial outcomes of placebo/nocebo effects are of equal relevance. A common pathway for such outcomes is the endogenous opioid system. This chapter describes the history of placebo/nocebo in medicine; delineates the current state of the literature related to placebo/nocebo in relation to pain modulation; summarizes research findings related to human performance in sports and exercise; discusses the implications of placebo/nocebo effects among diverse patient populations; and describes placebo/nocebo influences in research related to psychopharmacology, including the relevance of endogenous opioids to new lines of research on antidepressant pharmacotherapies.

Keywords Placebo · Nocebo · Pain · Endogenous opioids · Performance

Introduction

There is a substantial body of research and an accompanying wealth of literature involving placebo that pervades nearly all branches of science. Novel experimental conditions of interest to investigators—and to society—must be compared against known actors and against the passage of time itself. Such is the necessity of the scientific enterprise. However, in the rich history of science, the effort to design

P. L. Kerr (✉)
West Virginia University School of Medicine-Charleston, Charleston, WV, USA
e-mail: pkerr@hsc.wvu.edu

J. M. Gregg
Department of Surgery, VTCSOM, Blacksburg, VA, USA

P. L. Kerr et al. (eds.), *Endogenous Opioids*, Advances in Neurobiology 35,
https://doi.org/10.1007/978-3-031-45493-6_10

"controlled" experiments with inert comparison conditions to better understand the benefits of a new intervention is relatively young. Nonetheless, in a short time, research using placebo has clearly demonstrated that the central, peripheral, sympathetic, and parasympathetic nervous systems, immunological responses, and both pain and analgesic responses are activated by expectations almost as much as they are activated by substances.

Ross and Olson (1981) have provided a useful and succinct description of placebo: "a substance or procedure that is administered with suggestions that will modify a symptom or sensation but which, unknown to its recipient, has no specific pharmacological impact on the reaction in question. A standard placebo effect occurs when the administration of the placebo alters the recipient's state in accordance with its presumed impact" (p. 408). As a product of the era in which they were writing, Ross and Olson's (1981) description could not have captured the nuanced, diverse, and multifaceted dynamics that underpin placebo effects, many of which we have only recently begun to identify. Nonetheless, this definition is a helpful place from which to launch our discussion.

In the subsequent sections of this chapter, we delineate the mechanics involved in specific extensions of this definition, including both placebo and nocebo effects and placebo and nocebo responses. In particular, and in keeping with the trajectory of this book, we turn to the roles that endogenous opioids play in placebo and nocebo effects and responses across domains of scientific research, evidence-based clinical interventions, and the measurement and enhancement of human performance. As astutely observed by Benedetti (2002), "the placebo effect is a context effect" (p. 369). In this vein, much of the literature on placebo and nocebo effects upon which we elaborate in this chapter may be conceptualized as empirical and theoretical descriptions of context, a construct which we will revisit.

A Brief History of Placebo

Studies using placebo-like conditions were first published in the nineteenth century, though it was not until the early twentieth century that scientists conceptualized such conditions as a "placebo" (de Craen et al., 1999). Prior to that time, the concept of "placebo" was relegated to interventions mainly designed to placate patients. Two notable experiments by Haygarth (1800) and Flint (1863) introduced the concept of a placebo control in research. In what became a seminal work on the area of placebo, Haygarth (1800) described an experiment in which he compared the effects of metal rods ("Perkins Patent Tractors") to identically painted and sized wooden rods on alleviation of pain in a small group of patients:

> Five cases were chosen of chronick rheumatism, in the ankle, knee, wrist, and hip. (Haygarth, 1800, p. 3)

Haygarth's (1800) findings were as fascinating as the new research design he introduced: when given a sham treatment, i.e., the wooden "tractors," four of the five patients reported improvements in their pain, and this result was the same when the actual metal Perkins Tractors were used the following day. Two conclusions were drawn from this historical example. First, the metal tractors themselves were an inert intervention, a well-known historical fact that remains relevant in the modern era. Second, some extent of the outcomes of an intervention may be attributable in part to one's expectations of an outcome; as Haygarth noted in his report: "Such is the wonderful force of imagination" (Haygarth, 1800, p. 3).

Somewhat later in the nineteenth century, Flint (1863) reported on the results of his study of 13 patients diagnosed with "rheumatism" who were given "the placeboic remedy for rheumatism," viz., a heavily diluted extract of the quassia plant. His observation was that patients given this inert substance reported improvement in their symptoms, leading him to conclude that the symptoms of rheumatism "end from self-limitation." The conclusion from this second historical example is that something besides the direct effects of an active substance may contribute to a change in the course of a disease.

While Flint (1863) seemed to view his study as a method of observing the natural course of rheumatism, he inadvertently opened yet another early window into what would later come to be known as a *placebo effect*, i.e., "any effect that is not due to the specific, intended effects of a treatment" (Masters, 2013, p. 2926). The first documented conceptualization of a placebo effect was in a description of the results of a cold virus vaccine trial by Diehl et al. (1938) that used a placebo control group. The striking finding by Diehl and colleagues (1938) was not the inefficacy of the cold vaccine, but the ostensible efficacy of the placebo, which was comparable to the vaccine in their trial as well as that from previous trials. Those receiving placebo remained healthy at a rate that rendered differences between groups statistically nonsignificant. This later became known as a *placebo response*, i.e., "any beneficial effect that cannot be attributed to the specific intended effects" (Masters, 2013, p. 2926).

Clearly, the possibility that cognitive, psychological, cultural, environmental, and situational features of placebos and nocebos may have major effects on pain has long been recognized (Leslie, 1954; Beecher, 1955, 1961). Beecher observed that, in circumstances of warfare, many injured soldiers perceived little or no pain (Beecher, 1956). In 1965, Melzack and Wall proposed in their famous Gate Control Theory that noxious pain-responding neurons in the dorsal horn were powerfully controlled by supra spinal sites (Melzack & Wall, 1965). There were three subsequent key turning points in pain and behavioral research: first, the demonstration that placebo analgesia could be reversed by the administration of centrally acting opioid antagonists (Levine et al., 1978; Posner & Burke, 1985); second, the discovery and localization of a system of endogenous opioid receptor sites in the brain (Pert & Snyder, 1973); and third, the early identification of two distinct endogenous opioids, named endorphin and enkephalin (Hughes et al., 1975). In recent decades, researchers have shown interdependent links between endogenous opioids, chronic neuropathic pain (punishment, suffering), and reward (analgesia, maladaptive behavior) with substance abuse implications (Amanzio & Benedetti, 1999; Bodnar, 2017; Borsook, 2017; Nees et al., 2017; Ballantyne & Sullivan, 2017).

Neurobiological Underpinnings of Placebo Responses Across Conditions

The neurobiological pathways of placebo effects and responses are complex. While placebo analgesia has been clearly demonstrated to be associated with primarily endogenous opioid pathways, other neurochemicals and neuropeptides may be recruited during a placebo response. Indeed, a large meta-analysis ($k = 40$) of placebo effects in the pain research literature by Atlas and Wager (2014) found reliable decreases and increases in activation across several brain regions, noting consistent decreases in activation during expectation of reduced pain in the lateral prefrontal cortex, amygdala, insula, dorsal anterior cingulate cortex, amygdala, thalamus, and striatum. Conversely, reduced pain was consistently associated with increased activation in the left anterior insula, dorsolateral prefrontal cortex, ventromedial prefrontal cortex, rostral anterior cingulate cortex, midbrain periaqueductal gray (PAG), and striatum.

Atlas and Wager (2014) potentially found a common link through involvement of the striatum. Previous syntheses of the literature have also noted strong empirical evidence for the role of the ventral striatum in placebo responses (de la Fuente-Fernandez & Stoessl, 2004; Diederich & Goetz, 2007; Lidstone & Stoessl, 2007). This evidence has emerged consistently over the past two decades. For example, an early study by Goetz et al. (2000) demonstrated that dopaminergic activity in the striatum associated with improvements in motor function and performance was a mediator of placebo effects in patients with Parkinson's disease. Subsequently, in a small ($n = 6$) experimental study, de la Fuente-Fernandez and colleagues (2001) found that dopamine release in the striatum of patients with Parkinson's disease was proportional to the expectation of therapeutic effect, suggesting a causal relationship. The combined role of dopamine and endogenous opioids is further supported by the findings of Fulda and Wetter (2008). In a meta-analysis ($k = 36$) of RCTs of pharmacotherapies for restless leg syndrome (RLS), pooled placebo response across all outcomes was approximately 40%, indicating that much of the therapeutic response could be attributable to placebo effects. This is significant because previous empirical investigations have demonstrated that RLS symptomatology is equally responsive to both dopaminergic and opiatergic agonists. Not only does this indicate a dual-substrate pathophysiology for RLS, but it also points toward dual mechanisms of action for RLS interventions. More importantly, these data elucidate a unified neural pathway for placebo effects. Thus, from both a neurochemical and neuroanatomical perspective, there is evidence of a combinatorial effect of opioid-dopamine substrate interactions, especially within the limbic circuitry (de la Fuente-Fernandez & Stoessl, 2004; Kong & Benedetti, 2014).

Placebo and Nocebo Influences on Pain

The complex human experience of pain, with its primary components of sensory-discrimination, affective-motivational suffering, and cognitive-evaluative dimensions, is highly susceptible to manipulations by placebo and nocebo interventions (Benedetti, 2008; Moayedi, 2014; Vase et al., 2015). Research has demonstrated that placebo and nocebo effects related to pain are both mediated through central endogenous opioid mechanisms (Eippert et al., 2009; Colagiuri et al., 2015, Burns et al., 2017), the central focus of this book, and that these effects target the emotional (suffering) and cognitive-cultural components of pain with much less impact on pain sensory-discrimination (Singer, 2004, Jiang et al., 2016; Elman & Borsook, 2016; Gomtsian et al., 2018). This placebo-nocebo orientation for patients with chronically sensitized nervous systems has special relevance to sexual dimorphism in pain (Keogh, 2014; Mapplebeck et al., 2016) and reward-antireward systems that may influence endogenous opioid responses to pain and stress as well as treatment responses and addictions to opiate drugs (Benedetti et al., 2013; Borsook, 2017; Nees et al., 2017).

Pain-related placebos are inert substances and cognitive interventions whose primary clinical effects are to reduce (inhibit) pain, an *analgesia* effect. Nocebos are the mirror opposite of placebos and include inert substances and cognitive interventions whose primary clinical effects are to increase (stimulate) pain, a *hyperalgesia* effect. In evolutionary terms, placebo and nocebo effects coexist to favor survival from threats and dangerous events (nocebo effects) (Colloca, 2017) and also promote appetitive and safety behaviors (placebo effects) (Colloca et al., 2013). In cognitive terms, placebo and nocebo influences are opposing responses whereby nocebo pain intensifications are induced by negative expectations and placebo pain reductions are triggered by positive expectancies (Colloca & Benedetti, 2009; Petersen et al., 2014; Klinger et al., 2017).

A recent meta-analysis by Petersen et al. (2014) yielded findings that were both interesting and instructive. One of the most important findings was the comparable effect sizes of placebo and nocebo, which were $d = 0.81$ and $g = 0.62–1.01$, respectively. Additionally, Petersen et al. (2014) reported a slight advantage of procedures combining conditioning with verbal suggestion ($g = 0.76–1.17$) over verbal suggestion as a standalone nocebo induction procedure ($g = 0.64–0.87$). The data on the apparent additive value of conditioning in nocebo effects indicates two important factors to consider in nocebo effects. First, direct experience, viz., conditioning paradigms, matters and leads to more pronounced nocebo effects. Second, despite the added value of conditioning, verbal suggestion alone does lead to significant nocebo effects. The information people are provided regarding the potential for pain clearly affects their subjective interpretations of that experience and likely governs their experience of pain. As described in the subsequent sections of this chapter, these effects follow relatively well-defined neural pathways. One additional conclusion that can be drawn from this meta-analysis is that there may be equipoise when it comes to nocebo and placebo effects of noxious stimuli and analgesia. There is clear empirical evidence that expectations play a role in the subjective quality of the individual experience of pain and its relief.

Pain Research Targets

Research into pain and associated placebo and nocebo influences are traditionally subdivided into *experimental*, *acute* (nociceptive), and *chronic* (neuropathic) forms of pain (Treede et al., 2015). Each of these has contributed in important ways to our understanding of placebo and nocebo. We turn now to exploration of these respective subdivisions.

Research Target: Experimental Pain

Nonhuman animals at many levels of taxonomy display aversive responses to tissue damaging exposures and also appear to have levels of self-awareness and "consciousness" to experience the "suffering" component of pain (Attal et al., 1989; Bennett & Xie, 1988; Carlton et al., 1994). Many experimental animal models of pain continue to be useful to elucidate neurobiological effects of noxious stimulations and evaluate the potential for analgesic interventions (Birklein et al., 2018; Bobinski et al., 2018; Li et al., 2020). Research by Schafer et al. (2018) has attempted to bridge the cross-species gaps, using "dual process" models that compare central pathways and mechanisms of human placebo analgesia with pain modulation behavior in rodents. They have observed considerable overlap of human placebo to rodent pain modulation, including strong similarities of descending forebrain to brainstem and spinal cord pathways as well as similarities in opioid and cannabinoid receptor mechanisms. Nonetheless, well-reasoned arguments (scientific and philosophical) for movement away from animal experimentation in pain research are important to consider (Carbone, 2011), not the least of which is the low translational success across Phases II and III. Recent analyses suggest that over 80% of new drugs do not progress past Phase II and that only about half of those translate past Phase III (Arrowsmith, 2011a, b; Arrowsmith & Miller, 2013).

Human research, however, has extensively used *experimental models of pain* to study many aspects of pain among healthy volunteers, including possible responses to placebo and nocebo factors (Staahl et al., 2009; Vase et al., 2015). Prominent examples of human experimental pain models are the tourniquet ischemia test, the cold pressor test, cantharidin blister-base test, pressure algometer tolerances, noxious skin heat, electrically induced flair test, capsaicin skin stimuli, and ultraviolet B sunburn model (Anderson et al., 2015; Dusch et al., 2009; Fuchs et al., 2000; Gustorff et al., 2013; Keene, 1954; Klein et al., 2005; Shabes et al., 2016). Although much has been learned from research using these human pain models, there is existing evidence that neurotransmitter systems involved in placebo and nocebo effects in healthy humans may not be transferred directly to patients with chronic pain (Skyt et al., 2020).

Research Target: Acute Nociceptive Pain

In addition to experimental pain, many studies of placebo analgesia and nocebo hyperalgesia have been directed toward patients suffering from *acute* postoperative surgical, dental, and orofacial pain (Cooper et al., 2016; Johansen et al., 2014; Ong et al., 2004; Prosenz & Gustoff, 2017, Sessle, 2021; Slade et al., 2013) and transient inflammatory nociceptive conditions such as reversible infections and asthmatic and allergic conditions (Benedetti, 2008, 2014; Price et al., 2008, Kaptchuk & Miller, 2015).

Research Target: Chronic Neuropathic Pain—Sensitization

Studies of patients with chronic, sustained, or recurrent painful disease conditions have increasingly addressed placebo and nocebo influences (Benedetti, 2008; Petersen et al., 2012). Emphasis has been on patient populations with low back pain, irritable bowel syndrome (IBS), migraine headache, and Parkinson's pain (Klinger et al., 2017; Kaptchuk et al., 2010; Amanzio et al., 2009; Lidstone et al., 2010; de la Fuente-Fernandez, 2009) with lesser attention to female pelvic and male prostatic pain, RLS, and fibromyalgia (Kutch et al., 2017; Zhang et al., 2015; Fulda & Wetter, 2008; Clauw, 2014). Increasing attention has been directed toward studies of especially challenging patient populations with *chronic neuropathic pain* resulting from peripheral neurologic trauma (complex regional pain syndrome) (Vaso et al., 2014; Gregg, 2013; Drummond et al., 2018), degenerative inflammatory neuropathies (Zeigler et al., 2015; Demant et al., 2015) and brain trauma (chronic traumatic encephalopathy) with its long-term implications for post-traumatic stress disorder (PTSD) (Scioli-Salter et al., 2015; Harvold et al., 2018).

Chronic neuropathic pain is attributed to a hyperreactive state known as sensitization; these conditions may occur in both peripheral and central neural structures and may be modulated by placebo and nocebo factors (Apkarian et al., 2009). Peripheral nerves and their ganglia, when subjected to peripheral nerve injuries and recurrent or sustained inflammatory conditions such as peripheral neuropathies, postherpetic lesions, osteoarthritic vasculopathies, and neurotoxic exposures such as HIV, may undergo neuroplastic pathologic changes that result in spontaneous (ectopic) and triggered firing of the hyperactive peripheral structures (King et al., 2011, Haroutounian et al., 2014). Research by Sommer et al. (2018) has led to the conclusion that unresolved neuroinflammation may be a key common pathophysiology that drives the nervous system toward sensitization and chronic pain.

Patients with chronic *peripheral sensitization* display varying degrees of sensory nerve losses, spontaneous daily pain cycles, lowered pain detection and tolerance thresholds, and heightened pain responses to touch, percussions, and uncomfortable stimuli applied to nerve distributions—clinical states known as allodynia, hyperpathia, and hyperalgesia. At molecular levels peripheral sensitization is induced by nerve growth factor (NGF) through increases of transient receptor potential (TRPV1) in the peripheral nerves and sensory ganglia (Eskander et al., 2015; Obreja et al., 2018). If these peripheral neural changes are sustained, barrages of ectopic noxious impulses are propagated along damaged voltage-gated sodium channels (Devor, 2006; Black et al., 2008); this activity then projects to dorsal horn neurons in the spinal cord and brainstem to induce central sensitization at many levels of ascending and descending central pain pathways (Haroutounian et al., 2014; Obreja et al., 2018; Ringkamp & Raja, 2014; Sessle, 2005; Woolf & Salter, 2000). Research on placebo and nocebo influences on pain attributed to peripheral sensitization, however, is scant. More attention has been given to clinical conditions where chronic pain has been attributed to sensitization of the central nervous pain.

Central sensitization is a state of sustained and often spreading pain that is both spontaneous and evoked, presenting with lowered pain detection and tolerance thresholds (Woolf & Salter, 2000; Sessle, 2005; Latremoliere & Woolf, 2009; Woolf, 2011; Hansson, 2014). These phenomena are due to the upregulation of brain-derived neurotrophic factor (BDNF) and neuropeptides found in the spinal cord dorsal horn (Price & Inyang, 2015; Treede, 2016); these central changes are now known to be activated by microglia (Echeverry et al., 2017; Ji et al., 2016; Kronschlager et al., 2016; Piao et al., 2006; Sessle, 2007). The CNS may therefore be sensitized by inflammation, disease or injury initiated in the periphery and/or more directly by intracranial injuries to the brain and/or spinal cord (chronic traumatic encephalopathy), neurodegenerative disorders, (multiple sclerosis, Parkinson's disease, Alzheimer's disease), and neurovascular lesions (stroke, ischemia) (Apkarian et al., 2011; Baliki & Apkarian, 2015; Benedetti, 2008; Ji et al., 2016). Research has led to the conclusion that sustained neuroinflammation acting at many levels of the sensitized central pain system is a common activator of centralized chronic pain conditions and is the target of many therapeutic interventions including placebo analgesics that act through common central endogenous mechanisms (Elman & Borsook, 2016; Tinnermann et al., 2017; Sommer et al., 2018).

Pain Modulation: Endogenous Receptors

There is much evidence that pain modulation differs significantly between experimental, acute nociceptive and chronic neuropathic types of pain conditions, also markedly influencing the clinical expectations of cognitive factors, including placebo and nocebo interventions (Taylor et al., 2016).

In previous research with *normal patients* exposed to acute stressful events and acute nociceptive pain, the effects of endogenous opioid receptors mu and kappa have received primary attention; mu receptor activation has been shown to elicit pleasant, euphoric experiences and acute short-term pain relief (Ballantyne & Sullivan, 2017); kappa opioid receptors, by comparison, respond with stimulus aversiveness and dysphoria (unpleasantness) (Ikemoto, 2019; Corder et al., 2019).

However, with the development of *long-term neuropathic pain*, mu opioid receptor (MOR) activities become associated with distorted "rewards": diminishing analgesic effects, central and opioid-induced hyperalgesias, transient episodes of analgesia and euphoria, and opioid craving potentially leading to misuse and addictions (Ballantyne et al., 2018; Borsook, 2017; Le Merrer et al., 2009; Nees et al., 2017). Kappa opioid receptor (KOR) systems, in contrast to MOR activities, respond to ongoing stress and chronic neuropathic pain with the delayed emergence of "anti-reward" effects: escape aversive behaviors, reduced pain inhibition, increased pain hypersensitivities, and risk of depression and anhedonia (Corder et al., 2019; Ikemoto, 2019; Massaly et al., 2016a, b).

Both mu and kappa opioid systems act on peripheral nerves, cranial and dorsal root ganglia, and centrally in the spinal and medullary (trigeminal) and dorsal horns as well as numerous supraspinal pain processing sites (see Sessle, 2005, 2021). Opioid analgesia and hyperalgesia effects are largely mediated by GABA neurons, which trigger D_1 and D_2 dopaminergic receptors to influence the affective dimensions of pain (Wood, 2008; Scott et al., 2008).

In addition to the primary mu and kappa receptors, at least three other endogenous non-opioid endogenous receptor systems may co-activate with endogenous opioid systems to modulate pain and influence placebo-nocebo interactions (Colloca & Barsky, 2020). These include (1) estrogenic nonapeptide receptors responsive to the monoaminergics oxytocin and vasopressin (Alois & Bonafazi, 2006; Colloca et al., 2016; Gonzales & Condez-Lara, 2017; Tracy et al., 2015; Zakharian, 2017); (2) bidirectional neuron-immune cell interactions including regulatory T cell inhibition of neuropathic pain at peripheral nerve injury sites (Austin et al., 2012; Grace et al., 2014; Davoli-Ferreira et al., 2020; Scholtz & Woolf, 2007); and (3) endocannabinoid receptors with particular actions on stress-induced analgesia, fear-mediated pain, and neuroinflammation (Clayton et al., 2002; Maldonado et al., 2016; Rea et al., 2013).

The *endocannabinoid signaling system* (CBD) is a long-recognized pain-modulating immune cell system that acts in parallel with opioidergic systems to influence chronic inflammation and neuropathic pain (Benedetti et al., 2013; Nadal et al., 2013; Welch, 2009). At least two different cannabinoid receptors have been identified: CB_1R, highly expressed in central nervous centers including antinociceptive activities in the periaqueductal gray (PAG), nucleus raphe magnus (NMR), and basolateral amygdala (Benedetti et al., 2011; Finn et al., 2003; Rea et al., 2013), and CB_2R, localized in peripheral immune cells (Mitirattanakul et al., 2006). Considerable preclinical research has linked cannabinoid actions with serotonergic, noradrenergic, and endogenous opioids for acting synergistically on descending inhibitory pain pathways to modulate neuropathic pain and anxiety-like behavior (De Gregorio et al., 2019; Moldonado et al., 2016; Ossipov et al., 2014; Welch, 2009). With the recent liberalization of restrictions on clinical research, there has been accumulating evidence that the cannabinoids have mild-to-moderate analgesic efficacy as well as anxiety reduction for patients with chronic neuropathic pain (Lotsch et al., 2018; Maldonado et al., 2016; Mun et al., 2020).

The mechanisms and links between endogenous opioids, sex hormones, immune T cells, and endocannabinoids, however, remain unclear. More research is needed to clarify how and to what degree pain modulation and placebo/nocebo interventions are activated through the multiple coexisting receptor systems and how they influence reward and antireward central activities and clinical care.

Pain Modulation: Reward and Antireward

With ongoing stress, neuroinflammation, and neuropathic pain, endogenous mu opioid, kappa opioid, and endogenous cannabinoid receptor systems (MOR, KOR and CBD) become activated and expressed through coexisting bidirectional "reward and antireward" systems that primarily influence the aversive "emotional" component of pain (Borsook, 2017; Koob & Le Moel, 2008). There is increasing evidence that these systems also impact placebo and nocebo responses to pain (Benedetti et al., 2013; Porreca & Navratilova, 2017).

Pain Reward and MOR Systems

Mu opioid receptors (MOR) respond to *acute nociceptive pain* and stress in *normal patients* by activating mesolimbic brain systems that result in "pleasing" natural rewards associated with relief of pain relief (Elman & Borsook, 2016; Le Merrer et al., 2009). This is accomplished through a cascade of central actions of dopamine responding in key mesolimbic centers: amygdala, ventral tegmental area (VTA), nucleus accumbens (NAc), and periaqueductal gray (PAG) (Navratilova et al., 2015). The "reward hypothesis" further proposes that expectations of relief from pain are associated with feelings of reward (Leknes et al., 2011; Borsook et al., 2016; Navratilova et al., 2015) that correlate with increases in dopamine tone in key brain centers (Ballantyne & Sullivan, 2017; Borsook, 2017; de la Fuente-Fernandez et al., 2006, 2009, Lidstone et al., 2010; Navratilova et al., 2015).

However, with the development of *chronic neuropathic pain* and *sustained stress*, the MOR-dopamine links to pain relief reward systems may become dysfunctional, leading to losses in pain inhibition, perpetuation of chronic pain, and increasing craving and dependencies on exogenous opiate drugs (Koob & Le Moal, 2001; Lucher, 2016; Le Merrer et al., 2009; Nees et al., 2017; Volkow et al., 2017; Wightman & Robinson, 2002) (see Acree, chapter "Endogenous and Exogenous Opioids: Role in Substance Use Disorders", this volume). Much research into interrelationships between pain and mu opioid reward receptor activities and the role of dopamine tone has been facilitated by brain imaging (Baliki & Apkarian, 2015;

Diederich & Goetz, 2007). Mu receptor-driven reward activities are known to project bidirectionally with higher brain "reward system" centers of opioid-dopamine signaling centered in the nucleus accumbens (NAc), the amygdala, and the medial prefrontal cortex (PFC), especially the anterior cingulate cortex (ACC) (Baliki & Apkarian, 2015; Hyman et al., 2006; Ikemoto, 2019). Much pain reward activities have also been localized to the lower brainstem and spinal cord centers where MOR beta-endorphin signaling in midbrain PAG has been shown to increase dopamine tone facilitated by GABA to sustain pain inhibition (Jiang et al., 2016; Kong et al., 2013; Scott et al., 2008; Wood, 2004, 2008). With time and ongoing unrelieved pain and stress, however, mu opioid-dopamine reward systems become dysfunctional and give way to the dominant unpleasantness (dysphoria) of the kappa opioid receptor (KOR)-mediated "antireward" system (Jiang et al., 2016; Nees et al., 2017).

Pain Antireward and KOR Systems

Both acute nociceptive and chronic neuropathic pain aversiveness and unpleasantness are influenced by endogenous "antireward" system activities (Borsook et al., 2016; Corder et al., 2019; Ikemoto, 2019). Antireward pain aversiveness is initiated and driven by signaling increases in dynorphin-mediated kappa opioid receptors (KOR) concentrated in the basolateral and central nucleus of amygdala (CeA) leading to dysregulation of pain pathways at key brain levels of the mesolimbic pain response system (Koob, 2020; Massaly et al., 2016a, Navratilova et al., 2019; Phelps et al., 2019; Porreca & Navratilova, 2017). Neuroimaging studies have further shown that kappa opioid signaling in the central nucleus (CeA) of the amygdala also promotes ongoing painful hyperresponses to mechanical and thermal stimuli (allodynia and hyperalgesia) and enhances emotional and affective consequences of neuropathic pain (Chen et al., 2017; Corder et al., 2013, 2019; Liu et al., 2018; Navratilova et al., 2019).

Clinically, MOR and KOR receptor systems have opposite effects on experienced pain, whereby MOR produce transient clinical analgesia, euphoria, and opioid craving, and KOR/dynorphin activities, in contrast, are associated with dysphoria, aversive behavior, hyperalgesia, and hypersensitivities as well as anhedonia and depressive-like symptoms (Knoll & Carlezon, 2010;; Koob, 2020; Massaly et al., 2016b; Mukaetova-Ladinska et al., 2016). Clinical researchers have pointed out that the KOR/dynorphin antireward aversive state resembles and may have similar underlying mechanisms to ongoing opiate drug withdrawal symptoms (Bruchas et al., 2010; Massaly et al., 2016a, b; Phelps et al., 2019).

Balancing Pain Reward and Antireward: The Amygdala

The amygdala is a critical link controlling the emotional (aversive) component of pain, reward and antireward systems, and, by extension, placebo and nocebo effects (Zubieta & Stohler, 2009; Gandhi et al., 2020). Preclinical research has implicated kappa opioid signaling in the basolateral and central amygdala underlying the loss of pain inhibition in experimental animals (Phelps et al., 2019; Navratilova et al., 2019). The amygdala coordinates pain-suffering outputs through a descending cascade of activity on corticomesolimbic opioid receptors and dopamine release in inhibitory GABA centers in the nucleus accumbens (NAc), midbrain periaqueductal gray (PAG), and medullary rostroventromedial (RVM) regions that ultimately modulate pain through bidirectional excitability in the dorsal horn regions of the medulla and spinal cord (Bodnar, 2017; Jiang et al., 2016; Liu et al., 2018 Schwartz et al., 2014; Sessle, 2021).

Human studies have pointed to the amygdala's control of negative emotional states related to pain, increasing the perceived unpleasantness of pain without altering pain intensity from evoked stimuli (Navratilova et al., 2020) (see review by Bushnell et al., 2013). Jiang studied fMRI findings of 17 patients with chronic low back pain which showed amygdaloid effects in cognitive/affective networks (Jiang et al., 2016). These conclusions have been bolstered by other imaging research that have revealed abnormal gray matter densities in the amygdala among patients with chronic back pain (Cauda et al., 2014; Ung et al., 2012).

Amygdaloid pain-modulating influences are bidirectional, enhancing both reward (euphoria) and antireward (dysphoria), influencing comorbid anxiety-depressive disorders and the anhedonic decreases in motivation often found among patients with chronic pain (Gandhi et al., 2020; Leknes et al., 2011; Schwartz et al., 2014). Finally, amygdaloid controls have been shown to facilitate the delayed shift from nociceptive pain to chronic neuropathic pain (Hashmi et al., 2013; Gandhi et al., 2020) and appear to regulate the expression of clinical pain hypersensitivities of hyperalgesia and allodynia commonly present in patients with neuropathic pain (Corder et al., 2019; Liu et al., 2018).

Balancing Pain Reward and Antireward: Dopamine Effects

There is continuing controversy regarding the role of dopamine in pain reward and antireward and the variability in placebo and nocebo effects on neuropathic pain. Significantly, there are at least five variants of dopamine neurons located in many sites of the reward system, however, and dopamine neurons are anatomically and functionally very heterogeneous, exhibiting both excitatory and inhibitory actions (Fields, 2014; Navratilova et al., 2015). This may partly explain the variability found in placebo and nocebo effects among patients with neuropathic types of pain.

The functional relationships between dopamine and neuropathic pain have been extensively studied in patients with hyperalgesic postsurgical and post-nerve injury pain (Petersen et al., 2012, 2014; Skyt et al., 2018, 2020; Vase et al., 2005). In 2018, Skyt et al. studied dopamine placebo effects among 19 patients with neuropathic pain after thoracic surgery. They noted large placebo effects on patients' chronic neuropathic pain but found that neither increasing nor decreasing dopamine tone in the brain correlated with pain intensity levels. Their investigation used the dopamine antagonist haloperidol and the agonist levodopa and showed that, although dopamine markedly influenced the levels of desire and expectancy, there was no evidence of direct dopamine effects on neuropathic pain or direct placebo and nocebo outcomes. Other investigations have found similar disconnects between dopamine levels and direct placebo effects as well as both up and down dopamine placebo effects among patients with neuropathic pain (Bruchas et al., 2010; Bannister et al., 2015; Lucher, 2016; Massaly et al., 2016a, b; Le Merrer et al., 2009; Petersen et al., 2014; Scott et al., 2008; Vase et al., 2005; Volkow et al., 2012, 2017; Wightman & Robinson, 2002). It has also been argued that the variability found in studies of opioid-dopamine placebo effects on pain may be due to differences in research study populations; the effects found in healthy volunteers or those suffering from acute nociceptive pain may differ significantly from effects in patients with chronic neuropathic pain (Forsberg et al., 2017; Price et al., 2008; Taylor et al., 2016).

As we consider the conclusions regarding the role of bidirectional amygdaloid balances on pain and reward, the complex and controversial influences of the opioid-dopamine activities in pain modulation and reward/antireward activities, we must challenge the belief that the endogenous opioid-dopamine reward system is a simple and exclusive mechanism responsible for placebo analgesia and particularly nocebo effects on patients with widespread chronic neuropathic pain where maladaptive brain changes preexist (Price et al., 2008; Vase et al., 2015; Wood, 2004).

It would also appear that future studies should control for the presence of pain reward and antireward systems and the cognitive placebo-nocebo effects among patients with chronic neuropathic pain. Research and clinical care should also recognize the heterogeneity of dopamine neurons acting at differing levels of the brain reward system and consider that multiple neurotransmitter systems may act synergistically with dopamine as alternative pathways regarding pain and reward interactions (Navratilova et al., 2015; Welch, 2009).

Pain Modulation: Placebo Analgesia and Nocebo Hyperalgesia

The presence of potentially bidirectional inhibitory and excitatory reward and anti-reward central nervous systems sets the stage for understanding both *placebo analgesia effects* (Vase et al., 2015) and *nocebo hyperalgesia effects* (Bee & Dickenson, 2008; Phelps et al., 2019; Scott et al., 2008). These phenomena also help explain how both pain reductions (analgesia) or pain exacerbations (hyperalgesia) may be found at varying times for a single given patient with chronic neuropathic pain (Staud, 2012).

Sustained (chronic) neuropathic hyperalgesia is generally viewed as a failure of pain inhibition by receptors activating the diffuse pain modulation system (DPMS), traditionally mediated through descending (top-down) endogenous opioid inhibitory effects (Fields, 2014; Heinricher et al., 2009; Price & Prescott, 2015; Ren & Dubner, 2002). In addition, basic brain research using animal pain models has shown that aversive pain-modulating responses are also influenced by autonomic serotonergic and noradrenergic projection neurons in regions such as locus ceruleus that may have both inhibitory and excitatory effects on pain (Bannister et al., 2009, 2015; Bee & Dickenson, 2008; Fields et al., 1991, 1995; Llorca-Torralba et al., 2016; Porreca et al., 2002).

Pain Modulation: Placebo Analgesia

Placebo analgesic effects on pain are heavily dependent upon endogenous opioid molecules that influence inhibitory pain modulation (Kong et al., 2013; Petersen et al., 2012) with opioidergic, endocannabinoid, and dopaminergic interactions occurring at many brain levels to exert critical inhibitory effects on supraspinal, brainstem, and spinal dorsal horns (de la Fuente-Fernandez et al., 2006; Lidstone & Stoessl, 2007; Lidstone et al., 2010). Acting through a common descending pain modulation system, *placebo analgesic effects* may also substantially enhance exogenous opioid analgesic drug efficacy as well as abuse (Colloca et al., 2016; Eippert et al., 2009) acting through both endogenous opioid and non-opioid (endocannabinoid) agonists (Benedetti et al., 2011). There is other evidence that the ineffectiveness of exogenous opioid drugs, abuse, and possibly placebo analgesic "failures" are due to both pathological derangements of the endogenous descending inhibitory systems and nocebo hyperalgesic actions (Ballantyne, 2018; Finnerup et al., 2010; Kamermann et al., 2015; Price & Prescott, 2015).

Pain Modulation: Nocebo Hyperalgesia

Hyperalgesia is a neuropathic state of enhanced pain responses to normally innocuous or tolerable noxious stimuli; these include excess responses to mechanical, thermal, chemical, inflammatory, or environmental stimuli (see Fields, 2014; Jensen et al., 2001). Hyperalgesia is also characterized by sustained pain sensitization mediated through plastic neural responses to injury and inflammatory disease that occur in both peripheral and central nervous systems (for historical reference, see Amir & Devor, 1993; Chapman et al., 1998; Nystrom & Hagbarth, 1981; Wall & Gutnick, 1974; Wall & Devor, 1983). Clinical states of sustained hyperalgesia are commonly associated with nerve injuries and inflammatory diseases that drive central sensitization responses to immune cell actions (Fields, 2014; Salter & Stevens, 2017). Nerve growth factor (NGF) and other inflammatory cytochromes released by damaged immune cells (microglia, astroglia, Schwann cells) appear to play key roles in the induction and maintenance of hyperalgesic pain states (Chaumette et al., 2020; Hirth et al., 2013; Theodosiou et al., 1999; Tinnermann et al., 2017).

A second source of hyperalgesia is opioid induced (OIH) caused, paradoxically, by exogenous opioid drugs given to patients for management of both acute surgical pain (Fletcher & Martinez, 2014; Hayes et al., 2020; Hayes & Painter, 2017; Niedermayer et al., 2020; Servick & Rakola, 2016) and chronic neuropathic pain (Araldi et al., 2017; Bannister & Dickenson, 2010; Morasco et al., 2017). Recent preclinical research using the chronic constriction nerve injury male rodent model of established neuropathic pain confirms that morphine, oxycodone, and fentanyl measurably amplify established neuropathic pain (Green-Fulgham et al., 2019). This effect curiously was seen only when the drugs were administered 5–10 days after the nerve injury.

Clinically, OIH typically follows repeated exogenous mu opioid-active drug exposures, resulting in a state of increased spontaneous and evoked pain and diminished analgesic responses to these same drugs (Brush, 2012; Chu et al., 2008; Fields, 2014; Hay et al., 2009). Although OIH induction is most often associated with higher dose exogenous opioid exposures, even low-modest doses repeated after initial exposures appear to have the potential to induce hyperalgesic priming and drive the transition from acute to chronic forms of pain especially in young opioid-naive patients (Araldi et al., 2015; Corder et al., 2018). Imaging analyses have shown that OIH is associated with distorted endogenous opioids acting at many levels of the centrally sensitized pain endorphin pathways (Araldi et al., 2017; Corder et al., 2013).

Nocebo hyperalgesia, like placebo analgesia, appears to modulate pain (negatively) by acting within a cortico-subcortical-spinal descending pain network (rACC-PAG-spinal dorsal horn) that is modifiable by placebo and other cognitive actions (Benedetti et al., 2006, 2007; Colloca & Miller, 2011; Kong et al., 2008; Petersen et al., 2012, 2014; Tinnerman et al., 2017). Nocebo effects on neuropathic pain appear influenced by pain expectations as well as personal events, particularly mental health issues with anxiety and depression, and catastrophizing (Baliki & Apkarian, 2015; Hashmi et al., 2013). Hyperalgesic intensity is also responsive to *fear conditioning* (Colloca & Miller, 2011; Rea et al., 2013), particularly fear of high pain (Thomaidou et al., 2021). Progress in understanding the neural mechanisms underlying nocebo hyperalgesic effects has been aided by functional imaging studies (Benedetti et al., 2006; Kong et al., 2008). Questions remain, however, regarding the magnitude of nocebo effects among patients with chronic neuropathic pain (Petersen et al., 2014) and how that influences clinical practice (Colloca & Miller, 2011).

Although hyperalgesias in neuropathic pain clearly may be reduced by placebo and exacerbated by nocebo manipulations (Petersen et al., 2012; Tinnermann et al., 2017), more research is needed to clarify to what degree cognitive factors of expectations and emotional states influence placebo and nocebo effects on perceived pain levels among patients with concurrent nociceptive disease or injury-generated hyperalgesias (Petersen et al., 2014; Colloca et al., 2013, 2016).

Pain Modulation: Cognitive Factors and Empathy

Important research by Benedetti et al. (2013) has shown that placebo and nocebo cognitive interventions can change the meaning of pain from negative to positive for patients given verbal suggestions of positive expectations, hope, and beliefs. In their study, ischemic arm pain was induced in two subject groups: a first "negative" group was told that their pain was "to be expected"; the other "positive" group was told that their discomfort would be "beneficial to their muscles." Levels of pain tolerances in the positive group were found to be significantly higher than the negative group; and significantly, their higher pain tolerances were reversed by the administration of both the opioid antagonist drug naltrexone and also by the cannabinoid antagonist drug rimonabant. These findings lead to the conclusion that the positive (placebo) outcomes were determined by the co-activation of opioid and cannabinoid systems. Other clinical observations have revealed that differences in placebo pain perceptions may be heavily influenced by patient circumstances such as cancer pain, which is often associated with death, compared to postsurgical pain, usually associated with recovery and healing (Smith et al., 1998).

There are many other pain-modulating placebo effects known to be influenced by cognitive expectations and cultural, psychosocial, and religious practices (Jegindø et al., 2013; Kong et al., 2013; Wiech, 2016). Neuroimaging studies have further demonstrated cognitive-generated pain modulation in descending pain control network regions generated out of the dorsolateral prefrontal cortex (DLPFC), rostral anterior cingulate cortex, midbrain periaqueductal gray (PAG), and medullary raphe magnocellularis (RM) (Peirs & Seal, 2016; Vriens et al., 2011). Placebo analgesia involves an activation of these spinal cord inhibitory pathways including direct and indirect effects on sensitized dorsal horn neurons (Eippert et al., 2009; Peterson et al., 2014).

Central neural substrates mediating the expression of pain and fear have also been shown to overlap, with a critical linking element being the basolateral amygdala (BLA) which involves both emotional processing of painful stimuli and the expression of fear-conditioned analgesia (FCA) (Davis, 2001; Helmstetter & Bellgowan, 1994; Rea et al., 2013).

Pain Suffering and Empathy

We have cited the evidence that placebo analgesic interventions primarily target the affective-motivational (suffering) component of pain more than the sensory-discriminatory aspects (Elman & Borsook, 2016; Gomtsian et al., 2018; Singer,

2004) and that the main central pathways that mediate "antireward" activities associated with pain-suffering activities are centered in limbic systems originating in the medial forebrain, anterior cingulate cortex (ACC), insula, medial thalamus, hypothalamus, and especially amygdala which respond to external noxious stimuli and motivations (Bodnar, 2017; Neugebauer et al., 2009).

Humans expressing pain empathy appear to utilize these same pain-suffering brain regions' activities when research observers appear to "experience" physical pain when witnessing the suffering of another person in obvious physical pain. These observations, confirmed by brain imaging, contrast with brain imaging studies of somatosensory cortex and associated dorsolateral thalamic projections which are activated only when a person directly feels physical pain (Singer, 2004). Other researchers have shown that placebo analgesia is induced by face-to-face social observational learning but not by prerecorded patient images (Colloca & Benedetti, 2009; Hunter et al., 2014). Interestingly, brain imaging has also displayed signs of pain empathy behavior in nonhuman mammalian models of pain tied to the social bonding effects of oxytocin (Burkett et al., 2016). Thus, nonhuman animals may be similar to humans in their perceptions of pain in themselves and in others. Cross species research using pain models into how and to what extent placebo-like rewards, nocebo-like antirewards, and evidence reveal pain "empathy" in animals is important although obviously challenging.

From Great Expectations to Magic Feathers

Placebo and Nocebo Effects on Human Performance

One segment of placebo and nocebo effects that is almost exclusively in the domain of human research is task performance moderation (TPM), i.e., enhancement and detriment. TPM is commonly attributed to predictions of either desired or non-preferred outcomes whereby an individual develops an expectation. Expectations are comprised of anticipatory cognitions directed at a performance-related event or activity (Tracey, 2010). These cognitive constructs are often prompted by internal or external verbal stimuli aimed at motivation or dissuasion of effort for specific performance behavior. Similar to the effects of inert or sub-therapeutically active substances on remission of symptoms (or lack thereof), expectancy effects play a clear role in moderating performance.

A strong body of evidence supports the effects of expectation on the enhancement of motor performance, with much of this literature centered on sports. One of the earliest demonstrations of expectancy effect (Ariel & Saville, 1972) found that when athletes orally ingested a placebo pill that looked identical to a typical steroid pill, they exhibited significantly better performance in a series of weight-lifting tasks compared to pre-ingestion baseline. A consistent pattern of similar outcomes has been found in subsequent experimental investigations. Maganaris et al. (2000)

conducted a small (n = 11) study of professional powerlifters in which all participants were given a placebo that they were told was an anabolic steroid that would enhance their lifting performance. Performance improved significantly from baseline in Phase I of this experiment. Subsequently, 5 of the 11 participants were informed that they were actually given a placebo, while the other 6 were not; metrics of lifting performance subsequently significantly decreased in the subgroup who were told they received placebo. Using a similar cross-over design, Kalasountas and colleagues (2007) found comparable results with a slightly larger sample of nonathletes (N = 42). Participants who were given placebo but told they were given a performance-enhancing substance performed significantly better in weight-lifting exercises than those in the placebo condition. Following disclosure of placebo to half of participants (the placebo/no-placebo group), performance declined among those to whom this was disclosed, and performance was maintained among those to whom this was not disclosed. Thus, an expectancy effect was demonstrated in this experiment.

The previously described studies provide an initial glimpse of placebo effects and responses in human performance in sports and exercise. A clearer picture of the magnitude of placebo effects emerges from a meta-analysis (k = 14) by Berdi et al. (2011), which found an overall effect size for placebo of $g = 0.4$ (variance-weighted $g = 0.31$), with effects varying among subgroups, ranging from $g = 0.22$ (endurance exercises) to g = 0.48 (power exercises). These data indicate that across types of sports and exercise performance tasks, placebos yield small-to-moderate, yet significant, effects on performance outcomes. At the same time, the small quantity of eligible studies in this meta-analysis reflects the need for more rigorous work to be done in this area.

Similar to nocebo effects in pain described earlier, an additive effect of conditioning has been observed for placebo effects in physical performance. Carlino et al. (2014) found that administration of actual caffeine doses in an initial exercise trial (volitional leg extensor exercise) prior to administration of a non-caffeinated placebo before a subsequent exercise trial led to improved performance on the subsequent trial. Interestingly, this study found that verbal suggestion was only effective in eliciting a placebo response when participants were told there was at least a 50% chance of choosing a caffeinated beverage before exercise (versus no effect of 25% or 0% chance conditions). Again, direct experience provided by conditioning has a larger effect than verbal suggestion alone.

It is possible that placebos and nocebos may also modify fatigue effects (Carlino et al., 2014). Historically (e.g., Hill et al., 1924), fatigue states have been considered a natural "governor" of performance that rely on combined neuromuscular projections that integrate sensory, metabolic, and neurological signals to maximize work while limiting physical harm. Research suggests that there is a dual process that engages fatigue, comprised of excitatory/motivational and inhibitory systems operating along afferent and efferent pathways to evoke decreased or increased activity as warranted by the current signals (Noakes, 2012; Tanaka & Watanabe, 2012). Although there are inevitable physiological limits to performance that are insurmountable, Carlino and colleagues (2014) have proposed that placebo effects may delay cortical signaling deactivation of motor performance, while nocebo may accelerate such deactivation. This represents a potentially exciting future line of research, which will shape our understanding of the full range of human physical performance capabilities.

It is worth noting that, unlike in modern medicine, in the context of athletic performance, placebos may be more acceptable. Berdi and colleagues (2015) found that nearly half of their sample of professional athletes (n = 79) had previously experienced placebo effects on their athletic performance. Furthermore, a large majority (82%) reported believing that placebo effects could potentially enhance their performance, with most (67%) endorsing the acceptability of a placebo-related deception by a coach if such a deception could improve performance.

Within the realm of human performance in sports and exercise, there may be a place for the strategic use of placebo. Moderate but measureable improvements in performance can perhaps be achieved by activating the neural resources that are ostensibly inaccessible without the power of credible suggestion connected to a believable intervention. However, sustainable effects of placebo would likely require some form of cognitive restructuring conducive to integrating the person's new observations of their performance capabilities with existing belief structures about their performance limitations. Resolving the cognitive dissonance that would naturally arise from the discrepancy between belief in a particular performance limit and evidence of the ability to exceed that limit under conditions of mere expectation would be necessary. This is perhaps akin to the idea of the "magic feather" in the classic children's movie *Dumbo* (Disney et al., 1941), i.e., it was not the feather's "magic" but rather Dumbo's belief in his ability to fly that allowed him to do so. Thus, it is possible that placebo effects of many varieties, beyond human performance in sports and exercise, could be more meaningfully maintained if there was a system for organizing the new information inherently acquired in placebo responses, i.e., that improvement is possible with innate resources when those resources can be accessed and activated. How one induces such access and activation to internal resources is certainly a line of scientific inquiry unto itself and one worth pursuing.

Placebo and Nocebo Effects in Clinical Settings

A Biochemical-Psychosocial Matrix

As noted earlier in this chapter and elsewhere in this book, organismic responses to a pharmacological compound are proposed to be the direct result of biochemically based pharmacokinetic and pharmacodynamic properties of the substance, especially those substances that bind to opioid receptors. Thus, behavioral, cognitive, and emotional responses are commonly characterized as essentially biological or biochemical phenomena. The biochemical context of a drug's administration influences the matrix of responses that may be observed. However, biochemistry itself is characterized by a matrix of contexts. In humans, this matrix includes, among other things, physical conditions of the environment; the social structure of the environment; social cues and other demands for expected responses, outcomes, and behaviors in the environment; concurrent biochemical exposure; concurrent chemoreceptor activation; and learning history with all of these components that may affect predictions about (and, consequently, physiological responses to) the current situation.

Among humans there are intersecting biochemical and psychosocial contexts that influence the response matrix described above. In clinical applications of placebo, nocebo, and response expectancy in general, the construction of gender and sex differences has relevance. Sex and gender are differentiated; the former refers to aforementioned anatomical or hormonal differences, whereas the latter refers to social constructs that mediate interpersonal interactions (Johnson & Repta, 2012).

Clinical applications refer to the forms and forums of healthcare delivery, which occurs in a variety of social contexts that may affect treatment response (Benedetti, 2002). Those social contexts are comprised of interpersonal interactions that may be mediated by gender or sex, which may influence an individual's perception of the potential benefit she or he will derive from a pharmacological intervention delivered to them. As an example, if a patient interacting with a physician interprets the credibility or accuracy of information or competency of care he or she receives based on stereotypes about the physician's gender, the meaning ascribed to the medication or treatment may either enhance or diminish its perceived effects. Here, the meaning is conceptualized as either a placebo or nocebo effect; any consequence of that assigned meaning would be a placebo or nocebo response as described earlier. Both positive and null, or adverse, effects can be linked to expectations. The moderation of effects seen in nocebo effects may extend to adverse effects of medications or treatments and is a concerning potential outcome of discrimination in both healthcare settings and in daily life that may overflow into healthcare interactions.

As described by Benedetti (2002), the quality of the interpersonal interaction between physicians and patients is an essential piece of effective medical care and can impact treatment outcomes. A variety of interactions have been found that characterize the matrix described above: physicians who exhibit more patient-centered thinking are trusted more (Fiscella et al., 2004); patients are more satisfied with care, follow advice more consistently, and respond better to treatments from physicians they trust (Thom et al., 2002); trust predicts healthcare-seeking behavior, and extent of healthcare contact predicts greater trust in and satisfaction with physicians (Balkrishnan et al., 2003); and even minor characteristics of a healthcare encounter such as physician attire seem to matter—patients feel more trusting of professionally dressed physicians than those dressed casually, with the importance of attire on trustworthiness being significantly greater for female physicians than males (Rehman et al., 2005).

Of relevance to healthcare delivery, research using observational learning paradigms has found nocebo effects that vary by gender. For example, Klosterhalfen and colleagues (2009) found significantly greater nocebo effects of conditioning with nauseogenic suggestion in women than in men. Additionally, there is also evidence of [observer] × [target] × [gender] interaction effects in such paradigms. Swider and Babel (2013) reported that when male and female subjects were exposed to male and female demonstrators of painful stimuli, pain ratings were significantly higher following exposure to a male demonstrator in the experiment, regardless of the sex of the participant. Thus, the gender of the messenger and the recipient appears to matter as much as the message when it comes to nocebo. Such reciprocal interaction effects are likely to be continuously operating in healthcare environments and warrant further consideration.

Sex stereotypes are defined as a group of psychological characteristics attributed to males and females. Sex stereotypes fuel gender-role prescriptions—i.e., expectations related to one's perceived or expressed sex. Gender-role prescriptions for women and girls typically include passive, caretaking, or other-centered behaviors; when these expectations are violated, such behavior is more likely to be misinterpreted as aggressive, even though in men the same behavior may be considered normative due to different gender-role expectations. Research has identified deleterious effects of gender-role prescriptions for women in several domains (e.g., Eagly et al., 2007; Chin, 2002; Heilman, 2001; Schmitt & Allen, 2016; Wade, 2001).

The context of gender interactions becomes more complicated when examined outside of the heteronormative, cisgender lens. Research examining the role of sexual minority and gender identity minority status on healthcare experiences has found that compared to heterosexual patients and cisgender patients, patients who identify as lesbian, gay, bisexual, transgender, genderqueer, or questioning underutilize healthcare (Rounds et al., 2013; Shipherd et al., 2010) and are significantly more likely to report negative experiences during healthcare service delivery (Fish & Wilkinson, 2003; Fredriksen-Goldsen et al., 2013; Lane et al., 2008; O'Hanlan et al., 2004; Saphira & Glover, 2000). The experience of LGBTQ+ individuals when interacting with those who hold hostile, e.g., homophobic or transphobic, attitudes is commonly anxiety inducing (Rounds et al., 2013).

When discriminatory attitudes extend into the realm of healthcare, there are natural implications for nocebo effects in those services. If patients receive care from healthcare providers whom they perceive to be non-affirming or hostile based on sexual orientation or gender identity, healthcare that induces discomfort, e.g., invasive physical examinations, needle sticks for injections or blood draws, and medication side effects, may be experienced as intolerably uncomfortable, potentially leading to opting out of those procedures. Potential benefits of medical advice or recommendations may be muted in light of the perception that such prescriptions originated from a hostile source. The risk that true benefits will be discounted and true adverse effects will be amplified seems elevated in such circumstances. This represents an area of healthcare research in desperate need of systematic empirical study. With wider spread acceptance of same-sex sexual orientations, and wider recognition of the relative commonality of identifying as an LGTBQ+ person, the effect of being "out" to one's healthcare provider on the care received must be clarified as a matter of pragmatism. Trust in a healthcare provider is an essential and indelible part of the packaging in healthcare. Nocebo effects can only be more likely when that trust is missing because of perceived hostility.

Some Notes on Placebo Effects in Psychopharmacology

There is a wealth of literature on placebo effects in mental health treatment, which encompass both psychotherapy (e.g., "nonspecific factors" research; e.g., Kazdin, 1979; Beutler et al., 1986) and psychopharmacological research (e.g., Kirsch & Sapirstein, 1998; Kirsch et al., 2002). A full canvassing of this rich literature is beyond the scope of this chapter because of our emphasis on the endogenous opioid system. Nonetheless, some aspects and findings do merit consideration here.

Kirsch (2014) has delineated myriad problems with antidepressant research, noting that meta-analyses conducted by him and his colleagues have called into doubt on empirical grounds the purported benefits of newer medications, especially selective serotonin reuptake inhibitors (SSRIs). We will not rehash the heated debate (e.g., Klein, 1998) that ensued from the advent of Kirsch and colleagues' (1998, 2002, 2008) thorough, balanced, and unbiased re-analysis of SSRI clinical trial data. Instead, we will briefly focus on a lesser discussed question within the SSRI-placebo literature that is more relevant to this chapter, i.e., what else might SSRIs be doing besides selectively inhibiting serotonin reuptake?

The introduction of SSRIs was based on the relatively myopic serotonin model of depression (Kramer & Medawar, 1994). However, before the first SSRI—fluoxetine—had been on the market for even a decade, there were already questions about the legitimacy of the proposed mechanism of action and duration of its antidepressant effects (Barondes, 1994). Fourteen different 5-hydroxytryptophan receptor subtypes had already been identified by the early 1990s (see Siegel et al., 1999). Moreover, pharmacological blockade of other neurotransmitter reuptake inhibitors, viz., *serotonin-norepinephrine* reuptake inhibitors (SNRIs), was already in development. By this time, there were also data emerging in support of other potential mechanisms of antidepressant action, including serotonin receptor antagonism (not just reuptake inhibition) as observed in animal experiments with trazodone (e.g., Aprison & Hingtgen, 1981; Clements-Jewery et al., 1980; Hingtgen et al., 1984). Thus, the actual mechanism of action leading to antidepressant effects of serotonergic drugs was unclear both before and after fluoxetine went to market. What was clear early on is that the biological pathway of depression pathogenesis and its treatment was likely multifactorial.

In the proceeding decades, research identified the involvement of endogenous opioid activity in both depression (Bodkin et al., 1995; Hsu et al., 2015; Kennedy et al., 2006; Knoll & Carlezon, 2010) and anxiety disorders (Bruchas et al., 2010; Colasanti et al., 2011; Graeff, 2017; Sher, 1998). Blockade of opioid receptors (e.g., μ-opioid receptors) is associated with physiological anxiety symptoms and anhedonic mood states, while endogenous opioid activity is associated with placebo response in treatments of PTSD. Further evidence of the role of endogenous opioids in depression and anxiety comes from clinical and preclinical studies of pharmacotherapies. For example, in addition to reuptake blockade, SSRIs are also associated with upregulation of opioid peptides, e.g., proenkephalin 1 (Sillaber et al., 2008). In the same vein, there is an emerging line of research that has found μ-opioid agonists (e.g., buprenorphine) may be effective treatments for depression (Ehrich et al., 2015; Kosten, 2016; Schatzberg, 2016; Yovell et al., 2016). Experimental research in this line has identified that among nondepressed, healthy subjects, buprenorphine positively affects attentional bias, improves recall of pleasant emotional stimuli, reduces recognition of fear-related stimuli, and downregulates stress responses to stressful stimuli, i.e., people attend less to negatively valenced (possibly depressogenic) stimuli and recall more pleasant emotional stimuli in laboratory settings (Bershad et al., 2015, 2016; Ipser et al., 2013; Syal et al., 2015). A recent study by Bershad and colleagues (2018) shed further light on the potential cognitive mechanisms involved in the antidepressant effects of buprenorphine. In a study of 38 people with symptoms of depression, administration of 0.2 mg of buprenorphine led to significantly higher ratings of positivity for laboratory images related to social interactions compared to placebo. Similar to previous studies, Bershad et al. (2018) once again found significantly less recall of fear-related facial images compared with placebo.

The data related to the involvement of endogenous opioids in anxiety and depressive disorders, existing treatments, and emerging treatment options is instructive and exciting. Placebo responses, a main focus of this chapter, are known to account for over 80% of the effects seen in SSRI studies (Kirsch, 2014).

As we have elaborated extensively above, the endogenous opioid system (the main focus of this book) is also established as a substrate for placebo and nocebo responses. Furthermore, some of the most recent laboratory and clinical research suggests that depression and anxiety may not only be related to dysregulation of opioid receptors (Kennedy et al., 2006), but that opioid receptor agonists may target depression (and potentially anxiety) symptoms specifically (Bershad et al., 2018; Kosten, 2016). However, our knowledge about the engagement of the endogenous opioid system in these areas of human health has a relationship to what research has told us about exogenous opiates/opioids. The deleterious effects of opioid/opiate addiction, as described in this chapter as well as by Acree (chapter "Endogenous and Exogenous Opioids: Role in Substance Use Disorders", this volume), place any proposal for the use of opioids or opioid receptor agonists in a precarious position; it may be scientifically sound but socially or politically unacceptable. Nonetheless, it seems plausible at the time of this writing that we are at a crossroads in pharmacology. This crossroads presents us with the opportunity to move in a direction that is perhaps new, ostensibly unconventional, yet empirically grounded.

Conclusions

As discussed extensively in this chapter, placebo and nocebo effects have far-reaching implications for human health and human performance. The very existence of a placebo effect at all does suggest that the human mind is, as Haygarth put it centuries ago, a "wonderful force" indeed. As we have described, that force and its neurobiological mechanisms are just beginning to be understood.

As two sides of the same coin, the placebo-nocebo dyad is at once complimentary and reciprocally confirmatory. Given that the human nervous system is structured with so much reciprocity (e.g., sympathetic and parasympathetic; central and peripheral; afferent and efferent), there is a sensibility to this. It would make little sense for expectation to yield only positive effects while having little effect on adverse or aversive outcomes and vice versa. The nocebo effect is essentially confirmation that the power of expectation to influence subjective experience extends across the full range of experience types—pleasant to unpleasant and appealing to aversive. In this way, we may think of placebo and nocebo effects, and their neurobiological mechanisms, as reflecting a continuum.

As we have shown, placebo and nocebo effects also provide unique further opportunities to understand the endogenous opioid system. The discovery of placebo and nocebo effects has given us an organic comparison by which to control, so that we can understand more clearly the actual effects of opioidergic/opiatergic compounds. It has also demonstrated the interrelationships between endogenous opioids and many other molecules. The findings of attenuation of these effects by the blockade of opioid receptors tell us as much about innate opioids and their actions, as they do about the magnitude of effect yielded by cognition and emotion. These data support the contention that our interpretations of our experiences, and the meaning we ascribe to them, affect the shape and direction of those experiences, from the receptor level up.

There are some key implications of the data discussed in this chapter. First, one prominent theme to be extracted from these data strikes at the heart of this volume, i.e., the centrality of the endogenous opioid system. Placebo and nocebo data have already elucidated the pivotal roles played by the endogenous opioid system throughout the body. Research that aims to elicit placebo responses in particular will contribute further data on the interrelationships between opioid receptors, associated systems (e.g., endorphinergic, dopaminergic, noradrenergic), and their neuroanatomical correlates. These multisystem relationships must also be integrated with emerging genomic data. As with all complex biological responses, there is an intersection of genetic polymorphisms at the core of any phenotype, including the placebo response.

Another implication of these data is that the placebo effect and, by the same token, the nocebo effect are best thought of as "meaning effects," sometimes termed "meaning responses" (Moerman, 2002; Moerman & Jonas, 2002). Adding this terminology to descriptions of science may seem superfluous. However, we contend that this term adds clarity, rather than muddying the conceptual waters. While the placebo and nocebo effects observed in the foregoing research travel along a complex network of neural paths, those effects start at the point of interpretation. What one tells oneself about one's experience has much to do with the ultimate impact of that experience. Interestingly, the findings from placebo and nocebo research are consistent with data from emotion regulation research that has found strong evidence for the regulating effects of cognitive reappraisal of emotionally activating events (see Gross, 2014). So great is the influence of interpretation on emotion and physiology, that reappraisal has become a pillar of some major evidence-based psychological treatments (e.g., dialectical behavior therapy; Linehan, 2014).

A final theme of the data from placebo is that there is much untapped potential within humans. Whether it be performance of physical tasks, healing from disease, or responding adversely to pseudo-noxious stimuli, placebo and nocebo data suggest that the human brain and body move perhaps more synchronously than historically recognized. This should serve as only more credibility for a post-Descartian era. Moreover, these robust autogenic effects can potentially be better capitalized upon to enhance the human condition and alleviate suffering. Future researchers may do well to identify and develop methodologies to amplify placebo effects and to minimize nocebo effects. Additional concentrated lines of research devoted to understanding both the full potential and the inherent limits of these phenomena remain needed at this time.

References

Aloisi, A. M., & Bonafazi, M. (2006). Sex hormones, central nervous system and pain. *Normal Behavior, 50*, 1–7.

Amanzio, M., & Benedetti, F. (1999). Neuropharmacological dissection of placebo analgesia: Expectation-activated opioid systems versus conditioning-activated opioid systems versus conditioning-activated specific subsystems. *Journal Neuroscience, 19*, 484–494.

Amanzio, M., Corazzini, L. L., Vase, L., & Benedetti, F. (2009). A systematic review of adverse events in placebo groups of anti-migraine clinical trials. *Pain, 146*, 261–269.

Amir, R., & Devor, M. (1993). Ongoing activity in neuroma afferents bearing retrograde sprouts. *Brain Research, 630*(1–2), 283–288.

Andersen, S., Petersen, M. W., Svendsen, A. S., & Gazerani, P. (2015). Pressure pain thresholds assessed over temporals, master, and frontal muscles in healthy individuals, patients with tension-type headache, and those with migraine-a systematic review. *Pain, 156*(8), 1409–1423.

Apkarian, A. V., Baliki, M. N., & Geha, P. Y. (2009). Towards a theory of chronic pain. *Progress in Neurobiology, 87*, 81–97.

Apkarian, A. V., Hashmi, J. A., & Baliki, M. N. (2011). Pain and the brain: specificity and plasticity of the brain in clinical chronic pain. *Pain, 152*(3), S49–S64.

Aprison, M. H., & Hingtgen, J. N. (1981). Hypersensitive serotonergic receptors: A new hypothesis for one subgroup of unipolar depression derived from an animal model. In B. Haber, S. Gabay, M. R. Issidorides, & S. G. A. Alivisatos, (Eds). Serotonin. *Advances in Experimental Medicine and Biology, 133*(pp. 627–656). Springer, Boston, MA. Boston, MA.

Araldi, D., Ferrari, L. F., & Levine, J. D. (2015). Repeated mu-opioid exposure induces a novel form of the hyperalgesic priming model for transition to chronic pain. *Journal Neuroscience, 35*, 12502–12517.

Araldi, D., Ferrari, L. F., & Levine, J. D. (2017). Hyperalgesic priming (type II) induced by repeated opioid exposure: Maintenance mechanisms. *Pain, 158*(7), 1204–1216.

Ariel, G., & Saville, W. (1972). Effect of anabolic steroids on reflex components. *Journal of Applied Physiology, 32*(6), 795–797.

Arrowsmith, J. (2011a). Phase II failures: 2008-2010. *Nature Reviews Drug Discovery, 10*(5), 328–329.

Arrowsmith, J. (2011b). Phase III and submission failures: 2007–2010. *Nature Reviews Drug Discovery 10*, 1.

Arrowsmith, J., & Miller, P. (2013). Phase II and Phase III attrition rates 2011-2012. *Nature Reviews Drug Discovery, 12*(8), 569–570.

Atlas, L. Y., & Wager, T. D. (2014). A meta-analysis of brain mechanisms of placebo analgesia: Consistent findings and unanswered questions. *Handbook Experimental Pharmacology, 225*, 137–169. https://doi.org/10.1007/978-3-44519-8_3

Attal, N., Kayser, V., Jazat, F., & Guilbaud, G. (1989). Behavioral evidence for a bidirectional effect of systemic naloxone in a model of experimental neuropathy in the rat. *Brain Research, 494*, 276–284.

Austin, P. J., Kim, C. F., Perera, C. J., & Moalem-Taylor, G. (2012). Regulatory T cells attenuate neuropathic pain following peripheral nerve injury and experimental autoimmune neuritis. *Pain, 153*, 1916–1931.

Baliki, M. N., & Apkarian, A. V. (2015). Nociception, pain, negative moods, and behavior selection. *Neuron, 87*, 474–491.

Balkrishnan, R., Dugan, E., Camacho, F. T., & Hall, M. A. (2003). Trust and satisfaction with physicians, insurers, and the medical profession. *Medical Care, 41*(9), 1058–1064.

Ballantyne, J. C. (2018). The brain on opioids. *Pain, 159*, S24–S30.

Ballantyne, J. C., & Sullivan, M. D. (2017). Discovery of endogenous opioid systems: What it has meant for the clinician's understanding of pain and its treatment. *Pain, 158*(12), 2290–2300.

Ballantyne, J. C., Fishman, S. M., & Rathmell, J. P. (2018). Bonica's management of pain. Lippincott Williams & Wilkins.

Bannister, K., & Dickenson, A. H. (2010). Opioid hyperalgesia. *Current Opinions Supportive Palliative Care, 4*, 1–5.

Bannister, K., Bee, L. A., & Dickenson, A. H. (2009). Preclinical and early investigations related to monoaminergic pain modulation. *Neurotherapeutics, 6*, 703–712.

Bannister, K., Patel, R., Goncalves, L., Townson, L., & Dickenson, A. H. (2015). Diffuse noxious inhibitory controls and nerve injury: Monoamine inhibitions and facilitations. *Pain, 156*(9), 1803–1811.

Barondes, S. H. (1994). Thinking about prozac. *Science, 263*(5150), 1102–1103.

Bee, L. A., & Dickenson, A. H. (2008). Descending facilitation from the brainstem determines behavioral and neuronal hypersensitivity following nerve injury and efficacy of pregabalin. *Pain, 140*, 209–223.

Beecher, H. K. (1955). The powerful placebo. *JAMA, 159*, 1602–1606.

Beecher, H. K. (1956). The measurement of pain, prototype for the quantitative study of subjective responses. *Pharmacy Review, 9*, 59.

Beecher, H. K. (1961). Surgery as placebo: A quantitative study of bias. *JAMA, 176*, 1102–1107.

Benedetti, F. (2002). How the doctor's words affect the patient's brain. *Evaluation & the Health Professions, 25*(4), 369–386.

Benedetti, F. (2008). Mechanisms of placebo and placebo-related effects across diseases and treatments. *Annual Review Pharmacology, 48*, 33–60.

Benedetti, F. (2014). *Placebo effects: Understanding the mechanisms in health and disease* (2nd ed.). Oxford University Press.

Bennett, G. J., & Xie, Y. K. (1988). A peripheral mononeuropathy in rat that produces disorders of pain sensation like those seen in man. *Pain, 33*, 87–107.

Benedetti, F., Amanzio, M., Vighetti, S., & Asteggiano, G. (2006). The biochemical and neuroendocrine bases of the hyperalgesic nocebo effect. *Journal Neuroscience, 26*, 12014–12022.

Benedetti, F., Amanzio, M., Rosato, R., & Blanchard, C. (2011). Nonopioid placebo analgesia is mediated by CB1 cannabinoid receptors. *Nature Medicine, 17*, 1228–1230.

Benedetti, F., Lanotte, M., Lopiano, L., & Colloca, L. (2007). When words are painful: unraveling the mechanisms of the nocebo effect. *Neuroscience, 147*(2), 260–271.

Benedetti, F., Thoen, W., Blanchard, C., Vighetti, S., & Arduino, C. (2013). Pain as a reward: Changing the meaning of pain from negative to positive coactivates opioid and cannabinoid systems. *Pain, 154*, 361–367.

Bérdi, M., Köteles, F., Szabó, A., & Bárdos, G. (2011). Placebo effects in sport and exercise: a meta-analysis. *European Journal of Mental Health, 6*(2), 196.

Bérdi, M., Köteles, F., Hevesi, K., Bárdos, G., & Szabo, A. (2015). Elite athletes' attitudes towards the use of placebo-induced performance enhancement in sports. *European Journal of Sport Science, 15*(4), 315–321.

Bershad, A. K., Jaffe, J. H., Childs, E., & de Wit, H. (2015). Opioid partial agonist buprenorphine dampens responses to psychosocial stress in humans. *Psychoneuroendocrinology, 52*, 281–288.

Bershad, A. K., Seiden, J. A., & de Wit, H. (2016). Effects of buprenorphine on responses to social stimuli in healthy adults. *Psychoneuroendocrinology, 63*, 43–49.

Bershad, A. K., Ruiz, N. A., & de Wit, H. (2018). Effects of buprenorphine on responses to emotional stimuli in individuals with a range of mood symptomatology. *International Journal of Neuropsychopharmacology, 21*(2), 120–127.

Beutler, L. E., Crago, M., & Arizmendi, T. G. (1986). Therapist variables in psychotherapy process and outcome. *Handbook of Psychotherapy and Behavior Change, 3*, 257–310.

Birklein, F., Ibrahim, A., Schlereth, T., & Kingery, W. S. (2018). The rodent tibia fracture model: A critical review and comparison with the complex regional pain syndrome literature. *The Journal of Pain, 19*, 1102.e1–1102.e19.

Black, J. A., Nikolsen, L., Kroner, K., Jensen, T. S., & Waxman, S. G. (2008). Multiple sodium channel isoforms and mitogen-activated protein kinases are present in painful human neuromas. *Annals of Neurology, 64*, 644–653.

Bobinski, F., Teixeira, J. M., Sluka, K. A., & Santos, R. D. (2018). Interleukin-4 mediates the analgesia produced by low-intensity exercise in mice with neuropathic pain. *Pain, 159*(3), 437–450.

Bodnar, R. J. (2017). Endogenous opiates and behavior: 2015. *Peptides, 88*, 126–188.

Bodkin, J. A., Zornberg, G. L., Lukas, S. E., & Cole, J. O. (1995). Buprenorphine treatment of refractory depression. *Journal of Clinical Psychopharmacology, 15*(1), 49–57.

Borsook, D. (2017). Opioidergic tone and pain susceptibility: Interactions between reward systems and opioid receptors. *Pain, 158*(2), 185–186.

Borsook, D., Linnman, C., Faria, V., Strassman, A. M., Becerra, L., & Elman, I. (2016). Reward deficiency and anti-reward in pain chronification. *Neuroscience Biobehavioral Review, 68*, 282–297.

Bruchas, M. R., Land, B. B., & Chavkin, C. (2010). The dynorphin-kappa opioid system as a modulator of stress-induced and pro-addictive behaviors. *Brain Research, 1314*, 1–19.

Brush, D. E. (2012). Complications of long-term opioid therapy for the management of chronic pain: The paradox of opioid-induced hyperalgesia. *Journal Medical Toxicology, 8*, 387–392.

Burkett, J. P., Andari, E., Johnson, Z. V., Curry, D. C., der Waal, F. B. M., & Young, L. J. (2016). Oxytocin-dependent consolidation behavior in rodents. *Science, 351*(6271), 375–378.

Burns, J. W., Bruehl, S., France, C. R., Schuster, E., Orloskka, D., Buvanendran, A., Chont, M., & Gupta, R. (2017). Psychological factors predict opioid analgesia through endogenous opioid function. *Pain, 158*(3), 391–399.

Bushnell, M. C., Ceko, M., & Low, L. A. (2013). Cognitive and emotional control of pain and its disruption in chronic pain. *National Review Neuroscience, 14*, 502–511.

Carlton, S. M., Lekan, H. A., Kim, S. H., & Chung, J. M. (1994). Behavioral manifestations of an experimental model for peripheral neuropathy produced by spinal nerve ligation in the primate. *Pain, 56*, 155–166.

Carlino, E., Piedimonte, A., & Frisaldi, E. (2014). The effects of placebos and nocebos on physical performance. *Placebo*, 149–157.

Carbone, L. (2011). Pain in laboratory animals: the ethical and regulatory imperatives. *PLoS One, 6*(9), e21578.

Cauda, F., Palermo, S., Costa, T., Torta, R., Duca, S., Vercelli, U., Germiniani, G., & Torta, D. M. E. (2014). Gray matter alterations in chronic pain: A network-oriented meta-analytic approach. *Neuroimage Clinical, 4*, 676–686.

Chapman, V., Suzuki, R., & Dickenson, A. H. (1998). Electrophysiological characterization of spinal neuronal response properties in anaesthetized rats after ligation of spinal nerves L5-L6. *The Journal of Physiology, 507*(3), 881–894.

Chaumette, T., Delay, D., Barbier, J., Boudieu, L., Aissouni, Y., Meleine, M., Lashhermes, A., Legha, W., Antraigue, S., Carvalho, F. A., Eschalier, A., Ardid, D., Moqrich, A., & Marchand, F. (2020). c-Jun/p38MAPK/ASIC3 pathways specifically activated by nerve growth factor through TrkA are crucial for mechanical allodynia development. *Pain, 161*(5), 1109–1123.

Chen, Z., Chen, X., Liu, M., Dong, Z., Ma, L., & Yu, S. (2017). Altered functional connectivity of amygdala underlying the neuromechanism of migraine pathogenesis. *Journal Headache Pain, 18*, 7.

Chin, E. L. (2002). This side of doctoring: Reflections from women in medicine. SAGE: Thousand Oaks, CA.

Chu, L. F., Angst, M. S., & Clark, D. (2008). Opioid-induced hyperalgesia in humans: Molecular mechanisms and clinical considerations. *Clinical Journal Pain, 24*, 479–496.

Clauw, D. J. (2014). Fibromyalgia: A clinical review. *JAMA, 311*, 1547–1555.

Clayton, N., Marshall, F. H., Bountra, C., & O'Shaughnessy, C. T. (2002). CB1 and CB2 cannabinoid receptors are implicated in inflammatory pain. *Pain, 96*, 253–260.

Clements-Jewery, S., Robson, P. A., & Chidley, L. J. (1980). Biochemical investigations into the mode of action of trazodone. *Neuropharmacology, 19*(12), 1165–1173.

Colagiuri, B., Schenk, L. A., Kessler, M. D., Dorsey, S. G., & Colloca, L. (2015). The placebo effect: From concepts to genes. *Neuroscience, 307*, 171–190.

Colasanti, A., Rabiner, E. A., Lingford-Hughes, A., & Nutt, D. J. (2011). Opioids and anxiety. *Journal of Psychopharmacology, 25*(11), 1415–1433.

Colloca, L. (2017). Nocebo effects can make you feel pain. *Science, 6359*, 44.

Colloca, L., & Barsky, A. J. (2020). Placebo and nocebo effects. *New England Journal Medicine, 382*, 554–561.

Colloca, L., & Benedetti, F. (2009). Placebo analgesia induced by social observational learning. *Pain, 144*, 28–34.

Colloca, L., & Miller, F. G. (2011). The nocebo effect and its relevance for clinical practice. *Psychosomatic Medicine, 73*, 598–600.

Colloca, L., Klinger, R., Flor, H., & Bingel, U. (2013). Placebo analgesia: Psychological and neurobiological mechanisms. *Pain, 154*, 511–514.

Colloca, L., Pine, D. S., Ernst, M., Miller, F. G., & Grillon, C. (2016). Vasopressin boosts placebo analgesic effects in women: a randomized trial. *Biological Psychiatry, 79*, 794–802.

Cooper, S. A., Desjardins, P. J., Turk, D. C., Dworkin, R. H., Katz, N. P., Kehlet, H., Ballantyne, J. C., Burke, L. B., Carragee, E., Cowan, P., Croll, S., Dionne, R. A., Farrar, J. T., Gilron, I., Gordon, D. B., Iyengar, S., Jay, G. W., Kalso, E., Kerns, R. D., et al. (2016). Research design considerations for single-dose analgesic trials in acute pain. IMMPACT recommendations. *Pain, 157*, 288–301.

Corder, G., Doolen, S., & Donahue, R. R. (2013). Constitutive mu-opioid receptor activity leads to long-term endogenous analgesia and dependence. *Science, 341*, 1394–1399.

Corder, G., Castro, D. C., Bruchas, M. R., & Scherrer, G. (2018). Endogenous and exogenous opioids in pain. *Annual Review Neuroscience, 41*, 453–473.

Corder, G., Ahanonu, B., Grewe, B. F., Wang, D., Schnitzer, M. J., & Scherrer, G. (2019). An amygdalar neural ensemble that encodes the unpleasantness of pain. *Science, 363*, 276–281.

Davoli-Ferreira, M., de Lima, K. A., Fonseca, M. M., Guimaraes, R. M., Gomes, F. I., Cavallini, M. C., Quadros, A. U., Kusuda, R., Cunha, F. Q., Alves-Filho, J. C., & Cunha, T. M. (2020). Regulatory T cells counteract neuropathic pain through inhibition of the Th1 response at the site of peripheral nerve injury. *Pain, 161*(8), 1730–1743.

Davis, M. (2001). Fear-potentiated startle in rats. *Current Protocols in Neuroscience, 14*(1), 8–11.

De Craen, A. J., Kaptchuk, T. J., Tijssen, J. G., & Kleijnen, J. (1999). Placebos and placebo effects in medicine: historical overview. *Journal of the Royal Society of Medicine, 92*(10), 511–515.

De Gregorio, D., McLaughlin, R. J., Posa, L., Ochoa-Sanchez, R., Enns, J., Lopez- Canui, M., Aboud, M., Maione, S., Comai, S., & Gobbi, G. (2019). Cannabadiol modulates serotonergic transmission and anxiety-like behavior in a model of neuropathic pain. *Pain, 160*, 136–150.

de la Fuente-Fernandez, R. (2009). The placebo-reward hypothesis: Dopamine and the placebo effect. *Parkinsonism Related Disorders, 15*(suppl 3), S72–S74.

De la Fuente-Fernández, R., Ruth, T. J., Sossi, V., Schulzer, M., Calne, D. B., & Stoessl, A. J. (2001). Expectation and dopamine release: mechanism of the placebo effect in Parkinson's disease. *Science, 293*(5532), 1164–1166.

de la Fuente-Fernandez, R., & Stoessl, A. J. (2004). The biochemical basis of the placebo effect. *Science Engineering Ethics, 10*(1), 143–150.

de la Fuente-Fernandez, R., Lidstone, S., & Stoessl, A. J. (2006). Placebo effect and dopamine release. *Journal of Neural Transmission Supplementum*, 415–418. https://doi.org/10.1007/978-3-211-45295-0_62

Demant, D. T., Lund, K., Finnerup, N. B., Vollert, J., Maier, C., Segerdahl, M. S., Jensen, T. S., & Sindrup, S. H. (2015). Pain relief with lidocaine 5% patch in localized peripheral neuropathic pain in relation to pain phenotype: A randomized, double-blind, and placebo-controlled, phenotype panel study. *Pain, 156*, 2234–2244.

Devor, M. (2006). Sodium channels and mechanisms of neuropathic pain. *Journal Pain, 7*, 3–12.

Diederich, N. J., & Goetz, C. G. (2007). The placebo treatments in neurosciences: New insights from clinical and neuroimaging studies. *Neurology, 71*(9), 677–684.

Diehl, H. S., Baker, A. B., & Cowan, D. W. (1938). Cold vaccines: an evaluation based on a controlled study. *Journal of the American Medical Association, 111*(13), 1168–1173.

Disney, W., Armstrong, S., Ferguson, N., Jackson, W., Kinney, J., Roberts, B., ... & Elliotte, J. (1941). Dumbo [Motion picture]. EEUU: Walt Disney Home Video.

Drummond, P. D., Finch, P. M., Birklein, F., Stanton-Hicks, M., & Knudsen, L. F. (2018). Hemisensory disturbances in patients with complex regional pain syndrome. *Pain, 159*(9), 1824–1832.

Dusch, M., Schelm, M., Obreja, O., Force, E., Schmelz, M., & Rukwied, R. (2009). Comparison of electrically induced flare response patterns in human and pig skin. *Inflammation Research, 58*, 639–648.

Eagly, A. H., Carli, L. L., & Carli, L. L. (2007). Through the labyrinth: The truth about how women become leaders (Vol. 11). Boston, MA: Harvard Business School Press.

Echeverry, S., Xiang, Q. S., Yang, H., Wu, Y., Lorenzo, L. E., Perez-Sanchez, J., Bonin, R. P., Koninck, Y. D., & Zhang, J. (2017). Spinal microglia are required for long-term maintenance of neuropathic pain. *Pain, 158*(9), 1792–1801.

Ehrich, E., Turncliff, R., Du, Y., Leigh-Pemberton, R., Fernandez, E., Jones, R., & Fava, M. (2015). Evaluation of opioid modulation in major depressive disorder. *Neuropsychopharmacology, 40*(6), 1448–1455.

Eippert, F., Finsterbusch, J., Bingel, U., & Buchel, C. (2009). Direct evidence for spinal cord involvement in placebo analgesia. *Science, 326*, 404.

Elman, I., & Borsook, D. (2016). Common brain mechanisms of chronic pain and addiction. *Neuron, 89*, 11–36.

Eskander, M. A., Ruparel, S., Green, D. P., Chen, P. B., Por, E. D., Jeske, N. A. & Hargreaves, K. M. (2015). Persistent nociception triggered by nerve growth factor (NGF) is mediated by TRPV1 and oxidative mechanisms. *Journal Neuroscience, 35*, 8593–8603.

Fields, H. L. (2014). Neuroscience. More pain; less gain. *Science, 345*, 513–514.

Fields, H. H., Heinricher, M. M., & Mason, P. (1991). Neurotransmitters in nociceptive modulatory circuits. *Annual Review Neuroscience, 14*, 219–245.

Fields, H. H., Malick, A., & Burstein, R. (1995). Dorsal horn projection targets of ON and OFF cells in the rostral ventromedial medulla. *Journal of Neurophysiology, 1995*(74), 1742–1759.

Finn, D. P., Jhaveri, M. D., Beckett, S. R. G., Roe, C. H., Kendall, D. A., Marsden, A., & Chapman, V. (2003). Effects of direct periaqueductal grey administration of a cannabinoid receptor agonist on nociceptive and aversive responses in rats. *Neuropharmacology, 45*, 594–604.

Finnerup, N. B., Sindrup, S. H., & Jensen, T. S. (2010). The evidence for pharmacological treatment of neuropathic pain. *Pain, 150*, 573–581.

Fish, J., & Wilkinson, S. (2003). Understanding lesbians' healthcare behaviour: The case of breast self-examination. *Social Science & Medicine, 56*(2), 235–245.

Fiscella, K., Meldrum, S., Franks, P., Shields, C. G., Duberstein, P., McDaniel, S. H., & Epstein, R. M. (2004). Patient trust: is it related to patient-centered behavior of primary care physicians?. *Medical Care, 42*(11), 1049–1055.

Fletcher, D., & Martinez, V. (2014). Opioid-induced hyperalgesia in patients after surgery: A systematic review and meta-analysis. *British Journal Anaesthesia, 112*, 991–1004.

Flint, A. (1863). ART. I.--A contribution toward the Natural History of Articular Rhcumatism; consisting of a report of thirteen cases treated solely with palliative measures. *The American Journal of the Medical Sciences (1827-1924), 46*(91), 17.

Forsberg, J. T., Martinussen, M., & Flaten, M. A. (2017). The placebo analgesic effect in healthy individuals and patients: A meta-analysis. *Psychosomatic Medicine, 79*, 388–394.

Fredriksen-Goldsen, K. I., Kim, H. J., Barkan, S. E., Muraco, A., & Hoy-Ellis, C. P. (2013). Health disparities among lesbian, gay, and bisexual older adults: Results from a population-based study. *American Journal of Public Health, 103*(10), 1802–1809.

Fuchs, P. N., Campbell, J. N., & Meyer, R. A. (2000). Secondary hyperalgesia persists in capsaicin desensitized skin. *Pain, 84*, 141–149.

Fulda, S., & Wetter, T. C. (2008). Where dopamine meets opioids: A meta-analysis of the placebo effect in restless legs syndrome treatment studies. *Brain, 131*(4), 902–917.

Gandhi, W., Rosenek, N. R., Harrison, R., & Salomons, T. V. (2020). Functional connectivity of the amygdala is linked to individual difference in emotional pain facilitation. *Pain, 161*(2), 300–307.

Goetz, C. G., Leurgans, S., Raman, R., & Stebbins, G. T. (2000). Objective changes in motor function during placebo treatment in PD. *Neurology, 54*(3), 710–714.

Gomtsian, L., Bannister, K., Eyde, N., Robles, D., Dickerson, A. H., Porreca, F., & Navratilova, E. (2018). Morphine effects within the rodent anterior cingulate cortex and rostral ventromedial medulla reveal separable modulation of affective and sensory qualities of acute or chronic pain. *Pain, 159*(12), 2512–2521.

Gonzalez, A., & Condez-Lara, M. (2017). Peripheral oxytocin receptors inhibit the nociceptive input signal to spinal dorsal horn wide-dynamic-ranges neurons. *Pain, 158*(11), 2117–2128.

Grace, P. M., Hutchinson, M. R., Maier, S. F., & Watkins, L. R. (2014). Pathological pain and the neuroimmune interface. *National Review Immunology, 14*, 217–231.

Graeff, F. G. (2017). Translational approach to the pathophysiology of panic disorder: Focus on serotonin and endogenous opioids. *Neuroscience & Biobehavioral Reviews, 76*, 48–55.

Green-Fulgham, S. M., Ball, J. B., Kwilasz, A. J., Fabisiak, T., Maier, S. F., Watkins, L. R., & Grace, P. M. (2019). Oxycodone, fentanyl, and morphine amplify established neuropathic pain in male rats. *Pain, 160*(11), 2634–2640.

Gregg, J. M. (2013). Historical perspectives on trigeminal nerve injuries. In M. Miloro (Ed.), *Trigeminal nerve injuries* (pp. 1–16). Springer.

Gross, J. J. (Ed.). (2014). Handbook of emotion regulation, 2nd Edition. New York: Guilford publications.

Gustorff, B., Sycha, T., Lieb-Samal, D., Role, R., Treede, R. D., & Magerl, W. (2013). The pattern and time course of somatosensory changes in the human UVB sunburn model reveal the presence of peripheral and central sensitization. *Pain, 154*, 586–597.

Hansson, P. (2014). Translational aspects of central sensitization induced by primary afferent activity: What it is and what it is not. *Pain, 155*, 1932–1934.

Haroutounian, S., Nikolajsen, L., Bendtsen, T. F., Finnerup, N. B., Kristensen, A. D., Hasselstrom, J. B., & Jensen, T. S. (2014). Primary afferent input critical for maintaining spontaneous pain in peripheral neuropathy. *Pain, 155*, 1272–1279.

Harvold, M., MacLeod, C., & Vaegter, H. B. (2018). Attentional avoidance is associated with increased pain sensitivity in patients with chronic posttraumatic pain and comorbid posttraumatic stress. *Clinical Journal Pain, 34*, 22–29.

Hashmi, J. A., Baliki, M. N., Huang, L., Baria, A. T., Torbey, S., Hermann, K. M., Schnitzer, T. J., & Apkarian, A. V. (2013). Shape shifting pain: Chronification of back pain shifts brain representation from nociceptive to emotional circuits. *Brain, 136*, 2751–2768.

Hay, J. L., White, J. M., Bochner, F., Somogyi, A. A., Semple, T. J., & Rounsefell, B. (2009). Hyperalgesia in opioidmanaged chronic pain and opioid-dependent patients. *The Journal of Pain, 10*(3), 316–322.

Hayes, C. J., & Painter, J. T. (2017). A comprehensive clinical review of opioid-induced allodynia: Discussion of the current evidence and clinical implications. *Journal Opioid Management, 13*, 95.

Hayes, C. J., Kreegs, E. E., Hudson, T., Brown, J., Li, C., & Martin, B. C. (2020). Impact of opioid dose escalation on pain intensity: A retrospective cohort study. *Pain, 161*(5), 979–988.

Haygarth, J. (1800). Of the Imagination, as a Cause and as a Cure of Disorders of the Body; Exemplified by Fictitious Tractors, and Epidemical Convulsions. Read to the Literary and Philosophical Society of Bath. By John Haygarth, MD...... R. Cruttwell; and sold by Cadell and Davies, London.

Heilman, M. E. (2001). Description and prescription: How gender stereotypes prevent women's ascent up the organizational ladder. *Journal of Social Issues, 57*(4), 657–674.

Helmstetter, F. J., & Bellgowan, P. S. (1994). Effects of muscimol applied to the basolateral amygdala on acquisition and expression of contextual fear conditioning in rats. *Behavioral Neuroscience, 108*(5), 1005–1009.

Heinricher, M. M., Tavares, I., Leith, J. L., & Lumb, B. M. (2009). Descending control of nociception: Specificity, recruitment and plasticity. *Brain Research Review, 60*, 14–25.

Hill, A. V., Long, C. N. H., & Lupton, H. (1924). The effect of fatigue on the relation between work and speed, in contraction of human arm muscles. *The Journal of Physiology, 58*(4-5), 334–337.

Hirth, M., Rukwied, R., Gromann, A., Turnquist, B., Weinkauf, B., Francke, K., Albrecht, P., Rice, F., Hagglof, B., Ringkamp, M., Engelhardt, M., Schultz, C., Schmelz, M., & Obreja, O. (2013). Nerve growth factor induces sensitization of nociceptors without evidence for increased intrapidermal nerve fiber density. *Pain, 154*, 2500–2511.

Hingtgen, J. N., Hendrie, H. C., & Aprison, M. H. (1984). Postsynaptic serotonergic blockade following chronic antidepressive treatment with trazodone in an animal model of depression. *Pharmacology Biochemistry and Behavior, 20*(3), 425–428.

Hsu, D. T., Sanford, B. J., Meyers, K. K., Love, T. M., Hazlett, K. E., Walker, S. J., ... & Zubieta, J. K. (2015). It still hurts: altered endogenous opioid activity in the brain during social rejection and acceptance in major depressive disorder. *Molecular Psychiatry, 20*(2), 193–200.

Hughes, J., Smith, T. W., Kosterlitz, H. W., Fothergill, L. A., Morgan, B. A., & Morris, H. R. (1975). Identification of two related pentapeptides from the brain with potent opiate agonist activity. *Nature, 258*, 577–580.

Hunter, T., Siess, F., & Colloca, L. (2014). Socially induced placebo analgesia: A comparison of a pre-recorded versus live face-to-face observation. *European Journal of Pain, 18*(7), 914–922.

Hyman, S. E., Malenka, R. C., & Netler, E. J. (2006). Neural mechanisms of addiction: The role of reward-related learning and memory. *Annual Review Neuroscience, 29*, 565–598.

Ikemoto, S. (2019). Regulation of negative emotional behavior. *Science, 366*(6469), 1071–1072.

Ipser, J. C., Terburg, D., Syal, S., Phillips, N., Solms, M., Panksepp, J., ... & van Honk, J. (2013). Reduced fear-recognition sensitivity following acute buprenorphine administration in healthy volunteers. *Psychoneuroendocrinology, 38*(1), 166–170.

Jegindø, E. M. E., Vase, L., Skewes, J. C., Terkelsen, A. J., Hansen, J., Geertz, A. W., ... & Jensen, T. S. (2013). Expectations contribute to reduced pain levels during prayer in highly religious participants. *Journal of Behavioral Medicine, 36*, 413–426.

Jensen, T. S., Gottrup, H., Sindrup, S. H., & Bach, F. W. (2001). The clinical picture of neuropathic pain. *European Journal of Pharmacology, 429*(1-3), 1–11.

Ji, R. R., Chamessian, A., & Zhang, Y. Q. (2016). Pain regulation by non-neuronal cells and inflammation. *Science, 354*(6312), 572–577.

Jiang, Y., Oathes, D., Hush, J., Darnall, B., Charvat, M., Mackey, S., & Etkin, A. (2016). Perturbed connectivity of the amygdala and its subregions with the central executive and default mode networks in chronic pain. *Pain, 157*(9), 1970–1978.

Johnson, J. L., & Repta, R. (2012). Sex and gender: Beyond the binaries. In J. Oliffe & L. J. Greaves (Eds.). Designing and conducting gender, sex, and health research (pp. 17–37). SAGE Publications.

Johansen, A., Schirmer, H., Stubhaug, A., & Nielsen, C. S. (2014). Persistent post- surgical pain and pain sensitivity in the Tromso study; comorbid pain matters. *Pain, 155*, 341–348.

Kalasountas, V., Reed, J., & Fitzpatrick, J. (2007). The effect of placebo-induced changes in expectancies on maximal force production in college students. *Journal of Applied Sport Psychology, 19*(1), 116–124.

Kamerman, P. R., Wadley, A. L., Davis, K. D., Hietaharju, A., Parmanand, J., Kopf, A., Meyer, A. C., Raja, S. N., Rice, A. S. C., Smith, B. H., Trete, R.-D., & Wiffen, P. J. (2015). World Health Organization essential medicines lists: Where are the drugs to treat neuropathic pain? *Pain, 156*, 793–797.

Kaptchuk, T. J., & Miller, F. G. (2015). Placebo effects in medicine. *New England Journal Medicine, 373*, 8–9.

Kaptchuk, T. J., Friedlander, E., Kelley, J. M., & Sanchez, M. N. (2010). Placebos without deception: A randomized controlled trial in irritable bowel syndrome. *PLoS One, 5*, e15591.

Kazdin, A. E. (1979). Nonspecific treatment factors in psychotherapy outcome research. *Journal of Consulting and Clinical Psychology, 47*(5), 846–851.

Keene, K. D. (1954). Pain-sensitive tests; the pressure altimeter. *Lancet, 266*, 193–199.

Kennedy, S. E., Koeppe, R. A., Young, E. A., & Zubieta, J. K. (2006). Dysregulation of endogenous opioid emotion regulation circuitry in major depression in women. *Archives of General Psychiatry, 63*(11), 1199–1208.

Keogh, E. (2014). Gender differences in the nonverbal communication of pain: A new direction for sex, gender, and pain research? *Pain, 155*, 1927–1931.

King, T., Qu, C., Okun, A., Mercado, R., Ren, J., Brion, J. L., & Porreca, F. (2011). Contribution of afferent pathways to nerve injury-induced spontaneous pain and evoked hypersensitivity. *Pain, 152*, 1997–2005.

Kirsch, I. (2014). The emperor's new drugs: medication and placebo in the treatment of depression. *Placebo,* 291–303.

Kirsch, I., & Sapirstein, G. (1998). Listening to Prozac but hearing placebo: A meta-analysis of antidepressant medication. *Prevention & Treatment, 1*(2), Article 2a.

Kirsch, I., Moore, T. J., Scoboria, A., & Nicholls, S. S. (2002). The emperor's new drugs: An analysis of antidepressant medication data submitted to the US Food and Drug Administration. *Prevention & Treatment, 5*(1), 23a.

Kirsch, I., Deacon, B. J., Huedo-Medina, T. B., Scoboria, A., Moore, T. J., & Johnson, B. T. (2008). Initial severity and antidepressant benefits: a meta-analysis of data submitted to the Food and Drug Administration. *PLoS Medicine, 5*(2), e45.

Klein, D. F. (1998). Listening to meta-analysis but hearing bias. *Prevention & Treatment, 1*(2), Article 6c.

Klein, T., Mageri, W., Role, R., & Treede, R. D. (2005). Human surrogate model of neuropathic pain. *Pain, 115*, 227–233.

Klinger, R., Kothe, R., Schmitz, J., Kamping, S., & Flor, H. (2017). Placebo effects of a sham opioid solution: A randomized controlled study in patients with chronic low back pain. *Pain, 158*, 1893–1902.

Klosterhalfen, S., Kellermann, S., Braun, S., Kowalski, A., Schrauth, M., Zipfel, S., & Enck, P. (2009). Gender and the nocebo response following conditioning and expectancy. *Journal of Psychosomatic Research, 66*(4), 323–328.

Knoll, A. T., & Carlezon, W. A. (2010). Dynorphin, stress, and depression. *Brain Research, 1314*, 56–73.

Kong, J., & Benedetti, F. (2014). Placebo and nocebo effects: an introduction to psychological and biological mechanisms. *Placebo*, 3–15.

Kong, J., Gollub, R. L., Polich, G., Kirsch, I., Laviolette, P., Vangel, M., Rosen, B., & Kaptchuk, T. J. (2008). A functional magnetic resonance imaging study on the neural mechanisms of hyperalgesic nocebo effect. *The Journal of Neuroscience, 28*, 13354–13362.

Kong, J., Jensen, K., Loiotile, R., Cheetham, A., Way, H. Y., Tan, Y., Rosen, B., Smoller, J. W., Kaptchuk, T. J., & Gollub, R. L. (2013). Functional connectivity of the frontoparietal network predicts cognitive modulation of pain. *Pain, 154*(3), 459–467.

Koob, G. F. (2020). Neurobiology of opioid addiction: Opponent process, hyperkatifeia, and negative reinforcement. *Biological Psychiatry, 87*(1), 44–53.

Koob, G. F., & Le Moal, M. (2001). Drug addiction, dysregulation of reward, and allostasis. *Neuropsychopharmacology, 24*(2), 97–129.

Koob, G. F., & Le Moal, M. (2008). Addiction and the brain antireward system. *Annual Review Psychology, 59*, 29–53.

Kosten, T. R. (2016). An opioid for depression?. *American Journal of Psychiatry, 173*(5), 446–447.

Kramer, P. D., & Medawar, C. (1994). Listening to prozac. *Perspectives in Biology and Medicine, 37*(3), 460.

Kronschlager, M. T., Drdla-Schutting, R., Gassen, M., Honsek, S. D., Teichmann, H. L., & Sandkuhler, J. (2016). Gliogenic LTP spreads widely in nociceptive pathways. *Science, 354*(6316), 1144–1148.

Kutch, J. J., Ichesco, E., Hampson, J. J., Labus, J. S., Farmer, M. A., Martucci, K. T., Ness, K. T., Deutsch, G., Apkarian, V., Mackey, S. C., Klumpp, D. J., Schaeffer, J., Rodriguez, L. V., Kreder, K. J., Buchwald, D., Andriole, G. L., Lai, H. H., Mullins, C., Kusek, J. W., et al. (2017). Brain signature and functional impact of centralized pain: A multidisciplinary approach to the study of chronic pelvic pain (MAPP) network study. *Pain, 158*(10), 1979–1991.

Latremoliere, A., & Woolf, C. J. (2009). Central sensitization: A generator of pain hypersensitivity by central neural plasticity. *Journal Pain, 10*, 895–926.

Lane, T., Mogale, T., Struthers, H., McIntyre, J., & Kegeles, S. M. (2008). "They see you as a different thing": the experiences of men who have sex with men with healthcare workers in South African township communities. *Sexually Transmitted Infections, 84*(6), 430–433.

Leknes, S., Lee, M., Berna, C., Andersson, J., & Tracey, I. (2011). Relief as a reward: Hedonic and neural responses to safety from pain. *PLoS One, 6*, e17870.

Leslie, A. (1954). Ethics and practice of placebo therapy. *The American Journal of Medicine, 16*(6), 854–862.

Le Merrer, J., Becker, J. A., Befort, K., & Kieffer, B. L. (2009). Reward processing by the opioid system in the brain. *Physiological Reviews, 89*, 1379–1412.

Levine, J. D., Gordon, N. C., & Fields, H. L. (1978). The mechanism of placebo analgesia. *Lancet, 2*, 654–657.

Li, Y. D., Ge, J., Luo, Y. J., Xu, W., Wang, J., Lazarus, M., Hong, Z. Y., Qu, W. M., & Huang, Z. L. (2020). High cortical delta power correlates with aggravated allodynia by activating anterior cingulate cortex GABAergic neurons in neuropathic pain mice. *Pain, 161*(2), 288–299.

Lidstone, S. C., & Stoessl, A. J. (2007). Understanding the placebo effect: Contributions from neuroimaging. *Molecular Imaging Biology, 9*(4), 176–185.

Lidstone, S. C., Schulzer, M., Dinelle, K., Mak, E., Soessi, V., Ruth, T. J., de la Fuente-Fernandez, R., Phillips, A. G., & Stoessl, A. J. (2010). Effects of expectation on placebo-induced dopamine release in Parkinson's disease. *Archives General Psychiatry, 67*(8), 857–865.

Linehan, M. (2014). DBT® Skills training manual. Guilford Publications.

Liu, D., Tang, Q. Q., Yin, C., Song, Y., Liu, Y., Yang, J. X., Liu, H., Zhang, Y.-M., Wu, S.-Y., Song, Y., Juarez, B., Ding, H.-L., Han, M.-H., Zhang, H., & Cao, J.-L. (2018). Brain-derived neurotrophic factor- mediated projection-specific regulations of depressive-like and nociceptive behaviors in the mesolimbic reward circuitry. *Pain, 159*(1), 175–188.

Llorca-Torralba, M., Borges, G., Neto, F., Mico, J. A., & Barrocos, E. (2016). Noradrenergic locus coeruleus pathways in pain modulation. *Neuroscience, 338*, 93–113.

Lotsch, J., Weyer-Nenkhoff, I., & Tegeder, I. (2018). Current evidence of cannabinoid- based analgesia obtained in preclinical and human experimental settings. *European Journal Pain, 22*, 471–484.

Lucher, C. (2016). The emergence of a circuit model for addiction. *Annual Review of Neuroscience, 39*, 257–276.

Maganaris, C., Collins, D., & Sharp, M. (2000). Expectancy effects and strength training: Do steroids make a difference? *The Sport Psychologist, 14*, 272–278.

Maldonado, R., Banos, J. E., & Cabanero, D. (2016). The endocannabinoid system and neuropathic pain. *Pain, 157*(Supplement 1), S23–S32.

Mapplebeck, J. C. S., Beggs, S., & Salter, M. W. (2016). Sex differences in pain: A tale of two immune cells. *Pain, 157*, S2–S6.

Massaly, N., Moron, J. A., & Al-Hasani, R. (2016a). A trigger for opioid misuse: Chronic pain and stress dysregulate the mesolimbic pathway and kappa opioid system. *Frontiers in Neuroscience, 10*, 480.

Masters, S. (2013). Placebo. In G. F. Gebhart & R. F. Schmidt (Eds.). Encyclopedia of Pain. Springer, Berlin, Heidelberg. https://doi.org/10.1007/978-3-642-28753-4_3362

Massaly, N., Ream, A., Hipolito, L., Wilson-Poe, A., Walker, B., Bruchas, M., & Moron-Concepcion, J. (2016b). Kappa opioid receptors in the nucleus accumbens mediate pain induced decrease in motivated behavior. *Journal Pain, 17*, S56.

Melzack, R., & Wall, P. D. (1965). Pain mechanisms: A new theory. *Science, 150*, 971–979.

Mitrirattanakul, S., Ramakul, N., Guerro, A. V., Matsuka, Y., Ono, T., Iwase, H., Mackie, K., Fauli, K. F., & Spigelman, I. (2006). Site-specific increases in peripheral cannabinoid receptors and their endogenous ligands in a model of neuropathic pain. *Pain, 126*, 102–114.

Moayedi, M. (2014). All roads lead to the insula. *Pain, 155*, 1920–1921.

Moerman, D. E. (2002). The meaning response and the ethics of avoiding placebos. *Evaluation & the Health Professions, 25*(4), 399–409.

Moerman, D. E., & Jonas, W. B. (2002). Deconstructing the placebo effect and finding the meaning response. *Annals of Internal Medicine, 136*(6), 471–476.

Morasco, B. J., Yarborough, B. J., Smith, N. X., Dobscha, S. K., Deyo, R. A., Perrin, N. A., & Green, C. A. (2017). Higher prescription opioid dose is associated with worse patient- reported pain outcomes and more health care utilization. *Journal Pain, 18*, 437–445.

Mukaetova-Ladinska, E. B., Steel, M., Coppock, M., Cosker, G., Scully, J. P., & McNally, R. J. (2016). Dysphoria is a risk factor for depression in medically ill older people. *International Journal of Geriatric Psychiatry, 31*, 1233–1240.

Mun, C. J., Letzen, J. E., Peters, E. N., Campbell, C. M., Vandrey, R., Gajewski- Nemes, J., DiRenzo, D., Caulfield-Noll, C., & Finan, P. H. (2020). Cannabinoid effects on responses to quantitative sensory testing among individuals with and without clinical pain: A systematic review. *Pain, 161*(2), 244–260.

Nadal, X., La Porta, C., Andrea, B. S., & Maldonado, R. (2013). Involvement of the opioid and cannabinoid systems in pain control: New insights from knockout studies. *European Journal Pharmacology, 716*, 142–157.

Navratilova, E., Atcherley, C. W., & Porreca, F. (2015). Brain circuits encoding reward from pain relief. *Trends Neuroscience, 38*(11), 741–750.

Navratilova, E., Ji, G., Phelps, C., Qu, C., Hein, M., Yaknitsea, V., Neugebauer, V., & Porreca, F. (2019). Kappa opioid signaling in the central nucleus of the amygdala promotes disinhibition and aversiveness of chronic neuropathic pain. *Pain, 160*(4), 824–832.

Navratilova, E., Nation, K., Remeniuk, B., Neugebauer, V., Bannister, K., Dickenson, A. H., & Porreca, F. (2020). Selective modulation of tonic aversive qualities of neuropathic pain by morphine in the central nucleus of the amygdala requires endogenous opioid signaling in the anterior cingulate cortex. *Pain, 161*(3), 609–618.

Nees, F., Becker, S., Millenet, S., Banaschewski, T., Poustka, L., Bokde, A., Bromberg, U., Buchel, C., Conrod, P. J., Desrivieres, S., Frouin, V., Galinat, J., Garavan, H., Heinz, A., Ittermann, B.,

Martinot, J. L., Papadoipoulos, O. D., Paus, T., Smolka, M. N., et al. (2017). The IMAGEN Consortium I. Brain substrates of reward processing and the u-opioid receptor: A pathway into pain. *Pain, 158*, 212–219.

Neugebauer, V., Galhardo, V., Maine, S., & Mackey, S. C. (2009). Forebrain pain mechanisms. *Brain Research Reviews, 60*(1), 226–242.

Niedermayer, S., Heyn, J., Guenther, F., Kuchenhoff, H., & Luchting, B. (2020). Remifentanil for abdominal surgery is associated with unexpectedly unfavorable outcomes. *Pain, 161*(2), 266–273.

Noakes, T. D. (2012). Fatigue is a brain-derived emotion that regulates the exercise behavior to ensure the protection of whole body homeostasis. *Frontiers in Physiology, 3*(82), 1–13.

Nyström, B., & Hagbarth, K. E. (1981). Microelectrode recordings from transected nerves in amputees with phantom limb pain. *Neuroscience Letters, 27*(2), 211–216.

O'Hanlan, K. A., Dibble, S. L., Hagan, H. J. J., & Davids, R. (2004). Advocacy for women's health should include lesbian health. *Journal of Women's Health, 13*(2), 227–234.

Obreja, O., Rukwied, R., Nagler, L., Schmidt, M., Schmelz, M., & Namer, B. (2018). Nerve growth factor locally sensitizes nociceptors in human skin. *Pain, 159*(3), 416–426.

Ong, C. K., Seymour, R. A., & Tan, J. M. (2004). Sedation with midazolam leads to reduced pain after dental surgery. *Anesthesia Analgesia, 98*, 1289–1293.

Ossipov, M. H., Morimura, K., & Porreca, F. (2014). Descending pain modulation and chronification of pain. *Current Opinions Palliative Care, 8*, 143–151.

Peirs, C., & Seal, R. P. (2016). Neural circuits for pain: Recent advances and current views. *Science, 354*(6312), 578–584.

Pert, C. B., & Snyder, S. H. (1973). Opiate receptor: Demonstration in nervous tissue. *Science, 179*, 1011–1014.

Petersen, G. L., Finnerup, N. B., Norskov, K. N., Grosen, K., Pilegaard, H. K., Benedetti, F., Price, D. D., Jensen, T. S., & Vase, L. (2012). Placebo manipulations reduce hyperalgesia in neuropathic pain. *Pain, 153*, 1292–1300.

Petersen, G. L., Finnerup, N. B., Colloca, L., Amanzio, M., Price, D. D., Jensen, T. S., & Vase, L. (2014). The magnitude of nocebo effects in pain: A meta-analysis. *Pain, 155*, 1426–1434.

Phelps, C. E., Navratilova, E., Dickenson, A. H., Porreca, F., & Bannister, K. (2019). Kappa opioid signaling in the right central amygdala causes hind paw specific loss of diffuse inhibitory controls in experimental neuropathic pain. *Pain, 160*(7), 1614–1621.

Piao, Z. G., Cho, I. H., & Park, C. K. (2006). Activation of glia and microglia p38MAPK in medullary dorsal horn contributes to tactile hypersensitivity following trigeminal sensory nerve injury. *Pain, 121*, 219–231.

Porreca, F., & Navratilova, E. (2017). Reward, motivation, and emotion of pain and its relief. *Pain, 158*(suppl 1), S43–S49.

Porecca, F., Ossipov, M. H., & Gebhart, G. F. (2002). Chronic pain and medullary descending facilitation. *Trends Neuroscience, 25*, 319–325.

Posner, J., & Burke, C. A. (1985). The effects of naloxone on opiate and placebo analgesia in healthy volunteers. *Psychopharmacology, 87*, 468–472.

Price, T. J., & Inyang, K. E. (2015). Commonalities between pain and memory mechanisms and their meaning for understanding chronic pain. *Progress Molecular Biology Translational Science, 131*, 409–434.

Price, T. J., & Prescott, S. A. (2015). Inhibitory regulation of the pain gate and how its failure causes pathological pain. *Pain, 154*(5), 789–792.

Price, D. D., Finniss, D. G., & Benedetti, F. (2008). A comprehensive review of the placebo effect: Recent advances and current thought. *Annual Review Psychology, 59*, 565–590.

Prosenz, J., & Gustoff, B. (2017). Midazolam as an active placebo in 3 fentanyl- vallidated nociceptive pain models. *Pain, 158*, 1264–1271.

Rea, K., Weredeselam, M. O., Harhen, B., Kerr, D. M., Galligan, R., Fitzgerald, S., Moore, M., Roche, M., & Finn, D. P. (2013). Evidence for a role of GABAergic and glutamatergic signaling in the basolateral amygdala in endocannabinoid-mediated fear-conditioned analgesia in rats. *Pain, 154*, 576–585.

Rehman, S. U., Nietert, P. J., Cope, D. W., & Kilpatrick, A. O. (2005). What to wear today? Effect of doctor's attire on the trust and confidence of patients. *The American Journal of Medicine, 118*(11), 1279–1286.

Ren, K., & Dubner, R. (2002). Descending modulation in persistent pain: An update. *Pain, 100*, 1–6.

Ringkamp, M., & Raja, S. N. (2014). A sore spot: Central or peripheral generation of chronic neuropathic spontaneous pain. *Pain, 155*, 1189–1191.

Rounds, K., Burns Mcgrath, B., & Walsh, E. (2013). Perspectives on provider behaviors: A qualitative study of sexual and gender minorities regarding quality of care. *Contemporary Nurse, 44*(1), 99–110.

Ross, M., & Olson, J. M. (1981). An expectancy-attribution model of the effects of placebos. *Psychological Review, 88*(5), 408–437.

Salter, M. W., & Stevens, B. (2017). Microglia emerge as central players in brain disease. *Nature Medicine, 23*, 1018–1027.

Saphira, M., & Glover, M. (2000). New Zealand national lesbian health survey. *Journal of the Gay and Lesbian Medical Association, 4*, 49–56.

Schafer, S. M., Geuter, S., & Wager, T. D. (2018). Mechanisms of placebo analgesia: A dual-process model informed by insights from cross-species comparisons. *Progress in Neurobiology, 160*, 101–122.

Schatzberg, A. F. (2016). Opioids in psychiatric disorders: back to the future?. *American Journal of Psychiatry, 173*(6), 564–565.

Schmitt, C. M., & Allen, J. I. (2016). View from the top: perspectives on women in gastroenterology from society leaders. *Gastroenterology Clinics, 45*(2), 371–388.

Scholz, J., & Woolf, C. J. (2007). The neuropathic triad: Neurons, immune cells and glia. *Nature Neuroscience, 10*, 1361–1368.

Schwartz, N., Temkin, P., Jurado, S., Lim, B. K., Heifets, B. K., Polepalli, J. S., & Malenka, R. C. (2014). Decreased motivation during chronic pain requires long-term depression in the nucleus accumbens. *Science, 345*, 535–542.

Scioli-Salter, E. R., Forman, D. E., Otis, J. D., Gregor, K., Valovski, I., & Rasmusson, A. M. (2015). The shared neuroanatomy and neurobiology of comorbid chronic pain and PTSD: Therapeutic implications. *Clinical Journal Pain, 31*, 363–374.

Scott, D. J., Stohler, C. S., Egnatuk, C. M., Wang, H., Koeppe, R. A., & Zubieta, J. K. (2008). Placebo and nocebo effects are defined by opposite opioid and dopaminergic responses. *Archives General Psychiatry, 65*, 220–231.

Servick, K., & Rakola, M. (2016). Primed for pain: Do opioids make chronic users more sensitive to pain? *Science, 354*(6312), 560–571.

Sessle, B. J. (2005). Trigeminal central sensitization. *Reviews Analgesia, 8*, 85–102.

Sessle, B. J. (2007). Glia: Non-neural players in orofacial pain. *Journal Orofacial Pain, 21*(3), 169–170.

Sessle, B. J. (2021). Craniofacial pain: Mechanisms and modulatory processes. In P. L. Kerr, C. Sirbu, & J. M. Gregg (Eds.), *Endorphins: From molecular biology to modern medicine*. Springer Nature Switzerland AG.

Shabes, P., Schloss, N., Mageri, W., Schmahl, C., Treede, R. D., & Baumgartner, U. (2016). A novel human surrogate model of non-injurious sharp mechanical pain. *Pain, 157*(1), 214–224.

Sher, L. (1998). The role of the endogenous opioid system in the pathogenesis of anxiety disorders. *Medical Hypotheses, 50*(6), 473–474.

Shipherd, J. C., Green, K. E., & Abramovitz, S. (2010). Transgender clients: Identifying and minimizing barriers to mental health treatment. *Journal of Gay & Lesbian Mental Health, 14*(2), 94–108.

Siegel, A., Roeling, T. A., Gregg, T. R., & Kruk, M. R. (1999). Neuropharmacology of brain-stimulation-evoked aggression. *Neuroscience & Biobehavioral Reviews, 23*(3), 359–389.

Singer, T. (2004). Empathy for pain involves the affective but not sensory components of pain. *Science, 303*, 1157–1162.

Sillaber, I., Panhuysen, M., Henniger, M. S., Ohl, F., Kühne, C., Pütz, B., ... & Holsboer, F. (2008). Profiling of behavioral changes and hippocampal gene expression in mice chronically treated with the SSRI paroxetine. *Psychopharmacology, 200*, 557–572.

Skyt, I., Moslemi, K., Baastrup, C., Grosen, K., Benedetti, F., Petersen, G. L., Price, D. D., Hall, K. T., Kaptchuk, T. J., Svensson, P., Jensen, T. S., & Vase, L. (2018). Dopaminergic tone does not influence pain levels during placebo interventions in patients with chronic neuropathic pain. *Pain, 159*(2), 261–272.

Skyt, I., Lunde, S. J., Baastrup, C., Svensson, P., Jensen, T. S., & Vase, L. (2020). Neurotransmitter systems involved in placebo and nocebo effects in healthy participants and patients with chronic pain: A systematic review. *Pain, 161*(1), 11–23.

Slade, G. D., Smith, S. B., Zaykin, D. V., Tchivileva, I. A., Gibson, D. G., Yuryev, A., Mazo, I., Bair, E. Y., Fillingim, R., Ohrbach, R., Greenspan, J., Maixner, W., & Diatchenko, L. (2013). Facial pain with localized and widespread manifestations: Separate pathways of vulnerability. *Pain, 154*, 2335–2343.

Smith, W. B., Gracely, R. H., & Safer, M. A. (1998). The meaning of pain: Cancer patients' rating and recall of pain intensity and affect. *Pain, 78*, 123–129.

Sommer, C., Leiters, M., & Uceyler, N. (2018). Inflammation in the pathophysiology of neuropathic pain. *Pain, 159*(3), 595–602.

Staahl, C., Olesen, A. E., Andersen, T., Arendt-Nielsen, L., & Drewes, A. M. (2009). Assessing analgesic actions of opioids by experimental pain models in healthy volunteers - an updated review. *British Journal Pharmacology, 68*, 149–168.

Staud, R. (2012). Abnormal endogenous pain modulation is a shared characteristic of many chronic pain conditions. *Expert Review Neurotherapy, 12*, 577–585.

Świder, K., & Bąbel, P. (2013). The effect of the sex of a model on nocebo hyperalgesia induced by social observational learning. *PAIN®, 154*(8), 1312–1317.

Syal, S., Ipser, J., Terburg, D., Solms, M., Panksepp, J., Malcolm-Smith, S., ... & van Honk, J. (2015). Improved memory for reward cues following acute buprenorphine administration in humans. *Psychoneuroendocrinology, 53*, 10–15.

Tanaka, M., & Watanabe, Y. (2012). Supraspinal regulation of physical fatigue. *Neuroscience & Biobehavioral Reviews, 36*(1), 727–734.

Taylor, A. M., Becker, S., Schweinhardt, P., & Calill, C. (2016). Mesolimbic dopamine signaling in acute and chronic pain: Implications for motivation, analgesia, and addiction. *Pain, 157*, 1194–1198.

Theodosiou, M., Rush, R. A., Zhou, X. F., Hu, D., Walker, J. S., & Tracey, D. J. (1999). Hyperalgesia due to nerve damage: Role of nerve growth factor. *Pain, 81*, 245–255.

Thom, D. H., Kravitz, R. L., Bell, R. A., Krupat, E., & Azari, R. (2002). Patient trust in the physician: relationship to patient requests. *Family Practice, 19*(5), 476–483.

Thomaidou, M. A., Veldhuijzen, D. S., Meulders, A., & Evers, A. W. M. (2021). An experimental investigation into the mediating role of pain-related fear in boosting nocebo hyperalgesia. *Pain, 162*(1), 287–299.

Tinnermann, A., Geuter, S., Sprenger, C., Finsterbusch, J., & Buches, C. (2017). Interactions between brain and spinal cord mediate value effects in nocebo hyperalgesia. *Science, 358*, 105–108.

Tracey, I. (2010). Getting the pain you expect: mechanisms of placebo, nocebo and reappraisal effects in humans. *Nature Medicine, 16*(11), 1277–1283.

Tracy, L. M., Georglou-Karistianis, N., Gibson, S. J., & Giummarra, M. J. (2015). Oxytocin and the modulation of pain experience: Implications for chronic pain management. *Neuroscience Biobehavioral Review, 55*, 53–67.

Treede, R. D. (2016). Gain control mechanisms in the nociceptive system. *Pain, 157*(6), 1199–1204.

Treede, R. D., Rief, W., Barke, A., Aziz, Q., Bennett, M. I., Benoliel, R., Cohen, M., Evers, S., Nanna, B., Finnerup, M. B., First, M. B., Perrot, S., Scholz, J., Schug, S., Smith, B. H., et al. (2015). A classification of chronic pain for ICD-11. *Pain, 156*(6), 1003–1007.

Ung, H., Brown, J. E., Johnson, K. A., Younger, J., Hush, J., & Mackey, S. (2012). Mulitvariate classification of structural data defects in chronic low back pain. *Cerebral Cortex, 24*, 1037–1044.

Vase, L., Robinson, M. E., Verne, G. N., & Price, D. D. (2005). Increased placebo analgesia over time in irritable bowel syndrome (IBS) patients is associated with desire and expectation but not endogenous opioid mechanisms. *Pain, 115*, 338–347.

Vase, L., Amanzio, M., & Price, D. D. (2015). Nocebo vs placebo: The challenges of trial design in analgesia research. *Clinical Pharmacy Therapy, 97*, 143–150.

Vaso, A., Adahan, H. M., Gjika, A., Zahaj, S., Zhurda, T., & Vyshka, G. (2014). Peripheral nervous system origin of phantom limb pain. *Pain, 155*, 1384–1391.

Volkow, N. D., Wang, G. J., Fowler, J. S., & Tomasi, D. (2012). Addiction circuitry in the human brain. *Annual Review of Pharmacology and Toxicology, 52*, 321–336.

Volkow, N. D., Wise, R. A., & Baler, R. (2017). The dopamine motive system: Implications for drug and food addiction. *Nature Reviews Neuroscience, 18*(12), 741–752.

Vriens, J., Owsianik, G., Hofmann, T., Philipp, S. E., Stab, J., Chen, X., ... & Voets, T. (2011). TRPM3 is a nociceptor channel involved in the detection of noxious heat. *Neuron, 70*(3), 482–494.

Wade, M. E. (2001). Women and salary negotiation: The costs of self-advocacy. *Psychology of Women Quarterly, 25*(1), 65–76.

Wall, P. D., & Gutnick, M. (1974). Ongoing activity in peripheral nerves: the physiology and pharmacology of impulses originating from a neuroma. *Experimental Neurology, 43*(3), 580–593.

Wall, P. D., & Devor, M. (1983). Sensory afferent impulses originate from dorsal root ganglia as well as from the periphery in normal and nerve injured rats. *Pain, 17*(4), 321–339.

Welch, S. P. (2009). Interaction of the cannabinoid and opioid systems in the modulation of nociception. *International Review Psychiatry, 21*, 143–151.

Wiech, K. (2016). Deconstructing the sensation of pain: The influence of cognitive processes on pain perception. *Science, 354*(6312), 584–577.

Wightman, R. M., & Robinson, D. L. (2002). Transient changes in mesolimbic dopamine and their association with 'reward'. *Journal of Neurochemistry, 82*(4), 721–735.

Wood, P. B. (2004). Stress and dopamine: Implications for the pathophysiology of chronic widespread pain. *Medical Hypotheses, 62*, 420–424.

Wood, P. B. (2008). Role of central dopamine in pain and analgesia. *Expert Reviews Neurotherapy, 8*, 781–797.

Woolf, C. J. (2011). Central sensitization: Implications for the diagnosis and treatment of pain. *Pain, 152*(3), S2–S215.

Woolf, C. J., & Salter, M. W. (2000). Neuronal plasticity: Increasing the gain in pain. *Science, 288*, 1765–1769.

Yovell, Y., Bar, G., Mashiah, M., Baruch, Y., Briskman, I., Asherov, J., ... & Panksepp, J. (2016). Ultra-low-dose buprenorphine as a time-limited treatment for severe suicidal ideation: a randomized controlled trial. *American Journal of Psychiatry, 173*(5), 491–498.

Zakharian, E. (2017). Oxytocin modulates nociception as an agonist of pain-sensing TRPV1. *Cell Reports, 21*(6), 1681–1691.

Zhang, R., Chomistek, A. K., Dimitrakoff, J. D., Giovannucci, E. L., Willett, W. C., Rosner, B. A., & Wu, K. (2015). Physical activity and chronic prostatitis/chronic pelvic pain syndrome. *Medicine and Science in Sports and Exercise, 47*(4), 757–764.

Ziegler, D., Duan, W. R., Guohua, A., Thomas, J. W., & Nothaft, W. (2015). A randomized double-blind, placebo, and active-controlled study of T-type calcium channel blocker ABT-639 in patients with diabetic peripheral neuropathic pain. *Pain, 156*(10), 2013–2020.

Zubieta, J. K., & Stohler, C. S. (2009). Neurobiological mechanisms of placebo responses. *Annals NY Academy Science, 1156*, 198–210.

Physical Exercise as an Intervention for Depression: Evidence for Efficacy and Mu-Opioid Receptors as a Mechanism of Action

Colleen Pettrey, Patrick L. Kerr, and T. O. Dickey

Abstract Physical exercise is often cited as an important part of an intervention for depression, and there is empirical evidence to support this. However, the mechanism of action through which any potential antidepressant effects are produced is not widely understood. Recent evidence points toward the involvement of endogenous opioids, and especially the mu-opioid system, as a partial mediator of these effects. In this chapter, we discuss the current level of empirical support for physical exercise as either an adjunctive or standalone intervention for depression. We then review the extant evidence for involvement of endogenous opioids in the proposed antidepressant effects of exercise, with a focus specifically on evidence for mu-opioid system involvement.

Keywords Depression treatment · Exercise · Endogenous opioids · Mu-opioids · Endorphins

Introduction

Depression: Prevalence and Access to Treatment

Unipolar depression, including both major depressive disorder (MDD) and persistent depressive disorder (PDD; formerly referred to as dysthymic disorder), is a significant health problem for tens of millions of people globally. Depression is

C. Pettrey (✉)
Atrium Health Behavioral Health, Charlotte, NC, USA

P. L. Kerr
West Virginia University School of Medicine-Charleston, Charleston, WV, USA

T. O. Dickey
Behavioral Medicine & Psychiatry, WVU School of Medicine, Charleston, WV, USA

P. L. Kerr et al. (eds.), *Endogenous Opioids*, Advances in Neurobiology 35,
https://doi.org/10.1007/978-3-031-45493-6_11

highly prevalent with approximately 4.4% of people worldwide estimated to experience a depressive disorder of some form (World Health Organization, 2017). Research indicates that depression may be more common among some populations and in some settings. For example, a meta-analysis of depression identification in primary care settings found a rate of 21.9% (Mitchell et al., 2009). Data related to access of depression treatment by those who experience depression suggests that this varies widely, ranging from 15 to 60% (Maginn et al., 2004; Wittchen et al., 2001). More recently, Olfson et al. (2016) reported that 8.4% of their nationwide US sample (N = 46,417) screened positive for depression, but only 28.7% received treatment. Interestingly, less than a third (29.9%) of those reporting treatment screened positive for depression; this may point toward either effectiveness of treatment or some inefficiencies in the mental healthcare system.

In the USA, recent data from the 2020 report of the National Survey on Drug Use and Health conducted by the Substance Abuse and Mental Health Services Administration (SAMHSA) (2020) indicates that prevalence of major depressive episodes (MDEs) increased significantly ($p > 0.05$) between 2016 and 2019 for adolescents 12–17 years old (12.8% vs. 15.7%), adults18–25 years old (10.9% vs. 15.2%), and adults 26–49 (7.4% vs. 8.9%).

The prevalent and impairing nature of depression for adults and adolescents highlights the importance of depression treatment. Epidemiological data from SAMHSA's research over the past decade indicates that most adults who report experiencing a MDE receive treatment of some form (66.3% in 2019). However, these same data are more variable for adolescents' access to treatment for depression. Starting in 2005, the rate of depressed adolescents accessing treatment declined for several years, hitting a low of 34.6% in 2009, then returning to a rate of 40.9% in 2016, which was comparable to the rate in 2004 (40.3%). The most recent data indicate that in 2019, 43.3% of adolescents 12–17 years old reporting depression symptoms received treatment. This represents a 15-year high. While this trend is encouraging, it still suggests inadequate access to mental health services in general and does not speak to the quality or scientific validity of the services being accessed. In short, most adolescents experiencing depression do not receive treatment, and the quality of the treatment that is received is unclear.

The lack of treatment may account for the increased rates of serious suicidal ideations and suicide planning observed over the past decade. Both of these rates nearly doubled for 18–25-year-olds (serious suicidal ideations, 6.1% in 2009 vs. 11.8% in 2019; suicide planning, 2% in 2009 vs. 3.9% in 2019). These factors also increased significantly for 26–49-year-olds (serious suicidal ideations, 4.3% in 2009 vs. 5.3% in 2019; suicide planning, 1% in 2009 vs. 1.5% in 2019). Suicide attempts also increased significantly in young adults between 2009 and 2019 (1.1% vs. 1.8%). Using standardized mortality ratios, high-quality meta-analytic studies indicate that depression is associated with a 19.7 times greater risk of death by suicide than those who are not depressed (Chesney et al., 2014).

In sum, depression affects millions of people across the lifespan and is often associated with impairment in functioning. While the majority of adults who are depressed receive some form of treatment, nearly one-third do not; and the majority of adolescents (~56%) who are depressed receive no treatment at all. This may suggest a pattern of untreated adolescent depression progressing to continued depression in early adulthood (i.e., adolescents in 2009 would be young adults in 2019), where suicidal ideations, plans, and behaviors increased dramatically and significantly. This trend is concerning considering both the suffering inherent in depression and the risk of suicidal ideations and behaviors, as well as the risk of death by suicide. Treatment for depression that is both effective and accessible is thus essential in disrupting this pattern.

Physical Activity and Depression

A decrease in physical activity level is one sign and symptom of depression. Decreased physical activity is also often associated with disengagement from and lack of interest in typically enjoyable activities. Together, this pattern of symptoms may lead to more sedentary behavior, which may exacerbate developing depression symptoms. Indeed, a recent meta-analysis by Huang et al. (2020) suggests that sedentary behavior (especially when the activity is not cognitively engaging, e.g., passively watching television) may also increase the risk for depression (Huang et al., 2020). Consistent with this, a recent study by Long et al. (2019) in a diverse and nationally representative US sample ($N = 65{,}059$) of children ages 6–17 found that children who are depressed are significantly less likely to report physical activity. This risk appeared to be further increased for adolescents (12–17-year-olds were less likely to report physical activity than 6–11-year-olds). Though these data are clearly correlational and not causal, the activity-depression relationship is clear and points toward a potential target.

An earlier meta-analysis by Reed and Ones (2006) further demonstrates the potential connection between physical activity and mood and affect. Reed and Ones' extensive examination of pooled data across two decades found that across studies ($k = 158$; $N = 13{,}101$), exercise was associated with increases in positive affect. Their aggregated data were consistent with those of prior individual studies which identified lower baseline positive affect as a moderator of post-exercise positive affect. This meta-analysis also found a zone-limited, dose-dependent relationship, such that moderate (but not extreme/vigorous) exercise lasting up to (but not longer than) 35 minutes was associated with consistent increases in post-exercise positive affect. Reed and Ones (2006) also reported an effect duration of approximately 30 minutes. These analyses controlled for potential threats to internal validity and study quality and were robust to both, suggesting relative generalizability. Although mood differs from affect, and this meta-analysis did not focus on exercise as a treatment for depression per se, people diagnosed with depressive disorders do experience frequent and protracted negative affective states, leading to the depressing of their mood. Thus, these findings lay an important framework for understanding both the antidepressant effects of physical exercise and the potential limits we should expect of such effects.

The exact mechanism(s) through which any behavioral treatments for depression yield their effects remains unclear at this time and is beyond the scope of this chapter. However in this section, we review the best available data pertaining to the use of exercise as an intervention for depression. We then discuss the evidence for the involvement of endogenous opioid activity as an underlying mechanism, with an emphasis on the μ-opioid system.

Physical Exercise as a Treatment for Depression in Children and Adolescents

As noted in the preceding section, interventions for depression during childhood and adolescence are essential. Researchers have conducted several studies of physical exercise as an intervention for child and adolescent depression over the past three decades. This has led to a growing literature base from which some conclusions about efficacy may be drawn based on meta-analytic data.

Recently, multiple meta-analyses have been published evaluating the outcomes of the highest quality evidence for physical exercise as a treatment for depression in children and adolescents. Axelsdottir et al. (2020) conducted a meta-analysis of physical exercise interventions for depression in 12–18-year-olds ($k = 4$; $N = 159$). This study found a moderate effect of exercise interventions across trials (SMD = −0.59). However, the researchers noted the inconsistent evidence even in those studies that met the criteria for their analyses, leading to a low level of certainty based on their data.

Similarly, Oberste et al. (2020) conducted a meta-analysis of physical activity as a depression treatment in adolescents 12–18 years old ($k = 10$; $N = 491$). This analysis found a medium effect of physical activity across studies ($g = -0.47$) and larger effects associated with moderate ($g = -0.82$) and vigorous ($g = -0.51$) physical exercise intensity. No significant differences were found based on: (1) exercise session duration; (2) number of sessions; (3) length of intervention; or (4) standalone ($g = -0.58$) vs. combined treatment ($g = -0.34$) comparisons.

Finally, Wegner et al. (2020) recently published a systematic review of prior meta-analyses in children and adolescents (ages 5–20 years old), providing an even higher level of evidence to consider. Their meta-analysis of four meta-analytic datasets found an overall effect of $d = -0.51$, indicating a moderate effect of physical exercise on depression in children and adolescents.

Physical Exercise as a Treatment for Depression in Adults

An early meta-analysis by Craft and Landers (1998) of 30 studies of the effectiveness of physical exercise for the treatment of depression found a medium to large overall effect of $d = -0.72$ across studies. Higher-intensity exercise seemed to be associated with larger effects (e.g., $d = -0.096$ for running vs. −0.56 for walking), yet nonaerobic exercise also yielded a strong effect ($d = -0.82$). The largest effect was found for comparisons with a no-intervention/waitlist control group ($d = -0.77$), while no significant effects were found for comparisons with other behavioral interventions nor for group versus individual interventions.

A more recent meta-analysis of 23 randomized controlled trials ($N = 977$) by Kvam et al. (2016) found a moderate overall effect of physical exercise on depression ($g = -0.68$), but this effect was found to be diminished over follow-up intervals ($g = -0.22$). Compared to no treatment, physical exercise yielded a large effect ($g = -1.24$) and a medium effect for comparisons with realistic heterogeneous treatment-as-usual conditions ($g = -0.48$). Kvam reported no significant effects of physical exercise compared with psychological treatments and antidepressant medications, but a moderate effect when exercise was combined with antidepressant treatment (Kvam et al., 2016).

Similarly, Schuch et al. (2016a) found in their meta-analysis of 25 randomized controlled trials ($N = 1487$) that physical exercise produced a large statistically significant effect on depression symptoms (SMD = 1.11), with stronger effect for vigorous (SMD = 1.34) and moderate (SMD = 1.33) exercise compared to "light to moderate" levels (SMD = 0.58). Schuch et al. (2016a) also examined the clinical significance of physical activity effects, finding mean score decreases of 4.52 for the Hamilton Rating Scale for Depression (HAM-D) and 6.46 for the Beck Depression Inventory (BDI), respectively.

Josefsson et al. (2014) reported findings that were aligned with several studies mentioned previously. In their meta-analysis of 13 studies of people with depressive disorders, physical exercise was associated with a moderate-to-large effect ($g = -0.77$) on depression symptoms. Similar to previous authors, Josefsson commented on the overall suboptimal quality of studies, noting that effects were reduced in magnitude when only the most rigorous studies were included.

In a rigorous meta-meta-analysis, Rebar et al. (2015) examined the effects of physical exercise on both depression and anxiety symptoms. However, their analysis focused on adults without psychiatric diagnoses, and studies that measured anxiety and depression symptoms, but in which physical exercise was not a treatment for any diagnosed disorder, per se. The researchers reported that, across 6 meta-analyses and 92 studies ($N = 4310$), physical exercise produced a moderate effect on depression symptoms (SMD = −0.50).

In contrast to the previous work discussed, Krogh et al. (2017) meta-analysis ($k = 35$; $N = 2498$) found minimal effects due largely to low-quality data. Their overall reported effect of physical exercise on depression was a SMD of −0.66; however, when controlling for bias by reanalyzing only the studies with minimal risk of bias, they reported an overall low quality of evidence from this literature.

Some lines of research have focused on older adults experiencing depression. Meta-analyses for geriatric populations have found generally similar results. For example, meta-analyses by both Ghasemi Pirbalouti et al. (2019; $k = 6$) and Klil-Drori et al. (2020; $k = 9$) found a medium effect of 0.64. An earlier meta-analysis by Bridle et al. (2012) found a SMD of −0.34 in favor of exercise interventions across 9 studies. A recent meta-analysis by Miller et al. (2020; $k = 15$) compared the subtypes of physical exercise interventions and significant improvements over control conditions for “mind-body” exercises, e.g., tai chi and qigong (gs = −0.87 to −1.38); aerobic exercises, e.g., walking and dancing (gs = −0.051 to −1.02); and resistance exercises (gs = −0.41 to −0.92), while there were no significant differences in effects among interventions. The foregoing findings are further supported by Bigarella et al. (2021) recent meta-meta-analysis of 12 meta-analyses (k = 97 RCTs), which found a moderate effect range (SMD = −0.90 to −0.14) for exercise in geriatric populations, with a statistically significant pooled effect (OR = 2.24).

Although it is beyond the scope of this chapter, it is also worth noting that physical exercise has been found via meta-analyses to be effective for reducing depression in specific populations of people with medical conditions including cancer (d = −0.13 to −0.22; Brown et al., 2012; Craft et al., 2012; also see Hancock & Sirbu, chapter “Depression, Cancer, Inflammation, and Endogenous Opioids: Pathogenic Relationships and Therapeutic Options”, this volume), heart failure (e.g., Tu et al., 2014), and diabetes (g = −0.59; Narita et al., 2019) and those with multiple (i.e., two or more) medical comorbidities (SMD = −0.80; Bricca et al., 2020).

Summary of Meta-analyses of Physical Exercise as a Treatment for Depression

In aggregate, the published data suggest that for people of all ages experiencing depression, some physical activity is better than no physical activity for improving symptoms, whether done independently or with others. Combined, the data indicate that the effects of physical exercise may be enhanced if the exercise is slightly more intense, but not always, and that this effect may diminish linearly with age. Finally, physical exercise as a standalone intervention may be equally effective for depression compared with both psychological and psychopharmacological treatments, but it may enhance other treatments (especially pharmacotherapies) when integrated adjunctively.

Clearly, further research with more consistent and higher-quality methodology that is robust to bias is needed to fully understand the true effects of physical exercise on depression. Nonetheless, the findings to date clearly indicate a “signal” within the “noise,” and it is reasonable to conclude that there is indeed an effect of physical exercise on depression symptoms. We turn next to examining what neurobiological substrates may underpin this effect, with an emphasis on the role of the endogenous opioid system in the pathogenesis and alleviation of depression.

Endogenous Opioid System Involvement in the Pathogenesis of Depression

A critical part of understanding if physical exercise is an effective intervention for depression is identifying a valid, empirically sound theory that would explain the effects observed. The assumption for many years was that the stimulating and energizing effects of exercise naturally had an antidepressant effect. Historically, this has been simplistically attributed to the "release of endorphins." However, we now know that multiple peptides may be involved in a neurological response to physical activity. Moreover, as the research described above highlights, a prescription for exercise is not one-size-fits-all for type, intensity, or even duration. While the general principle of "some is better than none" may be derived from the empirical literature, the remaining question of "why" merits explanation. In this final section, we examine evidence for mu-opioid system activity as a pathway mediating the antidepressant effects of exercise.

Depression and the Mu-Opioid System

The activity of μ-opioid receptors (μ-OR) is one potential mediator of the effects of physical activity on depression symptoms. One important line of research in this regard is data pertaining to variations in μ-OR presence in relevant neuroanatomical locations. Research findings in this regard have been mixed. An early study by Zubieta and associates (2003) examining the relationship between experimentally induced sadness (via autobiographical recall) and μ-OR activity (via positron-emission tomography; PET) in non-depressed participants found reduced neurotransmission of μ-opioids in the amygdala, rostral anterior cingulate, ventral pallidum, and temporal cortex, all of which have an established relationship to emotionality and emotion regulation. Although not focused on depression per se, these preliminary findings pointed toward the involvement of μ-opioids in negative affect.

A subsequent study by Kennedy et al. (2006) also used mood induction and PET imaging to investigate μ-opioid activity. However, this research group examined the differences between depressed ($n = 14$) and non-depressed control ($n = 14$) participants. Similar to Zubieta, this study used a self-induction paradigm in which participants were asked to "vividly" imagine a significantly sad event, which was identified and rehearsed prior to study participation. In contrast to Zubieta et al. (2003), analyses by Kennedy et al. (2006) found no significant changes in opioidergic activity in response to sadness induction among the non-depressed controls. Additionally, they reported a significant increase in μ-opioid neurotransmission for participants diagnosed with depression. As further evidence that μ-OR dysregulation may be a differentiating factor or feature of depression, control participants exhibited deactivation (significant decreases) of μ-opioid activity in the rostral anterior cingulate when transitioning from the neutral condition to the sadness induction condition.

Hsu et al. (2013) reported opposite findings in their study of the effects of experimentally induced sadness (via a social rejection stressor) on μ-OR activity in non-depressed ($N = 18$) participants. Using this paradigm, Hsu reported *increased* μ-OR activity (measured by μ-OR availability using PET with [^{11}C]-carfentanil radiotracing) following mood induction.

Hsu et al. (2013) also conducted a small follow-up study in which they built upon the findings of their earlier research by comparing depressed ($n = 17$) participants to non-depressed participants ($n = 18$). This study also used a social rejection stressor to induce sadness and PET imaging again with [^{11}C]-carfentanil radiotracing to investigate μ-OR activity. Findings revealed a significant reduction in μ-OR activity for those diagnosed with major depressive disorder, but not for non-depressed control participants. Specifically, participants with depression exhibited significant mean deactivation in the left and right amygdala, whereas no such deactivation occurred for non-depressed control participants. Conversely, significant increases in μ-OR activity in the nucleus accumbens were observed in the non-depressed control group, but no significant increase was observed in the depression group. In sum, depressed people exhibited no increases and significant decreases in μ-OR activity when sadness was induced.

More recently, Nummenmaa et al. (2020) used PET to measure the relationship between μ-OR availability and symptoms of depression and anxiety in non-depressed/non-anxious participants. They reported an inverse association between μ-OR availability and depression symptoms in several neural structures that are integral to emotion regulation. Notably, these included the amygdala, thalamus, hippocampus, ventral striatum, and insular cortex. Thus, even subclinical depressive symptoms are associated with significantly less μ-OR availability.

Additionally, recent data also further elucidate some of the potential mechanisms through which μ-OR may operate. Al-Fadhel et al. (2019) found significantly higher serum levels of endogenous opioids in unmedicated male participants with MDD ages 14–70 ($n = 60$) compared to 30 age-matched male participants without MDD, including μ-ORs (4.06 pg/ml vs. 3.02 pg/ml; $p = 0.004$), beta-endorphin (βE; (34.15 pg/ml vs. 24.83 pg/ml; $p < 0.001$), and the proinflammatory cytokines interleukin-10 (IL-10; 9.52 pg/ml vs. 6.03 pg/ml; $p < 0.001$) and interleukin-6 (IL-6; 16.39 pg/ml vs. 11.44 pg/ml; $p = 0.046$). Interestingly, serum levels of endomorphin-2 (EM-2) were not significantly different (3.24 pg/ml vs. 3.17 pg/ml; $p = 0.417$). However, among participants with MDD, only μ-OR (but not βE or EM-2) was correlated with IL-10. This research group found similar results in a subsequent study (Al-Hakeim et al., 2019) in which they identified that, among men with MDD, μ-OR and βE were significantly correlated with IL-6 and IL-10. Interestingly, they also noted a significant correlation of IL-6 and IL-10 with κ-OR and dynorphin levels. These findings must be considered in the context of their sample limitations (i.e., males only); it is unclear how results may have differed if a more gender-inclusive sample had been used.

Other researchers have also found empirical links between cytokines and endogenous opioids in depression pathogenesis. Prossin et al. (2011) first identified the relationship between interleukin-18 (IL-18) and μ-OR activation. First, IL-18 levels were significantly higher in people with MDD than non-depressed control participants, suggesting another example of an immunological pathway for depression. Second, IL-18 was positively correlated with negative affect in non-depressed individuals, further establishing the relationship of this cytokine to depressogenesis. Third, IL-18 was positively correlated with μ-OR binding potential at baseline and with μ-OR activation in the ventral striatum and amygdala in response to experimentally induced sad mood.

In a follow-up study using mood induction and PET, Prossin et al. (2016) found that IL-18 was significantly increased during sad mood induction and significantly decreased during neutral mood induction. Consistent with their previous findings, Prossin et al. (2016) also reported that, for people with MDD, the increase in IL-18 was positively correlated with increases in μ-OR activation in the amygdala (bilaterally), the anterior cingulate (bilaterally), and the ventral pallidum (left).

Recently, Lutz and associates (2021) used a postmortem analysis of brain tissue paradigm to investigate the relationship between MDD and μ-OR expression and activation in specific neural regions. Postmortem brain tissue was derived from suicide decedents for whom psychiatric diagnoses and psychosocial history were well characterized and confirmed. Lutz et al., (2021) reported significantly higher levels of maximum μ-OR stimulation in cells from the insular cortex. Importantly, this effect remained robust when controlling for age, sex, history of child abuse, substance use disorder diagnosis, and substance use. These findings converge with and extend earlier data published by other research groups, which have found significantly more μ-OR density and activation in suicide decedents (e.g., thalamus, caudate nucleus; Gabilondo et al., 1995; Gross-Isseroff et al., 1990; Scarr et al., 2012; Zalsman et al., 2005).

The preceding data discussed above point toward evidence of a relationship between μ-OR activity and depression. These findings should portend clinical evidence that μ-OR activation is involved in mediating the antidepressant effects of interventions, including those observed in exercise and physical activity. However, clinically oriented data have been mixed.

While the picture is somewhat complex and nuanced, these data indicate that μ-OR activity plays a role in the pathogenesis of depression. The observations thus far point toward common substrates related to affective valence, processing of emotionally salient stimuli, and emotion regulation.

Mu-Opioid Activity and Physical Exercise

Studies of the relationship between exercise and μ-opioids span approximately four decades. Early research focused heavily on the effects of aerobic exercise on βE, which can act as a hormone, a neurotransmitter, and a neuromodulator. Initial findings bolstered support for the hypothesis that βE mediated the effects of exercise on mood, which contributed to the vernacular description of a “runner’s high” being attributable to “endorphin release” that persists today.

The empirical basis for a direct relationship between βE and exercise, and in turn βE modulation as the mechanism of therapeutic action for exercise on depression, is complex. Some initial findings did indeed indicate that intensive aerobic exercise increased βE levels in plasma and also improved mood (Farrell et al., 1982; Frailoli et al., 1980; Gambert et al., 1981; Goldfarb et al., 1987). Further support for this model was derived from early data indicating that administration of naloxone prior to exercise alters pain perception (higher pain threshold, higher pain tolerance) during and after exercise (e.g., Koltyn et al., 1996; Surbey et al., 1984).

Another key piece in the exercise-endorphin-mood would be the effects of μ-OR antagonism on post-exercise mood changes. However, data from research examining the effects of μ-OR antagonist administration prior to exercise on mood have not consistently supported the model. There are comparable volumes of studies that have found μ-OR antagonist (e.g., naloxone, naltrexone) attenuated or blocked altogether improvements in mood following exercise (Daniel et al., 2002; Janal et al., 1984; Jarvekulg & Viru, 2002) and those that have found no effect (Allen & Coen, 1987; Farrell et al., 1982; Grossman et al., 1984). With this piece missing, alternative explanations are necessary.

One alternative explanation is that μ-opioids interact with multiple neurotransmitter systems that may also be affected by exercise, e.g., dopamine and brain-derived neurotrophic factor (BDNF). As noted in Fig. 1, there are several potential pathways through which μ-opioids may co-influence mood in response to exercise, including dopaminergic, GABA-ergic, glutaminergic, and hormonal systems, among others. One key potential mediator is brain-derived neurotrophic factor (BDNF), which we discuss next.

Brain-Derived Neurotrophic Factor and Exercise

Brain-derived neurotrophic factor (BDNF) is a protein, found in various regions in the central nervous system and produced in peripheral tissues (Barde et al., 1982; Murer et al., 2001). Multiple neurological processes are influenced by BDNF, both at the biochemical (neurogenesis, dendritic growth, and synaptic plasticity) and clinical (cognitive improvement, metabolic regulation, and homeostasis) levels (Gorski et al., 2003; Vaynman et al., 2004; Matthews et al., 2009; Pedersen et al., 2009; Gomez-Pinilla et al., 2008). Decreased levels of BDNF have been shown to correlate with hippocampal atrophy, cognitive impairment, and depressive symptoms (Erickson et al., 2012; Murer et al., 2001; Guilloux et al., 2012; Karege et al., 2005).

BDNF can cross the blood-brain barrier and be stored in peripheral tissues. Additionally, there is some exploratory evidence suggesting that peripherally generated BDNF may have central effects (Erickson et al., 2012; Schmidt & Duman, 2010). While animal models are able to measure central BDNF, in vivo analysis of central BDNF on human subjects has been limited; however, it has been shown that

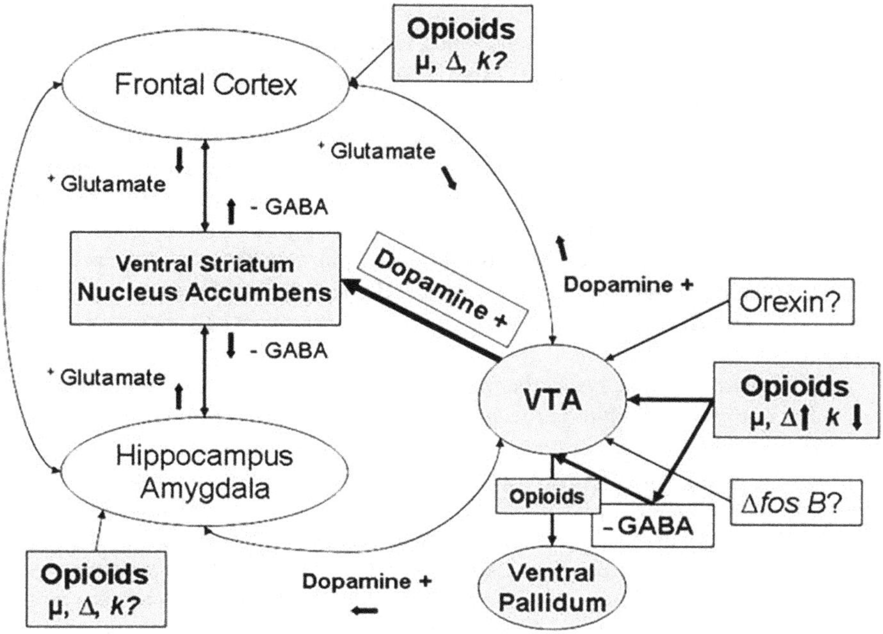

Fig. 1 Plausible opioid and early gene modulation of the mesolimbic dopamine system involved with motivation and pleasure. (Adapted from Dishman and O'Connor (2009). Used with permission)

peripheral and central levels of BDNF do correlate in some animal models (Angelucci et al., 2011).

BDNF was first linked to physical exercise in 1995 (Neeper et al., 1996). Since then, multiple studies on both humans and animals have attempted to answer the question of how this relates to brain function and what impact it may have on future treatment options. A meta-analysis by Szuhany et al. (2015) investigated the effect of exercise on BDNF levels in humans. Twenty-nine studies were analyzed ($N = 1111$ patients, mean age 42.1, 46.6% female), and it was found that both acute exercise bouts and regular exercise regimens had a positive effect on BDNF levels. While a single session of exercise (evidenced by 14 studies analyzed) yielded a moderate effect size, a regular exercise regimen intensified the effect of BDNF increase that occurs immediately post-exercise (evidenced by 8 studies analyzed). Furthermore, regular exercise increases resting levels of BDNF (evidenced by 13 studies analyzed). Interestingly, females did not exhibit as high of an increase in BDNF as did males (Szuhany et al., 2015).

Conversely, a meta-analysis by Schuch et al. (2016b) contradicts the effects reported above. Schuch et al. (2016b) reported that their pooled analysis of three studies of BDNF and exercise found no significant effects of exercise on serum levels of BDNF. Data from two studies analyzed found no effects of BDNF levels on reduction in depression symptoms following exercise.

Animal studies have revealed some potential mechanisms for how BDNF employs the beneficial effects described above. In addition to reversing hippocampal degeneration and enhancing synaptic neurotransmission, BDNF helps reduce the effects of oxidative stress and other chronic stress-mediating effects (Cotman & Berchtold, 2007; Cotman et al., 2007; Siette et al., 2013; Yang et al., 2014). Previous animal studies revealed that both forced and non-forced (treadmill versus activity wheel, respectively) activities increase hippocampal neurogenesis, cell proliferation, and dendritic branching, and even as little as 1 week of exercise improved learning in animals (Stranahan et al., 2007; van Praag et al., 1999; Vaynman et al., 2004).

BDNF interacts and activates the tropomyosin-related kinase B (TrKB) receptor, a member of the tyrosine kinase receptor family, resulting in phosphorylation of the tyrosine residue and facilitating adaptor binding. Subsequent pathways are then activated, including Ras-MAPK and PI3K, resulting in neuronal differentiation, cell proliferation and survival, and synaptic plasticity. A specific mechanism of interest has been the cAMP response element binding protein (CREB), a transcription factor responsible for long-term potentiation and synaptic plasticity. BDNF modulates the function of signaling systems within cells, including the calcium-calmodulin kinase II and mitogen-activated protein kinase, which result in the production and function of CREB (Minichiello, 2009; Shen et al., 2001).

BDNF, Pain, and Endogenous Opioids

Most evidence for the relationship between BDNF and endogenous opioid activity comes from research using rodent models. In an early study, Siuciak et al. (1994) found that infusion of BDNF into the midbrain of rats led to analgesia-like behaviors (i.e., increased latency in tail-flick following introduction of a noxious stimulus). This effect remained consistent for 6 days and was reversed with the administration of naloxone, further implicating a relationship between BDNF and the endogenous opioid system.

A subsequent study by this same research group (Siuciak et al., 1995) found that BDNF infusion increased βE levels by 63% and 97% in the periaqueductal gray (PAG)/dorsal raphe nucleus region and spinal cord, respectively, whereas met-enkephalin was not significantly increased. Similarly, Frank et al. (1997) reported onset of analgesia (increased tail-flick test latency) in rats following BDNF infusion in the midbrain, which increased from baseline to day 1, and from day 1 to day 6, and was sustained from day 6 to day 12 with continuous infusion. Subsequently, Cirulli et al. (2000) reported evidence of analgesia in rats (i.e., decreased pain sensitivity as indicated by increased latency to hindpaw licking when exposed to a hotplate stimulus) following a single intracerebral ventricular administration of BDNF. Consistent with these findings, a study by Guo et al. (2006) found that:

electrical stimulation of the PAG resulted in BDNF release; low-dose infusion of BDNF resulted in higher pain sensitivity; and high-dose infusion of BDNF produced sustained analgesia-like behavior, indicating lower pain sensitivity. Guo (2006) also found that these processes were mediated by descending projections from the PAG to the rostralventromedial medulla (RVM), through which BDNF downregulates TrkB neurons in the RVM.

Consistent with the early research findings by Siuciak et al. (1994, 1995) on the increased μ-opioid activity induced by BDNF induction, Zhang et al. (2006) reported that BDNF mRNA expression was upregulated by μ-opioids (βE, endomorphin-1, and endomorphin-2) and δ-opioids (met-enkephalin, leu-enkephalin). Notably, this was attenuated by naltrexone administration, which was also consistent with Siuciak et al. (1994).

Although beyond the scope of this chapter, the relationship of BDNF to other neurotransmitter systems relevant to depression merits mention. Most notably, research has found association between BDNF (i.e., levels, mRNA expression) and serotonergic activity (see Martinowich & Lu, 2008). This relationship is supported predominantly by animal research, but an emerging line of human research has also found converging evidence for this (Bhang et al., 2011; Fisher et al., 2017; Molteni et al., 2010).

In summary, BDNF availability is associated with pain sensitivity and thresholds. BDNF infusion is associated with the release of endogenous opioids, especially μ- and δ-opioids, which mediate BDNF mRNA expression. BDNF is also released by physical activity, including aerobic and anaerobic exercise. Together, the extant data suggest that the relationship between μ-opioids and antidepressant effects of exercise may be mediated by BDNF.

Conclusions and Future Directions

The preceding sections have established that μ-opioid system activity functionally influences mood and may be involved in depression pathogenesis. Understandably, this would suggest μ-ORs as a target for intervention. Indeed, as noted in Kerr and Gregg (this volume), an emerging body of pharmacotherapy research on opioid agonists seems promising in that regard. These findings may also explain the therapeutic effects of exercise for people with depression; however, this would be predicated on changes in μ-opioid system activity during exercise.

The role of μ-opioids in the observed antidepressant effects of physical exercise is nuanced. Instead of a direct path in which exercise "releases endorphins," which then alleviate depression symptoms, a more complex picture has emerged from empirical research. While μ-opioids likely have some relationship to the antidepressant effects of exercise, these molecules interact with a rich neurobiological ecosystem of neurotransmitters, hormones, proteins, and growth factors. The antidepressant effects of exercise may result from these dynamic interactions and others. Future research must extend previous findings to better understand these relationships.

It is worth noting that factors both known and unknown, both hypothesized and not contemplated, may affect relationships between mood, behavior, and neurobiology in ways we have yet to fully consider. An excellent example of this is a recent study by Sun et al. (2021), which found significant differences in μ-OR availability in humans ($N = 204$) based on season, i.e., spring and summer > autumn and winter. While such findings clearly require replication for any clinical utility to be extracted, they do highlight the potential influence of confounding variables on what would otherwise seem to be a straightforward relationship.

Finally, translational research with direct relevance to humans is lacking and needed. Much of what we think we know about exercise and depression actually derives from animal research. While such research has indeed laid a scientific foundation upon which to build, the clinical value is inherently limited. Here we must consider that preclinical (i.e., animal) studies of endogenous opioids have substantially limited generalizability and applicability to humans, due in part to the documented differences in receptor binding even between nonhuman species. For example, there is nearly twofold higher μ-OR binding in rats compared to guinea pigs; and κ-OR binding in guinea pigs is nine times higher than that found in rats (Thomasy et al., 2007). Differences between humans and rodents in OR binding have also been documented. For example, there is a higher κ-OR binding in human brains compared to rats, but a higher δ-OR binding in rat brains compared to humans (Pfeiffer et al., 1982). To truly advance the clinical science underpinning endogenous opioids and exercise in the treatment of depression in humans, future research must increasingly and, eventually, exclusively focus on empirical investigations of human participants.

References

Al-Fadhel, S. Z., Al-Hakeim, H. K., Al-Dujaili, A. H., & Maes, M. (2019). IL-10 is associated with increased mu-opioid receptor levels in major depressive disorder. *European Psychiatry, 57*, 46–51.

Allen, M. E., & Coen, D. (1987). Naloxone blocking of running-induced mood changes. *Annals of Sports Medicine, 3*, 190–195.

Angelucci, F., Gelfo, F., De Bartolo, P., Caltagirone, C., & Petrosini, L. (2011). BDNF concentrations are decreased in serum and parietal cortex in immunotoxin 192 IgG-Saporin rat model of cholinergic degeneration. *Neurochemistry International, 59*, 1–4.

Axelsdóttir, B., Biedilæ, S., Sagatun, Å., Nordheim, L. V., & Larun, L. (2020). Exercise for depression in children and adolescents–a systematic review and meta- analysis. *Child and Adolescent Mental Health, 26*(4), 347–356. https://doi.org/10.1111/camh.12438

Barde, Y. A., Edgar, D., & Thoenen, H. (1982). Purification of a new neurotrophic factor from mammalian brain. *The EMBO Journal, 1*(5), 549–553.

Bhang, S., Ahn, J. H., & Choi, S. W. (2011). Brain-derived neurotrophic factor and serotonin transporter gene-linked promoter region genes alter serum levels of brain-derived neurotrophic factor in humans. *Journal of Affective Disorders, 128*(3), 299–304.

Bigarella, L. G., Ballotin, V. R., Mazurkiewicz, L. F., Ballardin, A. C., Rech, D. L., Bigarella, R. L., & Selistre, L. D. S. (2021). Exercise for depression and depressive symptoms in older adults: An umbrella review of systematic reviews and meta-analyses. *Aging & Mental Health*, 1–11.

Bricca, A., Harris, L. K., Saracutu, M., Smith, S. M., Juhl, C. B., & Skou, S. T. (2020). Benefits and harms of exercise therapy in people with multimorbidity: A systematic review and meta-analysis of randomised controlled trials. *Ageing Research Reviews, 63*, 101166.

Bridle, C., Spanjers, K., Patel, S., Atherton, N. M., & Lamb, S. E. (2012). Effect of exercise on depression severity in older people: Systematic review and meta- analysis of randomised controlled trials. *The British Journal of Psychiatry, 201*(3), 180–185.

Brown, J. C., Huedo-Medina, T. B., Pescatello, L. S., Ryan, S. M., Pescatello, S. M., Moker, E., et al. (2012). The efficacy of exercise in reducing depressive symptoms among cancer survivors: A meta-analysis. *PLoS One, 7*(1), e30955.

Chesney, E., Goodwin, G. M., & Fazel, S. (2014). Risks of all-cause and suicide mortality in mental disorders: A meta-review. *World Psychiatry, 13*(2), 153–160.

Cirulli, F., Berry, A., & Alleva, E. (2000). Intracerebroventricular administration of brain-derived neurotrophic factor in adult rats affects analgesia and spontaneous behaviour but not memory retention in a Morris Water Maze task. *Neuroscience Letters, 287*(3), 207–210.

Cotman, C. W., & Berchtold, N. C. (2007). Physical activity and the maintenance of cognition: Learning from animal models. *Alzheimer's and Dementia, 3*(2), S30–S37.

Cotman, C. W., Berchtold, N. C., & Christie, L. A. (2007). Exercise builds brain health: Key roles of growth factor cascades and inflammation. *Trends in Neurosciences, 30*(9), 464–472.

Craft, L. L., & Landers, D. M. (1998). The effect of exercise on clinical depression and depression resulting from mental illness: A meta-analysis. *Journal of Sport and Exercise Psychology, 20*(4), 339–357.

Craft, L. L., VanIterson, E. H., Helenowski, I. B., Rademaker, A. W., & Courneya, K. S. (2012). Exercise effects on depressive symptoms in cancer survivors: A systematic review and meta-analysis. *Cancer Epidemiology and Prevention Biomarkers, 21*(1), 3–19.

Daniel, M., Martin, A. D., & Carter, J. (2002). Opiate receptor blockade by naltrexone and mood state after acute physical activity. *British Journal of Sports Medicine, 26*, 111–115.

Dishman, R. K., & O'Connor, P. J. (2009). Lessons in exercise neurobiology: The case of endorphins. *Mental Health and Physical Activity, 2*(1), 4–9.

Erickson, K. I., Miller, D. L., & Roecklein, K. A. (2012). The aging hippocampus: Interactions between exercise, depression, and BDNF. *Neuroscientist, 18(1), 82–97.*

Farrell, P. A., Gates, W. K., Maksud, M. G., & Morgan, W. P. (1982). Increases in plasma ß-endorphin/B-lipotropin immunoreactivity after treadmill running in humans. *Journal of Applied Physiology, 52*, 1245–1249.

Fisher, P. M., Ozenne, B., Svarer, C., Adamsen, D., Lehel, S., Baaré, W. F. C., et al. (2017). BDNF val66met association with serotonin transporter binding in healthy humans. *Translational Psychiatry, 7*(2), e1029–e1029.

Fraioli, F., Moretti, C., Paolucci, D., Alicicco, E., Crescenzi, F., & Fortunio, G. (1980). Physical exercise stimulates marked concomitant release of beta-endorphin and adrenocorticotropic hormone (ACTH) in peripheral blood in man. *Experientia, 36*, 987–989.

Frank, L., Wiegand, S. J., Siuciak, J. A., Lindsay, R. M., & Rudge, J. S. (1997). Effects of BDNF infusion on the regulation of TrkB protein and message in adult rat brain. *Experimental Neurology, 145*(1), 62–70.

Gabilondo, A. M., Meana, J. J., & García-Sevilla, J. A. (1995). Increased density of μ- opioid receptors in the postmortem brain of suicide victims. *Brain Research, 682*(1–2), 245–250.

Gambert, S. R., Garthwaite, T. L., Pontzen, C. H., Cook, E. E., Tristani, F. E., Duthie, E. H., et al. (1981). Running elevates plasma beta-endorphin immunoreactivity and ACTH in untrained human subjects. *Proceedings, Society of Experimental Biological Medicine, 168*, 1–4.

Ghasemi Pirbalouti, M., Shariat, A., & Ghazanfari, A. (2019). A meta-analysis of exercise therapy on reducing depression among older adults in Iran. *Quarterly Journal of Health Psychology, 8*(30), 69–80.

Goldfarb, A. H., Hatfield, B. D., Sforzo, G. A., & Flynn, M. G. (1987). Serum Beta-endorphin levels during a graded exercise test to exhaustion. *Medicine and Science in Sports and Exercise, 19*, 78–82.

Gomez-Pinilla, F., Vanyman, S., & Ying, Z. (2008). Brain-derived neurotrophic factor functions as a metabotrophin to mediate the effects of exercise on cognition. *European Journal of Neuroscience, 28*, 2278–2287.

Gorski, J. A., Zeiler, S. R., Tamowski, S., & Jones, K. R. (2003). Brain-derived neurotrophic factor is required for the maintenance of cortical dendrites. *Journal of Neuroscience, 23*, 6856–6865.

Gross-Isseroff, R., Dillon, K. A., Israeli, M., & Biegon, A. (1990). Regionally selective increases in μ opioid receptor density in the brains of suicide victims. *Brain Research, 530*(2), 312–316.

Grossman, A., Bouloux, P., Price, P., Drury, P. L., Lam, K. S., Turner, T., et al. (1984). The role of opioid peptides in the hormonal responses to acute exercise in man. *Clinical Science (Colch.), 67*, 483–491.

Guilloux, J. P., Douillard-Guilloux, G., Kota, R., Wang, X., Gardier, A. M., Martinowich, K., et al. (2012). Molecular evidence for BDNF-and GABA-related dysfunctions in the amygdala of female subjects with major depression. *Molecular Psychiatry, 17*(11), 1130–1142.

Guo, W., Robbins, M. T., Wei, F., Zou, S., Dubner, R., & Ren, K. (2006). Supraspinal brain-derived neurotrophic factor signaling: A novel mechanism for descending pain facilitation. *Journal of Neuroscience, 26*(1), 126–137.

Hsu, D. T., Sanford, B. J., Meyers, K. K., Love, T. M., Hazlett, K. E., Wang, H., et al. (2013). Social feedback activates the endogenous opioid system. *Molecular Psychiatry, 18*(11), 1147–1147.

Huang, Y., Li, L., Gan, Y., Wang, C., Jiang, H., Cao, S., & Lu, Z. (2020). Sedentary behaviors and risk of depression: A meta-analysis of prospective studies. *Translational Psychiatry, 10*(1), 1–10.

Janal, M. N., Colt, E. W., Clark, W. C., & Glusman, M. (1984). Pain sensitivity, mood and plasma endocrine levels in man following long-distance running: Effects of naloxone. *Pain, 19*, 13–25.

Jarvekulg, A., & Viru, A. (2002). Opioid receptor blockade eliminates mood effects of aerobic gymnastics. *International Journal of Sports Medicine, 23*, 155–157.

Josefsson, T., Lindwall, M., & Archer, T. (2014). Physical exercise intervention in depressive disorders: Metaanalysis and systematic review. *Scandinavian Journal of Medicine & Science in Sports, 24*(2), 259–272.

Karege, F., Vaudan, G., Schwald, M., Perroud, N., & La Harpe, R. (2005). Neurotrophin levels in postmortem brains of suicide victims and the effects of antemortem diagnosis and psychotropic drugs. *Molecular Brain Research, 136*(1), 29–37.

Kennedy, S. E., Koeppe, R. A., Young, E. A., & Zubieta, J. K. (2006). Dysregulation of endogenous opioid emotion regulation circuitry in major depression in women. *Archives of General Psychiatry, 63*(11), 1199–1208.

Klil-Drori, S., Klil-Drori, A. J., Pira, S., & Rej, S. (2020). Exercise intervention for late- life depression: A meta-analysis. *The Journal of Clinical Psychiatry, 81*(1), 6914.

Koltyn, K. F., Garvin, A. W., Gardiner, R. L., & Nelson, T. F. (1996). Perception of pain following aerobic exercise. *Medicine and Science in Sports and Exercise, 28*(11), 1418–1421.

Krogh, J., Hjorthøj, C., Speyer, H., Gluud, C., & Nordentoft, M. (2017). Exercise for patients with major depression: A systematic review with meta-analysis and trial sequential analysis. *BMJ Open, 7*(9), e014820.

Kvam, S., Kleppe, C. L., Nordhus, I. H., & Hovland, A. (2016). Exercise as a treatment for depression: A meta-analysis. *Journal of Affective Disorders, 202*, 67–86.

Long, S., Rogers, M. L., & Gjelsvik, A. (2019). The influence of depression status on weekly exercise in children ages 6 to 17 years. *Preventive Medicine Reports, 13*, 199–204.

Lutz, P. E., Almeida, D., Filliol, D., Jollant, F., Kieffer, B. L., & Turecki, G. (2021). Increased functional coupling of the mu opioid receptor in the anterior insula of depressed individuals. *Neuropsychopharmacology, 46*(5), 920–927.

Maginn, S., Boardman, A. P., Craig, T. K. J., Haddad, M., Heath, G., & Stott, J. (2004). The detection of psychological problems by general practitioners. *Social Psychiatry and Psychiatric Epidemiology, 39*(6), 464–471.

Martinowich, K., & Lu, B. (2008). Interaction between BDNF and serotonin: Role in mood disorders. *Neuropsychopharmacology, 33*(1), 73–83.

Matthews, V. B., Åström, M. B., Chan, M. H. S., Bruce, C. R., Krabbe, K. S., Prelovsek, O., et al. (2009). Brain-derived neurotrophic factor is produced by skeletal muscle cells in response to contraction and enhances fat oxidation via activation of AMP-activated protein kinase. *Diabetologia, 52*(7), 1409–1418.

Miller, K. J., Gonçalves-Bradley, D. C., Areerob, P., Hennessy, D., Mesagno, C., & Grace, F. (2020). Comparative effectiveness of three exercise types to treat clinical depression in older adults: A systematic review and network meta-analysis of randomised controlled trials. *Ageing Research Reviews, 58*, 100999.

Minichiello, L. (2009). TrkB signaling pathways in LTP and learning. *Nature Reviews Neuroscience, 10*(12), 850–860.

Mitchell, A. J., Vaze, A., & Rao, S. (2009). Clinical diagnosis of depression in primary care: A meta-analysis. *The Lancet, 374*(9690), 609–619.

Molteni, R., Cattaneo, A., Calabrese, F., Macchi, F., Olivier, J. D., Racagni, G., et al. (2010). Reduced function of the serotonin transporter is associated with decreased expression of BDNF in rodents as well as in humans. *Neurobiology of Disease, 37*(3), 747–755.

Murer, M. G., Yan, Q., & Raisman-Vozari, R. (2001). Brain-derived neurotrophic factor in the control human brain, and in Alzheimer's disease and Parkinson's disease. *Progress in Neurobiology, 63*(1), 71–124.

Narita, Z., Inagawa, T., Stickley, A., & Sugawara, N. (2019). Physical activity for diabetes-related depression: A systematic review and meta-analysis. *Journal of Psychiatric Research, 113*, 100–107.

Neeper, S. A., Gomez-Pinilla, F., Choi, J., & Cotman, C. W. (1996). Physical activity increases mRNA for brain-derived neurotrophic factor and nerve growth factor in rat brain. *Brain Research, 726*, 49–56.

Nummenmaa, L., Karjalainen, T., Isojärvi, J., Kantonen, T., Tuisku, J., Kaasinen, V., et al. (2020). Lowered endogenous mu-opioid receptor availability in subclinical depression and anxiety. *Neuropsychopharmacology, 45*(11), 1953–1959.

Oberste, M., Medele, M., Javelle, F., Lioba Wunram, H., Walter, D., Bloch, W., et al. (2020). Physical activity for the treatment of adolescent depression: A systematic review and meta-analysis. *Frontiers in Physiology, 11*(185), 1–17.

Olfson, M., Blanco, C., & Marcus, S. C. (2016). Treatment of adult depression in the United States. *JAMA Internal Medicine, 176*(10), 1482–1491.

Pedersen, B. K., Pedersen, M., Krabbe, K. S., et al. (2009). Role of exercise-induced brain-derived neurotrophic factor production in the regulation of energy homeostasis in mammals. *Experimental Physiology, 94*(12), 1153–1160.

Pfeiffer, A., Pasi, A., Mehraein, P., & Herz, A. (1982). Opiate receptor binding sites in human brain. *Brain Research, 248*(1), 87–96.

Prossin, A. R., Koch, A. E., Campbell, P. L., McInnis, M. G., Zalcman, S. S., & Zubieta, J. K. (2011). Association of plasma interleukin-18 levels with emotion regulation and μ-opioid neurotransmitter function in major depression and healthy volunteers. *Biological Psychiatry, 69*(8), 808–812.

Prossin, A. R., Koch, A. E., Campbell, P. L., Barichello, T., Zalcman, S. S., & Zubieta, J. K. (2016). Experimental sadness induces relevant interactions between central endogenous opioid activation and plasma IL-18 concentrations in depressed volunteers. *Molecular Psychiatry, 21*(2), 151–151.

Rebar, A. L., Stanton, R., Geard, D., Short, C., Duncan, M. J., & Vandelanotte, C. (2015). A meta-meta-analysis of the effect of physical activity on depression and anxiety in non-clinical adult populations. *Health Psychology Review, 9*(3), 366–378.

Reed, J., & Ones, D. S. (2006). The effect of acute aerobic exercise on positive activated affect: A meta-analysis. *Psychology of Sport and Exercise, 7*(5), 477–514.

Scarr, E., Money, T. T., Pavey, G., Neo, J., & Dean, B. (2012). Mu opioid receptor availability in people with psychiatric disorders who died by suicide: A case control study. *BMC Psychiatry, 12*(1), 1–10.

Schmidt, H. D., & Duman, R. S. (2010). Peripheral BDNF produces antidepressant-like effects in cellular and behavioral models. *Neuropsychopharmacology, 35*, 2378–2391.

Schuch, F. B., Vancampfort, D., Richards, J., Rosenbaum, S., Ward, P. B., & Stubbs, B. (2016a). Exercise as a treatment for depression: A meta-analysis adjusting for publication bias. *Journal of Psychiatric Research, 77*, 42–51.

Schuch, F. B., Deslandes, A. C., Stubbs, B., et al. (2016b). Neurobiological effects of exercise on major depressive disorder: A systematic review. *Neuroscience and Biobehavioral Reviews, 61*, 1–11.

Shen, H., Tong, L., Balazs, R., & Cotman, C. W. (2001). Physical activity elicits sustained activation of the cyclic AMP response element-binding protein and mitogen- activated protein kinase in the rat hippocampus. *Neuroscience, 107*(2), 219–229.

Siette, J., Westbrook, R. F., Cotman, C., et al. (2013). Age-specific effects of voluntary exercise on memory and the older brain. *Biological Psychiatry, 73*(5), 435–442.

Siuciak, J. A., Altar, A., Wiegand, S. J., & Lindsay, R. M. (1994). Antinociceptive effect of brain-derived neurotrophic factor and neurotrophin-3. *Brain Research, 633*(1–2), 326–330.

Siuciak, J. A., Wong, V., Pearsall, D., Wiegand, S. J., & Lindsay, R. M. (1995). BDNF produces analgesia in the formalin test and modifies neuropeptide levels in rat brain and spinal cord areas associated with nociception. *European Journal of Neuroscience, 7*(4), 663–670.

Stranahan, A. M., Khalil, D., & Gould, E. (2007). Running induces widespread structural alterations in the hippocampus and entorhinal cortex. *Hippocampus, 17*, 1017–1022.

Substance Abuse and Mental Health Services Administration. (2020). *Key substance use and mental health indicators in the United States: Results from the 2019 National Survey on drug use and health (HHS publication no. PEP20–07–01-001, NSDUH series H-55)*. Center for Behavioral Health Statistics and Quality, Substance Abuse and Mental Health Services Administration. Retrieved from https://www.samhsa.gov/data/

Sun, L., Tang, J., Liljenbäck, H., Honkaniemi, A., Virta, J., Isojärvi, J., et al. (2021). Seasonal variation in the brain μ-opioid receptor availability. *Journal of Neuroscience, 41*(6), 1265–1273.

Surbey, G. D., Andrew, G. M., Cervenko, F. W., & Hamilton, P. P. (1984). Effects of naloxone on exercise performance. *Journal of Applied Physiology, 57*(3), 674–679.

Szuhany, K. L., Bugatti, M., & Otto, M. W. (2015). A meta-analytic review of the effects of exercise on brain-derived neurotrophic factor. *Journal of Psychiatric Research, 60*, 56–64.

Thomasy, S. M., Moeller, B. C., & Stanley, S. D. (2007). Comparison of opioid receptor binding in horse, guinea pig, and rat cerebral cortex and cerebellum. *Veterinary Anaesthesia and Analgesia, 34*(5), 351–358.

Tu, R. H., Zeng, Z. Y., Zhong, G. Q., Wu, W. F., Lu, Y. J., Bo, Z. D., et al. (2014). Effects of exercise training on depression in patients with heart failure: A systematic review and meta-analysis of randomized controlled trials. *European Journal of Heart Failure, 16*(7), 749–757.

van Praag, H., Kempermann, G., & Gage, F. H. (1999). Running increases cell proliferation and neurogenesis in the adult mouse dentate gyrus. *Nature Neuroscience, 2*(3), 266–270.

Vaynman, S., Ying, Z., & Gomez-Pinilla, F. (2004). Hippocampal BDNF mediates the efficacy of exercise on synaptic plasticity and cognition. *European Journal of Neuroscience, 20*, 2580–2590.

Wegner, M., Amatriain-Fernández, S., Kaulitzky, A., Murillo-Rodriguez, E., Machado, S., & Budde, H. (2020). Systematic review of meta-analyses: Exercise effects on depression in children and adolescents. *Frontiers in Psychiatry, 11*(81), 1–12.

Wittchen, H. U., Höfler, M., & Meister, W. (2001). Prevalence and recognition of depressive syndromes in German primary care settings: Poorly recognized and treated? *International Clinical Psychopharmacology, 16*(3), 121–135.

World Health Organization. (2017). *Depression and other common mental disorders: Global health estimates. No. WHO/MSD/MER/2017.2*. World Health Organization.

Yang, J. L., Lin, Y. T., Chuang, P. C., Bohr, V. A., & Mattson, M. P. (2014). BDNF and exercise enhance neuronal DNA repair by stimulating CREB-mediated production of apurinic/apyrimidinic endonuclease 1. *Neuromolecular Medicine, 16*, 161–174.

Zalsman, G., Molcho, A., Huang, Y., Dwork, A., Li, S., & Mann, J. J. (2005). Postmortem μ-opioid receptor binding in suicide victims and controls. *Journal of Neural Transmission, 112*(7), 949–954.

Zhang, H., Torregrossa, M. M., Jutkiewicz, E. M., Shi, Y. G., Rice, K. C., Woods, J. H., et al. (2006). Endogenous opioids upregulate brain-derived neurotrophic factor mRNA through δ-and μ-opioid receptors independent of antidepressant-like effects. *European Journal of Neuroscience, 23*(4), 984–994.

Zubieta, J. K., Ketter, T. A., Bueller, J. A., Xu, Y., Kilbourn, M. R., Young, E. A., & Koeppe, R. A. (2003). Regulation of human affective responses by anterior cingulate and limbic μ-opioid neurotransmission. *Archives of General Psychiatry, 60*(11), 1145–1153.

The Endogenous Opioid System as a Pathway of Positive Emotions

Jennifer Barenz, Maeve O'Donnell, and Joey Smith

Abstract Pleasant emotions take a variety of forms and are a key part of the human experience. Although negative emotions have often been a focus of research, positive emotions, e.g., joy, pleasure, and love, have recently gained more attention. Each of these emotions is rich and complex in its own right. However, positive emotions appear to serve key evolutionary functions, which are mediated by complex biological substrates. This chapter summarizes key research and explores the biological underpinnings of positive emotions, with an emphasis on the roles that endogenous opioids play in the experience, expression, and development of positive emotions. The necessity of emphasizing positive emotions in research is also discussed.

Keywords Positive emotions · Endogenous opioids · Joy · Love

Introduction

Within the field of emotion research, the study of positive emotions is relatively new. Although negative emotions are distinctly adaptive and recognized as essential for survival, they are also attributed to myriad forms of suffering. Historically, the behavioral and psychological sciences have focused on understanding negative emotions for the sake of ameliorating the suffering they cause (Fredrickson, 1998). However, there is now wider recognition of the value in understanding the lighter quadrants of the emotional spectrum. Fredrickson and Cohn have contended that positive emotions also serve an evolutionarily adaptive function, albeit within a

J. Barenz (✉)
Tomah VA Medical Center, Tomah, WI, USA

M. O'Donnell
Seattle Children's & University of Washington School of Medicine, Seattle, WA, USA

J. Smith
Colorado State University, Fort Collins, CO, USA

P. L. Kerr et al. (eds.), *Endogenous Opioids*, Advances in Neurobiology 35,
https://doi.org/10.1007/978-3-031-45493-6_12

different timeframe; they contribute primarily to acquisition of the resources necessary for long-term survival (the "broaden-and-build" model), rather than facilitating the avoidance of impending danger (Fredrickson, 1998, 2001; Howell et al., 2007).

The experience of positive emotions, such as joy, love, and excitement, is empirically associated with both psychological well-being and improved physical health. Research indicates that those who experience more frequent positive emotions have better health outcomes, with accumulating evidence indicating that such outcomes may extend to improvements in mortality rates (Chida & Steptoe, 2008; Holt-Lunstad et al., 2010). Early meta-analytic research suggested that this may account for the protective effects of having meaningful social connections on morbidity and mortality, with more recent research confirming this directly (Corbett et al., 2006; Kok et al., 2013). Indeed, both social connectedness and the positive emotions it engenders may correlate with a common underlying neurobiological variable. Endogenous opioids, including endorphins, represent a likely candidate as this common link.

Despite substantial growth in research on endogenous opioids and positive emotions, the evidence for this association is still in a relatively nascent stage. Even newer is the line of research examining the relationship of specific positive emotions and endogenous opioids. As these literatures emerge, examination of the current state of the evidence for these associations is warranted. This chapter reviews the current evidence supporting the association between endogenous opioids and positive emotions in general. Additionally, this chapter reviews extant evidence for the role of endogenous opioids in those specific positive emotions that have borne out the most significant evolutionarily adaptive benefits across human and prehuman history, i.e., pleasure and love.

Endogenous Opioids and Positive Emotional States

Opioid receptors were the first types of neurotransmitter receptors discovered in the brain, which ushered in a new era of thinking about emotions and their underlying biological structures. Panksepp and colleagues accurately hypothesized early on that endogenous opioids alleviate psychological pain, e.g., through social bonding and attachment behaviors, in the same way that exogenous opiates alleviate physical pain. Several decades of research has generally supported this prediction (Panksepp, 2010). There is now a more nuanced understanding of the endogenous opioid system, endogenous opioid receptor subtypes, and their underlying neural paths.

Affective neuroscientific studies have identified seven primary emotional processes (Panksepp & Watt, 2011; Zubieta, et al., 2003). These include four positively valenced processes (i.e., seeking; play; care; lust) and three negatively valenced processes (i.e., rage; fear; panic/grief). Among the positively valenced processes, *seeking* is most closely linked to states of desire (e.g., survival resources, valued tools and other objects); *play* is most often associated with joy and happiness; *care* is commonly linked to nurturance and parental nurturing behaviors; and *lust* is associated with sexual desire and the physiology of sexual behaviors. Among negatively valenced processes, *rage* is associated with anger; *fear* is associated with the full range of fearful emotional states including anxiety; and *panic/grief* is associated with distress and loss, especially social loss. Furthermore, each process is defined and manifested by a unique set of criteria, including engagement of emotion-specific action patterns; precipitation by a restricted group of unconditioned stimuli; activation of affective-neural response that persists beyond the duration of activating stimuli; affective-neural activation that upregulates and/or downregulates afferent and efferent sensory neuron signals; regulation of learning processes; and eventual regulation of emotions via cortical processes (Panksepp & Watt, 2011). In addition to these distinguishing features, each process serves a specific evolutionary function.

Positive emotional states are most attributable to the seeking, care, lust, and play processes listed above. Interestingly, some of the key evidence for the role of endogenous opioids in positive emotional states comes from data related to the negative emotional states. For example, Zubieta et al. (2003) reported over a decade ago that low limbic opioid activity is predictive of sadness in humans. Additionally, as described by Kerr and Gregg (this volume), emerging research indicates the antidepressant action of μ-opioid receptor agonists and κ-opioid receptor antagonists (e.g., buprenorphine). Buprenorphine may also serve to reduce distress-related behaviors associated with panic/grief processes (Panksepp et al., 1978). This is consistent with research findings that indicate endogenous opioids—along with prolactin and oxytocin—inhibit arousal (activated by corticotropin-releasing factor and glutamate) in the dorsal periaqueductal gray-anterior cingulate circuit associated with panic/grief during distress related to separation loss. Similarly, following the activation of the rage process by glutamate and Substance P, endogenous opioids (along with gamma-aminobutyric acid; GABA) inhibit this activation, permitting a return to homeostasis, and stable organismic functioning. Endogenous opioids, and μ-opioids in particular, essentially act to neurobiologically downregulate emotional pain across multiple emotional states. A full discussion of the ways in which this research applies to other forms of psychopathology is beyond this chapter's emphasis. Nonetheless, there are clear implications of these excitatory/inhibitory relationships for depression, disordered eating, and addiction, which are addressed in other chapters in this book.

Beyond the inhibition of the processes involved in emotional pain, endogenous opioids are associated with neural circuits involved in pleasant emotions. As part of their meta-analysis of studies of endogenous opioids and human emotions in the Neurosynth database, Nummenmaa and Tuominen (2018) found a particularly strong relationship between μ-opioid receptors and positive emotions. Specifically, when examining data from neuroimaging studies employing blood oxygenation level-dependent (BOLD) response methodology, the strongest correlation between μ-opioid receptor distribution in the brain and neural substrates of emotional activity was with circuits associated with pleasant emotions ($\underline{r} = 0.44$). However, Nummenmaa and Tuominen (2018) report in a follow-up activation likelihood estimation (ALE) analysis of PET data from healthy participants in the Neurosynth database that, although μ-opioid receptor distribution and activation remained strongly and significantly related to emotional circuits, there was little differentiation between positive and negative emotions. μ-Receptors appear to equally influence emotional pleasure (e.g., happiness, reward) and pain (e.g., fear) (Nummenmaa & Tuominen, 2018).

Positive emotions are predicated on positive affect. Neurophysiologically, positive affective states are facilitated by both endogenous opioids and dopamine. Consistent with their biological mandate, endogenous opioids appear to facilitate appetitive feelings (e.g., enjoying an experience); dopamine appears to facilitate more active attracted feelings (e.g., desiring or wanting an experience or part of an experience; (Berridge & Robinson, 2003; Peciña et al., 2006)). Similar to the discussion of placebo by Kerr and Gregg (this volume) and addiction by Acree (this volume), the relationship between endogenous opioid activity and dopaminergic activity in the context of positive emotions is often inextricable. There are some emerging lines of research that suggest complex relationships among pleasant emotions and endogenous opioid activity (especially μ-opioids), and these are discussed further in the concluding chapter of this volume by Kerr, Gregg, and Sirbu.

Endogenous Opioids and Specific Positive Emotions

The preceding sections introduced evidence for the endogenous opioid system as a mechanism of action facilitating positive emotional states in general. Doing so is a prerequisite starting point for a discussion of the roles that endogenous opioids play in specific positive emotions. In this section, we review the empirical support for the roles of endogenous opioids in specific positive emotions. We turn now specifically to happiness/joy and love.

Pleasure: Happiness and Joy

Endogenous opioids, e.g., β-endorphin, have been associated with the activating emotional experiences associated with pleasure, including joy and happiness (Gale & Edwards, 2016; Jones & Ellis, 1996). For example, β-endorphin has been linked to pleasurable sensations including euphoria and "feelings of pleasantness" (Bird & Kuhar, 1977; Chemali et al., 2008; Schull et al., 1981; Szekely, 1983; Wildmann et al., 1986). This endorphin is synthesized and stored in the anterior pituitary gland from the precursor protein proopiomelanocortin (Guillemin et al., 1977; O'Riordan et al., 1988; Rose, 1985). When stimuli are presented, the hypothalamus is signaled by corticotropin-releasing hormone, which causes the pituitary to secrete β-endorphin and the associated pleasurable sensations including euphoria and "feelings of pleasantness" (Bird & Kuhar, 1977; Morley et al., 1991; Schull et al., 1981; Szekely, 1983; Wildmann et al., 1986).

Research in both humans and animal models has found that μ-opioid receptors are engaged during and by limbic system activity associated with pleasure and have a relationship to dopaminergic activity (Berridge & Kringelbach, 2013). Soderman and Unterwald (2009) found that administration of cocaine precipitated the release of a μ-opioid receptor binding peptide via the D_2 receptors within multiple limbic structures, including the ventral tegmental area and nucleus accumbens, caudate putamen, and frontal cortex in a rodent model. This is consistent with animal model data from Yoshida et al. (1999) finding nucleus accumbens and ventral tegmental area μ-opioid activity mediated by dopaminergic activity in these same regions. Colasanti et al. (2012) found that in healthy humans, high-dose amphetamine administration decreased exogenous opiate (i.e., carfentanil) binding in a similar pattern in the caudate, putamen, anterior cingulate, frontal cortex, and insula. This was consistent with data from Zubieta et al. (2003) indicating that μ-opioid activity measured using PET in the amygdala, ventral pallidum, and anterior cingulate mediates decreases in positive affect in response to experimental induction of negative mood in a laboratory task. Thus, in both humans and animals, endogenous opioids and specifically μ-opioids and receptors are integral to the regulation of pleasure-related emotional states.

Romantic Love, Compassion, and Self-Compassion

The definitions of love are expansive, the nuances inconceivable, and the neural research on love, at best, incomplete. Yet, recent scientific evidence suggests that there may be a discernible role of endogenous opioids and associated peptides with love. In this section, we will discuss three facets of love and the role of endogenous opioids in the experience of this emotion. First, we will discuss the experience of romantic love, defined as a state of intense desire for another (Acevedo et al., 2012). Next, we will discuss compassion, which is believed to encompass a deep awareness of other's suffering and a desire to relieve associated suffering (Cassell, 2009). Finally, self-compassion will be discussed, which is considered to be tripartite—kindness toward the self, awareness of common humanity, and mindfulness (Neff, 2003a, 2003b).

Hawkes contends that the pleasant emotional state, sometimes described in terms of intoxication as a "high," that is associated with romantic love is connected with increased endorphinergic activity, yet this has remained somewhat speculative (Hawkes, 1992). There is further evidence to suggest that not all romantic love processes are created equally, at least as far as the brain is concerned. "Falling in love" is considered a transitional period from not experiencing love to being in love, marked by motivation toward a romantic relationship, and characterized by intrusive thoughts and images of the object of your affections and behaviors intended to elicit a reciprocal response (Aron et al., 1995; Leckman, & Mayes, 1999). The experience of romantic love is conceptually different, marked by stability, reduced emotionality, and mutual interdependencies (Berscheid, 2010). Cortisol was found to be higher among those who had fallen in love in the last 6 months, as compared to people who were single or had been romantically linked with their partner for longer than 6 months, although there were no significant differences in cortisol at 12–24 month follow-up (Marazziti & Canale, 2004). It has been hypothesized that the cortisol responses observed in this study may represent a nonspecific effect of arousal and stress that exists while falling in love, but may not be representative when one is in love.

Other aspects of neurobiology have been discussed within the context of love (Esch & Stefano, 2005). Generally, "falling in love" has been associated with the release of neuropeptides, such as oxytocin (Esch & Stefano, 2005; Marazziti & Canale, 2004). At the same time, however, *falling* in love has been associated with serotonin depletion, while *being* in love has been associated with peak endorphins via an alkaloidal neurotransmitter, anandamide (Bloom & Kupfer, 1995; Kurup & Kurup, 2003). In addition, brain differences in people who have a tendency to fall in love versus those who are less likely to report falling in love have been observed. In one study, those with increased tendency to fall in love also had elevated levels of tyrosine catabolites, which have been associated with synthesis of endogenous morphine (Arun et al., 1998). Within animal models, there is research to suggest that sexual stimulation in rats and hamsters may lead to increased activity in the endogenous opioid systems (Hawkes, 1992). It is important to consider, too, that subjective experiences of romantic love and sex may be influential in biologically rewarding signaling molecules, such as endorphins and endocannabinoids (Kurup & Kurup, 2003).

Beyond falling and being in love with a romantic partner, there are many ways in which people exhibit caring behaviors toward non-romantic others, extending their reach into empathy, service, and compassion. Specifically, as a form of love, compassion evokes an affiliative sense—an understanding of another person's world and a desire to impact it positively. Oxytocin and endorphins are thought to play a key role in both the giving and receiving of compassion. It has been hypothesized that adaptations to the autonomic nervous system have occurred over time that regulate and override one's fight-or-flight response in order to experience others as not only rewarding but also soothing and physiologically regulating (Porges, 2007). Several studies have shown that compassionate or caring touch releases endorphins and oxytocin, inspiring the "rest and digest" properties of the parasympathetic nervous system and relating to lowered physiological markers of stress (Carter, 1998; Porges, 2007). Endorphins are considered a key part of compassionate love, especially in processes that require caregiving, such as a parent-child relationship (Hawkes, 1992). Models of compassion-focused therapy have been developed, in part, to capitalize on the parasympathetic nervous system activation and endorphin release which occurs when extending compassion or caring for others (Ashworth et al., 2011; Gilbert, 2014).

Although people notably and tangibly understand the benefits of acting in compassionate ways toward others, less is known about the neurobiological correlates of extending caring to the self. In its theoretical and empirical infancy, self-compassion is defined as reacting to the self with kindness, recognizing that one is not alone in navigating the trials of humanity, and holding thoughts and feelings in balanced awareness, especially in times of distress (Neff, 2003a, b). Akin to its theoretical cousin, self-compassion portends to provide a similar calming and soothing effect as compassion. However, the key difference lies in the availability for one to provide this resource for the self.

Gilbert has proposed that the soothing effects of self-compassion are mediated through a "safeness-soothing" emotional regulation system, which is linked to endogenous opioids, including endorphins (Gilbert, 2005, 2009). The "safeness-soothing" emotional regulation system is linked to contentedness, security, and well-being, with physical links to improved immune system functioning and higher pain threshold (Depue & Morrone-Strupinsky, 2005). Higher pain thresholds have been used as proxy measures of endogenous opioid release in prior research (Dunbar et al., 2011). Although yet to be empirically tested, some theories suggest that self-compassion confers psychological benefit, in part, through the release of endorphins and decreased responsiveness to non-harmful threat. For example, compassion-focused therapy is designed to engage safeness-soothing emotion regulation, thereby linking people to feeling safety and security, which, in turn, confers physiological benefit (Gilbert, 2009). More neurobiological research is needed in this area, specifically empirical tests of endorphin and neuropeptide functioning when exhibiting self-compassion in stressful situations.

Conclusions

As we have described, understanding the phenomenology and neurobiology of positive emotional states is among the newer of scientific endeavors. Direct study of positive emotions is emerging, and related research provides some direction to this line of inquiry. As described, there is a solid theoretical basis for attributing a central role of the endogenous opioid system in positive emotions, which is based on solid empirical science.

Emotional experiences exist along a continuum that spans from appetitive to aversive. Research has begun to elucidate the ways in which this full range of emotional experiences has likely been necessary for survival of species, both humans and others. Neural substrates underpinning this continuum converge and diverge at points that we are only beginning to understand. At a molecular level, it remains challenging to disentangle positive and negative emotions from one another, in part, because of their shared neurobiology. Indeed, they are separate yet clearly related phenomena.

The current research efforts devoted to fully understanding the functions and pathways of positive emotions are essential and must be extended and expanded. With the majority of emotion research over the past century having been devoted to aversive emotions, the time has come for a change in direction. After all, alleviating emotional pain necessitates amelioration of suffering and induction of competing pleasant emotions. The state of the current literature, though limited relative to the volume of pathocentric research, is intriguing, inspiring, and promising nonetheless. Pleasure is more than an absence of pain and merits equal consideration, attention, and resources for investigation.

References

Acevedo, B. P., Aron, A., Fisher, H. E., & Brown, L. L. (2012). Neural correlates of long-term intense romantic love. *Social Cognitive and Affective Neuroscience, 7*(2), 145–159.

Aron, A., Paris, M., & Aron, E. N. (1995). Falling in love: Prospective studies of self-concept change. *Journal of Personality and Social Psychology, 69*(6), 1102–1112.

Arun, P., Ravikumar, A., Leelamma, S., & Kurup, P. A. (1998). Endogenous alkaloids in the brain of rats loaded with tyrosine/tryptophan & in the serum of patients of neurodegenerative & psychiatric disorders. *Indian Journal of Medical Research, 107*, 231–238.

Ashworth, F., Gracey, F., & Gilbert, P. (2011). Compassion focused therapy after traumatic brain injury: Theoretical foundations and a case illustration. *Brain Impairment, 12*(2), 128–139.

Berridge, K. C., & Kringelbach, M. L. (2013). Neuroscience of affect: Brain mechanisms of pleasure and displeasure. *Current Opinion in Neurobiology, 23*(3), 294–303.

Berridge, K. C., & Robinson, T. E. (2003). Parsing reward. *Trends in Neurosciences, 26*(9), 507–513.

Berscheid, E. (2010). Love in the fourth dimension. *Annual Review of Psychology, 61*, 1–25.

Bird, S. J., & Kuhar, M. J. (1977). Iontophoretic application of opiates to the locus coeruleus. *Brain Research, 122*, 523–533.

Bloom, F. E., & Kupfer, D. J. (1995). Psychopharmacology. In *Fourth generation of progress*. Raven Press.

Carter, C. S. (1998). Neuroendocrine perspectives on social attachment and love. *Psychoneuroendocrinology, 23*(8), 779–818.

Cassell, E. J. (2009). Compassion. In S. J. Lopez & C. R. Snyder (Eds.), *Oxford library of psychology: Oxford handbook of positive psychology* (pp. 393–403). Oxford University Press.

Chemali, Z. N., Chahine, L. M., & Naassan, G. (2008). On happiness: A minimalist perspective on a complex neural circuitry and its psychosocial constructs. *Journal of Happiness Studies, 9*(4), 489–501.

Chida, Y., & Steptoe, A. (2008). Positive psychological Well-being and mortality: A quantitative review of prospective observational studies. *Psychosomatic Medicine, 70*(7), 741–756.

Colasanti, A., Searle, G. E., Long, C. J., Hill, S. P., Reiley, R. R., Quelch, D., et al. (2012). Endogenous opioid release in the human brain reward system induced by acute amphetamine administration. *Biological Psychiatry, 72*(5), 371–377.

Corbett, A. D., Henderson, G., McKnight, A. T., & Paterson, S. J. (2006). 75 years of opioid research: The exciting but vain quest for the holy grail. *British Journal of Pharmacology, 147*(S1), S153–S162.

Depue, R. A., & Morrone-Strupinsky, J. V. (2005). A neurobehavioral model of affiliative bonding: Implications for conceptualizing a human trait of affiliation. *Behavioral and Brain Sciences, 28*(3), 313–349.

Dunbar, R. I., Baron, R., Frangou, A., Pearce, E., van Leeuwin, E. J., Stow, J., et al. (2011). Social laughter is correlated with an elevated pain threshold. *Proceedings of the Royal Society of London B: Biological Sciences, 279*(1731), 1161–1167.

Esch, T., & Stefano, G. B. (2005). The neurobiology of love. *Neuroendocrinology Letters, 26*(3), 175–192.

Fredrickson, B. L. (1998). What good are positive emotions? *Review of General Psychology, 2*(3), 300–319.

Fredrickson, B. L. (2001). The role of positive emotions in positive psychology: The broaden-and-build theory of positive emotions. *American Psychologist, 56*(3), 218–226.

Gale, A., & Edwards, J. A. (2016). *Individual differences and psychopathology: Physiological correlates of human behaviour* (Vol. 3). Academic Press.

Gilbert, P. (Ed.). (2005). *Compassion: Conceptualisations, research and use in psychotherapy*. Routledge.

Gilbert, P. (2009). Introducing compassion-focused therapy. *Advances in Psychiatric Treatment, 15*(3), 199–208.

Gilbert, P. (2014). The origins and nature of compassion focused therapy. *British Journal of Clinical Psychology, 53*(1), 6–41.

Guillemin, R., Vargo, T., Rossier, J., Minick, S., Ling, N., Rivier, C., Vale, W., & Bloom, F. (1977). β-Endorphin and adrenocorticotropin are secreted concomitantly by the pituitary gland. *Obstetrical & Gynecological Survey, 33*(3), 184–185.

Holt-Lunstad, J., Smith, T. B., & Layton, J. B. (2010). Social relationships and mortality risk: A meta-analytic review. *PLoS Medicine, 7*(7), e1000316.

Howell, R. T., Kern, M. L., & Lyubomirsky, S. (2007). Health benefits: Meta-analytically determining the impact of Well-being on objective health outcomes. *Health Psychology Review, 1*(1), 83–136.

Jones, R. A., & Ellis, G. D. (1996). Effect of variation in perceived risk on the secretion of β-endorphins. *Leisure Sciences, 18*(3), 277–291.

Kok, B. E., Coffey, K. A., Cohn, M. A., Catalino, L. I., Vacharkulksemsuk, T., Algoe, S. B., et al. (2013). How positive emotions build physical health: Perceived positive social connections account for the upward spiral between positive emotions and vagal tone. *Psychological Science, 24*(7), 1123–1132.

Kurup, R. K., & Kurup, P. A. (2003). Hypothalamic digoxin, hemispheric dominance, and neurobiology of love and affection. *International Journal of Neuroscience, 113*(5), 721–729.

Leckman, J. F., & Mayes, L. C. (1999). Preoccupations and behaviors associated with romantic and parental love. Perspectives on the origin of obsessive-compulsive disorder. *Child and Adolescent Psychiatric Clinics of North America, 8*(3), 635–665.

Marazziti, D., & Canale, D. (2004). Hormonal changes when falling in love. *Psychoneuroendocrinology, 29*(7), 931–936.

Morley, J. E., Benton, D., & Solomon, G. F. (1991). The role of stress and opioids as regulators of the immune response. In J. A. McCubbin, P. G. Kaufmann, & C. B. Nemeroff (Eds.), *Stress, neuropeptides, and systemic disease* (pp. 221–231). Academic Press.

Neff, K. D. (2003a). The development and validation of a scale to measure self-compassion. *Self and Identity, 2*(3), 223–250.

Neff, K. (2003b). Self-compassion: An alternative conceptualization of a healthy attitude toward oneself. *Self and Identity, 2*(2), 85–101.

Nummenmaa, L., & Tuominen, L. (2018). Opioid system and human emotions. *British Journal of Pharmacology, 175*(14), 2737–2749.

O'Riordan, H. L. H., Malan, P. G., & Gould, R. P. (Eds.). (1988). *Essentials of endocrinology* (2nd ed.). Blackwell Scientific.

Panksepp, J. (2010). Affective neuroscience of the emotional BrainMind: Evolutionary perspectives and implications for understanding depression. *Dialogues in Clinical Neuroscience, 12*(4), 533–545.

Panksepp, J., & Watt, D. (2011). What is basic about basic emotions? Lasting lessons from affective neuroscience. *Emotion Review, 3*(4), 387–396.

Panksepp, J., Herman, B., Conner, R., Bishop, P., & Scott, J. P. (1978). The biology of social attachments: Opiates alleviate separation distress. *Biological Psychiatry, 13*(5), 607–618.

Peciña, S., Smith, K. S., & Berridge, K. C. (2006). Hedonic hot spots in the brain. *The Neuroscientist, 12*(6), 500–511.

Porges, S. W. (2007). The polyvagal perspective. *Biological Psychology, 74*(2), 116–143.

Rose, R. M. (1985). Psychoendocrinology. In J. D. Wilson & D. W. Foster (Eds.), *Textbook of endocrinology* (7th ed., pp. 653–681). W. B. Saunders.

Schull, J., Kaplan, H., & O'Brien, C. P. (1981). Naloxone can alter experimental pain and mood in humans. *Physiological Psychology, 9*(3), 245–250.

Soderman, A. R., & Unterwald, E. M. (2009). Cocaine-induced mu opioid receptor occupancy within the striatum is mediated by dopamine D2 receptors. *Brain Research, 1296*, 63–71.

Szekely, J. I. (1983). The most characteristic in vivo effects of opiates. In J. I. Szekely & A. Z. Ronai (Eds.), Opioid peptides (Vol. I, pp. 29–118). : CRC Press.

Wildmann, J., Kruger, A., Schmole, M., Neimann, J., & Matthaie, J. (1986). Increase of circulating beta-endorphin-like immunoreactivity correlates with the change in feeling of pleasantness after running. *Life Sciences, 38*, 997–1003.

Yoshida, Y., Koide, S., Hirose, N., Takada, K., Tomiyama, K., Koshikawa, N., & Cools, A. R. (1999). Fentanyl increases dopamine release in rat nucleus accumbens: Involvement of mesolimbic mu-and delta-2-opioid receptors. *Neuroscience, 92*(4), 1357–1365.

Zubieta, J. K., Ketter, T. A., Bueller, J. A., Xu, Y., Kilbourn, M. R., Young, E. A., & Koeppe, R. A. (2003). Regulation of human affective responses by anterior cingulate and limbic μ-opioid neurotransmission. *Archives of General Psychiatry, 60*(11), 1145–1153.

Endogenous Opioids and Volunteering: On the Evolutionary Significance of Helping Others

Alina Simona Rusu

Abstract The human tendency to help others in need has been subject to trans-, inter-, and multidisciplinary studies (e.g., anthropology, neurobiology, evolutionary psychology, economy), within the frame of studying the mechanisms and adaptive significance of human prosocial behavior. Volunteering directed to unrelated and unfamiliar individuals is one common form of such helping behavior. Helping others may be adaptive for a species at a macro-level, which in turn is mediated by neurobiological mechanisms. A key target for analysis of the neurobiological underpinnings of volunteering is the endogenous opioid system (EOS). This chapter discusses EOS activity as a potential mediator of volunteering behavior. Evidence of the congruence between EOS involvement in social group behavior and social bonding and the role of these phenomena in volunteerism is reviewed. Models and empirical evidence of the mechanisms and adaptive value of helping unrelated others are discussed and integrated, including the mammalian caregiving system, the neurobiological model of prosocial behavior, synchrony promoting social bonding, and stress-driven motivation of prosocial action in immediate needs.

Keywords Formal volunteering · Helping behavior · Endorphins · Adaptive value

Introduction

The human tendency to help conspecifics (i.e., those of the same species) in need, but also heterospecific beings, such as animals, has been subject to many trans-, inter-, and multidisciplinary studies (e.g., social psychology, anthropology, neurobiology, evolutionary psychology, economics, and sociology). Across scientific disciplines, such studies have commonly focused on the mechanisms and adaptive

A. S. Rusu (✉)
Department of Fundamental Sciences, Faculty of Animal Science and Biotechnologies, University of Agricultural Sciences and Veterinary Medicine, Cluj-Napoca, Romania
e-mail: alina.rusu@usamvcluj.ro

P. L. Kerr et al. (eds.), *Endogenous Opioids*, Advances in Neurobiology 35,
https://doi.org/10.1007/978-3-031-45493-6_13

significance of human prosocial behavior (e.g., Brown et al., 2009; Brown & Brown, 2015; Rossi, 2001; Tomasello et al., 2012). This chapter addresses several physiological, neurobiological, and evolutionary aspects of volunteering behavior (*note: volunteering is seen here as a form of prosocial behavior directed to unrelated others*). Throughout the chapter, research that identifies the potential roles of the endogenous opioid system (EOS) within the functionality of volunteering behavior is synthesized, with an emphasis on the positive dimensions of this form of cooperative behavior.

In the general field of psychology, a common definition of prosocial behavior is "a broad category of acts that are defined by some significant segment of society and/or one's social group as generally beneficial to other people" (Penner Dovidio et al., 2005, p. 366). From an evolutionary perspective, prosocial behavior is considered a form of behavior that brings fitness benefits to the recipient and diminishes the fitness of the helper (West et al., 2007).

In the evolutionary history of socially living mammals, communication through direct positive physical contacts, such as grooming, has been challenged by several selective pressures, such as time constraints and the increasingly large social networks (Launay et al., 2016). Hence, in the evolutionary history of our species, cognitive and social behavioral strategies emerged to buffer individuals against the stress of social life, to maintain social cohesion and to allow bonding between multiple individuals simultaneously (Dunbar et al., 2009; Launay et al., 2016).

Life in bonded social groups of humans is known to be associated with a diversity of inclusive fitness-related benefits, such as longevity and subjective well-being, as well as costs, such as the stress associated with the increased size of social networks of unrelated individuals with whom a person can interact (Dunbar & Shultz, 2010; Launay et al., 2016). Inclusive fitness is commonly defined as the ability of an individual to further pass her/his genes on to the next generations (Hamilton, 1964). In primates, social bonding is primarily supported by specific forms of social behavior, such as grooming (e.g., Dunbar, 1991; Dunbar & Lehmann, 2013), which is known to be mechanistically underpinned by the release of endogenous opioids (Keverne et al., 1989; Martel et al., 1995).

The costs of helping others (e.g., caregiving, informal and formal volunteering) that are emphasized in the literature include psychological and physiological correlates of caregiver distress. Most of the studies addressing this phenomenon fail to distinguish between the stress associated with helping behavior per se and the feelings (e.g., compassion, sadness) about the recipient (Brown & Brown, 2006, 2015). However, recent evidence from large-scale studies suggests that informal caregivers, as well as formal volunteers, experience positive states associated with helping others and several improvements in health and psychological well-being (e.g., Brown et al., 2009; Jenkinson et al., 2013).

Most of the studies documenting the neurobiological substrate of successful social interactions, such as offering and receiving help from others in situations of need, point toward the rewarding value of these types of interactions for humans (Krach et al., 2010). As mentioned above, dopamine is one of the most commonly suggested candidates for the potential explanation of the phenomenon often referred as *volunteer's high* or *helper's high* (Luks, 1988), commonly described as a sensation of pleasure and subjective happiness associated with the participation in volunteering and charity activities.

The connection between EOS activity and social bonding activities is strongly supported in nonhuman primates. For example, in talapoin monkeys, the administration of an endorphin receptor antagonist in males was associated with increased rates of dyadic grooming (Meller et al., 1980; Fabre-Nys et al., 1982). Also, in the same species, direct measurement of beta-endorphins in the central nervous system indicated that there were higher levels of this molecule following social grooming (Fabre-Nys et al., 1982).

There is a large body of literature supporting grooming's function as a reinforcing behavior for social cohesion and peaceful relations in primate groups. A recent evolutionary perspective has emerged regarding the possibility that the EOS might have been co-opted in the evolutionary history of primates to support the human need for the enhanced social bond (Launay et al., 2016). Based on this perspective, one could interpret the functions of prosocial behavior like volunteering as "grooming-like behaviors" (or to "groom at a distance"; Dunbar, 2012, p. 1843), without direct physical contact with the recipient(s). Thus, from an evolutionary perspective, volunteering may promote social connectedness among large communities in the same ways as grooming. Moreover, the existence of coordinated programs of volunteering means that this type of grooming-like behavior can be simultaneously directed to multiple members of the same community (e.g., activities in which health services and food and/or sanitary packages are simultaneously offered in a coordinated manner to a whole community by volunteers belonging to the same organization).

One of the most common definitions of a volunteer is "…someone who contributes time to helping others with no expectation of pay or other material benefit to herself" (Wilson & Musick, 1999, p. 141). Two types of volunteering are generally referred to in the literature: *formal volunteering*, which is done through some association and organizations, and *informal volunteering*, which is done directly, with no associations. Formal volunteering implies organized activities based on the identification of specific needs of the recipients, while informal volunteering is most commonly a spontaneous activity, often driven by immediate needs in a current situation (Wilson & Musick, 1999). Both types of volunteering are predominantly directed to unrelated persons.

Less direct evidence of the connection between the EOS and social behavior exists in humans, especially between EOS and volunteering. However, in their summary of the literature, Launay et al. (2016) identified areas of the brain with high concentrations of opioid receptors that are responsive to social rejection and acceptance (e.g., Hsu et al., 2013) and that experimental interference with EOS activity in humans can affect the perception of positive social stimuli (Chelnokova et al., 2014). Also, several investigations using physiological methods to operationalize the pain threshold – commonly considered a proxy measure of endorphin release – indicate connections of endorphin release with synchronized physical human activities in social contexts, e.g., laughter while watching humorous videos (Dunbar,

2012; Nummenmaa et al., 2016), and increased exertive activities in the presence of other people (e.g., Cohen & Ejsmond-Frey, 2010).

Physiological Value of Volunteering

Several studies suggest that volunteering can bring significant benefits to volunteers at subjective levels of well-being that have physiological underpinnings (Calvo et al., 2012; Wheeler et al., 1998; see also Bekkers et al., 2016 for a comprehensive review on the physiological correlates of volunteering). There is an increasing body of research, including experimental and quasi-experimental studies (e.g., awareness programs and coordinated activities for students aiming to increase their civic participation), indicating that volunteers report more positive affects and a higher self-esteem compared to non-volunteering control group participants (Bekkers et al., 2016; Konrath, 2014; Konrath & Brown, 2012).

Nonetheless, research on the neurophysiological mechanisms hypothesized to underlie the reported benefits of volunteering is currently lacking. For example, dopamine is a neurotransmitter often associated with reward and pleasant emotions. The neural regions that produce dopamine have been found to be those that are most active when someone makes charitable donations (Moll et al., 2006), contributing to an assumption that dopamine may be responsible for the "warm glow" experience of giving (Andreoni, 1990); however, there is no direct evidence for this. Conversely, there is evidence that those inclined to give may experience more subjective pleasure from charitable behavior than those who are not inclined to give (Ferguson et al., 2012).

Hormones are also proposed as a neurophysiological mechanism of volunteering benefits. Oxytocin is a neuropeptide implicated in positive social interactions, which would be logically associated with volunteering. Although there are studies indicating that nasally administrated oxytocin (compared to a placebo) was associated with an increase in motivation to donate in male participants (Barazza et al., 2011), there are not yet studies examining a direct connection between oxytocin release and volunteering behavior (Bekkers et al., 2016). Research on vasopressin, a neuropeptide involved in social behavior, has yielded emerging evidence for a relationship with aspects of positive social behavior, e.g., such as perception of the level of friendliness of human faces (Thompson et al., 2006). However, no studies directly examining vasopressin in volunteering behaviors have been conducted. Cortisol, a hormone activated by the hypothalamic-pituitary-adrenocortical axis in response to stress, was examined in adolescents in one study, which reported no significant effect of a four-month volunteering program on participants' cortisol levels compared to the waitlist group (Schreier et al., 2013). Similarly, research on testosterone and generosity has yielded contradictory findings, with testosterone administration being associated with both increases and decreases in generosity or neither (Bekker et al., 2016). No studies have focused on the effects of volunteering on testosterone levels. More research is needed to elucidate the neurophysiological processes that lead to the subjectively reported benefits of volunteering.

Volunteering as Activation of Caregiving Motivation

In their attempt to provide an integrative explanation of research on the connections between helping others and health benefits for the helper, Brown and Brown (2015) propose a neurobiological model of prosocial (helping) behavior aiming to identify the neuronal substrates (events and circuitry) behind the health-related benefits of helping behavior and of the adaptive significance of motivation to help. The model is referred to as a *caregiving model*, which, at its primary evolutionary significance, like the parental caregiving system (Bowlby, 1969), functions on a fitness-improvement logic (i.e., in the direction of achieving a safe social environment for the preservation of community).

Based on the well-documented prosocial behavioral parallels between humans and other mammalian species (Preston, 2013), and the evidence of the cross-species conservation of the mammalian nervous system (Finlay & Darlington, 1995), the bedrock of Brown and Brown's (2015) caregiving model is the neuroscience of mammalian parenting (Numan, 2006). This offers a useful tool for understanding the potential neural map of prosocial motivation in humans. Thus, Brown and Brown (2015) describe a set of neurocontingencies that might have their evolutionary roots in providing offspring care, which, through genetic and cultural evolution, might have offered a "...regulatory infrastructure to support a more flexible and generalized system behavior, not restricted to biological kin and not readily exploited by cheaters" (Brown & Brown, 2015, p. 3).

Brown and Brown (2015) propose three evolutionarily based explanations for the redirection of parent-like helping behaviors to unrelated individuals: (1) indirect reciprocity (Alexander, 1987; Nowak, 2006; Trivers, 1971), i.e., a beneficial act whose return comes from individual(s) other than the act's recipient (Tullberg, 2004); (2) the fitness interdependences, i.e., the bidirectional correlation of the helper's reproductive success with that of the recipient (Brown & Brown, 2006; Roberts, 2005); and (3) the idea of considering helping unrelated others as an evolutionary "misfire," meaning that the helping response, which is assumed to be evolutionarily shaped to help individuals in need who are either kin or reciprocators (direct and indirect ones), could be activated in the current social environment by others in need, regardless of relatedness and of helping reciprocity (Dawkins, 1989, as cited in Brown & Brown, 2015).

Regarding the health effects of helping others in need, Brown and Brown (2015) hypothesize that prosocial motivation (caregiving motivation) is part of a chain of interconnected biochemical events that function in the direction of stress and inflammation reduction. According to the neurobiological caregiving model, oxytocin, which reduces stress and inflammation through interaction with other hormones that regulate immunological functioning, modulates the neural circuitry behind caregiving (active help) motivation and behavior.

The flow of the components of the model begins with the helper's perception of distress, followed by recognition of need and/or vulnerability of the other individual. Subsequently, caregiving motivation is activated and acted upon. The model ends with the health and longevity benefits for the helper (for a detailed description of the model, see Brown & Brown, 2015). Regarding neural circuitry, the most important component in the Brown and Brown's (2015) neurobiological caregiving model appears to be the medial preoptic area of the hypothalamus (MPOA), which is hypothesized to regulate the motivation for caregiving, in interconnection with the amygdala and other brain regions, as well as other neural and hormonal substrates (e.g., progesterone and oxytocin). Brown and Brown (2015) specified that this model, which is actually the extended version of the original model (Brown et al., 2012), does not offer an all-encompassing theory of helping behavior, but is rather a proposed model of *other-directed motivation* that can favor active helping of individuals in need.

Besides the perception of others in need and the potential stress- and inflammatory-regulating effects of other-directed motivation and helping behavior, research on prosocial behavior provides evidence connecting helping and health benefits for the helpers, i.e., the rewarding value of positive and successful social interactions. However, when considering the positive dimensions of volunteering as successful social interaction, it is important to take into account that there are also health-related risks associated with working with suffering persons in a helping capacity (Thieleman & Cacciatore, 2014). Several studies indicate that volunteers and professionals working with traumatized beings (humans and animals) may experience negative cognitive, emotional, and physiological effects. These adverse effects are commonly referred as compassion fatigue, which encompasses the concepts of secondary (vicarious) trauma and burnout (Adams et al., 2006). Positive intragroup interactions, such as those experienced in a supportive workplace, may help reduce risk for some aspects of compassion fatigue (Adams et al., 2006).

Volunteering as a Reward

Krach et al. (2010) took into account the involvement of dopamine in the prediction and mediation of rewarding stimuli (Schultz, 2001) in their comprehensive literature analysis of the reward of social interactions, noting the fact that "…successful social interactions comprise some of the most potent rewarding stimuli for human beings" (Krach et al., 2010, p. 2). Besides dopamine, the neuropeptide oxytocin has been often mentioned in the context of animal models for social rewards (Liu et al., 2010). Skuse and Gallagher (2009) recently hypothesized that an interaction between dopamine and oxytocin might actually underlie positive (and rewarding) social interactions in humans.

It is well documented that endorphins play an important role in social bonding with unrelated conspecifics in nonhuman primates (Meller et al., 1980) and that the system might be activated by physiological arousal. This physiological arousal might be associated with the interpretation of suffering and need for help in other persons, including those unrelated to the prospective volunteer. In other words, one could consider that volunteering has the potential to decrease the level of distress associated with the presence of other suffering beings (i.e., negative reinforcement function of volunteering).

These findings are consistent with the perception-action model of empathy (Preston & De Waal, 2002), which predicts that the variety of multimodal cues processed by the observers might activate neural patterns of activity in the direction of helping or not a person in need, including the observers' personal representation of stress from previous experiences – the more empathic observers may show more physiological resonance with the person in need (Hofelich & Preston, 2011; Buchanan & Preston, 2014). It could be that the level of familiarity (and/or relatedness) with the person in need mediates the intensity of arousal resonance.

Volunteering as Synchrony

Recent studies (Launay et al., 2016; Mogan et al., 2017) suggest that human behaviors that involve synchronized movements, such as dance, singing, rituals, etc., can be understood as evolutionary behavioral strategies developed to exploit existing neurobiological and psychological mechanisms related to social bonding. This exploitation is considered to take place in the direction of allowing humans to experience and promote social connection in larger social groups, e.g., synchronization may have primarily evolved to save time when socially bonding, but there is evidence that synchronization during singing can encourage social bonding in human groups larger than 150 individuals, regardless of their level of genetic relatedness (Weinstein et al., 2015).

Synchronous activity among humans is considered an adaptive mechanism for large-scale human social bonding (Launay et al., 2016; Mogan et al., 2017). Thus, a *model of synchrony* might also be an explanation for the communality of coordinated volunteering behavior in human species.

There is evidence that endorphins play a significant role in social bonding with nonrelatives in nonhuman primates (Launay et al., 2016). Starting from this type of evidence, it has been suggested that the capacity of experiencing endorphin-related positive affects in social settings might motivate subsequent interactions with individuals who are present, regardless of whether they are related or not; hence, the EOS may promote social bonding (Launay et al., 2016).

One key question remains whether the communality of formal volunteering activities might be explained by the endorphin-based effects of the synchronized movements in humans. This type of volunteering specifically refers to a structured format of volunteering, which implies synchronized decisions and activities of the group members of associations and organizations providing help to persons in need. Prior scientific investigations have found that these activities may promote social bonding (Launay et al., 2016).

Launay et al. (2016) proposed that there is emerging evidence of the connection between synchronized human activities and a greater increase in pain threshold (release of endorphins), compared to solo and/or unsynchronized activities. These authors also argue that synchronization is essentially mimicry involving temporal prediction of the movements of the other co-participants and that the human tendency to synchronize often happens unintentionally (e.g., Issartel et al., 2007). Moreover, it appears that prosocial human individuals demonstrate more spontaneous synchronization than self-oriented individuals (Lumsden et al., 2012) and that people tend to perceive synchrony as an indicator of social closeness (e.g., Miles et al., 2009). Prior studies have also found that synchronization even among strangers facilitates social bonding, even when people have no visual contacts with the others, e.g., synchronization with the sounds produced by another person (Launay et al., 2016).

Considering the connection between synchronization and social closeness on one hand and the role of EOS in promoting social bonding among groups of people on the other hand, it is possible that coordinated and organized volunteering behavior might be explained by the model of synchrony as an adaptive mechanism for large-scale human social bonding. A recent meta-analysis (Mogan et al., 2017) of 42 studies in which experimentally manipulated synchrony conditions were compared to control conditions highlights the importance of taking into account group size when discussing the positive effect of synchrony on prosocial behaviors in humans. Across studies, synchrony in larger groups increased positive affect and prosocial behaviors; however, the relationship between synchrony, social bonding, and social cognition was not affected by group size (Mogan et al., 2017). Mogan et al. (2017) suggest that distinct process mechanisms (neurocognitive versus affective) might underpin the dimensions of synchrony's effects as a function of group size and that synchrony in large groups might drive prosocial effects through affective mechanisms, viz., the process termed by Durkheim collective effervescence. These pathways are not necessarily based on the conscious perception and evaluation of matching behaviors (Mogan et al., 2017). In this vein, at a larger scale, coordination is improved by operating independently of social prediction, and affective responses may facilitate cooperation in an automatic manner, i.e., without relying on partners' specific expectations (Bulbulia, 2012; Mogan et al., 2017). In this context, the EOS may be a proxy for the proposed neurobiological mechanism of affective responses prompted by synchronous behavior (Launay et al., 2016; Tarr et al., 2015; Mogan et al., 2017). Further investigation is needed to confirm these hypothesized mechanisms.

Volunteering as a Stress Reduction Mechanism

Recent studies on the mechanisms of prosocial decisions and behaviors, such as the effects on decision-making of social environment and of the social stress perceived at individual level, indicate that there are situations where stress promotes prosocial behaviors, such as empathic behavior and altruism (Taylor et al., 2000; de Waal et al., 2000). In their comprehensive and multidimensional analysis (neurobiological, behavioral, and evolutionary) of the idea that stress can lead to prosocial action in immediate need, Buchanan and Preston (2014) start from evidence indicating that the physiological stress response has evolved as an adaptive way to motivate behavior and release metabolic energy in acute need situations (e.g., Sapolsky et al., 2000). Sapolsky et al. (2000) proposed that this adaptive function is connected to the stressful experience of social living in humans, i.e., unpredictable and uncontrollable situations that an individual can encounter while interacting with conspecifics (Buchanan & Preston, 2014). Hence, depending on the adaptive value of the individual response in context, stress in a social context can not only lead to aggressive reactions among individuals but also foster affiliative responses, such as prosocial behavior (Buchanan & Preston, 2014). This idea has support from studies of crowding in nonhuman primates, in which monkeys displayed an increased rate of affiliative gestures and grooming when crowded (de Waal et al., 2000). This is contrary to the presumption based on evolutionary principles that crowding might unilaterally produce violence and aggression due to competition for space.

Similarly, there is evidence that stress can be associated with prosocial behavior even in male humans, who, like males across other species, tend to exhibit a "fight-or-flight" response to stress, while females, based on their maternal abilities, tend to be characterized by more prosocial types of responses, such as "tend-and-befriend" (Preston, 2013). However, recent studies on humans exposed to stressful tasks, i.e., the Trier Social Stress Test, indicate that stressful conditions lead to more altruistic behaviors in both male and female participants who reported increased positive affects in response to increased stress, while those participants who showed a greater stress-induced cortisol response made more selfish/self-protective decisions (Starcke et al., 2011). These results can be interpreted based on the importance of interindividual differences in perception of others in need, i.e., individuals who find a stressor to be more of a challenge and report more positive affect might respond in a more prosocial manner, while those who consider the stressor to be a threat might respond in a more self-protective manner (Lazarus & Folkman, 1984).

Buchanan and Preston (2014) have proposed possible mechanisms for stress-induced altruistic aid, i.e., helping others in immediate need regardless of relation. These hypothesized mechanisms are predicated on the mammalian neural circuitry supporting offspring care, which largely overlap with those associated with the reward processes and motivational decisions in the mammalian mesolimbocortical system (Brown et al., 2012; Preston et al., 2007). In sum, by extending the neural system for offspring care to altruism in humans, i.e., helping unrelated others at a current cost to the helper, one might better understand a spontaneous prosocial behavioral decision under stressful conditions, e.g., self-sacrificing acts of saving strangers from fire or from icy waters.

Volunteering As Evolutionarily Stable Strategy

Volunteering behavior is also relevant to discussions of the evolution of cooperation (He et al., 2014; Taylor & Nowak, 2007). Taylor and Nowak (2007) propose that, besides mutation and selection, cooperation can be considered a fundamental principle of evolutionary dynamics. Starting from the two possible actions among interacting individuals as defined by the prisoner's dilemma, i.e., "cooperation" vs. "defection," Taylor and Nowak (2007) discuss five specific mechanisms by which natural selection leads to cooperation: (1) direct reciprocity (i.e., the actions of an individual depend on the previous encounters with the same actor); (2) indirect reciprocity (i.e., the actions of an individual depend on the actions he/she has received from the actor with whom he/she has interacted and on what that specific actor has done to others); (3) kin selection (i.e., implies interactions between genetic relatives); (4) group selection (i.e., refers to cooperation between individuals and between groups); and (5) network reciprocity (i.e., refers to clusters of cooperators that can prevail over defectors within a given population). All of these strategies have been mathematically analyzed from the perspective of costs and benefits (evolutionary game dynamics), by taking into consideration the fact that the fitness of an individual depends on the relative abundance (frequency) of various strategies (phenotypes) in a population. Among these five strategies, indirect reciprocity is the key element for understanding the evolution of prosocial behaviors in humans, including the form of volunteering behavior in which the players might face the dilemma of producing public goods at a cost to themselves and/or when some of the players are weaker than others, i.e., mixed population (for a detailed description of the results of this analysis, see Taylor & Nowak, 2007). Based on recent reconsiderations of the evolutionarily based interpretations that public good can only be produced by stronger players in an asymmetric game, He et al. (2014) indicate that volunteering can function as an evolutionarily stable strategy, i.e., a strategy that, if adopted by a population, cannot be invaded by any competing alternative strategy (Maynard-Smith & Price, 1973). This is because both stronger and weaker players of a population can produce public goods depending on the initial conditions, i.e., phenotype of initial strategy of individuals. While the method of evolutionary game theory is commonly used to analyze several aspects of the evolution of cooperation, this methodology cannot be considered either direct evidence or a mechanism for the evolution of prosocial behaviors in humans and other species (Taylor & Nowak, 2007).

The decision to volunteer, seen by some authors as the decision to participate in the prisoner's dilemma or stay a loner (Hauert et al., 2002), remains a challenging topic of research. The possibility that endogenous opioid neuropeptides play a role in this decision could be considered based on the social rewarding value of volunteering, stress-reduction function, and other fitness-related benefits addressed in this chapter.

Summary and Conclusions

There are still many individual, group, and contextual variables to be taken into account in order to understand humans' decisions to help complete strangers. It is possible that observing pain and suffering in others might activate in a prosocial manner (or in an avoidant self-protective one) the common neural patterns developed through the personal past experiences of the observers of persons in immediate need (Singer et al., 2004; Buchanan & Preston, 2014). Recent studies using standard methods in human stress research indicate that physiological stress can resonate between humans, and this happens in relation to the level of trait empathy of the observers of those experiencing stress (Buchanan et al., 2012).

Both synchronous behavior and subjective distress activate the EOS. There are also rewarding consequences for successful social interactions, which are associated with EOS activity. Therefore, when conceptualized both as a form of synchrony and as a response to personal distress activated by the distress of others in need, we can conclude that endorphins might be placed in this picture as a potential connecting component between these explanatory models of the human tendency to help unrelated others.

The emergence of coordinated activities of helping others, such as group-based volunteering, appears to be quite recent in human evolutionary history (e.g., volunteering in formal volunteer service programs dates back only to the mid-1800s; Harris et al., 2016). Due to the development of media, most humans are continuously exposed to suffering and threats to survival of other beings, e.g., a lack of medical supplies, famine, domestic violence, etc. In most cases, although the immediate needs of the individuals are perceived, e.g., famine in some regions of the world, help cannot be offered by coming into direct contact with the persons in need, e.g., a person from Europe who decides to help children from another continent. In this context, volunteering might be one of the adaptive mechanisms evolved by humans to promote social connectedness in large groups of unrelated individuals. Such mechanisms could build and reinforce the neurocircuitry that might favor the probability of proactive social behavior at a community level, such as receiving help, e.g., resources, assistance, etc., when human individuals (and their offspring) are in need in the future, or may stimulate actions to prevent the situations of being in need. In the same line of interpretation, Taylor and Nowak (2007) point out that all five mechanisms for the evolution of cooperation (direct reciprocity, indirect reciprocity, kin selection, group selection, and network reciprocity) appear to have a unifying principle, viz., assortment—i.e., cooperators are more likely to cooperate with each other than with defectors. While assortment is considered a consequence of a mechanism for the evolution of cooperation, and not a mechanism itself, further questions remain to be answered on how assortment is achieved. The EOS might be an optimal candidate for understanding the neurobiological underpinnings of assortment in the evolution of behavioral forms of cooperative behavior, including volunteering.

References

Adams, R. E., Boscarino, J. A., & Figley, C. R. (2006). Compassion fatigue and psychological distress among social workers: A validation study. *American Journal of Orthopsychiatry, 76*, 103–108.

Alexander, R. D. (1987). *The biology of moral systems*. Aldine de Gruyter.

Andreoni, J. (1990). Impure altruism and donations to public goods: A theory of warm glow giving. *Economic Journal, 100*, 464–477.

Barraza, J. A., McCullogh, M. E., Ahmadi, S., & Zach, P. J. (2011). Oxytocin infusion increases charitable donations regardless of monetary resources. *Hormones and Behavior, 60*, 148–151.

Bekkers, R., Konrath, S. H., & Smith, D. H. (2016). Physiological correlates of volunteering. In D. H. Smith, R. A. Stebbins, & J. Grotz (Eds.), *The Palgrave handbook of volunteering, civic participation, and nonprofit associations* (pp. 541–579). Palgrave Macmillan.

Bowlby, J. (1969). *Attachment and loss* (Vol. 1). Basic Books.

Brown, S. L., & Brown, R. M. (2006). Target article: Selective investment theory: Recasting the functional significance of close relationships. *Psychological Inquiry, 17*, 1–29.

Brown, S. L., & Brown, R. M. (2015). Connecting prosocial behavior to improved physical health: Contributions from the neurobiology of parenting. *Neuroscience and Behavioral Reviews, 55*, 1–17.

Brown, S. L., Smith, D. M., Schulz, R., Kabeto, M. U., Ubel, P. A., Poulin, M., & Langa, K. M. (2009). Caregiving behavior is associated with decreased mortality risk. *Psychological Science, 20*, 488–494.

Brown, S. L., Brown, R. M., & Preston, S. (2012). A model of human caregiving motivation. In S. L. Brown, R. M. Brown, & L. A. Penner (Eds.), *Moving beyond self interest: Perspectives from evolutionary biology, neuroscience and social sciences*. Oxford University Press.

Buchanan, T. V., & Preston, S. D. (2014). Stress leads to prosocial action in immediate need situations. *Frontiers in Behavioral Neuroscience, 8*(5), 1–6.

Buchanan, T. W., Bagley, S. L., Stansfield, R. B., & Preston, S. D. (2012). The empathic, physiological resonance of stress. *Social Neuroscience, 7*, 191–201. https://doi.org/10.1080/17470919.2011.588723

Bulbulia, J. (2012). Spreading order: Religion, cooperative niche construction, and risky coordination problems. *Biology and Philosophy, 27*, 1–27.

Calvo, R., Zheng, Y., Kumar, S., Olgiati, A., & Berkman, L. (2012). Well-being and social capital on planet earth: Cross-national evidence from 142 countries. *PLoS One, 7*, e42793.

Chelnokova, O., Laeng, B., Eikemo, M., Riegels, J., Løseth, G., Maurud, H., et al. (2014). Rewards of beauty: The opioid system mediates social motivation in humans. *Molecular Psychiatry, 19*, 746–747.

Cohen, E. E. A., & Ejsmond-Frey, R. (2010). Rowers' high: Behavioural synchrony is correlated with elevated pain thresholds. *Biology Letters, 6*, 106–108.

Dawkins, R. (1989). *The selfish gene*. Oxford University Press.

de Waal, F. B., Aureli, F., & Judge, P. G. (2000). Coping with crowding. *Scientific American, 282*, 76–81.

Dunbar, R. I. M. (1991). Functional significance of social grooming in primates. *Folia Primatologica, 57*, 121–131.

Dunbar, R. I. M. (2012). Bridging the bonding gap: The transition from primates to humans. *Philosophical Transactions of the Royal Society B: Biological Sciences, 367*, 1837–1846.

Dunbar, R. I. M., & Lehmann, J. (2013). Grooming and social cohesion in primates: A comment on Grueter et al. *Evolution and Human Behavior, 34*, 453–455.

Dunbar, R. I. M., & Shultz, S. (2010). Bondedness and sociality. *Behaviour, 147*, 775–803.

Dunbar, R. I. M., Korstjens, A. H., & Lehmann, J. (2009). Time as an ecological constraint. *Biological Reviews, 84*, 413–429.

Fabre-Nys, C., Meller, R. E., & Keverne, E. B. (1982). Opiate antagonists stimulate affiliative behaviour in monkeys. *Pharmacology Biochemistry and Behavior, 16*, 653–659.

Ferguson, E., Atsma, F., De Kort, W., & Veldhuizen, I. (2012). Exploring the pattern of blood donor beliefs in first-time, novice, and experienced donors: Differentiating reluctant altruism, pure altruism, impure altruism, and warm glow. *Transfusion, 52*(2), 343–355.

Finlay, B. L., & Darlington, R. B. (1995). Linked regularities in the development and evolution of mammalian brains. *Science, 268*, 1578–1584.

Hamilton, W. D. (1964). The genetic evolution of social behavior. *Journal of Theoretical Biology, 7*, 1–52.

Harris, B., Morris, A., Ascough, R. S., Chikoto, G. L., Elson, P. R., McLoughlin, J., Muukkonen, M., et al. (2016). History of associations and volunteering. In D. H. Smith, R. A. Stebbins, & J. Grotz (Eds.), *Palgrave research handbook on volunteering and nonprofit associations*. Springer.

Hauert, C. A., DeMonte, S., Hofbauer, J., & Sigmund, K. (2002). Volunteering as red queen mechanism for cooperation in public goods game. *Science, 296*, 1129–1132.

He, J. Z., Wang, R. W., & Li, Y. T. (2014). Evolutionary stability in the asymmetric volunteer's dilemma. *PLoS One, 9*(8), 1–6.

Hofelich, A. J., & Preston, S. D. (2011). The meaning in empathy: Distinguishing conceptual encoding from facial mimicry, trait empathy and attention to emotion. *Cognition & Emotion, 26*, 119–128.

Hsu, D. T., Sanford, B. J., Meyers, K. K., Love, T. M., Hazlett, K. E., Wang, H., et al. (2013). Response of the opioid system to social rejection and acceptance. *Molecular Psychiatry, 18*, 1211–1217.

Issartel, J., Marin, L., & Cadopi, M. (2007). Unintended interpersonal co-ordination: 'Can we march to the beat of our own drum?'. *Neuroscience Letters, 411*, 174–179.

Jenkinson, C. E., Dickens, A. P., Jones, K., Thompson-Coon, J., Taylor, R. S., Rogers, M., et al. (2013). Is volunteering a public health intervention? A systematic review and meta-analysis of the health and survival of volunteers. *BMC Public Health, 13*, 773.

Keverne, E. B., Martensz, N. D., & Tuite, B. (1989). Beta-endorphin concentrations in cerebrospinal fluid of monkeys are influenced by grooming relationships. *Psychoneuroendocrinology, 14*, 155–161.

Konrath, S. (2014). The power of philanthropy and volunteering. In F. A. Huppert & C. L. Cooper (Eds.), *Interventions and policies to enhance wellbeing: A complete reference guide* (Vol. VI). Wiley.

Konrath, S., & Brown, S. L. (2012). The effects of giving on givers. In N. Roberts & M. Newman (Eds.), *Handbook of health and social relationships*. American Psychological Association.

Krack, S., Paulus, F. M., Bodden, M., & Kircher, T. (2010). The rewarding nature of social interactions. *Frontiers in Behavioral Neuroscience, 4*(22), 1–3.

Launay, J., Tarr, B., & Dunbar, R. I. M. (2016). Synchrony as an adaptive mechanism for large-scale human social bonding. *Ethology, 122*, 1–11.

Lazarus, R. S., & Folkman, S. (1984). *Stress, appraisal and coping*. Springer.

Liu, Y., Aragona, B. J., Young, K. A., Dietz, D. M., Nestler, E. J., & Wang, Z. (2010). Nucleus accumbens dopamine mediates amphetamine-induced impairment of social bonding in a monogamous rodent species. *Proceedings of the National Academy of Sciences U.S.A., 107*, 1217–1222.

Luks, A. (1988). Doing good: Helper's high. *Psychology Today, 22*(10), 34–42.

Lumsden, J., Miles, L. K., Richardson, M. J., Smith, C. A., & Macrae, C. N. (2012). Who syncs? Social motives and interpersonal coordination. *Journal of Experimental Social Psychology, 48*, 746–751.

Martel, F. L., Nevison, C. M., Simpson, M. J., & Keverne, E. B. (1995). Effects of opioid receptor blockade on the social behavior of rhesus monkeys living in large family groups. *Developmental Psychobiology, 28*, 71–84.

Meller, R. E., Keverne, E. B., & Herbert, J. (1980). Behavioural and endocrine effects of naltrexone in male talapoin monkeys. *Pharmacology Biochemistry and Behavior, 13*, 663–672.

Miles, L. K., Nind, L. K., & Macrae, C. N. (2009). The rhythm of rapport: Interpersonal synchrony and social perception. *Journal of Experimental Social Psychology, 45*, 585–589.

Mogan, R., Fischer, R., & Bulbulia, J. A. (2017). To be in synchrony or not? A meta-analysis of synchrony's effects on behavior, perception, cognition and affect. *Journal of Experimental Social Psychology, 72*, 13–20.

Moll, J., Krueger, F., Zahn, R., Pardini, M., de Oliveira-Souza, R., & Grafman, J. (2006). Human fronto–mesolimbic networks guide decisions about charitable donation. *Proceedings of the National Academy of Sciences, 103*(42), 15623–15628.

Nowak, M. (2006). Five rules for the evolution of cooperation. *Science, 314*, 1560–1563.

Numan, M. (2006). Hypothalamic neural circuits regulating maternal responsiveness toward infants. *Behavioral and Cognitive Neuroscience Review, 5*, 163–190.

Nummenmaa, L., Tuominen, L., Dunbar, R. I. M., Hirvonen, J., Manninen, S., Arponen, E., et al. (2016). Social touch modulates endogenous μ-opioid system activity in humans. *NeuroImage, 138*, 242–247.

Penner, L. A., Dovidio, J. F., Piliavin, J. A., & Schroeder, D. A. (2005). Prosocial behavior: Multilevel perspectives. *Annual Reviews of Psychology, 56*, 365–392.

Preston, S. D. (2013). The origins of altruism in offspring care. *Psychological Bulletin, 139*, 1305–1341.

Preston, S. D., & de Waal, F. B. (2002). Empathy: Its ultimate and proximate bases. *Behavioral Brain Sciences, 23*, 1–20.

Preston, S. D., Bechara, A., Damasio, H., Grabowski, T. J., Stansfield, R. B., Mehta, S., & Damasio, A. R. (2007). The neural substrates of cognitive empathy. *Social Neuroscience, 2*(3–4), 254–275.

Roberts, G. (2005). Cooperation through interdependence. *Animal Behavior, 70*, 901–908.

Rossi, A. S. (2001). *Caring and doing for others: Social responsibility in the domains of family, work, and community*. University of Chicago Press.

Sapolsky, R. M., Romero, L. M., & Munck, A. U. (2000). How do glucocorticoids influence stress response? Integrating permissive, suppressive, stimulatory and preparative actions. *Endocrinological Review, 21*, 55–89.

Schreier, H. M. C., Schonert-Reichl, K. A., & Chen, E. (2013). Effect of volunteering on risk factors for cardiovascular disease in adolescents: A randomized controlled trial. *JAMA Pediatrics, 167*, 327–537.

Schultz, W. (2001). Reward signaling by dopamine neurons. *The Neuroscientist, 7*, 293–302.

Singer, T., Seymour, B., O'Doherty, J., Kaube, H., Dolan, R. J., & Frith, C. D. (2004). Empathy for pain involves the affective but not sensory components of pain. *Science, 303*, 1157–1162.

Skuse, D. H., & Gallagher, L. (2009). Dopaminergic-neuropeptide interactions in the social brain. *Trends in Cognitive Sciences, 13*, 27–35.

Smith, J. M., & Price, G. R. (1973). The logic of animal conflict. *Nature, 246*, 15–18.

Starcke, K., Polzar, C., Wolf, O. T., & Brand, M. (2011). Does stress alter every day moral decision-making? *Psychoneuroendocrinology, 36*, 201–219.

Tarr, B., Launay, J., Cohen, E., & Dunbar, R. (2015). Synchrony and exertion during dance independently raise pain threshold and encourage social bonding. *Biology Letters, 11*, 20150767.

Taylor, C., & Nowak, M. A. (2007). Transforming the dilemma. *Evolution, 61*, 2281–2292.

Taylor, S. E., Klein, L. C., Lewis, B. P., Gruenewald, T. L., Gurung, R. A., & Updegraff, J. A. (2000). Biobehavioral responses to stress in females: Tend-and-befriend, not fight-or-flight. *Psychological Review, 107*, 411–429.

Thieleman, K., & Cacciatore, J. (2014). Witness to suffering: Mindfulness and compassion fatigue among traumatic bereavement volunteers and professionals. *Social Work, 59*, 34–41.

Thompson, R. R., George, K., Walton, J. C., Orr, S. P., & Benson, C. J. (2006). Sex-specific influences of vasopressin on human social communication. *Proceedings of the National Academy of Sciences, 103*, 7889–7894.

Tomasello, M., Melis, A. P., Tennie, C., Wyman, E., & Herrmann, E. (2012). Two key steps in the evolution of human cooperation: The interdependence hypothesis. *Current Anthropology, 53*, 673–692.

Trivers, R. L. (1971). The evolution of reciprocal altruism. *QUARTERLY REVIEW OF BIOLOGY, 46*, 35–57.

Tullberg, J. (2004). On indirect reciprocity: The distinction between reciprocity and altruism, and a comment on suicide terrorism. *The American Journal of Economics and Sociology, 63*, 1193–1212.

Weinstein, D., Launay, J., Pearce, E., Dunbar, R. I. M., & Stewart, L. (2015). Music, pain thresholds, and social connections: The effect of singing in small and large groups. *Evolution and Human Behavior, 37*, 152–158.

West, S. A., Griffin, A. S., & Gardner, A. (2007). Social semantics: Altruism, cooperation, mutualism, strong reciprocity and group selection. *Journal of Evolutionary Biology, 20*, 415–432.

Wheeler, J. A., Gorey, K. M., & Greenblatt, B. (1998). The beneficial effects of volunteering for older volunteers and the people they serve: A meta-analysis. *International Journal of Aging and Human Development, 47*, 69–79.

Wilson, J., & Musick, M. (1999). The effects of volunteering on the volunteers. *Law and Contemporary Problems, 62*, 141–168.

Endogenous and Exogenous Opioids: Role in Substance Use Disorders

Lindsay Acree

Abstract Opioid use disorders have become an epidemic in recent years with rates nearly quadrupling since 1999 according to the US Centers for Disease Control and Prevention (Centers for Disease Control, Wide-ranging online data for epidemiologic research (WONDER). CDC, National Center for Health Statistics, Atlanta. Retrieved December 19, 2017, from http://wonder.cdc.gov, 2016). To understand substance use disorder (SUD) as a disease, many aspects must be studied including the circuitry in the brain, adaptations to neuronal circuitry and neurotransmitters, genetic variations increasing the risk for SUD, and treatments available for SUD. The mechanism in which an exogenous opioid may cause SUD is nearly identical to the mechanism of an endogenous opioid. This chapter reviews the clinical and epidemiological aspects of opioid use disorder, as well as the interactions between endogenous and exogenous opioids. Additionally, this chapter discusses current scientific data regarding genetic variations and mechanisms within brain circuitry and the role of endogenous opioids in substance use disorders generally (and opioid use disorder specifically). Future applications of these data to treatment of substance use disorders are also discussed.

Keywords Addiction · Exogenous opioids · Endogenous opioids · Opioid receptor · Naloxone · Substance use disorder

Endogenous and Exogenous Opioids

Introduction

Nearly everyone has been affected by substance use disorder (SUD, colloquially referred to as "addiction") in some way, with little to no discrimination between cultural groups and social status. Many struggle with the concept of addiction or

L. Acree (✉)
Department of Pharmacy Practice, University of Charleston School of Pharmacy, Charleston, WV, USA
e-mail: lindsayacree@ucwv.edu

P. L. Kerr et al. (eds.), *Endogenous Opioids*, Advances in Neurobiology 35,
https://doi.org/10.1007/978-3-031-45493-6_14

SUD. Is it a disease or a choice? Scientific evidence exists indicating that SUD is a disease-like process that is associated with specific changes in brain circuitry due to reward and conditioning and genetic propensity for substance use identified through differences in gene expression and heredity (Volkow et al., 2012; Khokhar et al., 2010). Consistent with the disease/disorder model, the American Psychiatric Association (APA) delineates several diagnostic criteria for diagnosing SUDs (APA, 2013).

Criteria for several types and subtypes of SUDs are listed in the *Diagnostic and Statistical Manual of Mental Disorders, Fifth Edition* (*DSM-5*; APA, 2013). Ten areas of SUDs are identified (Nathan et al., 2016). In previous editions of the *DSM*, a distinction was made between substance use and dependence reflecting different levels of severity and/or impairment; however, this approach to differentiation has been removed from the *DSM-5* (Nathan et al., 2016). The *DSM-5* also separated substance-related disorders from substance-induced disorders. Substance-related disorders are defined as "a cluster of cognitive, behavioral, and physiological symptoms indicating that the individual continues using the substance despite significant substance-related problems" (APA, 2013). Substance-induced disorders include withdrawal, intoxication, and other drug-related mental disorders (Nathan et al., 2016). Eleven criteria or symptoms are identified for use in diagnosis of substance use disorders. These criteria are ultimately the same, but each substance use disorder is diagnosed separately based upon drug-specific symptoms (Nathan et al., 2016). The 11 criteria identified are in relation to the following: (1) using larger amounts of the substance or using for a more extended period of time than intended; (2) unsuccessful attempts to decrease substance use; (3) increasing time commitment to obtaining the substance; (4) drug cravings; (5) difficulty carrying out certain responsibilities due to use of the drug; (6) continued substance use despite adverse social or interpersonal consequences; (7) continued substance use despite adverse social or interpersonal consequences; (8) using in dangerous environments or situations; (9) stopping enjoyable activities; (10) tolerance as indicated by increasing amounts needed for original effects; and (11) withdrawal symptoms when not using the substance (APA, 2013). To diagnose based upon these criteria, a patient must have two to three symptoms for mild substance use disorders, four to five symptoms for moderate substance use disorders, and six or more for severe substance use disorders (APA, 2013).

Several pathways within the brain circuitry are hypothesized to lead to SUD; however, the most accepted pathway centers on the increase in dopamine within the nucleus accumbens (NAc) leading to the rewarding effects of several substances (Volkow et al., 2012). However, reward is not the only factor that may lead to substance use disorder. Conditioning of specific signals may also play a role in SUD and relapse (Volkow et al., 2012). Conditioning is a repeated association with cues in the environment that the brain connects with use or rewarding effects of the drug. It is often a location, person, or instrument used in the process of obtaining or using the drug. In addition to conditioning, the speed that the drug reaches the brain and provides the rewarding effects may impact the risk of a substance use disorder due to the almost immediate reward (Volkow et al., 2012). A more intense high occurs when the drug is administered through smoking or intravenous route due to the quick delivery of the drug to the receptors and the subsequent increase in dopamine reinforcing that behavior (Volkow et al., 2012).

Currently, several pharmacotherapy treatment options are available for patients with SUDs. Medications often used in the treatment of opioid use disorder are buprenorphine, buprenorphine/naloxone, methadone, and naltrexone. With the exception of naltrexone, the medications used to treat opioid use disorder attenuate the withdrawal symptoms, reducing or eliminating the discomfort of withdrawal. Naltrexone is an opioid antagonist, which is used to deter use of opiates or opioids once the drug is no longer in the system and the individual has completed detoxification. Naltrexone may also lessen the cravings associated with substance use disorders. Each medication has a place in therapy, and the choice of medication should be patient specific Alkermes, Inc., 2010).

Basic History of Opioids

Opium, which is derived from the poppy plant (*Papaver somniferum*), was the first opiate to appear in history and was documented as early as 300 B.C. Some authors have even suggested that opium may have been used as early as 3000 B.C. (Pathan & Williams, 2012; Van Ree et al., 1999). Prior to the discovery of morphine, opium was often used as a tincture to treat a number of ailments (Van Ree et al., 1999). Opium use spread throughout the globe, making its way into China. The illegal export of opium into China led to the Opium Wars, the first of which began in 1840. Addiction to opium was beginning to become an issue in China, and in 1839, therefore, the government seized and destroyed a shipment of opium owned by the British. Hostilities continued and eventually led to the war beginning in 1840, which ended with China ceding land and additional ports to the British. The second Opium War in 1856 resulted in the legalized trade of opium in China, leading to further proliferation of opium use (Pletcher & Encyclopedia Brittanica, Inc, 2018).

A pharmacist named Friedrich Sertürner isolated morphine in 1806, which would lead to more universal use in the medical community after the hypodermic needle was invented in 1853 (Pathan & Williams, 2012). Once morphine was isolated and available for use, several other opioid alkaloids were identified from the opium plant including codeine, papaverine, and thebaine (Pathan & Williams, 2012). It was quickly realized that morphine was just as addictive as opium, so the push began to find less addictive or nonaddictive opioids for use. Heroin was introduced in 1898 as a nonaddictive alternative to morphine and opium. Of course, it was later discovered that heroin was even more addictive than morphine (Van Ree et al., 1999). Many other opioids were later developed, such as meperidine in 1939 and methadone in 1946. Research into the activity of morphine eventually led to the development of opiate antagonists such as naloxone in the 1960s (Van Ree et al., 1999).

Endogenous opioids, or opioid peptides that are naturally occurring in the body, were not identified until the mid-1970s (Charbogne et al., 2014). These opioid peptides include enkephalins, dynorphins, neoendorphins, endomorphins, and β-endorphins. These endogenous opioids are produced through the cleavage of several endogenous peptides in response to stimuli, including proopiomelanocortin (POMC), prodynorphin (PDYN), and proenkephalin (PENK) (Merrer et al., 2009; Trigo et al., 2010). Endogenous opioids are able to bind to opioid receptors due to a specific N-terminal sequence of Tyr-Gly-Gly-Phe, which is expressed by all endogenous opioids, with the exception of endomorphins (Merrer et al., 2009; Trigo et al., 2010).

Pharmacology

Opioid Receptors

Shortly after the discovery of the endogenous opioid peptides, opioid receptors were identified as μ (mu), κ (kappa), and δ (delta). Each receptor type induces specific activity at the receptor binding with affinity to specific exogenous and endogenous opioids (Van Ree et al., 1999). The three types of opioid receptors are distributed throughout the brain, central nervous system, and peripheral nervous system. Opioid receptors are G protein-coupled receptors (GPCR) and can be bound by both endogenous and exogenous opioids (Merrer et al., 2009). Although mostly concentrated in the cortex, μ-, κ-, and δ-opioid receptors are also found in the brain stem and limbic system (Merrer et al., 2009). Typically, binding sites for all three receptor types overlap in most areas; however, in some structures, a higher expression, or only one type of receptor, may be detected. μ-Receptors are the most expressed binding site in the certain areas of the amygdala, thalamus, brain stem, and mesencephalon. The μ-opioid receptors are detected alone in the posterior thalamus, lateral geniculate thalamus, ventrolateral thalamus, ventromedial thalamus, sensory trigeminal nucleus, and nucleus ambiguus (Merrer et al., 2009). A higher expression of κ-opioid receptors is detected in the anterior forebrain (claustrum and endopiriform cortex), striatum including the nucleus accumbens and caudate putamen, olfactory tubercle, hypothalamus, pituitary, and preoptic area (Merrer et al., 2009). Expression of κ-binding sites is found alone in areas of the brain associated with the stress axis. These areas include the claustrum, arcuate nucleus, supraoptic nucleus, pituitary, paraventricular hypothalamus, and central nucleus of the amygdala (Merrer et al., 2009). Expression of δ-opioid receptors is limited in several areas of the brain including the hypothalamus, thalamus, mesencephalon, and brain stem. The δ-opioid receptor has the highest expression in the olfactory tract, neocortex, regions of the amygdala, and striatum (Merrer et al., 2009).

Endogenous Opioids

Enkephalins were the first endogenous opioid peptides to be discovered, followed by β-endorphins and dynorphins. PENK-containing neurons are widely distributed throughout the brain and contain both short- and long-projection neurons (Van Ree et al., 1999). Met- and leu-enkephalins are generated from PENK and typically bind the δ-opioid receptors with the highest affinity. PDYN-containing neurons are widely distributed throughout the CNS and also contain both short- and long-projection neurons. Dynorphins (dynorphin A, dynorphin B, α-neoendorphin, and β-neoendorphin) are generated from PDYN and bind to the κ-opioid receptor with the highest affinity and to the μ-opioid receptor with a low affinity. β-Endorphins (α-endorphin, β-endorphin, and γ-endorphin) are generated from POMC and bind to the μ-opioid receptor with the highest affinity (Trigo et al., 2010; Van Ree et al., 1999). Endomorphins were not discovered until 1997 and contain a different amino acid sequence compared to other opioid receptors (Table 1).

Table 1 Endogenous opioids with relation to opioid receptors

Precursor molecules	Endogenous opioids	Opioid peptides	Biosynthesis and distribution	Preferential receptor
Proenkephalin (PENK)	Enkephalins	Met-enkephalin Leu-enkephalin	Brain (wide distribution) Local circuits and long-projection neurons	δ-Opioid receptor
Prodynorphin (PDYN)	Dynorphins	Dynorphin A Dynorphin B α-Neoendorphin β-Neoendorphin	CNS (wide distribution) Short- and long-projection neurons	κ-Opioid receptor
Proopiomelanocortin (POMC)	β-Endorphin	β-Endorphin α-Endorphin γ-Endorphin Non-opioid peptides	Pituitary gland (biosynthesis of POMC) Arcuate nucleus of the hypothalamus Nucleus tractus solitaries Brain (widespread projections of neurons)	μ-Opioid receptor
Unknown	Endomorphins	Endomorphin 1 Endomorphin 2	Brain and upper brain stem Spinal cord and lower brain stem. Also present in primary afferent fibers of nervous cells (from which it originates), somatic and visceral sensory ganglia, immune system, and gastrointestinal tract	μ-Opioid receptor

Van Ree et al. (1999); Trigo et al. (2010); Gu et al. (2017); Przewlocki and Almeida (2017)

Exogenous Opioids

Opioids and opiates are a group of substances that bind to opioid receptors in the brain and spinal cord to induce antinociceptive effects within these sites. *Opioids* reduce pain through the inhibition of pain transmission in the dorsal horn as well as inhibition of excitatory neurotransmitters from the primary afferent (Schumacher et al., 2015). *Opiates* are considered direct derivatives from the poppy plant, and opioids are originated synthetically. Examples of opiates include papaverine, thebaine, morphine, and codeine (Merrer et al., 2009). Most other medications in this class are referred to as *opioids* including hydrocodone, oxycodone, and fentanyl. Although opioids are highly effective in reducing pain, they also are extremely addictive. Repeated use, especially in high doses, use of long-acting opioids, or use of high-potency opioids may quickly potentiate the symptoms of SUD, as well as tolerance to these medications, leading to opioid use disorder.

Opiates and opioids may be classified through chemical subgroups or through the action at the receptor (Pathan & Williams, 2012). The chemical subgroups are used to classify synthetic compounds based upon the opiate in which it is derived. These subgroups are as follows: morphinan derivatives, diphenylheptane derivatives, benzomorphan derivatives, and phenylpiperidine derivatives (Pathan & Williams, 2012). Rather than referring to chemical derivation, opiates and opioids are most often referred to based upon drug action at the receptor.

These drugs may be a full agonist, partial agonist, agonist-antagonist, antagonist, or inverse agonist. A full agonist binds the opioid receptor and provides a full response. In other words, the maximum stimulation at the receptor will occur (Pathan & Williams, 2012). An example of a full agonist would be morphine or fentanyl (most opioids). Many of the opioids used to treat pain are full agonists. Methadone, a long-acting full agonist, is also used in the treatment of opioid use disorder.

Opioid antagonists bind to the receptor but do not elicit a response. Rather, antagonists prevent the binding of other opioids (endogenous or exogenous) to the receptor. Opioid antagonists are especially beneficial in the treatment of opioid overdose as well as in the treatment for opioid use disorder as a maintenance medication after withdrawal. Examples of opioid antagonists are naltrexone and naloxone. There are two types of antagonists: neutral antagonists, which block agonist activity, and inverse agonists, which produce the opposite or negative effect at the receptor (Cruz & Granados-Soto, 2015; Sirohi et al., 2009). In cells that have previous exposure to opioid agonists, inverse agonists may increase intracellular cAMP (Sirohi et al., 2009). Partial agonists produce only a partial response at the receptor site. An example of a partial agonist is buprenorphine, which is often used in the treatment of opioid use disorder. When comparing the actions of a full agonist vs. a partial agonist, a full agonist would be the "glass is completely full" analogy, while the partial agonist would be the "glass is half full." Both partial agonists and inverse agonists act as an antagonist when an agonist is present (Cruz & Granados-Soto, 2015). Each classification of opioids has a place in therapy, which may take advantage of varying properties of the drugs. For instance, both methadone (full agonist) and buprenorphine (partial agonist) are used to treat opioid use disorder. Both medications are long-acting, which modifies the substance-seeking behavior in patients with opioid use disorder (Pathan & Williams, 2012).

Safety of Opioids

In addition to the risk of opioid use disorder, there are many concerns related to opioids. These medications or illicit substances are CNS (central nervous system) depressants. The main role of an opioid is to provide analgesia through inhibition of the ascending pathway from the site of pain to the brain as well as activation of the inhibitory or descending pathway leading from the brain to the dorsal horn (Cruz & Granados-Sotos, 2015). The most life-threatening CNS effect of opioids is respiratory depression. Respiratory depression may lead to respiratory arrest and death. Typically, the body recognizes the increasing levels of carbon dioxide, which is regulated or compensated through breathing. Opioids reduce the response in the brain stem to the higher levels of carbon dioxide; therefore, respiratory compensation does not occur (Cruz & Granados-Soto, 2015).

Brain Circuitry in Substance Use Disorder

Dopamine Theory

Much of the research in reward and positive reinforcement associated with drugs of misuse is in regard to the mesocorticolimbic dopamine system (Koob & Le Moal, 2001). The most significant areas of the brain connected in this dopamine circuitry are the ventral tegmental area (VTA) and the basal forebrain including the NAc, olfactory tubercle, amygdala, and the limbic cortices (Koob & Le Moal, 2001). A significant amount of the research in the drug reward system is related to the dopamine projections from the VTA to the NAc, opioid receptors and peptides within these regions of the brain, GABA, glutamate, and serotonin (Koob & Le Moal, 2001). With the use of many addictive drugs, dopamine is increased in the NAc, which leads to positive reinforcement or reward (Trigo et al., 2010; Lucher, 2016). However, the mechanism in which dopamine is increased varies between each drug. For example, opioids are involved in a disinhibition or dysregulation pathway in which an opioid receptor is bound, causing a reduction in the release of GABA. GABA typically inhibits the release of dopamine, but in a situation in which the GABA is reduced, dopamine release is increased (Lucher, 2016; Niikura et al., 2010). There are five dopamine receptors that may be involved in the reward pathway; however D1 and D2 receptors are the most understood. Typically, D1 receptors are activated through large increases in dopamine, and D2 receptors are activated through smaller changes in dopamine release. D1 receptors are associated with drug reward and conditioning cues, whereas D2 receptors are only involved in drug reward (Volkow et al., 2012).

Within the NAc, the rewarding effects are originated through the binding of both the μ- and δ-opioid receptors. Opioid agonists binding the μ-opioid receptors have the highest potential for misuse and possess the most rewarding effects (Trigo et al., 2010). Although the exact role of δ-opioid receptors in opioid use disorder is not completely understood, animal models demonstrate that self-administration is approximately 100 times lower in δ-opioid agonists. Dopamine release is also 1000 times lower with δ-opioid agonists when compared to selective μ-opioid agonists (Trigo et al., 2010). However, κ-opioid receptors are associated with an anti-reward system (Lutz & Kieffer, 2013; Niikura et al., 2010). Dynorphin (an endogenous opioid) acts upon the κ-receptor in the NAc to regulate extracellular dopamine. In a state in which dopamine is elevated (drug use), it is theorized that the dynorphin/κ--opioid receptor may decrease dopamine regulating these higher levels (Bruchas et al., 2010). If dopamine is decreased due to the lack of drug, the dynorphin/κ--opioid receptor plays a role in the aversive state associated in withdrawal (Niikura et al., 2010; Trigo et al., 2010). The anti-reward system strengthens the drug-seeking behavior to prevent the aversive effects. To avoid withdrawal, the individual feels as though the substance is needed (i.e., compulsive behavior).

After repeated exposure to opioids, tolerance and dependence occur due to desensitization of the μ-opioid receptor and changes in dopamine release. This mechanism is not completely identified and can vary based upon the drug administered and area in the brain (Trigo et al., 2010). In regard to downregulation of opioid receptors, evidence is conflicting indicating that downregulation of the opioid receptor is not solely responsible for tolerance and dependence. Varying studies reporting upregulation, downregulation, and no change in the opioid receptors challenge this theory (Trigo et al., 2010). One mechanism that may play a role in desensitization is through the endocytosis and recycling of opioid receptors, specifically μ-opioid receptors. Opioid receptors (G protein-coupled receptors) are recycled through a process in which the receptor is phosphorylated by G protein-coupled receptor kinases (GRKs). Phosphorylation activates the receptor and changes the conformation to increase affinity for cytosolic β-arrestin (Trigo et al., 2010). β-Arrestin causes an uncoupling of the receptor, which leads to internalization through endocytosis. However, some evidence indicates that internalization of the receptor is a mechanism to recycle the reactivated receptor to the surface (Trigo et al., 2010). Opioid agonists that induce endocytosis, such as fentanyl and methadone, typically stimulate less tolerance than medications that are considered "non-internalizing opioids [morphine]" (Trigo et al., 2010; Johnson et al., 2017). Morphine tends to activate G protein with little activation of the β-arrestin. Due to this poor activation of β-arrestin, morphine does not induce endocytosis or recycling of the receptor to the same extent as endogenous opioids and smaller molecules such as fentanyl and methadone. This can also be known as "bias" towards the G protein and is thought to lead to a higher tolerance. The bias for signaling the G protein is even more pronounced with buprenorphine when comparing to morphine, which leads to a higher tolerance (Johnson et al., 2017). Following repeated exposure to opioids, the δ-opioid receptor may assist to maintain the rewarding properties of opioids due to activation and translocation to neuronal membranes (Trigo et al., 2010).

Tolerance may also occur due to the downregulation of striatal D2 receptors. After repeated use of a substance, the number of D2 receptors decreases, increasing the demand for the drug. Once tolerance occurs, the brain has already learned the reward pathway and is driven to obtain the drug, regardless of consequences. The user is also conditioned by cues to be motivated and "aroused" when drug seeking or during the act of using. Often, this leads to less interest in activities not leading to the procurement of the drug, including natural pathways of reward (Volkow et al., 2012).

The demand for the drug may be to use more, chasing the high that they once had. The first high that a person experiences typically will never occur again. Once the brain has learned the reward, the pathway is set, and the individual will do what is necessary to experience the reward. However, withdrawal symptoms also potentiate drug-seeking behavior. As a person who continually uses substances becomes tolerant to the substance, the individual may not feel "normal" when not using, and they may require the substance to attain physiological homeostasis. The user is in a state of withdrawal or sickness when the drug cannot be obtained. Thus, seeking out and using the substance may in fact not be to feel the "high," but rather to prevent or soothe withdrawal symptoms.

Pharmacokinetics of the substances leading to SUD may increase the level of misuse potential based upon the speed at which the body experiences the increase in dopamine, as well as the bioavailability of the substance (Trigo et al., 2010). For example, opioids will bind to the receptors on the brain much quicker when administered via inhalation (smoking) or intravenous methods. When taken orally, the opioid will be metabolized first before reaching the site of action. This may take as little as several minutes, but the individual may not feel the peak effect for up to an hour depending upon the opioid used. The effect can be felt almost immediately when injected, which will increase the reinforcing properties of the drug. Speed of onset may also influence the conditioning aspect of SUD in which specific cues may induce the same effect of the drug (release of dopamine).

Each drug also has a specific bioavailability for oral vs. intravenous use. The "high" that a person experiences from oral use will be intensified when the same dose is injected. Essentially, by changing the route of administration, the dose is increased exponentially putting the individual at risk of an overdose. This risk is compounded by variability in the content and potency of substances, especially opioids, due to contamination, e.g., with fentanyl, in which there is a well-established association with overdose death (Carroll et al., 2017; Daniulaityte et al., 2019; Dowell et al., 2017; Klar et al., 2016; Park et al., 2018).

The reward and reinforcement of the increase in dopamine is essentially a learned process. When dopamine is released in response to a drug, the brain remembers this reward and reinforces this action so it can be repeated. Once the brain has established the expected reward with the use of a drug, the increase of dopamine is no longer needed. However, with substance use disorders, often the dopamine is still released in response to a drug but to a lesser extent. This process essentially rewires the brain to need this reward above all others (compulsive and substance-seeking behavior; Lucher, 2016). Some studies have demonstrated that the increase in dopamine is in response to the cue (conditioning) rather than the drug itself. It is thought that the increase in dopamine shifts from the drug itself to the act of using the drug or these conditioned cues. Cues can be anything from drug paraphernalia, the act of using, people, or an environment (Volkow et al., 2012). Cravings associated with drug cues may be due to increases of dopamine not only in the NAc but also in the ventral tegmental area (VTA), anterior cingulate cortex (ACC), orbitofrontal cortex, and insula (Volkow et al., 2012).

There is some evidence to suggest that the number of dopamine receptors, specifically D2 receptors, located in the prefrontal cortex (PFC) may potentiate substance use disorder. In one study of patients with a family history of alcohol use disorder (AUD), subjects not suffering from AUD had an increased availability of striatal D2 receptors compared to the typical quantity that was expected. It is theorized that the higher availability of D2 receptors in high-risk individuals may have allowed for normal metabolism and adequate function in the PFC, protecting these subjects from alcohol use disorder (Volkow et al., 2012). The PFC is also responsible for self-regulation in specific behaviors including use of drugs (George & Koob, 2010). It is postulated that disruption in the PFC may potentiate the symptoms of SUD through inadequate function, negative mood, reduced dopamine reward after long-term use, preference over immediate reward vs. delayed reward, and alterations in decision-making (Volkow et al., 2012). Changes or failure in the self-regulation function may predict later SUD potential, increased drug-seeking behavior, increased relapse, and overall risk for SUD (George & Koob, 2010).

Stress-Induced Behaviors

In addition to the dopamine theory, stress is hypothesized to induce many substance-related behaviors, including substance use, substance seeking, and substance use relapse. The dynorphin/κ-receptor activation is shown to maintain dependence to several substances and may even potentiate acute withdrawal symptoms in a stress-induced state (Bruchas et al., 2010). Activation of the dynorphin/κ-receptor reduces the levels of dopamine during substance use to manage the levels of dopamine. Bruchas et al. (2010) discussed the role of the dynorphin/κ-receptor through several studies on three separate animal models. In these studies, the dynorphin/κ-receptor activation through a selective agonist restored substance-seeking behaviors and administration of the drug. However, when a selective antagonist was administered at the κ-receptor, these behaviors were not inhibited, which is contradictory to what would be expected (Bruchas et al., 2010).

Corticotropin-releasing factor (CRF) is another stress-related factor in regard to substance use. Withdrawal to most substances that are misused, including opioids, releases CRF along with other hormones including adrenocorticotropic hormones (ACTH) and corticosterone. Therefore, it could be concluded that CRF would impact relapse and may further potentiate withdrawal effects in a stress-induced state. Several studies have been used to determine the exact role of CRF in relapse and SUD. In one animal model, CRF inserted into the VTA was shown to instigate relapse through reinstatement of conditioned place preference (CPP) and drug-seeking behavior, specifically implicating the CRF_1-R (Bruchas et al., 2010). Another study analyzed the circuitry involved in the stress-induced state through electrophysiology and demonstrated an alteration of both CRF_1-R and CRF_2–2 following drug administration, which in this case was cocaine. Based upon these studies, several investigators are looking further into the role of CRF in drug-seeking behavior and relapse.

It appears through review of these studies that a clear connection exists between the dynorphin/κ-receptor activation and CRF in regard to relapse and drug-seeking behaviors. Although the response at each of these receptors may vary based upon the drug used, both of these mechanisms could be considered future targets for pharmacotherapies for SUDs. Currently only a small number of medications are used to treat SUDs; novel drug targets may lead to more options for those with a substance use disorder.

Role of Endogenous Opioids in Substance Use Disorders

In several studies involving knockout mice, the role of endogenous opioids at each receptor site was tested. In μ-opioid receptor knockout mice, self-administration for most drugs including opioids, alcohol, delta-9-tetrahydrocannabinol (THC), and nicotine was decreased or completely eliminated. These results confirm the theory that the μ-opioid receptor is responsible for the rewarding effects of most drugs of abuse along with memory and adapted response associated with these behaviors (Merrer et al., 2009). δ-Opioid receptors play a different role in the body compared to μ-opioid receptors. Knockout mice with the δ-mutation demonstrated increased anxiety, depression, and motor impulsivity. This data validates that emotional behavior and drug reinforcement may be regulated by δ-opioid receptor activity in response to opioid agonists (Merrer et al., 2009). In κ-opioid receptor knockout mice, studies indicated that the κ-opioid receptor was responsible for the aversive effects of drug use as well as regulation of extracellular dopamine (Merrer et al., 2009).

Opioid use disorder is essentially an exploitation of the endogenous opioid system, which alters pathways intended to respond to natural reinforcement in order to respond to an exogenous stimulus (Merrer et al., 2009). Opioid receptors and endogenous ligands are widely spread throughout the body for many reasons. The endogenous opioid system (EOS) is responsible for a variety of functions including nociception, memory, and emotion (Trigo et al., 2010). Active endogenous peptides that bind the opioid receptors include β-endorphins, met-enkephalins, leu-enkephalins, dynorphins, and neoendorphins. Typically, dynorphins bind κ-opioid receptors to regulate dopamine levels and potentiate withdrawal symptoms. Met-enkephalins and leu-enkephalins bind with the highest affinity to δ-opioid receptors and bind to μ-opioid receptors to a lesser extent. β-Endorphins have a higher affinity for μ-opioid receptors than for δ- or κ-opioid receptors (Trigo et al., 2010). Endogenous opioids act very similarly to exogenous opioids in regard to binding of the receptors with subsequent release of dopamine (Leeman & Potenza, 2013).

Genetics Variations

Many genetic variations occur in regard to susceptibility to substance use disorders, including variations in opioid receptor coding genes and dopamine receptors (Khokhar et al., 2010). Two common variants within the μ-opioid receptor gene (*OPRM1*) are the 118A > G variant and the 17C > T SNP. Both variants within the coding region have been associated with a higher risk of opioid use disorder across several populations. Although the μ-opioid receptor is the primary binding site for most addictive drugs, variations within the κ-opioid receptor gene (*OPRK1*) and the δ-opioid receptor gene (*OPRD1*) may also increase the risk of substance use disorder (Khokhar et al., 2010).

Polymorphisms within the dopamine (D2) receptor gene (*DRD2*) include *Taq*1A and *Taq*1B. The substitution of C > T in *Taq*1A decreases the quantity of D2 receptors located within the striatum (Khokhar et al., 2010). With less dopamine receptors to bind dopamine released from binding of either endogenous or exogenous opioids, there is an increased risk to compensate for a lower dopamine through the use of drugs or other addictive behaviors (Khokhar et al., 2010). In studies evaluating patients at a high risk of SUD due to family history, those individuals with higher levels of D2 receptors in the striatum were less likely to have a substance use disorder concluding that higher levels of D2 receptors allowed these patients to overcome the alterations in brain chemistry that often leads to SUD (Volkow et al., 2012).

Many other genetic variations exist for not only opioid use disorder but also SUDs involving other substances, including alcohol, cocaine, and methamphetamine (Ozburn et al., 2015). In regard to the dopamine theory, alterations to the opioid receptors that will lead to the release of dopamine and alterations to the dopamine receptors receiving these signals may be the key to SUD.

Treatment

Treatment of Substance Use Disorders

Approximately seven medications are currently used to treat SUD; however, only three of those are used to treat opioid use disorder. These medications include buprenorphine (with or without naloxone), methadone, and naltrexone (Koob & Mason, 2016). Each drug has a specific mechanism in which cravings and drug-seeking behaviors are reduced. As previously stated, for pharmacotherapy to be effective, it must be combined with counseling and behavior therapy. Each medication binds to the opioid receptor and either acts as a full agonist, partial agonist, or antagonist.

Methadone, a full agonist, was the first treatment used in treating opioid use disorder and was approved for this purpose in 1972 (Koob & Mason, 2016). Methadone is a long-acting μ-opioid receptor agonist with some NMDA antagonist properties. The half-life of methadone is approximately 8–59 hours for adults (George et al., 2012) and may range higher or lower (3.5–60 h) for children (Berde et al., 1991). Because of this extended half-life, withdrawal symptoms are minimized without a high frequency of dosing. When withdrawal occurs from methadone, it is typically more prolonged and less severe than with other full agonists, including morphine. Without the withdrawal effects stimulating drug seeking, this behavior is minimized when methadone is used for treatment. Although methadone has been shown to be effective, there are several warnings and concerns associated with this medication. As with any opioid, misuse potential, diversion, accidental ingestion, overdose, and respiratory depression may occur. Methadone may also increase the QT interval, which can cause a fatal arrhythmia known as torsades de pointes (Stringer et al., 2009). This effect on QT prolongation may be additive with other medications carrying the same risk. If considering this medication, it is important to complete a medication review prior to initiation.

Another agonist used in the treatment of opioid use disorder is buprenorphine. However, buprenorphine is a partial agonist at the μ-opioid receptor and an antagonist at the κ-opioid receptor (Lutfy & Cowen, 2004). A partial agonist will bind the receptor with low efficacy so the individual does not experience the same rush of dopamine as with full agonists of the same opioid receptor. Buprenorphine also binds with high affinity for the μ-opioid receptor, which will antagonize full agonists (Koob & Mason, 2016). Therefore, substance use is deterred through antagonism, and withdrawal symptoms are minimized through partial agonism, reducing the substance-seeking behaviors and aversive symptoms seen in SUD. Similar to methadone, buprenorphine is also long-acting with a half-life of approximately 24–42 hours (Indivior, Inc., 2017). Naloxone, an opioid antagonist, is combined with some of the sublingual formulations of buprenorphine. When naloxone is ingested via oral or sublingual route, absorption is reduced compared with intravenous or intranasal administration. Naloxone in combination with buprenorphine is intended to be used as a deterrent for injection (Orman & Keating, 2009). However, buprenorphine has a much higher affinity for the μ-opioid receptor and will often block the effects of naloxone regardless of the route of administration. In addition, the half-life of naloxone is much shorter than with buprenorphine; therefore, if an individual does experience the effects of naloxone, it will be short-lived (Blazes & Morrow, 2020). As with each medication, some warnings exist with use of this medication. Although buprenorphine is a partial agonist, abuse potential, diversion, and accidental ingestion may still occur. Although overdose and respiratory depression occur to a lesser extent with buprenorphine alone, it is still a risk of the medication and may be increased when used in combination with other central nervous system (CNS) depressants. If overdose occurs with buprenorphine, naloxone may be less effective due to the high affinity for buprenorphine at the receptor. Therefore, multiple doses and repeated administration of naloxone may be required (SAMHSA, 2018).

The last medication to be discussed in the treatment of opioid use disorder is naltrexone. Naltrexone acts competitively as an antagonist at the μ-opioid receptor (Kjome & Moeller, 2011). Treatment initiation is substantially different than treatment with a partial or full agonist. Complete detoxification must occur prior to initiation of treatment. Detoxification may take seven or more days, and this medication does not attenuate the withdrawal symptoms. In fact, naltrexone will induce withdrawal in patients currently taking opioids or opiates. Naltrexone is available as a tablet or an injection, with a longer half-life observed in the injectable formulation (approximately 5–10 days) (Alkermes, Inc., 2010; Kjome & Moeller, 2011). Efficacy in opioid use disorder treatment is derived from blockade of the receptors deterring use of drugs and reduction of cravings. The major warning associated with this drug is hepatotoxicity and should not be used in patients with acute hepatitis or liver failure (Kjome & Moeller, 2011).

Overdose

Opioid overdose is occurring more frequently, especially with new drugs emerging into the market. Some of these newer drugs of abuse include carfentanil and many other synthetic opioids such as fentanyl analogs (acetylfentanyl, butyrfentanyl, and furanylfentanyl) much more potent than heroin or morphine (Howard & Hornsby-Myers, 2018). Overdose rates have quadrupled since 1999 and continue to increase each year (CDC, 2016). SUD can lead to opioid overdose; however, others are at a high risk as well including children, older individuals, patients using higher doses and long-acting opioids, those exiting jails or prisons, and patients receiving treatment (medication or abstinence-based) for SUD. Although occupational absorption of opioids through either inhalation or transdermal routes is unlikely to cause fatal overdose, several agencies have issued recommendations to protect first responders and those that may be exposed to high-potency opioids (Howard & Hornsby-Myers, 2018; Moss et al., 2017). Such recommendations have been issued through the

National Institute for Occupational Safety and Health (NIOSH), the Drug Enforcement Administration (DEA), the White House Safety Recommendations for First Responders, the American College of Toxicology, and the American Academy of Clinical Toxicology (Howard & Hornsby-Myers, 2018).

Inhalation of fentanyl or fentanyl analogs would require a significant amount to be inhaled over an extended amount of time to produce a toxic dose. Carfentanil and remifentanil are suspected as the drugs used in a weaponized aerosolized formulation at a theater in Moscow in 2002, in which 125 deaths occurred from exposure (Howard & Hornsby-Myers, 2018; Moss et al., 2017). Although fentanyl is lipophilic and may be absorbed through the skin, it takes several minutes to hours to reach a therapeutic concentration and even longer for peak concentration with optimal circumstances (through transdermal patch). Other factors may increase absorption of aqueous forms of fentanyl (moisture needs to be present for optimal absorption) such as broken skin, application of heat, and larger surface area of application (Moss et al., 2017). If powder comes in contact with the skin, absorption would not be optimal and can easily be washed from the skin with soap and water. Soap can solubilize fentanyl and decrease surface attraction to allow for removal from the skin with water. Hand sanitizers should be avoided as it may spread the opioid to additional surface area and potentially increase likelihood of absorption (Howard & Hornsby-Myers, 2018; Moss et al., 2017). Another important route of absorption may be through mucus membranes, in which bioavailability may be as much as 65% with administration through buccal or sublingual routes. Therefore, protection of the mucous membranes is important for those at a high risk of exposure. Fentanyl or other potent opioids can be transferred from gloves to mucous membranes through contact (touching eyes, nose, or mouth with contaminated gloves). Gloves should be changed frequently, and contact with mucous membranes should be avoided (Howard & Hornsby-Myers, 2018; Moss et al., 2017).

In addition, children are more frequently overdosing due to accidental ingestion. Prescription medications pose a high risk to children due to accessibility. Often children or adolescents obtain prescriptions from a friend or relative or from the medicine cabinet. In addition, younger children are often susceptible to mistaking prescription medications for harmless substances. Pills are often mistaken for candy, and fentanyl patches may appear as a sticker to a young child. Fentanyl is also available as other formulations, such as a lollipop for buccal administration. To a young child, this formulation may appear harmless but have deadly consequences. Prescription opioid-related hospitalizations increased between 1997 and 2012 for those less than 19 years of age with the highest increase in those less than 5 years of age (Gaither et al., 2016; Kane et al., 2018). According to a study by Kane et al., in children aged 1–17, the number of opioid-related admissions increased by 39%, while the admissions to PICU increased by 35% between the years of 2004 and 2015 (Kane et al., 2018). Deaths in this age group were approximately 1.6% and averaged 1.4% in those aged 1–5 (Kane et al., 2018). Although this study did show a decrease in mortality of this time frame, toxicities and fatal overdoses are still of major concern (Kane et al., 2018). Proper disposal of medication, keeping all medications out of the reach of children, educating children and adults regarding safe use of medication, and storing medications in a locked container may reduce the risk of toxicities or fatal overdoses in children and adolescents. Instructions and resources for the proper disposal of medications are available through the US Food and Drug Administration.

The elderly may be at a higher risk due to misuse or unintentional overdose. In the elderly, many factors may play into the increased risk of overdose. Reduced kidney function and liver function may cause opioids to become more toxic at therapeutic doses. Most opioids are metabolized in the liver and excreted by the kidneys. Therefore, impairment of either organ function may increase the concentrations of opioids in the body and may lead to more adverse effects, including respiratory depression. In addition, cognitive decline may also lead to unintentional overdose (taking medication too frequently or taking an incorrect dose). Multiple drug interactions may also be a factor in opioid overdoses in the elderly. Therefore, elderly patients should be screened prior to initiation of therapy, and caution should be used when prescribing opioids.

As this chapter discusses, changes in tolerance may lead to an increased demand for the drug. Increased demand may cause the user to increase intake, which will often lead to overdose. However, recovering individuals are also at a high risk of an overdose based upon a reduction of tolerance and rate of relapse. Currently, naloxone and nalmefene are available to treat suspected opioid overdoses and are becoming more available to the community for use. Naloxone has been approved for nonprescription use and is now available over-the-counter in both a 3mg and 4mg intranasal product.

Naloxone is an opioid antagonist and is available in several formulations including nasal sprays and injectable devices. This medication will competitively bind the opioid receptors, removing the opioid from the receptor in most cases, and should be used in a suspected overdose. Signs of opioid overdose include a bluish tint to the lips or fingertips, unresponsiveness, miosis, and respiratory depression (Substance Abuse and Mental Health Services Administration [SAMHSA], 2018). Naloxone may not be as effective in the event of an overdose with a partial agonist due to the high affinity binding of partial agonists (LoVecchio et al., 2007). Studies indicate that the half-life of naloxone may range approximately from 1 to 2 h, depending upon the route of administration (Howland, 1998). The implication of this half-life is such that most opioids have a half-life longer than 1–2 hours. After administration of naloxone, the individual should have immediate medical treatment to ensure their safety and that respiratory depression does not occur again after the levels of naloxone in the body have diminished (SAMHSA, 2018). Nalmefene is also an opioid antagonist, similar to naloxone, however the major differences between the two are that nalmefene has an extended half-life of approximately 11 hours and an almost five-fold higher affinity for the opioid receptor compared with naloxone (Infante et al., 2024). Many programs exist to supply first responders and community members with opioid antagonists to help prevent deaths from opioid overdoses, which may be a step in battling the epidemic.

Conclusion

Although many studies have focused on the mechanisms, causes, and genetic variants influencing SUD, there is still much to be learned regarding this disease. Most of the data regarding reward and reinforcement identifies the dopaminergic pathway as the most prominent mechanism in SUD. However, through additional research it is evident that dopamine is not solely responsible for substance use and behavioral disorders. The endogenous opioid system is intended to manage pain, emotion, and memory, but with the addition of exogenous opioids and other substances, this system has been altered to facilitate SUD. Exogenous and endogenous opioids elicit a similar effect in the body, suggesting an association in the mechanism of their effects. Further studies in endogenous opioids and identification of genetic variants may lead to additional and more efficacious treatments for SUD.

References

Alkermes, Inc. (2010). *Vivitrol (naltrexone for extended-release injectable suspension) package insert*. Retrieved December 1, 2017, from https://www.accessdata.fda.gov/drugsatfda_docs/label/2010/021897s015lbl.pdf

American Psychiatric Association. (2013). *Diagnostic and statistical manual of mental disorders* (5th ed.). American Psychiatric Publishing.

Berde, C. B., Beyer, J. E., Bournaki, M. C., Levin, C. R., & Sethna, N. F. (1991). Comparison of morphine and methadone for prevention of postoperative pain in 3- to 7-year-old children. *The Journal of Pediatrics, 119*(1 Pt 1),136–41.

Blazes, C. K., & Morrow, J. D. (2020). Reconsidering the usefulness of adding naloxone to buprenorphine. *Frontiers in Psychiatry, 11*, 549272.

Bruchas, M. R., Land, B. B., & Chavkin, C. (2010). The dynorphin/kappa opioid system as a modulator of stress-induced and pro-addictive behaviors. *Brain Research, 1314*, 44–55.

Carroll, J. J., Marshall, B. D., Rich, J. D., & Green, T. C. (2017). Exposure to fentanyl-contaminated heroin and overdose risk among illicit opioid users in Rhode Island: A mixed methods study. *International Journal of Drug Policy, 46*, 136–145.

Centers for Disease Control. (2016). *Wide-ranging online data for epidemiologic research (WONDER)*. CDC, National Center for Health Statistics. Retrieved December 19, 2017, from http://wonder.cdc.gov

Charbogne, P., Kieffer, B. L., & Befort, K. (2014). 15 years of genetic approaches *in vivo* for addiction research: Opioid receptor and peptide gene knockout in mouse models of drug abuse. *Neuropharmacology, 76*, 204–217.

Cruz, S. L., & Granados-Soto, V. (2015). Opioids and opiates: pharmacology, abuse, and addiction. Neurosci 21st century. New York: Springer, 3625–57.

Daniulaityte, R., Juhascik, M. P., Strayer, K. E., Sizemore, I. E., Zatreh, M., Nahhas, R. W., et al. (2019). Trends in fentanyl and fentanyl analogue-related overdose deaths–Montgomery County, Ohio, 2015–2017. *Drug and Alcohol Dependence, 198*, 116–120.

Dowell, D., Noonan, R. K., & Houry, D. (2017). Underlying factors in drug overdose deaths. *JAMA, 318*(23), 2295–2296.

Gaither, J. R., Leventhal, J. M., Ryan, S. A., & Camenga, D. R. (2016). National trends in hospitalizations for opioid poisonings among children and adolescents, 1997 to 2012. *JAMA Pediatrics, 170*(12), 1195–1201.

George, O., & Koob, G. F. (2010). Individual differences in prefrontal cortex function and the transition from drug use to drug dependence. *Neuroscience and Biobehavioral Reviews, 35*(2), 232–247.

George, M., Kitzmiller, J. P., Ewald, M. B. et al. (2012). Methadone toxicity and possible induction and enhanced elimination in a premature neonate. *Journal of Medical Toxicology, 8,* 432–435.

Gu, Z. H., Wang, B., Kou, Z. Z., Bai, Y., Chen, T., Dong, Y. L., & Li, Y. Q. (2017). Endomorphins: Promising endogenous opioid peptides for the development of novel analgesics. *Neurosignals, 25*, 98–116.

Howard, J., & Hornsby-Myers, J. (2018). Fentanyls and the safety of first responders: Science and recommendations. *American Journal of Industrial Medicine, 61*, 633–639.

Howland, M. A. (1998). Antidotes in depth. In *Goldfrank's toxicologic emergencies* (pp. 998–1009). Appleton and Lange.

Infante, A. F., Elmes, A. T., Gimbar, R. P., Messmer, S. E., Neeb, C., & Jarrett, J. B. (2024). Stronger, longer, better opioid antagonists? Nalmefene is NOT a naloxone replacement. *The International Journal on Drug Policy, 4*(104323). Retrieved May 7, 2024, from https://pubmed.ncbi.nlm.nih.gov/38232438/#:~:text=The%20U.S.%20Food%20and%20Drug,of%20naloxone%20for%20opioid%20reversal.

Johnson, T. A., Milan-Lobo, L., Che, T., Ferwerda, M., Lambu, E., McIntosh, N. L., et al. (2017). Identification of the first marine-derived opioid receptor "balanced" agonist with a signaling profile that resembles the endorphins. *ACS Chemical Neuroscience, 8*, 473–485.

Kane, J. M., Colvin, J. D., Bartlett, A. H., & Hall, M. (2018). Opioid-related critical care resource use in U.S. children's hospitals. *Pediatrics, 141*(4), e20173335.

Khokhar, J. Y., Ferguson, C. S., Zhu, A. Z. X., & Tyndale, R. F. (2010). Pharmacogenetics of drug dependence: Role of gene variations in susceptibility and treatment. *Annual Review of Pharmacology and Toxicology, 50*, 39–61.

Kjome, K. L., & Moeller, F. G. (2011). Long-acting injectable naltrexone for the management of patients with opioid dependence. *Substance Abuse: Research and Treatment, 5*, 1–9.

Klar, S. A., Brodkin, E., Gibson, E., Padhi, S., Predy, C., Green, C., & Lee, V. (2016). Furanyl-fentanyl overdose events caused by smoking contaminated crack cocaine—British Columbia, Canada, July 15–18, 2016. *Health Promotion and Chronic Disease Prevention in Canada: Research, Policy and Practice, 36*(9), 200–201.

Koob, G. F., & Le Moal, M. (2001). Drug addiction, dysregulation of reward, and allostasis. *Neuropsychopharmacology, 24*(2), 97–129.

Koob, G. F., & Mason, B. J. (2016). Existing and future drugs for the treatment of the dark side of addiction. *The Annual Review of Pharmacology and Toxicology, 56*, 299–322.

Leeman, R. F., & Potenza, M. N. (2013). A targeted review of the neurobiology and genetics of behavioral addictions: An emerging area of research. *Canadian Journal of Psychiatry, 58*(5), 260–273.

LoVecchio, F., Pizon, A., Riley, B., Sami, A., & D'Incognito, C. (2007). Onset of symptoms after methadone overdose. *The American Journal of Emergency Medicine, 25*(1), 57–59.

Lucher, C. (2016). The emergence of a circuit model for addiction. *Annual Review of Neuroscience, 39*, 257–276.

Lutfy, K., & Cowan, A. (2004). Buprenorphine: a unique drug with complex pharmacology. *Curr Neuropharmacol, 2*(4), 395–402.

Lutz, P. E., & Kieffer, B. L. (2013). The multiple facets of opioid receptor function: Implications for addiction. *Current Opinion in Neurobiology, 23*, 473–479.

Merrer, J. L., Becker, J. A. J., Befort, K., & Keiffer, B. L. (2009). Reward processing by the opioid system in the brain. *Physiological Reviews, 89*(4), 137–1412.

Moss, M. J., Warrick, B. J., Nelson, L. S., McKay, C. A., Dube, P., Gosselin, S., Palmer, R. B., Stolbach, A. I., & American Academy of Clinical Toxicology. (2017). ACMT and AACT position statement: Preventing occupational fentanyl and fentanyl analog exposure to emergency responders. *Clinical Toxicology, 56*(4), 297–300.

Nathan, P. E., Conrad, M., & Skinstad, A. H. (2016). History of the concept of addiction. *Annual Review of Clinical Psychology, 12*, 29–51.

Niikura, K., Narita, M., Butelman, E. R., Kreek, M. J., & Suzuki, T. (2010). Neuropathic and chronic pain stimuli downregulate central μ-opioid and dopaminergic transmission. *Trends in Pharmacological Sciences, 31*(7), 299–305.

Orman, J. S., & Keating, G. M. (2009). Buprenorphine/naloxone. *Drugs, 69*(5), 577–607.

Ozburn, A. R., Janowsky, A. J., & Crabbe, J. C. (2015). Commonalities and distinctions among mechanisms of addiction to alcohol and other drugs. *Alcoholism, Clinical and Experimental Research, 39*(10), 1863–1877.

Park, J. N., Weir, B. W., Allen, S. T., Chaulk, P., & Sherman, S. G. (2018). Fentanyl-contaminated drugs and non-fatal overdose among people who inject drugs in Baltimore, MD. *Harm Reduction Journal, 15*(1), 1–8.

Pathan, H., & Williams, J. (2012). Basic opioid pharmacology: An update. *British Journal of Pain, 6*(1), 11–16.

Pletcher, K., & Encyclopedia Brittanica, Inc. (2018). *Opium wars*. Encyclopedia Brittanica. Retrieved January 4, 2019, from: https://www.britannica.com/topic/Opium-Wars

Przewlocki, R. & Almeida, O. (2017). 1.08 - Stress and Opioid Systems, Editor(s): Donald W. Pfaff, Marian Joëls, Hormones, Brain and Behavior (Third Edition), Academic Press, 2017, Pages 225–260.

Schumacher, M. A., Basbaum, A. I., & Naidu, R. K. (2015). Opioid agonists & antagonists. In B. G. Katzung & A. J. Trevor (Eds.), *Basic & clinical pharmacology*. McGraw-Hill. Retrieved December 10, 2017, from http://0-accesspharmacy.mhmedical.com.library.acaweb.org/content.aspx?bookid=1193&sectionid=69108439

Sirohi, S., Dighe, S. V., Madia, P. A., & Yoburn, B. C. (2009). The relative potency of inverse opioid agonists and a neutral opioid antagonist in precipitated withdrawal and antagonism of analgesia and toxicity. *The Journal of Pharmacology and Experimental Therapeutics , 330*(2), 513–519.

Stringer, J., Welsh, C., & Tommasello, A. (2009). Methadone-associated QT interval prolongation and torsades de pointes. *American Journal of Health-System Pharmacy, 66*(9), 825–833.

Substance Abuse and Mental Health Services Administration. (2018). *SAMHSA opioid overdose prevention toolkit. HHS Publication No. (SMA) 18–4742*. Rockville, MD.

Trigo, J. M., Martin-Garcia, E., Berrendero, F., Robledo, P., & Maldonado, R. (2010). The endogenous opioid system: A common substrate in drug addiction. *Drug and Alcohol Dependence, 108*, 183–194.

Volkow, N. D., Wang, G. J., Fowler, J. S., & Tomasi, D. (2012). Addiction circuitry in the human brain. *Annual Review of Pharmacology and Toxicology 52*(1), 321–336. https://doi.org/10.1146/pharmtox.2012.52.issue-110.1146/annurev-pharmtox-010611-134625

Van Ree, J. M., Gerrits, M. A. F. M., & Vanderschuren, L. J. M. J. (1999). Opioids, reward and addiction: An encounter of biology, psychology, and medicine. *Pharmacological Review; The American Society for Pharmacology and Experimental Therapeutics, 51*(2), 341–396.

Endogenous Opioid Activity as the Mechanism of Action for *Mitragyna speciosa* (Kratom): The Current State of the Evidence

Adina Bowe and Patrick L. Kerr

Abstract Kratom (*Mitragyna speciosa*) is a substance derived from botanical compounds native to Southeast Asia. This substance has been cultivated predominantly in Thailand, Malaysia, Vietnam, and Myanmar, where it has historically been used in traditional medicine as a near panacea for several health problems. Such ritualistic use of kratom has been present for centuries; however, recreational use appears to have increased globally, especially in the United States. Pharmacodynamic and pharmacokinetic studies have found that kratom demonstrates a unique parabolic, dose-dependent pattern of effects ranging from stimulation to opioid and analgesic effects. Pharmacological research indicates that kratom is both a mu opioid receptor (μ-OR; MOR) and a kappa opioid receptor (κ-OR; KOR) agonist, which mediates its analgesic effects. Other research suggests that kratom may simultaneously act on dopaminergic and serotonergic receptors, which mediate its stimulant effects. This chapter reviews the literature related to the structural, functional, and cultural characteristics of kratom use. We begin with an overview of current and historical patterns of kratom, followed by a review of data on the pharmacodynamics and pharmacokinetics of kratom thus far.

Keywords Kratom · *Mitragyna speciosa* · Endogenous opioids · Pharmacodynamics · Pharmacokinetics

A. Bowe (✉) · P. L. Kerr
West Virginia University School of Medicine-Charleston, Charleston, WV, USA
e-mail: adbowe@hsc.wvu.edu

P. L. Kerr et al. (eds.), *Endogenous Opioids*, Advances in Neurobiology 35,
https://doi.org/10.1007/978-3-031-45493-6_15

Introduction

The uses of kratom have varied over time and geography. This plant-derived substance is well known in some regions of the world, where it has been used medicinally for centuries; and it is just now gaining popularity in other regions as it is introduced for a combination of recreational and medicinal purposes. Indeed, research over the past two decades has identified a wide range of uses for kratom among users (Kerrigan & Basiliere, 2022). From an early era, the medicinal uses of kratom have included alleviation of pain and discomfort for some people while also functioning as a stimulant for others. Such is the duality of kratom: well known to some and yet novel to many; stimulating at one dose, yet pain relieving at another; medicinal and also recreational.

The ostensible analgesic properties of kratom suggest that it operates on the endogenous opioid system, but this has only recently started to be understood. In this chapter, we describe the history of kratom and its cultivation; the patterns of kratom use; and the current state of scientific knowledge about the pharmacodynamics and pharmacokinetics of kratom. We emphasize the role of endogenous opioids and opioid receptors in facilitating the mechanisms of action that produce the primary effects of kratom.

Kratom: A Historical Context

Kratom (*Mitragyna speciosa; M. speciosa*), a genus in the Rubiaceae (i.e., coffee) family, is a tropical medicinal plant in Southeast Asia (Gong et al., 2012; Hassan et al., 2013; Saingam et al., 2016; Singh et al., 2016). This psychoactive plant has long been used in Southeast Asian countries, with reports of its use dating to as early as 1836 (Swogger et al., 2015). Kratom grows well in Malaysia, Thailand, the Philippines, and Myanmar, as well as in parts of Africa (Shellard, 1989; Tanguay, 2011). It is commonly known as "biak-biak," "bia," or "kratom" in Malaysia and "kratom," "tom," or "thom" in Thailand (Reanmongkol et al., 2007; Burkill, 1935). Dutch botanist Pieter Willem Korthals named the genus *Mitragyna*, due to the resemblance of the shape of kratom leaves to a bishop's ceremonial hat, called a miter or mitre (Warner et al., 2016).

Botanical Features of Kratom

Hassan et al. (2013) report that *M. speciosa* trees can grow to a height of 4–9 m and a width of 5 m. Some plants can reach a height of up to 15–30 m. The stem is erect and branching. The glossy green leaves of *M. speciosa* can grow to over 18 cm long and 10 cm wide with an oval or ovate-lanceolate shape and tapered ends. The flowers are a deep yellow and arranged in globular clusters attached to the leaf axils on long stalks. The florets can be up to 120 in number. The fruit produced by the kratom plant is round with oblong protruding vessels that contain many small, flat seeds (Hassan et al., 2013). The veins of kratom leaves are greenish-white or red in color. Interestingly, these plants differ in potency depending on their veins.

Studies have suggested that the "strain" or vein type of kratom corresponds to the age of the leaf (Braley & Hondrogiannis, 2020). For example, the red vein type is juvenile and is more potent than the mature green vein type. This results in differing pharmacological effects (Braley & Hondrogiannis, 2020). The leaves fall abundantly during the dry season of the year, and new growth is produced during the rainy season. The tree grows best in wet, humid, fertile soil, with medium to full sun exposure in areas protected from strong winds (Hassan et al., 2013). See Fig. 2 for examples of the botanical features and habitats of kratom (Fig. 1).

Fig. 1 The plant *Mitragyna speciosa* Korth. (**a**) Leaves of the plant, (**b**) naturally occurring trees, (**c**) and (**d**) cultivated plants. (Hassan et al., 2013)

Uses of Kratom Across Time, Geography, and Culture

The earliest scientific reports of kratom use were in the early nineteenth century. A report in 1836 described a way in which kratom leaves were used by people in Malaysia in place of opium (Burkill, 1935). Historically, people in Thailand ingested kratom to ease opioid withdrawal, with documented use for this purpose dating back to the 1940s (Tanguay, 2011).

Hassan et al. (2013) note that kratom's use in Malaysia and Thailand has been primarily for two broad purposes: as a stimulant to increase work efficiency and as a traditional homeopathic medicine. In the past, kratom was used predominantly by men, and restricted to those engaging in hard labor, such as agriculturalists and unskilled laborers, because of the belief of its effect on endurance (Chittrakarn et al., 2005). Increased work efficiency, endurance, and tolerance of hot and humid climate conditions made kratom an appealing option. These laborers were often farmers, rubber tapers, machine drivers, and fishermen who found that kratom increased their ability to work in excessively hot climates without tiring (Hassan et al., 2013; Suwanlert, 1975; Vicknasingam et al., 2010). Most male workers learned how to use kratom from other users in the community (Suwanlert, 1975).

Vicknasingam et al. (2010) study of people in Malaysia found self-reported use of kratom as an aphrodisiac. In this context, it was reportedly used to heighten sexual desire (Vicknasingam et al., 2010). Sweating, increased body temperature, perceived improvements in sexual performance, and subjective feelings of euphoria were also reported by people using kratom. These effects usually last between one and six hours (Ahmad & Aziz, 2012).

As a homeopathic medicine, women in villages in Southeast Asia use kratom's leaves to treat pain, cough, fevers, diabetes, hypertension, and anxiety. Modern products that include formulations as topical application of liniments, balms, or tinctures may provide effective alternatives for treatment of certain types of pains (Brown et al., 2017). It is also applied as a wound dressing (potentially because of its analgesic effects) and is believed to be a deworming agent and appetite suppressor (Assanangkornchai et al., 2007; Burkill, 1935; Burkill & Haniff, 1930; Lee, 1957; Saingam et al., 2016).

In addition to medicinal functions, kratom is sometimes used in cultural rituals. Kratom has also become a fixture in some spiritual practices and ceremonies as an offering to gods or spirits in exchange for fulfilling their desires or vows (Saingam et al., 2016). Kratom is also consumed in teashops or as a drink alternative by individuals restricted from alcohol consumption due to their religious beliefs. There is some evidence that people who practice Islam in Malaysia and South Thailand have used kratom as a substitute for alcohol because alcohol consumption is against Islamic teachings (Tanguay, 2011; Tungtananuwat & Lawanprasert, 2010).

A decade ago, Vicknasingam et al. (2010) noted two emerging trends in patterns of kratom usage. Their findings from a cross-sectional study in Malaysia showed that many people in the northern region of the country (close to the Thai border) who use illicit drugs were using kratom as a relatively inexpensive method to suppress opioid withdrawal symptoms and reduce the use of opioids (Vicknasingam et al., 2010). This same study found that people using kratom in northern Malaysian states relied on it for opioid withdrawal because it was more economically feasible and easily available and precluded the need to approach government facilities that might expose their identities (Vicknasingam et al., 2010). These trends in Southeast Asia were further established in a review of scientific literature several years later by this same research group (Singh et al., 2017), which noted data indicating that kratom's use for managing opioid withdrawal appeared to be more common in rural areas.

A decade later, there is further evidence that kratom is used for management of substance withdrawal. In a recent study by Singh et al. (2020), people currently using (n = 142) and previously using (n = 62) opioids reported using kratom at a comparable rate (approximately 900 ml/day). However, people who currently used opioids were nearly six times more likely to report using kratom for managing opioid withdrawal symptoms than those who no longer used opioids. Those who were no longer using opioids were significantly more likely to report using kratom to induce intoxication. This evidence that kratom is used to manage opioid withdrawal symptoms also suggests that this substance operates via one or more receptors within the endogenous opioid system. We will return to this topic in greater detail later in this chapter.

Kratom misuse has gained significant attention in Malaysia and Thailand (Chan et al., 2005). Researchers have noted a trend toward the use of kratom as a narcotic and/or substitute for alcohol, predominantly among younger people in more urban areas of Malaysia and Thailand (Ahmad & Aziz, 2012; Singh et al., 2016; Tanguay, 2011; Tungtananuwat & Lawanprasert, 2010). In Thailand, a homemade 4 × 100 kratom cocktail (referred to colloquially as "sii koon roi") has gained popularity among adolescents and young adults. This mixed drink consists of brewed kratom tea, cough syrup (e.g., diphenhydramine), and Coca-Cola. In addition to this, an assortment of substances such as anxiolytics, antidepressants, and analgesics may also be added based on preference, thus further enhancing the effects of kratom (Tanguay, 2011; Tungtananuwat & Lawanprasert, 2010). The use of the kratom cocktail with multiple drugs simultaneously may exacerbate the risk of morbidity and mortality via the compounding effects of multidrug toxicity (Tanguay, 2011). Similarly, in Malaysia, this kratom drink ("koroi") has become popular among people across multiple age groups who use kratom (Tanguay, 2011).

In the United States, an anonymous cross-sectional online survey of current kratom users (N = 8049) was conducted by the American Kratom Association to assess self-reported perceived benefits of kratom. This study documented analgesia, mood elevation, and anxiety reduction as perceived benefits (Grundmann, 2017). Interestingly, the study found what appeared to be a dose-dependent opioid-like effect of kratom. Consistent with this, participants also reported viewing kratom as beneficial for reducing or managing symptoms of opioid withdrawal associated with prescription opioid use (Swogger et al., 2015). These data are limited due to the potential for biased sampling and the lack of measurement precision. Nonetheless, it is noteworthy that even in self-report data, there is a correlation between reported volume of use and perceived benefits. These data also comport with the emerging trends noted by Vicknasingam et al. (2010) several years ago.

Routes of Administration for Kratom

Kratom has multiple routes of administration, which include smoking or chewing the leaves of the plant, brewing the leaves into a tea, ingesting powdered leaves into capsule or pill form, and making a kratom extract into a liquid dose (Hassan et al., 2013). Historically, individuals who used it for its stimulant effects used it to help reduce fatigue when carrying out manual labor on rubber plantations and seafaring. They are referred to as "chewers." These individuals typically start chewing kratom from the age of about 25 years. Nearly 70% of "chewers" are males, and their day-to-day consumption averages from 10 to 60 leaves.

More recently in Thailand, the use of a mixture produced from the broth of boiled kratom leaves with cola drinks and other substances, such as benzodiazepine and cough syrup, has become popular among adolescents and young adults (Saingam et al., 2013). The combination of cough syrup is thought to enhance and intensify the euphoric effect (Singh et al., 2014).

Kratom is consumed in Malaysia in the form of kratom juice or tea (Singh et al., 2016; Vicknasingam et al., 2010). Kratom traders often prefer to buy older kratom leaves from kratom cultivators when preparing kratom juice. The fresh kratom leaves are usually washed, boiled, and then left to brew at a lower temperature for up to 4 h, being stirred occasionally to prevent the leaves from sticking to the pot and burning. The brewed kratom juice is cooled and then packed into plastic bags (containing 275–360 ml) that are sold. The tea can last up to 3 days if properly preserved with ice once packaged in the plastic bags, which are sold for approximately RM5 (USD = $1.20) each (Singh et al., 2014, 2016; Vicknasingam et al., 2010). Kratom juice is bitter in taste; consequently, some users prefer to mix it with sweet beverages, e.g., Coca-Cola, Pepsi, etc.

Dried kratom leaves are often crushed, and the resulting powder may be inserted into gel capsules or prepared as a hot tea (Hassan et al., 2013). Plant ashes or baking soda is frequently added to help extract plant alkaloids prior to consumption. Sugar and honey are sometimes added due to the bitterness of the tea. The powder can also be cooked to yield a syrup-like consistency, which is then compressed into tablets (Hassan et al., 2013). Smoking of the dried leaf is less common but occasionally reported in Malaysia (Hassan et al., 2013).

Over a decade ago, it was recognized that kratom use had expanded from traditional and ceremonial uses to increased recreational use. Kratom gained popularity over the past several years across Southeast Asia and especially in Thailand. The tea-based cocktail known as 4 × 100 kratom preparations was among the most commonly used by high school students at a similar rate to cannabis (2.3–4.9%; Assanangkornchai et al., 2007).

The sale and supply of kratom are banned by the Malaysian government, whereas the legality of kratom is highly variable across much of the world. In Malaysia, kratom is normally sold in a clandestine environment in the community. The average daily intakes of kratom among Malaysian users are three glasses for a frequency of three times a day (Singh et al., 2016; Vicknasingam et al., 2010).

Prevalence and incidence data regarding kratom use is relatively scarce. No reliable data from North America (United States, Canada, Mexico), Europe, Africa, South America, or Australia have been published. In Asia, some data are available, though these too are scant. According to a 2008 national survey in Thailand, 1,078,152 people reported using kratom, and up to 70% of the male population use kratom daily in several southern districts (Tanguay, 2011).

Estimates of kratom use prevalence in the United States range from 10 to 15 million, but reliable evidence is lacking (Singh et al., 2020).

Some anecdotal reports initially suggested that people who immigrated from Southeast Asian countries first imported kratom into the United States in the 1980s and 1990s, with an expansion of use in the United States within the past decade (Henningfield et al., 2018; Kruegel & Grundmann, 2018). However, the path to kratom being used more widely is more complicated and intersectional than that. Global and local spread of kratom use has been attributed to online advertisements and social networks and easy transportation from the cultivation areas (Schmidt et al., 2011). Today, kratom is increasingly being sold through the Internet and local smart shops (Cinosi et al., 2015; Singh et al., 2016). Although commercial *M. speciosa* products are now widely available on the Internet (Hillebrand et al., 2010; Schmidt et al., 2011), quality and content vary, and preparations may not always be *M. speciosa* products (Hanna, 2012). These commercial *M. speciosa* products are offered as resin, dried leaf, or powder under the names "kratom," "mitragyna," "concentrated kratom," or "plant sample kratom" and many more.

Kratom is easily available and sold online, as well as at a variety of commercial outlets (e.g., specialty herbal stores, tobacco shops, and "head shops"). It is commonly marketed as an herbal or traditional medicine for various health problems ranging from pain to anxiety to opioid use disorder. Kratom is also frequently promoted as a "legal high" or "natural high" (Cinosi et al., 2015; Prozialeck et al., 2012; Singh et al., 2016). In some instances, kratom has been marketed in similar attractive packaging as many synthetic drugs potentially contributing to its sales success (Rosenbaum et al., 2012). Kratom is often thought of as a legal, safer, psychoactive alternative to other sedative and stimulant-type drugs which adds to its popularity (Babu et al., 2008).

Molecular, Pharmacokinetic, and Pharmacodynamic Properties of Kratom

Molecular Structure

Mitragynine was first isolated in 1921 (Grewal, 1932). Zacharias and colleagues (1965) identified the structure of mitragynine using X-ray crystallography, which was further elaborated by Liu et al. (2010) in a computational study. Mitragynine has been synthesized using a variety of methodologies over the past three decades, including an alcohol-based enzymatic hydrolytic process of racemic acetate and ketone enantioselective reduction (Takayama et al., 1995) and via Mori-Ban-Hegedus indole synthesis preparations and bis(cyclooctadiene)nickel/Pictet-Spengler reaction preparations of 4-methoxytryptophan. The molecular structures of *M. speciosa* (see Fig. 2) alkaloids were found to be either indoles with a methoxy group in the C19 position and an open E ring with substitution at C9 position or oxindoles with a closed E ring and no substitution at C9 position (Beckett et al., 1966; Shellard & Phillipson, 1966; Takayama et al., 2002).

The IUPAC name of mitragynine, the major indole alkaloid constituent of *M. speciosa*, is methyl (E)-2-[(2S,3S,12bS)-3-ethyl-8-methoxy-1,2,3,4,6,7,12,12b-octahydroindolo[2,3-a]-quinolizin-2-yl]-3-methoxyprop-2-enoate. Mitragynine has a CAS registry number of 4098-40-2, a molecular weight of 398.50 g/mol, a LogP (partition coefficient) of 1.73, and a pKa of 8.11 ± 0.11 (Ramanathan et al., 2015). It is structurally similar to yohimbine and voacangine (Hassan, et al., 2013; Prozialeck, et al., 2012). Mitragynine has a molecular composition of C23H30N2O4 (Ramanathan et al., 2015). Mitragynine crystals with inchoate structures demonstrate a polarization rotation ([α]D = −126 (c = 0.66) and −128 (c = 1.2), $CHCl_3$) (Kerschgens et al., 2012) with a melting point between 102 and 106 °C. The hydrochloride salt of mitragynine melts at 243 °C (EMCDDA, 2021). The relative lipophilicity of mitragynine makes the free base insoluble in both water and basic media (EMCDDA).

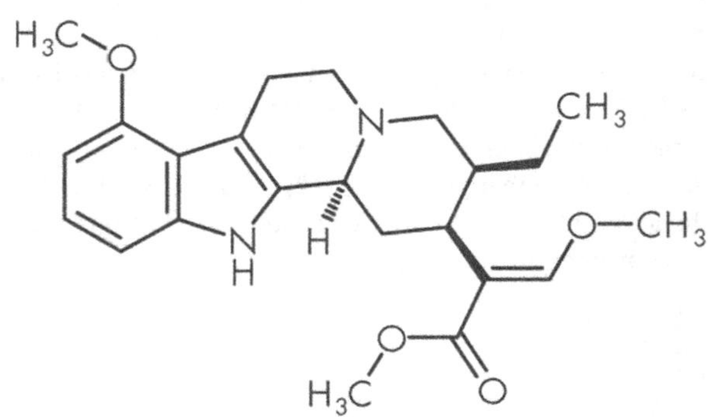

Fig. 2 Molecular structure: mitragynine Molecular formula: $C_{23}H_{30}N_2O_4$ Molecular weight: 398.50 g/mol European Monitoring Centre for Drugs and Drug Addiction (2021)

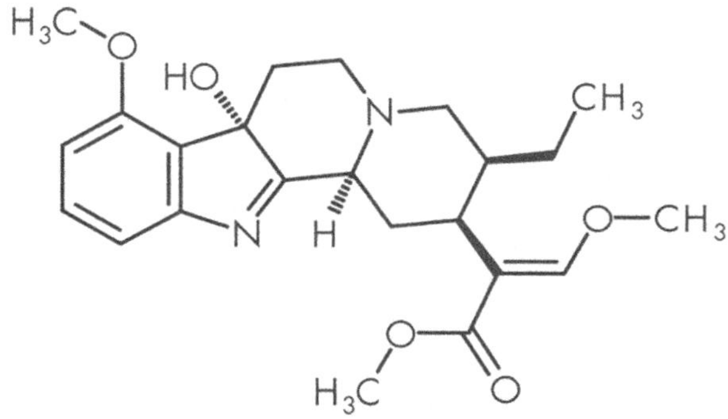

Fig. 3 Molecular structure: 7-hydroxymitragynine Molecular formula: $C_{23}H_{30}N_2O_5$ Molecular weight: 414.50 g/mol European Monitoring Centre for Drugs and Drug Addiction (2021)

Mitragynine is a white amorphous powder that is soluble in alcohol, chloroform, and acetic acid and is chemically known as 9-methoxy-corynantheidine. After its structural elucidation, three total syntheses, along with one formal synthesis of (−)-mitragynine, were reported (Kerschgens et al., 2012; Takayama et al., 1995; Ma et al., 2007; Sun & Ma 2011). Moreover, various semisynthetic approaches toward more potent mitragynine derivatives such as 7-hydroxymitragynine (7-HMG; see Fig. 3) and mitragynine pseudoindoxyl were also reported.

Among the compounds, mitragynine, 7-HMG, and mitragynine pseudoindoxyl render the potent morphine-like effects (Raffa 2014). Even though 7-HMG is a minor alkaloid in kratom compared to mitragynine, 7-HMG has a higher potential for abuse (Hemby et al., 2019). However Boyer et al. recognized that these compounds alone have failed to produce the powerful analgesic effects achieved by ingesting the whole leaves (Boyer et al., 2008).

Mitragynine is a weak base (pKa = 8.1) and is lipophilic (Log P = 1.73) (Ramanathan et al., 2015). Mitragynine has aqueous solubility limits of 130 μM at pH 4 and 83 μM at pH 7 (Han et al., 2019). This compound at 37 °C was moderately stable at neutral pH (˜3.5% degradation after 3 h) but degraded at pH 1.2 (by 26% degradation after 1–2 h) (Manda et al., 2014; Ramanathan et al., 2015). Mitragynine being a basic drug is highly solubilized and ionized in the stomach, thus making it less bioavailable. In addition, poor oral bioavailability was also likely to be caused by its degradation in the highly acidic gastric juice. There are no reported data of human oral bioavailability.

Pharmacokinetics: Absorption

Kong et al. (2017) and Manda et al. (2014) studied the in vitro absorption of mitragynine. These studies included in vitro absorption rates evaluating fluxes through the phospholipid bilayer, coherent Caco-2 cell monolayers, and MDR-MDCK cell monolayers. Substance permeability (i.e., between donor and acceptor compartments through an artificial phospholipid membrane) can be determined via a methodology known as the parallel artificial membrane permeability assay (PAMPA; Kansy, 1998). The artificial membrane used in PAMPA permits the observation of passive permeability because it has no transporters or efflux system. Consequently, PAMPA facilitates the classification of compounds that are passively transported using a valid and reliable in vitro model of passive transcellular permeation (Kansy, 1998; Ottaviani et al., 2006). Mitragynine fluxes through the phospholipid bilayer at pH 4 and 7.4 were 0.23×10^{-6} and 11×10^{-6} cm/s, respectively (Kong et al., 2017); these values suggested that mitragynine permeated as the unionized form.

Mitragynine absorption has been estimated using the Caco-2 monolayer, a continuous line of heterogeneous human epithelial colorectal adenocarcinoma cells commonly employed in pharmacological research to model the small intestinal tract of humans and to predict the absorption of orally administered drugs (Fogh, 1977). Using this methodology, Manda et al. (2014) reported that forward and reverse fluxes were 25×10^{-6} and 27×10^{-6} cm/s and independent of concentration. Mitragynine flux ratios were approximately 1, which suggests passive diffusion through membranes. Manda (2014) also reported forward and reverse flux symmetry of mitragynine in a model of the blood-brain barrier (i.e., with MDR-MDCK cell monolayers), yielding 15×10^{-6} and 17×10^{-6} cm/s, respectively. Thus, there is clear evidence that mitragynine, and most likely kratom itself, is absorbed through the small intestine and can pass through the blood-brain barrier.

Differences in stability of mitragynine have been found for gastric versus intestinal fluid. In simulated gastric fluid, mitragynine demonstrated a degradation of 26% while remaining stable in simulated intestinal fluid. Similarly, Manda et al. (2014) found that 7-HMG demonstrated a degradation of as much as 27% in simulated gastric fluid. Manda and colleagues have proposed that this degradation rate may, at least in part, explain the process through which 7-HMG is concerted to mitragynine (23%). Notably, simulated intestinal fluid was associated with only a 6% degradation. Conversely, Manda (2014) reports that mitraphylline (a mitragynine alkaloid) was stable in simulated gastric fluid but unstable in simulated intestinal fluid (13.6% degradation). When examining stability in the hepatic system, mitragynine demonstrated metabolic stability both in human liver microsomes and in S9 fractions (Kong et al., 2017; Manda et al., 2014). However, mitragynine alkaloids 7-HMG and mitraphylline demonstrated rapid hepatic metabolism, with half-lives of 24 min and 50 min, respectively, when exposed to human liver microsomes.

An animal study by Jagabalan et al. (2019) found comparable rates of intestinal absorption for mitragynine (1.11×10^{-4} cm/s), propranolol (1.12×10^{-4} cm/s), and atenolol (0.41×10^{-4} cm/s). This would suggest that mitragynine is absorbed passively and rapidly. That conclusion is further supported by Jagabalan's analyses indicating that P-glycoprotein (P-gp) or CYP3A4 inhibitors (azithromycin and ciprofloxacin, respectively) did not affect mitragynine fluxes (1.13×10^{-4} cm/s and 1.17×10^{-4} cm/s, respectively).

Pharmacokinetics: Protein Binding

Plasma protein binding is determined by equilibrium dialysis. All three compounds exhibited a high plasma protein binding (> 90%). By equilibrium dialysis, about 95% of mitragynine at tested concentration range of 5–15 μM bound to plasma protein after 24 h at 37 °C in human plasma (Manda et al., 2014). About 85% of mitragynine at a tested concentration of 10 μM binds to plasma proteins after 1 h at 37 °C in human plasma, determined by ultrafiltration (Kong et al., 2017).

P-gp is a plasma membrane protein that acts as a localized drug transport mechanism, actively exporting drugs out of the cell. The activity of mitragynine on P-gp was investigated by three studies (Manda et al., 2014, 2017; Meyer et al., 2015). Those studies concluded that mitragynine can induce and inhibit P-gp; thus, coadministration of mitragynine with drugs which are P-gp substrated may possibly lead to drug-herbal interaction. Notably, when protein binding is considered, free mitragynine concentration and by extension the extracellular levels are low. Thus, these actions on drug metabolism are mostly irrelevant to therapeutic or even toxic doses of mitragynine.

Metabolism

Most of mitragynine's metabolism occurs in the liver. Both phase I and phase II are involved in the metabolism of this psychoactive alkaloid. There were several studies investigating the roles of phase I and phase II using a system with liver microsomes or liver S9 fractions (Kong et al., 2017; Manda et al., 2014). Philipp et al. (2009) measured urinary metabolites by LC–MS/MS after oral dosing of 40 mg/kg mitragynine to rats, and urine samples were collected after 24 h. Seven phase I mitragynine metabolites were identified, and a further five conjugates as one sulfonate and four glucuronides were presumed to be phase II metabolites. In human urine, three sulfonates and four glucuronides were found (Philipp et al., 2009).

In terms of metabolism and excretion in humans, Trakulsrichai et al. (2015) conducted a study of ten chronic kratom users. In this human study, mitragynine was administered orally daily via kratom tea. The researchers found only 0.14% of the administered mitragynine in urine and proposed hepatic metabolism as its main metabolic transformation in humans (Trakulsrichai et al., 2015). While this study is small and not necessarily generalizable to all people who use kratom (e.g., the metabolism of people who use kratom chronically may differ from those who are kratom naive), it is generally consistent with preclinical animal studies described earlier.

Mitragynine is metabolized in humans via phase I and phase II mechanisms. The parent undergoes hydrolysis at the side chain methyl ester in position 16 (Hassan et al., 2013; Philipp et al., 2009; Prozialeck et al., 2012). O-Demethylation then takes place at the 9- and 17-methoxy groups. Oxidative and reductive transformations proceed to the intermediate aldehydes, which yield carboxylic acids and alcohols, respectively. A final step involves glucuronide and sulfate conjugate formation as a result of phase II metabolism which is excreted with the urine (Philipp et al., 2009; Hassan et al., 2013; Prozialeck et al., 2012).

In vitro experiments using isolated CYP450 enzymes indicate that kratom extracts inhibit various CYP enzymes, notably CYP 3A4, 2D6, and 1A2. This may lead to clinically significant interactions with other drugs given that there are substrates for these CYP enzymes (Kong et al., 2011).

In humans, the effects of kratom are felt 30–60 min after ingestion, although onset can be noticeable within about 10–20 min. Certain conditions (e.g., having recently eaten, using kratom capsules instead of teas/cocktails) may delay the initial effects. The effects of kratom typically last about 5–7 h, with the strongest effects occurring about 2–4 h following ingestion; some mild "aftereffects" may be noticeable up to 1 day later (Maruyama et al., 2009; Prozialeck et al., 2012; Rosenbaum et al., 2012; Scott et al., 2014). The half-life of mitragynine is 3.5 h, and the half-life of 7-HMG is approximately 2.5 h. Both are eliminated from the body primarily via urinary excretion (Neerman et al., 2013; Prozialeck et al., 2012). Current pharmacokinetic data from either animal or human research is limited, and mitragynine pharmacokinetics varies widely within and between species.

Dosage

The average weight of a *Mitragyna speciosa* leaf is 1.7 g when picked and 0.43 g when dried. Twenty leaves contain ~17 mg of mitragynine (Chan et al., 2005). However, the concentrations of alkaloids may vary among batches, as the time course of alkaloid degradation in the fresh and dried forms is unknown, as well as the manufacture and storage of the products.

Approximately 1–5 g of raw leaves, considered a low-to-moderate dose, produces mild stimulant effects. The onset of euphoric effects is experienced in about 10 min after using a few grams of dried leaves. People using kratom will likely feel happy, motivated, and strong (Suwanlert, 1975). At this dose, the user may also exhibit normal to slightly contracted pupils and skin blushing. There are usually minimal undesirable side effects at this dose; however, anxiety and internal agitation have been described (Prozialeck et al., 2012). Individuals using from 5 to 15 g of leaves are said to exhibit opioid-type effects. This may include euphoria, sleepiness, and increased analgesia similar to an opiate (Kruegel & Grundmann, 2018). Diarrhea may be a side effect at this dose.

Ingestion of higher doses (i.e., those exceeding 15 g) of kratom leaves leads to opioid mimetic effects, such as sedation, respiratory depression, and a feeling of being in a dreamlike state (Prozialeck et al., 2012). High doses are commonly associated with sweating, dizziness, nausea, and dysphoria initially. These effects quickly subside and are followed by calmness and a dreamlike state. Most people use kratom in a dose of less than 8 g, delivering 120–180 milligrams of mitragynine.

Mechanism of Action

Most of the psychoactive and opioid-like activities of kratom have been attributed to mitragynine and 7-HMG (Babu et al., 2008; Warner et al., 2016; Kruegel et al., 2019; Kruegel & Grundman, 2018). Mitragynine and its analogues, such as mitraphylline and 7-HMG, are indole alkaloids of the *Corynanthe*-type possessing a monoterpene (iridoid) moiety (Ingsathit et al., 2009). Other alkaloids in kratom also include raubasine and some yohimbe alkaloids (Chittrakarn et al., 2010; Raffa 2014; Prozialeck et al 2012; Takayama et al., 2002). The mitragynine contents in kratom vary with location and season. The difference in the alkaloidal content depends on several factors, such as the particular variety and age of the plant, environmental factors, and the time of harvest (Leon et al., 2009). When kratom is grown in Southeast Asia, the content of mitragynine tends to be higher, but in elsewhere it tends to be low or nonexistent (Ward et al., 2011).

Opioid Activity

Mitragynine and 7-HMG both exhibit dose-related antinociceptive effects and stimulate supraspinal μ- and δ-opioid receptors in vivo and in vitro models (Rosenbaum et al., 2012; Hassan et al., 2013; EMCDDA, 2015; Prozialeck et al., 2012; Babu et al., 2008; Thongpradichote et al., 1998; Tohda et al., 1997; Matsumoto et al., 1996a). Mitragynine and 7-HMG are selective and full agonists of μ-opioid subtype receptors (Rosenbaum et al., 2012; Hassan et al., 2013; EMCDDA, 2015; Prozialeck et al., 2012) and competitive antagonists at κ- and δ-opioid receptors (KOR and DOR) (Kruegal & Grundman, 2018). Although these alkaloids activate opioid and monoaminergic receptors, they differ from opioids in structure (Stolt et al., 2014).

While polarity is increased due to the additional hydroxyl group on 7-HMG as compared to mitragynine, the increased activity of 7-HMG is otherwise not well understood (Prozialeck et al., 2012). The presence of this hydroxyl group at C-7 increases the potency of 7-HMG (Horie et al., 2005; Takayama, 2004; Utar et al., 2011). Although found in a much lower concentration in kratom leaves, 7-HMG is 46-fold more potent than mitragynine and 13-fold stronger than morphine as an antinociceptive compound (Hassan et al., 2013).

Studies have suggested that both mitragynine and 7-HMG have a high affinity for opioid receptors with 7-HMG having the higher affinity for μ-OR (Chear et al., 2021; Yamamoto et al., 1999; Watanabe et al., 1997; Prozialeck et al., 2012). Mitragynine and 7-HMG differ in binding affinities to the various opioid receptor subtypes. These differences in binding affinity may be due to the differences in interaction between polar structures of mitragynine with a set of N-termini and carboxyl (COOH) transmembrane 4 and extracellular loops 2 and 3 located at the membrane which differentiate between the kappa, mu, and delta opioid receptors

(Taufik Hidayat et al., 2010). Studies differ in their report with respect to binding affinity among the receptor subtype (Han et al., 2019; Taufik Hidayat et al., 2010), but, in general, studies show that mitragynine and 7-HMG bind to opioid receptors with nM affinity (μ-OR > κ-OR > δ-OR) and efficacy (Han et al., 2019; Raffa et al., 2018).

Kruegel et al. (2019) reported that mitragynine acted as either a competitive antagonist or a partial agonist at the mouse μ-OR depending on the assay used (Kruegel et al., 2019). Conversely, Varadi et al. reported that mitragynine acted as a partial agonist of the μ-OR (EC50 = 203 nM; Emax = 65%) using a [^{35}S]GTPγS displacement assay (Varadi et al., 2016). As these studies showed differences in receptor modulation, it has been proposed that there is an active metabolite of mitragynine, generated after the first-pass metabolism that may also be involved in μ-OR agonism besides mitragynine itself (Han et al., 2019; Kruegel et al., 2019; Váradi et al., 2016). Multiple studies have found that the oral and intraperitoneal administration of mitragynine yields a more potent analgesic effect compared to subcutaneous route administration in rodent models (Kruegel et al., 2019; Macko et al., 1972; Sabetghadam et al., 2013). However, intracerebroventricular administration of mitragynine also produces analgesia, suggesting that no active metabolites are involved. Nevertheless, the detailed molecular mechanisms for the induction of analgesic effects have yet to be elucidated and remain a target for future research.

In animal studies using noxious or painful stimuli, extracts of kratom prolonged the latency of nociceptive responses using hot plate and tail-flick tests (Mossadeq et al., 2009; Reanmongkol et al., 2007; Sabetghadam et al., 2010). These responses were reduced by naloxone, supporting the theory of an opioid agonist mechanism of action. Caffeine coadministration enhanced the antinociceptive effects of a kratom alkaloid extract, as did the use of acetaminophen (Hassan et al., 2013).

Despite being a μ-OR partial and full agonist, mitragynine and 7- HMG present with a unique mechanism of action and pharmacology that is distinct from classical opioids such as morphine or heroin. Mitragynine and 7-HMG bind to the μ-OR causing recruitment and activation of the G protein-coupled signaling cascade; however, unlike classical opioids, this binding does not lead to recruitment of β-arrestin 2. β-Arrestin 2 is associated with side effects similar to those resulting from opioid receptor activation, e.g., respiratory depression, constipation, and physiological dependence (Kruegel et al., 2019; Varadi et al., 2016). Mitragynine and 7-HMG are classified as *atypical opioids* due to attenuated respiratory depression and reduced inhibition of gastrointestinal transit compared to classical opioids. For this reason, kratom is speculated to be a safer alternative to classic opioids (Varadi et al., 2016). However, recent evidence suggests that this compound is still associated with multiple health risks (Bowe & Kerr, 2020; Kerrigan & Basiliere, 2022).

Anti-inflammatory Effects

The anti-inflammatory effects of *Mitragyna speciosa* have also been studied by several researchers (Hassan et al., 2013; Utar et al., 2011; Mossadeq et al., 2009). Hassan and colleagues note that the administration of kratom extract inhibited acute limb edema, while long-term kratom use prevented the growth of granuloma tissue in animal studies (Hassan et al., 2013). Prior research has established that cyclooxygenase (COX) isoforms (COX-1 and COX-2) contribute to inflammation leading to the formation of prostaglandin E2 (PGE2; Meireles et al., 2019). PGE2 is a strong mediator of inflammation. Utar et al. (2011) found that mitragynine inhibits COX-2 mRNA and protein expression and thus also inhibits PGE2 formation. This same research group reported that the inhibitory effects of mitragynine may be dose dependent for COX-1, with little to no effect at lower doses, but may yield an effect at higher doses (Utar et al., 2011). The hypothesized anti-inflammatory effects may be mediated by the combined inhibition of pro-inflammatory molecule release and vascular permeability, as discussed earlier (Meireles et al., 2019).

Pharmacological Effects

Mitragyna speciosa produces a variety of pharmacologic effects, both in vivo and in vitro. Among those reported have been the inhibition of vas deferens contraction (Matsumoto et al., 2005) and the inhibition of gastric acid secretion (Tsuchiya et al., 2002). At the cellular level, mitragynine blocks neuronal Ca2+ channels, which partially inhibits the release of neurotransmitters from the axonal terminals in the vas deferens. This may account for the perception of "improved" sexual performance among men who use kratom. This would also be consistent with data on the involvement of endogenous opioids in sexual performance in general (see Khajehei, chapter "Endorphins, Sexuality, and Reproduction" in this volume). The neuronal Ca2+ channel-blocking effect of mitragynine is believed to be a general mechanism for other physiological effects (Matsumoto et al., 2005).

Further evidence for endogenous opioid activity in kratom comes from research examining induced muscle (ileum) contraction in animals (Watanabe et al., 1997; Horie et al., 2005). Researchers investigated the use of a crude kratom extract on electrically induced contraction in the ilea of guinea pigs. The crude kratom extract inhibited the ileum contraction; and this effect was reversed (inhibition of inhibitory action) by naloxone, further indicating μ-OR activity. Specific extracted alkaloids (7-HMG, mitragynine, SC, PAY, and SG) also inhibited induced ileum contractions in a dose-dependent pattern (Horie et al., 2005).

Kratom induces constipation through opioid and non-opioid mechanisms. A methanolic extract of kratom reduced defecation frequency and fecal weight in a castor oil-induced diarrhea model and slowed intestinal transit in rats (Hassan et al., 2013). Naloxone pre-administration only partially ameliorated these gastrointestinal effects. Intestinal transit time was also suppressed when subcutaneous 7-HMG was administered to mice (Hassan et al., 2013). Paynantheine, speciociliatine, and speciogynine, other alkaloids found in kratom, impede intestinal smooth muscle function (Hassan et al., 2013; Prozialeck et al., 2012). Speciociliatine inhibits acetylcholine release from presynaptic nerves, while paynantheine and speciogynine inhibit muscarinic receptors in ileal smooth muscles (Prozialeck et al., 2012).

Non-opioid Activity

Mitragynine has been shown to bind to non-opioid receptors in the CNS (central nervous system) and displays activity on descending noradrenergic and serotonergic systems and a suppressive action on the mechanical and thermal stimuli-induced nociceptive responses (Hassan et al., 2013; Thonpradichote et al., 1998; Tsuchiya et al., 2002). Binding to non-opioid receptors includes α-2 adrenergic receptor, adenosine A2a receptor, dopamine (D2), and the 5-HT2C and 5-HT7 serotonin receptors (Kruegel et al., 2019; Boyer et al., 2008). The analgesic effects of mitragynine are inhibited by idazoxan (an α-2 adrenergic receptor antagonist) and cyproheptadine (a nonspecific 5-HT antagonist) (Matsumoto et al., 1996b).

Mitragynine stimulates postsynaptic α-2 adrenergic receptors (Prozialeck et al., 2012; Giovannitti et al., 2015). Examples of medications with a similar effect are dexmedetomidine and clonidine. Dexmedetomidine is α-2 adrenergic agonist used for sedation. Clonidine is an α-2 adrenergic agonist used to manage pain, anxiety, attention deficit disorder, and symptoms of withdrawal (e.g., from opioids, benzodiazepines, alcohol, tobacco). Kratom has been used to manage opioid withdrawal symptoms (Boyer et al., 2008), and it was postulated that these are managed in part through the action on a-2 receptors. Kratom through a-2 agonism may have additive sedative and hypnotic effects with agents that act similarly (Prozialeck et al., 2012).

Other studies have further revealed that kratom's pharmacology is complex (Stolt et al., 2014). The pharmacological mechanisms responsible for stimulant activity are yet to be clearly established (Ujvary, 2014). Further work to investigate the mechanisms of action of other kratom alkaloids, such as paynantheine, speciogynine, and speciociliatine, may be important in delineating the non-opioid-related effects present in kratom (Takayama, 2004).

Raffa et al. (2018) summarized opioid pharmacology research findings to date on mitragynine and 7-HMG as follows:

- Mitragynine and 7-HMG bind to opioid receptors with nM affinity (μ-OR > δ-OR and κ-OR) and efficacy (agonist action).
- Mitragynine and 7-OH-MG produce antinociceptive effect when administered orally, s.c., or directly into brain ventricles.

- The antinociceptive effect is antagonized by naloxone and the more μ-OR-selective antagonist cyprodime (but less so by the more δ-OR- and κ-OR-selective antagonists naltrindole and nor-binaltorphimine respectively).
- Antinociceptive tolerance develops with 7-OH-MG, as does cross-tolerance to morphine.
- Naloxone precipitates withdrawal signs in 7-OH-MG-tolerant animals, but the withdrawal is milder than that from morphine.

Raffa et al. (2018) summarized the non-opioid pharmacology of mitragynine and 7-HMG as follows:

- Mitragynine interacts indirectly with α_2-adrenoceptors, a known mechanism of analgesic action.
- Mitragynine suppresses serotonin-induced head-twitch response in mice, suggestive of a 5-HT_{2A}-related action.
- Mitragynine is moderately active in antidepressant tests in mice.
- A role for descending noradrenergic and serotonergic inhibitory systems has been postulated for mitragynine-induced antinociception.

There are no published human pharmacologic or drug interaction studies on kratom or mitragynine making it impossible to fully understand kratom's therapeutic potential and risks and the populations most likely to benefit or experience harm from its use (Hassan et al., 2013; Prozialeck et al., 2012; Ulbricht et al., 2013; Tanguay, 2011).

Antidepressant and Anxiolytic Activity

Animal studies suggest that mitragynine may have antidepressant and anxiolytic-like effects in rodents (Kumarnsit et al., 2007; Moklas et al., 2013; Yusoff et al., 2016; Hazim et al., 2014). A single administration of *M. speciosa* extract reduced the duration of immobility in a forced swim depressomimetic paradigm indicating that the extract may have potential for functioning as an antidepressant (Kumarnsit et al., 2007). Additionally, other researchers have observed significant reductions in corticosterone levels in mice exposed to the forced swim test and tail suspension test, and have proposed that restoration of monoamine neurotransmitter levels (i.e. serotonin, noradrenaline and dopamine) and/or interaction with neuroendocrine hypothalamic-pituitary-adrenal axis systems, may mediate the antidepressant-like action of mitragynine (Farah Idayu et al., 2011). Yusoff et al. (2016) also demonstrated the anxiolytic effects of mitragynine in a plus-maze experiment paradigm (Yusoff et al., 2016).

Side Effects

Chronic use of kratom is associated with multiple adverse health consequences, including insomnia, dry mouth, constipation, weight loss, insomnia, micturition, dehydration, fatigue, skin hyperpigmentation, and low libido (Assanangkornchai et al., 2006; Saingam et al., 2016; Suwanlert, 1975; Vicknasingam et al., 2010). Kratom users report tolerance and in the absence of kratom experience withdrawal symptoms (Saingam et al., 2016; Prozialeck et al., 2012). These include muscle aches and spasms, anorexia, weight loss, decreased sexual drive, insomnia, runny nose, diarrhea, and watery eyes (Burkill & Haniff, 1930; Singh et al., 2014). Psychological withdrawal symptoms commonly reported are nervousness, restlessness, tension, anger, hostility, aggression, and sadness (Suwanlert, 1975; Singh et al., 2014).

Studies have also highlighted transplacental transfer of kratom to neonates after newborns suffered withdrawal symptoms from exposure in utero (Post et al., 2019). Several cases of neonatal abstinence syndrome due to maternal kratom use have been reported (Eldridge et al., 2018; Davidson et al., 2019; Mackay & Abrahams, 2018).

In a cross-sectional survey of 293 people who reported regular kratom use in three northern states of Peninsular Malaysia, Singh et al. (2014) found no evidence of major impairments in social functioning, despite using chronic kratom for prolonged periods.

Kratom Toxicity

Although kratom has the reputation of being minimally toxic (Henningfield et al., 2018), it has been implicated in an increasing number of emergency room visits and calls to poison control centers (Forrester, 2013; Trakulsrichai et al., 2013). Prior to 2015, no studies had been conducted to determine the blood concentration in patients, and it was proposed that future approach should consider this point in order to prevent overdose (Holler et al., 2011; Ramanathan et al., 2015). In 2019, Papsun described a validated analytical method of separating and identifying mitragynine from its isomers and other related natural products. A short series of case reports was presented that provided examples of apparent adverse events and the associated range of mitragynine concentrations in serum. Papsun et al. suggested that mitragynine concentrations between 100 and 500 ng/ml may need to be viewed as contributory, while concentrations >1000 ng/ml are more likely to be associated with fatalities and may be more causative in nature (Papsun et al., 2019).

Eggleston et al. (2019) conducted a retrospective study of kratom exposures reported to the National Poison Data System to determine the toxicities associated with kratom use. Kratom was listed as a cause or contributing factor in the death of four decedents identified by the county medical examiner's office (Eggleston et al.,

2019). A review of the literature reveals that kratom toxicity most commonly caused agitation, tachycardia, drowsiness, vomiting, and confusion. Serious effects of seizure, withdrawal, hallucinations, respiratory depression, liver toxicity, intrahepatic cholestasis, coma, and cardiac or respiratory arrest were also reported (Eggleston et al., 2019; Forrester, 2013; Kapp et al., 2011; Kupferschmidt, 2011; Post et al., 2019; Roche et al., 2008).

McIntyre et al. (2015) noted that the pharmacological properties of kratom and its potential toxicity are not completely clear. While kratom can produce negative side effects, it appeared to be primarily dangerous when users engage in polydrug use. Kronstrand et al. (2011) documented deaths associated with krypton, a substance that combined kratom with O-desmethyltramadol (Kronstrand et al. 2011). Furthermore, Karinen et al. (2014) discussed the death of a Norwegian man who had traces of mitragynine, 7-OH, zopiclone, citalopram, and lamotrigine in his system. Another case indicated that a mix of kratom, over-the-counter cold medications, and benzodiazepines was responsible for the death of a 17-year-old boy (Holler et al., 2011). McIntyre et al. (2015) reported a kratom-related fatality. A 24-year-old man found with a medical history significant for alcohol abuse and depression was found unresponsive in bed, and his postmortem evaluation detected kratom, together with venlafaxine, diphenhydramine, and mirtazapine (McIntyre et al., 2015).

In Thailand, there is an increase in community concern about contamination of kratom mixtures. Some kratom cocktails are suspected of containing other drugs such as benzodiazepines, as well as household consumer products, including fluorescent tubes, powdered mosquito coils, road paint, and pesticides (Taguay, 2011). Similar to the contamination of illicit opioid and stimulant supplies in the United States (e.g., with fentanyl; Carroll et al., 2017; Macmadu et al., 2017), the adulterants that are mixed with kratom may increase the risk of using.

Detection

Mitragynine presents an analytical challenge requiring an appropriate method of separating and identifying mitragynine itself from its isomers and other similar naturally occurring compounds (Papsun et al., 2019). Mitragynine and 7-HMG are not routinely detected in most drug testing or screening procedures in the clinical and forensic toxicology setting (Prozialeck et al., 2012). A range of methods has been developed for the analysis of the plant material and other kratom-containing substances including numerous chromatographic techniques (Limsuwanchote et al., 2014). High-performance liquid chromatography (HPLC), the most common of chromatographic techniques, and other LC techniques coupled with either ultraviolet (UV) or mass spectrometer (MS) detectors (e.g., electrospray) may be used to detect the active alkaloids in kratom leaves (Limsuwanchote et al., 2014; Philipp et al., 2010; Rosenbaum et al., 2012).

Kratom: The Current Legal and Regulatory Status

The full scope of the legality of kratom is complex, highly technical from a legal standpoint, and thus is beyond the scope of this chapter. Nonetheless, the regulatory landscape has some relevance to our ability to adequately study and understand any substance, including kratom. Therefore, we will provide a general overview of the current legal status of kratom as of the time of this writing in the early 2020s.

The restriction and regulation of kratom vary widely by region, country, and in some cases by state. The regulation of kratom appears to be rapidly changing in multiple directions, with some countries tightening restrictions and others loosening them. Kratom is illegal, banned, or tightly regulated in many Asian countries (including Indonesia, Japan, Thailand, Malaysia, Myanmar, Vietnam, Singapore, Taiwan, South Korea), Australia, and European countries (including the United Kingdom, France, Italy, Ireland, Finland, Romania, Denmark, Lithuania, Poland, Russia, Lithuania, Latvia). The status of kratom regulations in most African nations is not well known. However, kratom is mostly unregulated in countries in South America, Central America, and the Caribbean. In North America, kratom is unregulated in Canada and Mexico. In the United States, kratom is unregulated in most states, although a number of states and even some counties have passed or attempted to pass legislation to regulate and/or ban kratom.

In the United States, kratom is legal at the federal level but regulated at the state level in a patchwork of laws. The US Food and Drug Administration (FDA) introduced import restrictions on kratom in 2014 based on uncertainty relating to regulatory status and availability of safety data, although it should be noted that for the period from 2008 through March 2015, the FDA MedWatch system did not receive any serious adverse events (US FDA, 2014). Prior to 2016, within the United States, only five states controlled kratom, and the DEA listed kratom only as a drug of concern (Griffin & Webb, 2018). That began to change on July 29, 2016, with a report disseminated by the Centers for Disease Control and Prevention (CDC). Within that report, Anwar et al. (2016) noted that, during the time period of 2010–2015, poison control centers in the United States received 660 calls related to kratom exposures.

On August 30, 2016, the DEA announced its intention to place the chemicals mitragynine and 7-HMG into Schedule I of the CSA. However, on October 12, 2016, the DEA withdrew the intent to schedule notice from the federal register and stated their intent to gather more input from interested parties before making any further decisions. To date, kratom remains unscheduled in the United States at the federal level.

Conclusions

Although kratom has been cultivated and used for centuries, this is a compound that we are only beginning to understand. Its potential dual effects of stimulation and analgesia make this a unique compound. The wider availability of kratom and growth of kratom use over the past decade globally beyond its geographic roots in Southeast Asia has piqued the interest of both the general public and scientists. The historical and contemporary use of kratom for so many health conditions points toward some potential medicinal value that must be examined, while the nascent data on potential harms must also be considered. These emerging lines of evidence suggest that both caution and science are needed for kratom.

References

Ahmad, K., & Aziz, Z. (2012). Mitragyna speciosa use in the northern states of Malaysia: A cross-sectional study. *Journal of Ethnopharmacology, 141*(1), 446–450.

Anwar, M., Law, R., & Schier, J. (2016). Notes from the field: Kratom (Mitragyna speciosa) exposures reported to poison centers—United States, 2010–2015. *MMWR. Morbidity and Mortality Weekly Report, 65*(29), 748–749.

Assanangkornchai, S., Muekthong, A., Sam-Angsri, N., & Pattanasattayawong, Y. (2006). The use of mitragynine **speciosa** ("krathom"), an addictive plant, in Thailand. *Substance Use & Misuse, 42*, 2145–2157.

Assanangkornchai, S., Muekthong, A., Sam-Angsri, N., & Pattanasattayawong, U. (2007). The use of Mitragynine speciosa ("Krathom"), an addictive plant, in Thailand. *Substance Use & Misuse, 42*(14), 2145–2157.

Babu, K. M., McCurdy, C. R., & Boyer, E. W. (2008). Opioid receptors and legal highs: Salvia divinorum and Kratom. *Clinical Toxicology, 46*(2), 146–152.

Beckett, A. H., Shellard, E. J., Phillipson, J. D., & Lee, C. M. (1966). The Mitragyna species of Asia. VII. Indole alkaloids from the leaves of Mitragyna speciosa Korth. *Planta Medica, 14*(3), 277–288. https://doi.org/10.1055/s-0028-1100055

Bowe, A., & Kerr, P. L. (2020). A complex case of kratom dependence, depression, and chronic pain in opioid use disorder: Effects of buprenorphine in clinical management. *Journal of Psychoactive Drugs, 52*(5), 447–452.

Boyer, E. W., Babu, K. M., Adkins, J. E., McCurdy, C. R., & Halpern, J. H. (2008). Self-treatment of opioid withdrawal using kratom (Mitragynia speciosa korth). *Addiction (Abingdon, England), 103*(6), 1048–1050. https://doi.org/10.1111/j.1360-0443.2008.02209.x

Braley, C., & Hondrogiannis, E. M. (2020). Differentiation of commercially available kratom by purported country of origin using inductively coupled plasma–mass spectrometry. *Journal of Forensic Sciences, 65*(2), 428–437.

Brown, P. N., Lund, J. A., & Murch, S. J. (2017). A botanical, phytochemical and ethnomedicinal review of the genus Mitragyna korth: Implications for products sold as kratom. *Journal of Ethnopharmacology, 202*, 302–325. https://doi.org/10.1016/j.jep.2017.03.020

Burkill, I. H. (1935). *A Dictionary of the economic products of the malay peninsula, 2, 1480–1483.*

Burkill, I. H., & Haniff, M. (1930). Malay village medicine. *The Gardens' Bulletin Straits Settlements, 6*, 165–207.

Carroll, J. J., Marshall, B. D., Rich, J. D., & Green, T. C. (2017). Exposure to fentanyl-contaminated heroin and overdose risk among illicit opioid users in Rhode Island: A mixed methods study. *International Journal of Drug Policy, 46*, 136–145.

Chan, K. B., Pakiam, C., & Rahim, R. A. (2005). Psychoactive plant abuse: The identification of mitragynine in ketum and in ketum preparations. *Bulletin on Narcotics, 57*(1–2), 249–256.

Chear, N. J. Y., Leó̀n, F., Sharma, A., Kanumuri, S. R. R., Zwolinski, G., Abboud, K. A., et al. (2021). Exploring the chemistry of alkaloids from Malaysian Mitragyna speciosa (Kratom) and the role of oxindoles on human opioid receptors. *Journal of Natural Products, 84*(4), 1034–1043.

Chittrakarn, S., Keawwongsri, P., Kumarnsit, E., Sawangjaroen, K., & Keawpradub, N. (2005). Botanical description and pharmacological activity. In S. Assanangkornchai, A. Sirivongs, & N. Ayudhya (Eds.), *Krathom plant in Thai society: Culture, behavior, health, science, laws* (pp. 97–112). Ministry of Justice.

Chittrakarn, S., Keawpradub, N., Sawangjaroen, K., Kansenalak, S., & Janchawee, B. (2010). The neuromuscular blockadeproduced by pure alkaloid, mitragynine and methanol extract of kratom leaves (Mitragyna speciosa Korth.). *Journal of Ethnopharmacology, 129*, 344e9.

Cinosi, E., Martinotti, G., Simonato, P., Singh, D., Demetrovics, Z., Roman-Urrestarazu, A., et al. (2015). Following "the roots" of Kratom (Mitragyna speciosa): The evolution of an enhancer from a traditional use to increase work and productivity in Southeast Asia to a recreational psychoactive drug in western countries. *BioMed Research International, 2015*(968786), 1–11.

Davidson, L., Rawat, M., Stojanovski, S., & Chandrasekharan, P. (2019). Natural drugs, not so natural effects: Neonatal abstinence syndrome secondary to 'kratom'. *Journal of Neonatal-Perinatal Medicine, 12*(1), 109–112.

Eggleston, W., Stoppacher, R., Suen, K., Marraffa, J. M., & Nelson, L. S. (2019). Kratom use and toxicities in the united states. *Pharmacotherapy, 39*(7), 775–777. https://doi.org/10.1002/phar.2280

Eldridge, W. B., Foster, C., & Wyble, L. (2018). Neonatal abstinence syndrome due to maternal kratom use. *Pediatrics, 142*(6), e20181839.

European Monitoring Centre on Drugs and Drug Addiction. (2021). *Kratom drug profile*. Retrieved from: https://www.emcdda.europa.eu/publications/drug-profiles/kratom

Fogh, J., Wright, J. C., & Loveless, J. D. (1977). Absence of HeLa cell contamination in 169 cell lines derived from human tumours. *Journal of the National Cancer Institute, 21*, 393–408.

Forrester, M. B. (2013). Kratom exposures reported to Texas poison centers. *Journal of Addictive Diseases, 32*(4), 396–400.

Giovannitti, J. A., Jr., Thoms, S. M., & Crawford, J. J. (2015). Alpha-2 adrenergic receptor agonists: A review of current clinical applications. *Anesthesia Progress, 62*(1), 31–38.

Gong, F., Gu, H. P., Xu, Q. T., & Kang, W. Y. (2012). Genus Mitragyna: Ethnomedicinal uses and pharmacological studies. *Phytopharmacology, 3*(2), 263–272.

Grewal, K. S. (1932). Observations OX the pharmacology of Mitragynine. *Journal of Pharmacology and Experimental Therapeutics, 46*(3), 251–271.

Griffin, O. H., & Webb, M. E. (2018). The scheduling of kratom and selective use of data. *Journal of Psychoactive Drugs, 50*(2), 114–120.

Grundmann, O. (2017). Patterns of kratom use and health impact in the US—Results from an online survey. *Drug and Alcohol Dependence, 176*, 63–70.

Han, C., Schmitt, J., & Gilliland, K. M. (2019). DARK classics in chemical neuroscience: kratom. *ACS Chemical Neuroscience, 11*(23), 3870–3880.

Hanna, J. (2012). Bogus Kratom market exposed. *Entheogen Review, 12*(1), 26–28.

Hassan, Z., Muzaimi, M., Navaratnam, V., Yusoff, N. H., Suhaimi, F. W., Vadivelu, R., Vicknasingam, B. K., Amato, D., von Hörsten, S., Ismail, N. I., Jayabalan, N., Hazim, A. I., Mansor, S. M., & Müller, C. P. (2013). From Kratom to mitragynine and its derivatives: physiological and behavioural effects related to use, abuse, and addiction. *Neuroscience and biobehavioral reviews, 37*(2), 138–151. https://doi.org/10.1016/j.neubiorev.2012.11.012

Hazim, A. I., Ramanathan, S., Parthasarathy, S., Muzaimi, M., & Mansor, S. M. (2014). Anxiolytic-like effects of mitragynine in the open-field and elevated plus-maze tests in rats. *The Journal of Physiological Sciences, 64*(3), 161–169.

Hemby, S. E., McIntosh, S., Leon, F., Cutler, S. J., & McCurdy, C. R. (2019). Abuse liability and therapeutic potential of the Mitragyna speciosa (kratom) alkaloids mitragynine and 7-hydroxymitragynine. *Addiction Biology, 24*(5), 874–885. https://doi.org/10.1111/adb.12639

Henningfield, J. E., Fant, R. V., & Wang, D. W. (2018). The abuse potential of kratom according the 8 factors of the controlled substances act: Implications for regulation and research. *Psychopharmacology, 235*(2), 573–589.

Holler, J. M., Vorce, S. P., McDonough-Bender, P. C., Magluilo, J., Jr., Solomon, C. J., & Levine, B. (2011). A drug toxicity death involving propylhexedrine and mitragynine. *Journal of Analytical Toxicology, 35*(1), 54–59.

Hillebrand, J., Olszewski, D., & Sedefov, R. (2010). Legal highs on the Internet. *Substance use & misuse, 45*(3), 330–340. https://doi-org.wvu.idm.oclc.org/10.3109/10826080903443628

Horie, S., Koyama, F., Takayama, H., Ishikawa, H., Aimi, N., Ponglux, D., et al. (2005). Indole alkaloids of a Thai medicinal herb, Mitragyna speciosa, that has opioid agonistic effect in Guinea-pig ileum. *Planta Medica, 71*(03), 231–236.

Idayu, N. F., Hidayat, M. T., Moklas, M. A. M., Sharida, F., Raudzah, A. N., Shamima, A. R., & Apryani, E. (2011). Antidepressant-like effect of mitragynine isolated from Mitragyna speciosa Korth in mice model of depression. *Phytomedicine, 18*(5), 402–407.

Ingsathit, A., Woratanarat, P., Anukarahanonta, T., Rattanasiri, S., Chatchaipun, P., Wattayakorn, K., et al. (2009). Prevalence of psychoactive drug use among drivers in Thailand: A roadside survey. *Accident Analysis & Prevention, 41*(3), 474–478.

Jagabalan, J. Y., Murugaiyah, V., Zainal, H., Mansor, S. M., & Ramanathan, S. (2019). Intestinal permeability of mitragynine in rats using in situ absorption model. *Journal of Asian Natural Products Research, 21*(4), 351–363.

Kansy, M. (1998). Physicochemical high throughput screening: Parallel artificial membrane permeation assay in the description of passive absorption processes. *Journal of Medicinal Chemistry, 41*, 1007–1010.

Kapp, F. G., Maurer, H. H., Auwärter, V., Winkelmann, M., & Hermanns-Clausen, M. (2011). Intrahepatic cholestasis following abuse of powdered kratom (Mitragyna speciosa). *Journal of Medical Toxicology, 7*(3), 227–231.

Karinen, R., Fosen, J. T., Rogde, S., & Vindenes, V. (2014). An accidental poisoning with mitragynine. *Forensic Science International, 245*, e29–e32.

Kerrigan, S., & Basiliere, S. (2022). Kratom: A systematic review of toxicological issues. *Wiley Interdisciplinary Reviews: Forensic Science, 4*(1), 1–29.

Kerschgens, I. P., Claveau, E., Wanner, M. J., Ingemann, S., van Maarseveen, J. H., & Hiemstra, H. (2012). Total syntheses of mitragynine, paynantheine and speciogynine via an enantioselective thiourea-catalysed Pictet–Spengler reaction. *Chemical Communications, 48*(100), 12243–12245.

Kong, W. M., Chik, Z., Ramachandra, M., Subramaniam, U., Aziddin, R. E. R., & Mohamed, Z. (2011). Evaluation of the effects of Mitragyna speciosa alkaloid extract on cytochrome P450 enzymes using a high throughput assay. *Molecules, 16*(9), 7344–7356.

Kong, W. M., Mohamed, Z., Alshawsh, M. A., & Chik, Z. (2017). Evaluation of pharmacokinetics and blood-brain barrier permeability of mitragynine using in vivo microdialysis technique. *Journal of Pharmaceutical and Biomedical Analysis, 143*, 43–47.

Kronstrand, R., Roman, M., Thelander, G., & Eriksson, A. (2011). Unintentional fatal intoxications with mitragynine and O-desmethyltramadol from the herbal blend Krypton. *Journal of Analytical Toxicology, 35*(4), 242–247.

Kruegel, A. C., & Grundmann, O. (2018). The medicinal chemistry and neuropharmacology of kratom: A preliminary discussion of a promising medicinal plant and analysis of its potential for abuse. *Neuropharmacology, 134*, 108–120.

Kruegel, A. C., Uprety, R., Grinnell, S. G., Langreck, C., Pekarskaya, E. A., Le Rouzic, V., Ansonoff, M., Gassaway, M. M., Pintar, J. E., Pasternak, G. W., Javitch, J. A., Majumdar, S., & Sames, D. (2019). 7-Hydroxymitragynine is an active metabolite of mitragynine and a key mediator of its analgesic effects. *ACS Central Science, 5*(6), 992–1001. https://doi.org/10.1021/acscentsci.9b00141

Kumarnsit, E., Keawpradub, N., & Nuankaew, W. (2007). Effect of Mitragyna speciosa aqueous extract on ethanol withdrawal symptoms in mice. *Fitoterapia, 78*(3), 182–185. https://doi.org/10.1016/j.fitote.2006.11.012

Kupferschmidt, H. (2011, July). Toxic hepatitis after Kratom (Mitragyna sp.) consumption. *Clinical Toxicology, 49*(6), 532–532.

Lee, C. T. (1957). Addiction to **Mitragyna speciosa**. Malaya, University Malaya, Kuala Lumpur. *Proceedings of the Alumni Association, 10*, 322–324.

León, F., Habib, E., Adkins, J. E., Furr, E. B., McCurdy, C. R., & Cutler, S. J. (2009). Phytochemical characterization of the leaves of Mitragyna speciosa grown in USA. *Natural Product Communications, 4*(7), 907–910.

Limsuwanchote, S., Wungsintaweekul, J., Keawpradub, N., Putalun, W., Morimoto, S., & Tanaka, H. (2014). Development of indirect competitive ELISA for quantification of mitragynine in Kratom (Mitragyna speciosa (Roxb.) Korth.). *Forensic Science International, 244*, 70–77.

Liu, H., McCurdy, C. R., & Doerksen, R. J. (2010). Computational study on the conformations of mitragynine and mitragynaline. *Journal of Molecular Structure: THEOCHEM, 945*(1–3), 57–63.

Ma, J., Yin, W., Zhou, H., & Cook, J. M. (2007). Total synthesis of the opioid agonistic indole alkaloid mitragynine and the first total syntheses of 9-methoxygeissoschizol and 9-methoxy-N b-methylgeissoschizol. *Organic Letters, 9*(18), 3491–3494.

Mackay, L., & Abrahams, R. (2018). Novel case of maternal and neonatal kratom dependence and withdrawal. *Canadian Family Physician, 64*(2), 121–122.

Macko, E., Weisbach, J. A., & Douglas, B. (1972). Some observations on the pharmacology of mitragynine. *Archives internationales de pharmacodynamie et de therapie, 198*(1), 145–161.

Macmadu, A., Carroll, J. J., Hadland, S. E., Green, T. C., & Marshall, B. D. (2017). Prevalence and correlates of fentanyl-contaminated heroin exposure among young adults who use prescription opioids non-medically. *Addictive Behaviors, 68*, 35–38.

Manda, V. K., Avula, B., Ali, Z., Khan, I. A., Walker, L. A., & Khan, S. I. (2014). Evaluation of in vitro absorption, distribution, metabolism, and excretion (ADME) properties of mitragynine, 7-hydroxymitragynine, and mitraphylline. *Planta Medica, 80*(07), 568–576.

Manda, V. K., Avula, B., Dale, O. R., Ali, Z., Khan, I. A., Walker, L. A., & Khan, S. I. (2017). PXR mediated induction of CYP3A4, CYP1A2, and P-gp by Mitragyna speciosa and its alkaloids. *Phytotherapy Research, 31*(12), 1935–1945.

Maruyama, T., Kawamura, M., Kikura-Hanajiri, R., Takayama, H., & Goda, Y. (2009). The botanical origin of kratom (Mitragyna speciosa; Rubiaceae) available as abused drugs in the Japanese markets. *Journal of Natural Medicines, 63*(3), 340–344.

Matsumoto, K., Mizowaki, M., Suchitra, T., Murakami, Y., Takayama, H., Sakai, S. I., et al. (1996a). Central antinociceptive effects of mitragynine in mice: Contribution of descending noradrenergic and serotonergic systems. *European Journal of Pharmacology, 317*(1), 75–81.

Matsumoto, K., Mizowaki, M., Suchitra, T., Takayama, H., Sakai, S. I., Aimi, N., & Watanabe, H. (1996b). Antinociceptive action of mitragynine in mice: Evidence for the involvement of supraspinal opioid receptors. *Life Sciences, 59*(14), 1149–1155.

Matsumoto, K., Horie, S., Takayama, H., Ishikawa, H., Aimi, N., Ponglux, D., et al. (2005). Antinociception, tolerance and withdrawal symptoms induced by 7-hydroxymitragynine, an alkaloid from the Thai medicinal herb Mitragyna speciosa. *Life Sciences, 78*(1), 2–7.

McIntyre, I. M., Trochta, A., Stolberg, S., & Campman, S. C. (2015). Mitragynine 'Kratom'related fatality: A case report with postmortem concentrations. *Journal of Analytical Toxicology, 39*(2), 152–155.

Meireles, V., Rosado, T., Barroso, M., Soares, S., Gonçalves, J., Luís, Â., et al. (2019). Mitragyna speciosa: Clinical, toxicological aspects and analysis in biological and non-biological samples. *Medicine, 6*(1), 35–56.

Meyer, M. R., Wagmann, L., Schneider-Daum, N., Loretz, B., de Souza Carvalho, C., Lehr, C. M., & Maurer, H. H. (2015). P-glycoprotein interactions of novel psychoactive substances–stimulation of ATP consumption and transport across Caco-2 monolayers. *Biochemical Pharmacology, 94*(3), 220–226.

Moklas, M. A. M., Suliman, N. A., Taib, C. N. M., Hidayat, M. T., Baharuldin, S. F., Zakaria, F. N., et al. (2013). Sedative, cognitive impairment and anxiolytic effects of acute Mitragyna Speciosa in rodents. *Differences, 22*(11), 313–317.

Mossadeq, W. S., Sulaiman, M. R., Mohamad, T. T., Chiong, H. S., Zakaria, Z. A., Jabit, M. L., et al. (2009). Anti-inflammatory and antinociceptive effects of Mitragyna speciosa Korth methanolic extract. *Medical Principles and Practice, 18*(5), 378–384.

Neerman, M. F., Frost, R. E., & Deking, J. (2013). A drug fatality involving Kratom. *Journal of Forensic Sciences, 58*, S278–S279.

Ottaviani, G., Martel, S., & Carrupt, P. A. (2006). Parallel artificial membrane permeability assay: A new membrane for the fast prediction of passive human skin permeability. *Journal of Medicinal Chemistry, 49*(13), 3948–3954.

Papsun, D. M., Chan-Hosokawa, A., Friederich, L., Brower, J., Graf, K., & Logan, B. (2019). The trouble with kratom: Analytical and interpretative issues involving mitragynine. *Journal of Analytical Toxicology, 43*(8), 615–629.

Philipp, A. A., Wissenbach, D. K., Zoerntlein, S. W., Klein, O. N., Kanogsunthornrat, J., & Maurer, H. H. (2009). Studies on the metabolism of mitragynine, the main alkaloid of the erbal drug Kratom, in rat and human urine using liquid chromatography-linear ion trap mass spectrometry. *Journal of Mass Spectrometry, 44*(8), 1249–1261.

Philipp, A. A., Wissenbach, D. K., Weber, A. A., Zapp, J., & Maurer, H. H. (2010). Phase I and II metabolites of speciogynine, a diastereomer of the main Kratom alkaloid mitragynine, identified in rat and human urine by liquid chromatography coupled to low- and high-resolution linear ion trap mass spectrometry. *Journal of Mass Spectrometry : JMS, 45*(11), 1344–1357. https://doi.org/10.1002/jms.1848

Post, S., Spiller, H. A., Chounthirath, T., & Smith, G. A. (2019). Kratom exposures reported to United States poison control centers: 2011–2017. *Clinical Toxicology, 57*(10), 847–854.

Prozialeck, W. C., Jivan, J. K., & Andurkar, S. V. (2012). Pharmacology of kratom: an emerging botanical agent with stimulant, analgesic and opioid-like effects. *The Journal of the American Osteopathic Association, 112*(12), 792–799.

Raffa, R. B. (Ed.). (2014). *Kratom and other mitragynines: the chemistry and pharmacology of opioids from a non-opium source*. CRC Press.

Raffa, R. B., Beckett, J. R., Brahmbhatt, V. N., Ebinger, T. M., Fabian, C. A., Nixon, J. R., Orlando, S. T., Rana, C. A., Tejani, A. H., & Tomazic, R. J. (2015). Short overview of Mitragynines. In R. B. Raffa (Ed.), *Kratom and other mitragynines: the chemistry and pharmacology of opioids from a non-opium source (pp. 9–21). CRC Press.*

Raffa, R. B., Pergolizzi, J. V., Taylor, R., Ossipov, M. H., & NEMA Research Group. (2018). Nature's first "atypical opioids": Kratom and mitragynines. *Journal of Clinical Pharmacy and Therapeutics, 43*(3), 437–441.

Ramanathan, S., Parthasarathy, S., Murugaiyah, V., Magosso, E., Tan, S. C., & Mansor, S. M. (2015). Understanding the physicochemical properties of mitragynine, a principal alkaloid of Mitragyna speciosa, for preclinical evaluation. *Molecules, 20*(3), 4915–4927.

Reanmongkol, W., Keawpradub, N., & Sawangjaroen, K. (2007). Effects of the extracts from Mitragyna speciosa Korth. leaves on analgesic and behavioral activities in experimental animals. *Songklanakarin Journal of Science and Technology, 29*(Suppl 1), 39–48.

Roche, K. M., Hart, K., Sangalli, B., Lefberg, J., & Bayer, M. (2008, January). Kratom: A case of a legal high. *Clinical Toxicology, 46*(7), 598–598.

Rosenbaum, C. D., Carreiro, S. P., & Babu, K. M. (2012). Here today, gone tomorrow… and back again? A review of herbal marijuana alternatives (K2, Spice), synthetic cathinones (bath salts), kratom, Salvia divinorum, methoxetamine, and piperazines. *Journal of Medical Toxicology, 8*(1), 15–32.

Sabetghadam, A., Ramanathan, S., & Mansor, S. M. (2010). The evaluation of antinociceptive activity of alkaloid, methanolic, and aqueous extracts of Malaysian Mitragyna speciosa Korth leaves in rats. *Pharmacognosy Research, 2*(3), 181–185.

Sabetghadam, A., Navaratnam, V., & Mansor, S. M. (2013). Dose–response relationship, acute toxicity, and therapeutic index between the alkaloid extract of M itragyna speciosa and its main active compound mitragynine in mice. *Drug Development Research, 74*(1), 23–30.

Saingam, D., Assanangkornchai, S., Geater, A. F., & Balthip, Q. (2013). Pattern and consequences of krathom (Mitragyna speciosa Korth.) use among male villagers in southern Thailand: A qualitative study. *International Journal of Drug Policy, 24*(4), 351–358.

Saingam, D., Assanangkornchai, S., Geater, A. F., & Lerkiatbundit, S. (2016). Factor analytical investigation of Krathom (Mitragyna speciosa Korth.) withdrawal syndrome in Thailand. *Journal of Psychoactive Drugs, 48*(2), 76–85.

Schmidt, M. M., Sharma, A., Schifano, F., & Feinmann, C. (2011). "Legal highs" on the net—Evaluation of UK-based websites, products and product information. *Forensic Science International, 206*(1–3), 92–97.

Scott, T. M., Yeakel, J. K., & Logan, B. K. (2014). Identification of mitragynine and O-desmethyltramadol in Kratom and legal high products sold online. *Drug Testing and Analysis, 6*(9), 959–963.

Shellard, E. J. (1989). Ethnopharmacology of kratom and the Mitragyna alkaloids. *Journal of Ethnopharmacology, 25*(1), 123–124.

Singh, D., Müller, C. P., & Vicknasingam, B. K. (2014). Kratom (Mitragyna speciosa) dependence, withdrawal symptoms and craving in regular users. *Drug and Alcohol Dependence, 139*, 132–137.

Singh, D., Narayanan, S., & Vicknasingam, B. (2016). Traditional and non-traditional uses of Mitragynine (Kratom): A survey of the literature. *Brain Research Bulletin, 126*, 41–46.

Singh, D., Narayanan, S., Vicknasingam, B., Corazza, O., Santacroce, R., & Roman-Urrestarazu, A. (2017). Changing trends in the use of kratom (Mitragyna speciosa) in Southeast Asia. Abstract. *Human Psychopharmacology: Clinical and Experimental, 32*(3). https://doi.org/10.1002/hup.2582

Singh, D., Brown, P. N., Cinosi, E., Corazza, O., Henningfield, J. E., Garcia-Romeu, A., et al. (2020). Current and future potential impact of COVID-19 on Kratom (Mitragyna speciosa Korth) supply and use. *Frontiers in Psychiatry, 11*, 1355.

Stolt, A. C., Schröder, H., Neurath, H., Grecksch, G., Höllt, V., Meyer, M. R., et al. (2014). Behavioral and neurochemical characterization of kratom (Mitragyna speciosa) extract. *Psychopharmacology, 231*(1), 13–25.

Sun, X., & Ma, D. (2011). Organocatalytic approach for the syntheses of corynantheidol, dihydrocorynantheol, protoemetinol, protoemetine, and mitragynine. *Chemistry–An Asian Journal, 6*(8), 2158–2165.

Suwanlert, S. (1975). A study of kratom eaters in Thailand. *Bulletin on Narcotics, 27*(3), 21–27.

Swogger, M. T., Hart, E., Erowid, F., Erowid, E., Trabold, N., Yee, K., et al. (2015). Experiences of kratom users: A qualitative analysis. *Journal of Psychoactive Drugs, 47*(5), 360–367.

Takayama, H. (2004). Chemistry and pharmacology of analgesic indole alkaloids from the rubiaceous plant, Mitragyna speciosa. *Chemical and Pharmaceutical Bulletin, 52*(8), 916–928.

Takayama, H., Maeda, M., Ohbayashi, S., Kitajima, M., Sakai, S. I., & Aimi, N. (1995). The first total synthesis of (−)-mitragynine, an analgesic indole alkaloid in Mitragyna speciosa. *Tetrahedron Letters, 36*(51), 9337–9340.

Takayama, H., Ishikawa, H., Kurihara, M., Kitajima, M., Aimi, N., Ponglux, D., et al. (2002). Studies on the synthesis and opioid agonistic activities of mitragynine-related indole alkaloids: Discovery of opioid agonists structurally different from other opioid igands. *Journal of Medicinal Chemistry, 45*(9), 1949–1956.

Tanguay, P. (2011). Kratom in Thailand: Decriminalisation and community control. *Transnational Institute/International Drug Policy Consortium Series on Legislative Reform of Drug Policies, 13*, 1–16.

Taufik Hidayat, M., Apryani, E., Nabishah, B. M., Moklas, M. A. A., Sharida, F., & Farhan, M. A. (2010). Determination of mitragynine bound opioid receptors. *Advances in Medical and Dental Sciences, 3*(3), 65–70.

Thongpradichote, S., Matsumoto, K., Tohda, M., Takayama, H., Aimi, N., Sakai, S., & Watanabe, H. (1998). Identification of opioid receptor subtypes in antinociceptive actions of supraspinally-administered mitragynine in mice. *Life sciences, 62*(16), 1371–1378. https://doi.org/10.1016/s0024-3205(98)00075-7

Tohda, M., Thongpraditchote, S., Matsumoto, K., Murakami, Y., Sakai, S. I., Aimi, N., et al. (1997). Effects of mitragynine on cAMP formation mediated by δ-opiate receptors in NG108-15 cells. *Biological and Pharmaceutical Bulletin, 20*(4), 338–340.

Trakulsrichai, S., Tongpo, A., Sriapha, C., Wongvisawakorn, S., Rittilert, P., Kaojarern, S., & Wananukul, W. (2013). Kratom abuse in Ramathibodi Poison Center, Thailand: A five-year experience. *Journal of Psychoactive Drugs, 45*(5), 404–408.

Trakulsrichai, S., Sathirakul, K., Auparakkitanon, S., Krongvorakul, J., Sueajai, J., Noumjad, N., et al. (2015). Pharmacokinetics of mitragynine in man. *Drug Design, Development and Therapy, 9*, 2421.

Tsuchiya, S., Miyashita, S., Yamamoto, M., Horie, S., Sakai, S. I., Aimi, N., et al. (2002). Effect of mitragynine, derived from Thai folk medicine, on gastric acid secretion through opioid receptor in anesthetized rats. *European Journal of Pharmacology, 443*(1–3), 185–188.

Tungtananuwat, W., & Lawanprasert, S. (2010). Fatal 4x100; home-made kratom juice cocktail. *Journal of Health Research, 24*(1), 43–47.

Ujváry, I. (2014). Psychoactive natural products: Overview of recent developments. *Annali dell'Istituto superiore di sanita, 50*, 12–27.

Ulbricht, C., Costa, D., Dao, J., Isaac, R., LeBlanc, Y. C., Rhoades, J., & Windsor, R. C. (2013). An evidence-based systematic review of kratom (Mitragyna speciosa) by the natural standard research collaboration. *Journal of Dietary Supplements, 10*(2), 152–170.

Utar, Z., Majid, M. I. A., Adenan, M. I., Jamil, M. F. A., & Lan, T. M. (2011). Mitragynine inhibits the COX-2 mRNA expression and prostaglandin E2 production induced by lipopolysaccharide in RAW264. 7 macrophage cells. *Journal of Ethnopharmacology, 136*(1), 75–82.

Váradi, A., Marrone, G. F., Palmer, T. C., Narayan, A., Szabó, M. R., Le Rouzic, V., et al. (2016). Mitragynine/corynantheidine pseudoindoxyls as opioid analgesics with mu agonism and delta antagonism, which do not recruit β-arrestin-2. *Journal of Medicinal Chemistry, 59*(18), 8381–8397. https://doi.org/10.1021/acs.jmedchem.6b00748

Vicknasingam, B., Narayanan, S., Beng, G. T., & Mansor, S. M. (2010). The informal use of ketum (Mitragyna speciosa) for opioid withdrawal in the northern states of peninsular Malaysia and implications for drug substitution therapy. *International Journal of Drug Policy, 21*(4), 283–288.

Ward, J., Rosenbaum, C., Hernon, C., McCurdy, C. R., & Boyer, E. W. (2011). Herbal medicines for the management of opioid addiction. *CNS Drugs, 25*(12), 999–1007.

Warner, M. L., Kaufman, N. C., & Grundmann, O. (2016). The pharmacology and toxicology of kratom: from traditional herb to drug of abuse. *International Journal of Legal Medicine, 130*(1), 127–138. https://doi-org.wvu.idm.oclc.org/10.1007/s00414-015-1279-y.

Watanabe, K., Yano, S., Horie, S., & Yamamoto, L. T. (1997). Inhibitory effect of mitragynine, an alkaloid with analgesic effect from Thai medicinal plant Mitragyna speciosa, on electrically stimulated contraction of isolated Guinea-pig ileum through the opioid receptor. *Life Sciences, 60*(12), 933–942.

Yamamoto, L. T., Horie, S., Takayama, H., Aimi, N., Sakai, S., Yano, S., Shan, J., Pang, P. K., Ponglux, D., & Watanabe, K. (1999). Opioid receptor agonistic characteristics of mitragynine pseudoindoxyl in comparison with mitragynine derived from Thai medicinal plant Mitragyna speciosa. *General Pharmacology, 33*(1), 73–81. https://doi.org/10.1016/s0306-3623(98)00265-1

Yusoff, N. H., Suhaimi, F. W., Vadivelu, R. K., Hassan, Z., Rümler, A., Rotter, A., et al. (2016). Abuse potential and adverse cognitive effects of mitragynine (kratom). *Addiction Biology, 21*(1), 98–110.

Zacharias, D. E., Rosenstein, R. D., & Jeffrey, G. A. (1965). The structure of mitragynine hydroiodide. *Acta Crystallographica, 18*(6), 1039–1043.

Endogenous Opioids in the Homeostatic Regulation of Hunger, Satiety, and Hedonic Eating: Neurobiological Foundations

Marcela Rodriguez Flores and Sylvana Stephano Zúñiga

Abstract This chapter (part one of a trilogy) summarizes the neurobiological foundations of endogenous opioids in the regulation of energy balance and eating behavior, dysregulation of which translates to maladaptive dietary responses in individuals with obesity and eating disorders, including anorexia, bulimia, and binge eating disorder. Knowledge of these neurobiological foundations is vital to researchers' and clinicians' understanding of pathophysiology as well as the science-based development of multidisciplinary diagnoses and treatments for obesity and eating disorders. We highlight mechanisms of endogenous opioids in both homeostatic and hedonic feeding behavior, review research on the dysregulation of food reward that plays a role in a wide array of obesity and disordered eating, and the clinical implications of neurobiological responses to food for current science-based treatments for obesity and eating disorders.

Keywords Endogenous opioids · Eating disorders · Eating · Hunger · Obesity

Homeostatic Regulation of Hunger and Satiety

Food is essential for survival; its purpose is to satisfy nutritional needs. Several mechanisms regulate the way we eat. First, this faculty requires an efficient system of starting and braking to promote nutrition and avoid consumption in certain circumstances. However, eating behaviors also comprise complex phenomena

---

M. R. Flores (✉)
Obesity and Eating Disorders Clinic, Instituto Nacional de Ciencias Médicas y Nutrición Salvador Zubirán, Tlalpan, Mexico

Department of Brain Sciences, Imperial College London, Hammersmith Hospital, London, UK

S. S. Zúñiga
Obesity and Eating Disorders Clinic, Instituto Nacional de Ciencias Médicas y Nutrición Salvador Zubirán, Tlalpan, Mexico

P. L. Kerr et al. (eds.), *Endogenous Opioids*, Advances in Neurobiology 35,
https://doi.org/10.1007/978-3-031-45493-6_16

beyond the basic nutritional needs. The initiation and interruption of feeding integrates the neuroendocrine pathways of hunger and satiety with signals of reward, memory, decision-making, and learning by reciprocal connections in the Central Nervous System (CNS).

The most important brain areas where crucial energy homeostasis functions lie are found in the brainstem and hypothalamus (10–15% of the neuron population). In the hypothalamus, the paraventricular nucleus (PVN), arcuate nucleus (ARC), ventromedial nucleus (VMN), dorsomedial nucleus (DMN), and lateral hypothalamus (LH) mediate autonomic, hormonal, and metabolic control of hunger and satiety (Yu & Kim, 2012). These systems interact, integrate, and respond to the interoceptive and exteroceptive information that drives us to eat. The homeostatic regulation of food consists of a ceaseless dialogue between the brain and the body, as in the pre- and postprandial states. This dialogue causes that after energy expenditure, hunger increases (orexigenic effect), and after a meal, it is suppressed (anorexigenic effect). In ARC, a subset of neurons express the anorectic peptides proopiomelanocortin (POMC) and cocaine and amphetamine-related transcript (CART), and a group that co-expresses the orexigenic neuropeptide Y (NPY) and Agouti-related peptide (AgRP) (Fig. 1).

POMC neurons send axonal projections to second-order neurons in the PVN and LH, releasing α-melanocyte-stimulating hormone (α-MSH), which binds with melanocortin receptors 3 and 4 (MCR3 and MCR4). These signals have a powerful catabolic effect, reduce food intake, and increase energy expenditure (thermogenesis). AgRP is an endogenous melanocortin receptor antagonist that is released from neurons that produce NPY/AGRP. It stimulates appetite and has anabolic effects through its competition with α-MSH in the MCR3 and MCR4. Serotonin also acts on POMC neurons through its 5-hydroxytryptamine 2C receptor (5-HT-2CR) to promote satiety (Hall, 2011).

Roles of Endogenous Opioids in Eating Behavior and Physiology

Another fragment of POMC, β-endorphin, stimulates food intake due to autoinhibition of POMC (Crespo et al., 2014). Endogenous opioids – endorphins, enkephalins, dynorphins and endomorphines – fulfill their roles through specific receptors: mu (μ-), delta (δ-), and kappa (κ-) opioid receptors. β-endorphin acts via μ-opioid receptors and has the most important role in the regulation of appetite and food intake. Enkephalin receptors are μ-opioid receptors and δ-opioid receptors, dynorphin acts via κ-opioid receptors, and endomorphines via μ-opioid receptors. These receptors can also be activated by the administration of exogenous opiates, such as morphine and its derivatives. Opioid receptors responsible for the modulation of homeostatic signals are located in the hypothalamus (Katarzyna Pomorska et al., 2016).

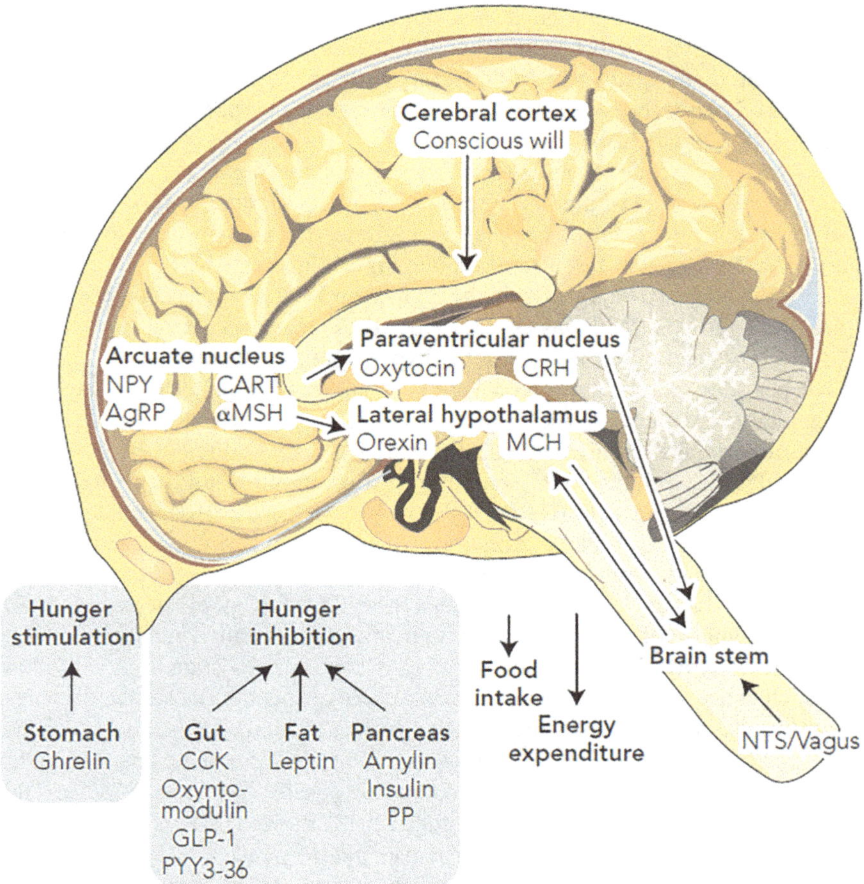

AgRP = agouti-related peptide. CART = cocaine and amphetamine-regulated transcript. CCK = cholecystokinin. CRH = corticotropin-releasing hormone. GLP-1 = glucagon-like peptide. MCH = melanin-concentrating hormone. αMSH = alpha melanocyte-stimulating hormone. NPY = neuropeptide Y. NTS = nucleus of the tractus solitarius. PP = pancreatic polypeptide. PYY = peptide YY. ◆

Fig. 1 Selected pathways involved in body weight regulation. (Adapted from: Proietto, 2011)

In parallel, peripheral regulation of food intake occurs in three main sites (adipose tissue, pancreatic β cells, and cells of the digestive tract) through their metabolic signals: leptin, insulin, intestinal peptide secretion, and activation of the autonomic nervous system.

Leptin is produced by adipocytes in proportion to fat mass. Together with insulin, it provides feedback to the CNS on the status of long-term energy reserves. It activates its receptors in the nucleus accumbens (NAc) in response to food consumption, inhibiting the production of NPY/AgRP and favoring the activation of α-MSH, which stimulates satiety and energy expenditure (Harris, 2013). It also influences other metabolic functions, insulin secretion, and the sympathetic nervous system (SNS; Grassi, 2014; Yau et al., 2014).

In addition to its function as a digestive and absorptive organ, the gastrointestinal tract regulates food consumption by promoting or responding to food intake. The main communication pathway between the gastrointestinal tract and first-order neurons in the NAc occurs through metabolic signals generated by nutrients at the nucleus of the solitary tract (NTS) in the brain stem via the vagus nerve, also known as the brain-gut axis (Williams et al., 2000). These signals slow gastric emptying and activate second-order neurons in the hypothalamus. The best known gastrointestinal peptides, ghrelin, peptide YY (PYY), cholecystokinin (CCK), pancreatic polypeptide (PP), glucagon-like peptide 1 (GLP-1), and oxynytomodulin (OXM) regulate food intake in an acute fashion. Fats, PYY, CCK, PP, GLP-1, and leptin stimulate POMC, inhibit AgRP, amplify MCR4 neurotransmission, and inhibit gamma amino butyric acid (GABA), causing a decrease in food intake and an increase in energy expenditure (Konturek et al., 2003; Vrang et al. 2006). In contrast, ghrelin, released by endocrine cells in the stomach, stimulates the orexigenic pathway by secreting GABA, and activates the NPY pathway by increasing AgRP, an inverse agonist of MCR4, and through interaction with dopamine signaling (Romero-Picó et al., 2013) Physiologically, its concentrations are high while fasting, which increase food intake and decrease energy expenditure, and they decrease after eating. Ghrelin promotes the rewarding effect of foods rich in fat or carbohydrates, and its effects are antagonized by activation of GLP-1 receptors in the CNS (Zwirska-Korczala et al., 2007). Treatments for obesity have tried blocking this pathway without success. Romero-Pico et al. (2013) showed that ghrelin stimulated food intake in rats using the κ-opioid receptor pathway independent of the AMP-activated protein kinase pathway (Romero-Picó et al., 2013). Orexin A is another neuropeptide secreted in the lateral hypothalamus with orexigenic effects. Experimental studies have indicated that its interaction with opioids is involved in both homeostatic and hedonic aspects of food intake (Álvarez-Crespo et al., 2013) (Fig. 2).

Hunger vs. Appetite

The decision to eat is not only influenced by the internal state of metabolic depletion, but by non-homeostatic factors, such as the palatability of food, environmental cues, and everyday sociocultural influences (anticipatory depletion). Hunger can be defined as the demand for calories (such as after starvation), while appetite is the demand for a particular food (what is craved). The former promotes conditioned responses and internal rhythms that generate hormonal, metabolic, and neurological adaptations, including learning and motivation in the latter. These motivated

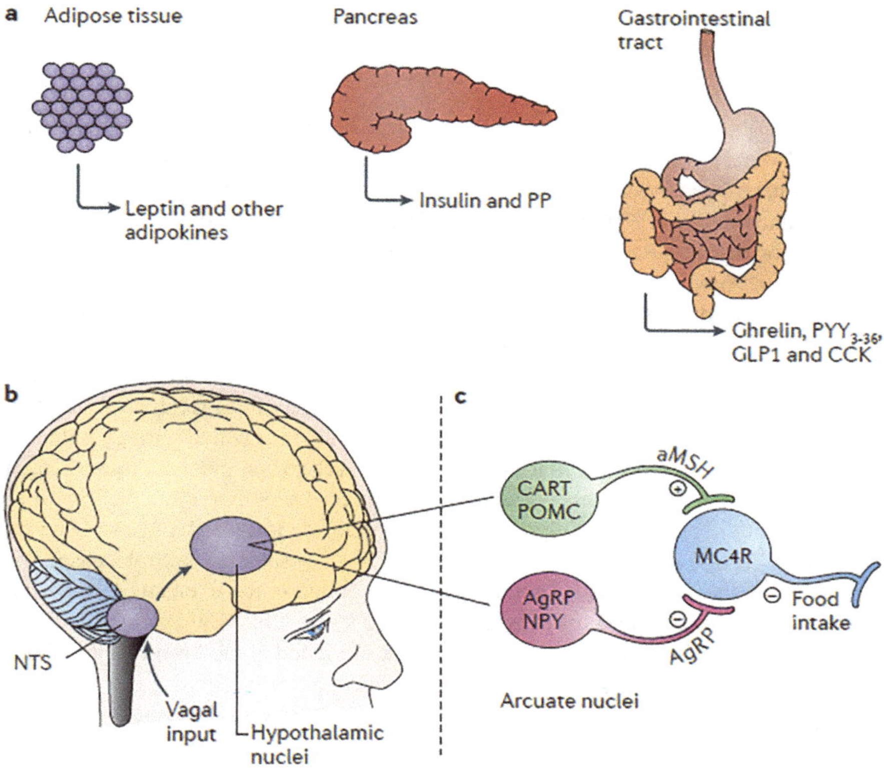

Fig. 2 Overview of homeostatic feeding circuits. (Adapted from: Kenny, 2011)

behaviors are consistent with the nature of the human brain to respond and seek immediate rewards through the dopamine (DA) pathways (Suzuki et al., 2012; Volkow et al., 2011). The availability of a variety of highly palatable foods is a stimulus to consume in response to appetite rather than hunger. Despite homeostatic regulation, when highly appetizing foods are available, not only as a source of nutrition, but as a reward, their consumption is stimulated, and the ability to resist this stimulus relies on self-control. This phenomena is configured individually starting in the fetal period and meshed along an individual's life. The main anatomical areas within the CNS that contain these functions are: The limbic and neocortical areas modulate memory and learning; the ventral tegmental area (VTA), striatum, and substantia nigra generate reward signals; and the prefrontal cortex directs decision-making (Hall, 2011).

Role of Endogenous Opioids as the Backstage of Metabolic Eating

The precise mechanism that governs the effects of opioids on food intake is not completely understood. The effects of endogenous opioids and the role of their receptors have been investigated in experimental and clinical studies, using opioid agonists and opioid receptor antagonists. The administration of opiates, such as morphine or butorphanol in rats has been shown to stimulate food intake (Gosnell et al., 1990). Agonists of opioid receptors, particularly μ- and κ-opioid receptors, have also been found to increase food intake (Gosnell et al., 1985; Morley et al., 1983). Among the opioid receptors, μ-opioid receptors play the most important role in feeding behavior. The activation of μ-opioid receptors increases food intake and may contribute to the development of obesity induced by high fat diet (Katarzyna Pomorska et al., 2016). κ-opioid receptor activation also influences energy balance. It has been shown that stimulation of κ-opioid receptors induced hyperphagia (Cooper et al., 1985). In another study, κ-opioid receptor-deficient mice exposed to a high fat diet did not gain weight and increased their locomotor activity and energy expenditure. κ-opioid receptor (KOR) deficiency was associated with deceased triglycerides synthesis and increased fatty acid β-oxidation in the liver, suggesting that κ-opioid receptor activation may favor fat storage in the liver (Czyzyk et al., 2010). δ-opioid receptor-induced orexigenic effects are mediated at the level of hypothalamus, by D_2 prostaglandin, which stimulate food intake through the activation of NPY system (Katarzyna Pomorska et al., 2016). Even though the complex mechanism of action of opioids is not completely clarified, research has identified that opioids' effects are due to the modulation of the activity of several neuropeptides. Stress and negative affect are reduced by food, and an inability to control stress may lead to increased food intake and obesity (Larsen & Hunter, 2006). Research has found that the endogenous opioid system is involved in stress and affective responses (Larsen & Hunter, 2006; Slochower et al., 1981) and may also mediate consummatory behavior (Berthoud, 2002). Extant data support an interaction between central opioid and melanocortin systems. NPY neurons also interact with the opioid system. The administration of naloxone decreases food intake caused by NPY. The receptors that mediate the action of this neuropeptide are μ-opioid receptors and KORs (reviewed in Nogueiras et al., 2012).

The central opioid and melanocortin systems interact to modulate food intake. Proopiomelanocortin (POMC) encodes for alpha-melanocyte-stimulating hormone (alfa-MSH) and beta-endorphin. POMC neurons express MORs inhibiting appetite; also, the activation of opioid receptors in GABAergic neurons inhibits presynaptic POMC neurons (Nogueiras et al., 2012). Proopiomelanocortin neurons express postsynaptic MORs, which respond to agonist hyperpolarizing and inhibiting action potential firing. Activation of opioid receptors (MOR, KOR, DOR) present in the GABAergic terminal inhibits the presynaptic POMC neurons. Blockade of both MOR and KOR receptors suppresses agouti-related peptide inducing food intake (Snyder & Gao, 2013). Opioid receptor antagonism with naloxone blocks the effect of AgRP, stimulating feeding responses (Saper et al., 2002).

Hedonic Eating

Learning how to eat and what to eat involves reward, aversion, cognition, and repetition. Through experience, we associate cues to these traits driving us to eat or reject food (Johnson, 2013). Palatability of food, a significant cue, stimulates taste and smell sensory organs sending exteroceptive information to the central nervous system making food rewarding or aversive. Reward systems allow the reinforcement of responses that have no homeostatic value, and activation of aversive symptoms by the limbic system conditions us to avoid food that can make us sick (Berridge et al., 2009; Volkow et al., 2011).

Peripheral homeostatic peptides such as leptin, insulin, ghrelin, and peptide YY also regulate eating behavior through rewarding properties of food and cognitive control over food intake. Insulin and leptin attenuate the response of limbic and cortical regions in the human brain to food stimuli. Peptide YY in high concentrations activates the orbitofrontal cortex by food stimuli decreasing intake, and ghrelin increases activation in the amygdala, orbitofrontal cortex, anterior insula, and striatum, increasing motivation to eat (Volkow et al., 2011). Both ghrelin and endogenous opioids contribute to reward signals as independent systems (Johnson, 2013; Nogueiras et al., 2012) (Fig. 3).

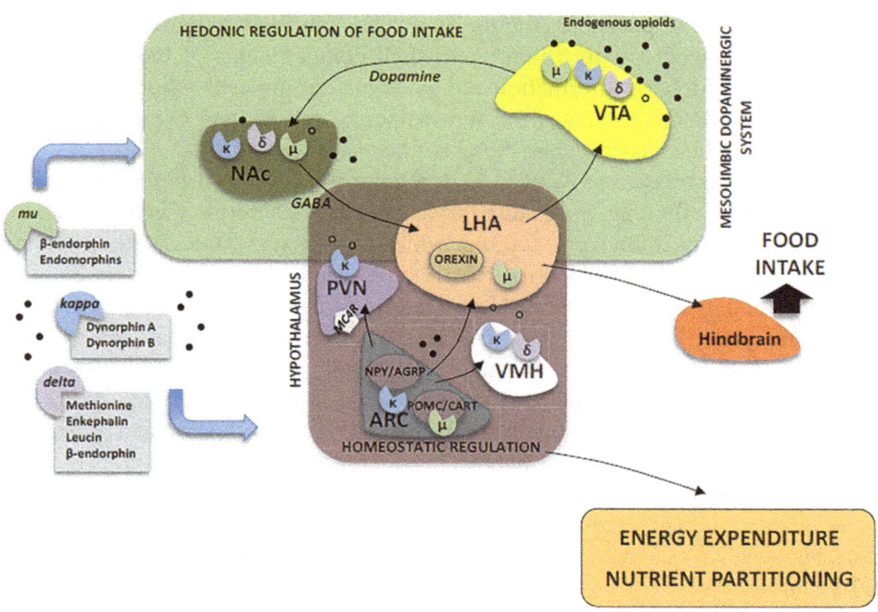

Fig. 3 Effects of the opioid system on energy balance. (Adapted from: Nogueiras et al., 2012)

Role of Endogenous Opioids in (The Backstage of) Hedonic Eating

Endogenous Opioids and Interactions with Dopamine, Serotonin, and GABA Systems

Eating is rewarding; liking food is part of what makes us get it. Dopamine, serotonin, GABA, the endocannabinoid, and opioid systems and their interactions are involved in this process (Valbrun & Zvonarev, 2020). Endogenous opioids have a role in the metabolic and hedonic pathways of eating. Endorphins, enkephalins, dynorphins, and endorphins act as neurotransmitters through the three types of receptors – that is, MOR, DOR, and KOR – which belong to the G protein-coupled receptor family (Novelle & Diéguez, 2018). These receptors are distributed through the central nervous system and participate in reward-aversion networks regulating mood, stress, and hedonic response (Darcq & Kieffer, 2018).

Activation of MORs expressed in the mesocorticolimbic network (VTA, NA, or ventral striatum, with the densest zone in the brain is the medial habenula) by VTA GABAergic interneurons relieves the local inhibitory tone and disinhibits DA neurons, which release DA to signal reward. Mu agonism reinforces the anticipatory reward response to food stimuli (Cottone et al., 2008). The habenular complex is a center for aversive processing and acts as an anti-reward system by inhibiting DA neuron activity. MOR in the MHb neurons may regulate dislike responses (Darcq & Kieffer, 2018), and association to previous learning could condition anticipatory aversive response to palatable stimuli. They also contribute with decision-making through coordinated activity in core and shell compartments of the NAc, which integrate reward-related information and guide to goal-directed actions. The dorsal raphe nucleus (DRN) regulates the serotoninergic release and expresses MORs; endogenous opioids may regulate mood by this pathway.

Kappa opioid receptors (KOR) and dynorphin function as a highly responsive stress system facilitating a negative affective state. Dynorphin is produced by GABAergic D-1 medium spiny neurons, which also express KOR. KOR-dynorphin signaling inhibits DA release in the NAc and prefrontal cortex, disrupting emotion regulation and enabling dysphoria. Likewise, glutamatergic projection from basolateral amygdala (BLA) expresses KOR and in the central amygdala (CeA) KOR inhibits local GABAergic function and interacts with stress hormone corticotropin-releasing factor (CFR) and decreases 5HT release in the DRN (Darcq & Kieffer, 2018). These mechanisms enable negative regulation of motivation and abstinence in drug addiction. The same may happen in patients with food addiction (Erbs et al., 2015; Volkow et al., 2011).

KOR systemic activation disrupts inhibitory control and may underlie interactions between stress and compulsivity. KOR activation in dopamine neurons in the VTA can decrease dopamine release in the PFC, dysregulating dopamine signals received in the corticostriatal network disrupting behavioral inhibition (Abraham et al. 2018). Distorted activation of the KOR-dynorphin system could be an underlying alteration in eating disorders, especially in binge-eating behavior and stress eating.

DORs participate in emotional responses reducing anxiety, depression, and promoting motivated behaviors. Contrary to KOR, agonism of DORs alleviate the negative mood associated with acute withdrawal and abstinence and increases inhibitory controls (Darcq & Kieffer, 2018). DORs in GABAergic forebrain neurons mediate locomotor function inhibiting D1-stimulated hyperactivity (Chung et al., 2015) (Fig. 4).

The opioid system and orexins modulate both homeostatic and hedonic pathways; mainly the effects of the opioid system are involved in the reinforcement of the pleasurable properties of food. Endogenous opioid systems regulate the hedonic value of food intake independently from the ongoing metabolic needs of the individual (Hayward et al., 2002).

Berridge et al. (2009) studied the components of reward in food intake and the structures that regulate these components: liking, wanting, and learning (Berridge et al., 2009). The circuits involved are the limbic system (amygdala, hippocampus), thalamus, reward system (VTA, NAc), and prefrontal cortex, mainly orbitofrontal cortex projections.

The NAc integrates the salience assessment of food with executive and cognitive processes receiving interoceptive information from taste and gastrointestinal organs via the vagus nerve and nucleus of the solitary tract conveying internal homeostasis from the arcuate nucleus (ARC) and lateral hypothalamus (LH) and input from cortico-limbic areas associated with cognitive and emotional processing. The core of the NAc sends projections to the basal ganglia motor control circuits and the shell to the ventral pallidum and lateral hypothalamus to regulate energy balance and eating behavior (Bouarab et al., 2019; Mesulam, 1998; Nathan & Bullmore, 2009).

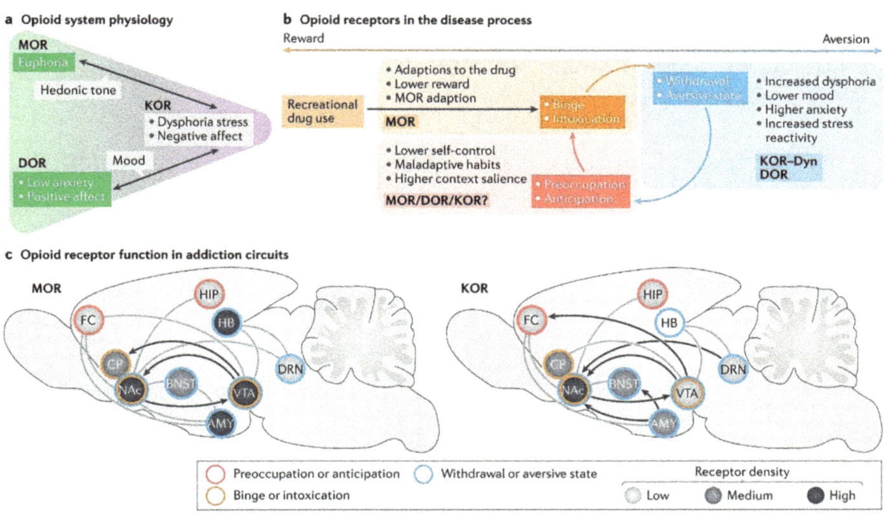

Fig. 4 Opioid receptors in physiology and addiction. (Adapted from: Darcq & Kieffer, 2018)

Within these regions, there are essential areas for pleasure responses modulated by endogenous opioids. Via mu receptors, the rostral dorsal quadrant shell of the NAc and the posterior subregion of the ventral pallidum enhanced liking (Berridge et al., 2009). Stimulation of mu receptors of GABAergic interneurons of the ventral tegmental leads to inhibition of GABAergic interneurons, increasing dopamine release in the shell of the NAc, giving the sensation of pleasure (Nathan & Bullmore, 2009). In the ventral tegmental area, dopamine release increases the palatability of food. This interaction between the opioid and dopaminergic systems is critical for goal-directed motivational behavior (Fig. 5a, b).

MORs are abundant in brain areas that receive DA neuron projections and among NAc neurons that project back to the VTA regulating motivational aspects of

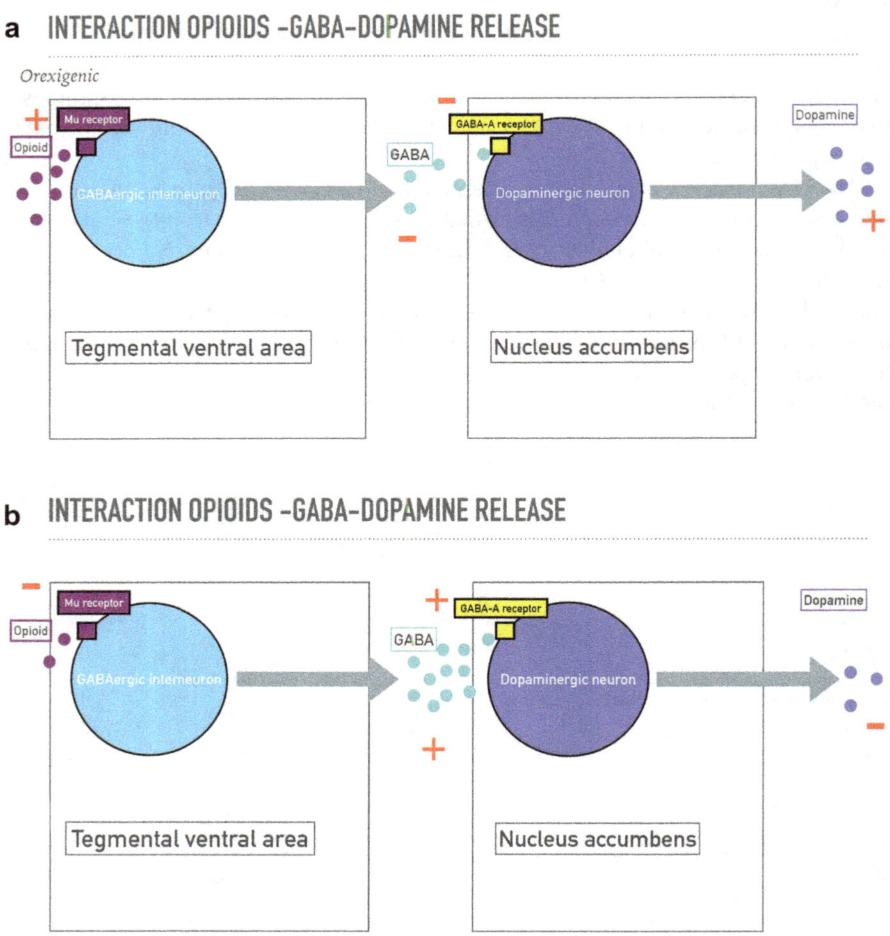

Figs. 5 (**a, b**) Interactions between endogenous opioids, GABA, and dopamine. (Adapted from: Nathan & Bullmore, 2009)

behavior, also contribute to decision-making through coordinated activity in core and shell compartments of the NAc, which integrate reward-related information and guide to goal-directed actions. Neurons co-expressing MOR/DOR are detected in hypothalamic and amygdaloidal areas inhibiting feeding through avoidance and positive affect, also are present in basal ganglia contributing to the integration of information that controls motor responses to auditory, visual, or olfactory stimuli (Erbs et al., 2015). Food deprivation, which enhances the hedonic response to food, also increases the motivational value of nonfood rewards. Opioidergic modulation of feeding increases carbohydrate intake in carbohydrate referrers and fat intake in fat referrers (Gosnell et al., 1990).

References

Abraham, A. D., Fontaine, H. M., Song, A. J., Andrews, M. M., Baird, M. A., Kieffer, B. L., et al. (2018). κ-Opioid receptor activation in dopamine neurons disrupts behavioral inhibition. *Neuropsychopharmacology, 43*(2), 362–372. https://doi.org/10.1038/npp.2017.133

Álvarez-Crespo, M., Martínez-Sánchez, N., Ruíz-Pino, F., Garcia-Lavandeira, M., Alvarez, C. V., Tena-Sempere, M., et al. (2013). The orexigenic effect of orexin-A revisited: Dependence of an intact growth hormone axis. *Endocrinology, 154*(10), 3589–3598. https://doi.org/10.1210/en.2013-1251

Berridge, K. C., Robinson, T. E., & Aldridge, J. W. (2009). Dissecting components of reward: 'Liking', 'wanting', and learning. *Current Opinion in Pharmacology, 9*(1), 65–73. https://doi.org/10.1016/j.coph.2008.12.01

Berthoud, H. R. (2002). Multiple neural systems controlling food intake and body weight. *Neuroscience & Biobehavioral Reviews, 26*(4), 393–428.

Bouarab, C., Thompson, B., & Polter, A. M. (2019). VTA GABA neurons at the interface of stress and reward. *Frontiers in Neural Circuits, 13*, 78. https://doi.org/10.3389/fncir.2019.00078

Chung, P. C. S., Keyworth, H. L., Martin-Garcia, E., Charbogne, P., Darcq, E., Bailey, A., et al. (2015). A novel anxiogenic role for the delta opioid receptor expressed in GABAergic forebrain neurons. *Biological Psychiatry, 77*(4), 404–415. https://doi.org/10.1016/j.biopsych.2014.07.033

Cooper, S. J., Jackson, A., & Kirkham, T. C. (1985). Endorphins and food intake: kappa opioid receptor agonists and hyperphagia. *Pharmacology Biochemistry and Behavior, 23*(5), 889–901.

Cottone, P., Sabino, V., Steardo, L., & Zorrilla, E. P. (2008). Opioid-dependent anticipatory negative contrast and binge-like eating in rats with limited access to highly preferred food. *Neuropsychopharmacology, 33*(3), 524–535. https://doi.org/10.1038/sj.npp.1301430

Crespo, C., Cachero, A., Jiménez, L., Barrios, V., & Ferreiro, E. (2014). Peptides and food intake. *Frontiers in Endocrinology, 5*, 1–13.

Czyzyk, T. A., Nogueiras, R., Lockwood, J. F., McKenzie, J. H., Coskun, T., Pintar, J. E., et al. (2010). kappa-Opioid receptors control the metabolic response to a high-energy diet in mice. *The FASEB Journal, 24*(4), 1151–1159.

Darcq, E., & Kieffer, B. L. (2018). Opioid receptors: Drivers to addiction? *Nature Reviews Neuroscience, 19*(8), 499–514. https://doi.org/10.1038/s41583-018-0028-x

Erbs, E., Faget, L., Scherrer, G., Matifas, A., Filliol, D., Vonesch, J. L., et al. (2015). A mu–delta opioid receptor brain atlas reveals neuronal co-occurrence in subcortical networks. *Brain Structure and Function, 220*(2), 677–702. https://doi.org/10.1007/s00429-014-0717-9

Gao, W. J., & Snyder, M. A. (2013). NMDA hypofunction as a convergence point for progression and symptoms of schizophrenia. *Frontiers in Cellular Neuroscience, 7*(31), 1–12. https://doi.org/10.3389/fncel.2013.00031

Gosnell, B. A., Romsos, D. R., Morley, J. E., Levine, A. S. (1985). Opiates and medial hypothalamic knife cuts cause hyperphagia through direct mechanisms. *Behavioral Neuroscience, 99*(6):1181–1191.

Gosnell, B. A., Krahn, D. D., & Majchrzak, M. J. (1990). The effects of morphine on diet selection are dependent upon baseline diet preferences. *Pharmacology Biochemistry and Behavior, 37*(2), 207–212. https://doi.org/10.1016/0091-3057(90)90322-9

Grassi, G. (2014). Leptin, the sympathetic nervous system and blood pressure: The tale is still without an end. *Journal of Hypertension, 32*(4), 738–739. https://doi.org/10.1097/HJH.0000000000000123

Hall, J. E. (2011). *Guyton and Hall textbook of medical physiology* (12th ed.). Elsevier Saunders.

Harris, R. B. (2013). Evidence that leptin-induced weight loss requires activation of both forebrain and hindbrain receptors. *Physiology & Behavior, 120*, 83–92. https://doi.org/10.1016/j.physbeh.2013.07.004

Hayward, M. D., Pintar, J. E., & Low, M. J. (2002). Selective reward deficit in mice lacking β-endorphin and enkephalin. *Journal of Neuroscience, 22*(18), 8251–8258. https://doi.org/10.1523/jneurosci.22-18-08251.2002

Johnson, A. W. (2013). Eating beyond metabolic need: How environmental cues influence feeding behavior. *Trends in Neurosciences, 36*(2), 101–109. https://doi.org/10.1016/j.tins.2013.01.002

Katarzyna Pomorska, D., do-Rego, J., do-Rego, J., Zubrzycka, M., & Janecka, A. (2016). Opioid and cannabinoid system in food intake. *Current Pharmaceutical Design, 22*(10), 1361–1370.

Kenny, P. J. (2011). Common cellular and molecular mechanisms in obesity and drug addiction. *Nature Reviews Neuroscience, 12*(11), 638–651.

Konturek, S. J., Pepera, J., Zabielski, K., Konturek, P. C., Pawlik, T., Szlachcic, A., & Hahn, E. G. (2003). Brain-gut axis in pancreatic secretion and appetite control. *Journal of Physiology and Pharmacology, 54*(3), 293–317.

Larsen, P. J., & Hunter, R. G. (2006). The role of CART in body weight homeostasis. *Peptides, 27*(8), 1981–1986.

Mesulam, M. M. (1998). From sensation to cognition. *Brain: A Journal of Neurology, 121*(6), 1013–1052. https://doi.org/10.1093/brain/121.6.1013

Morley, J. E., Levine, A. S., Gosnell, B. A., Kneip, J., & Grace, M. (1983). The kappa opioid receptor, ingestive behaviors and the obese mouse (ob/ob). *Physiology & Behavior, 31*(5), 603–606.

Nathan, P. J., & Bullmore, E. T. (2009). From taste hedonics to motivational drive: Central μ-opioid receptors and binge-eating behaviour. *International Journal of Neuropsychopharmacology, 12*(7), 995–1008. https://doi.org/10.1017/S146114570900039X

Nogueiras, R., Romero-Picó, A., Vazquez, M. J., Novelle, M. G., López, M., & Diéguez, C. (2012). The opioid system and food intake: Homeostatic and hedonic mechanisms. *Obesity Facts, 5*(2), 196–207. https://doi.org/10.1159/000338163

Novelle, M. G., & Diéguez, C. (2018). Food addiction and binge eating: Lessons learned from animal models. *Nutrients, 10*(1), 71. https://doi.org/10.3390/nu10010071

Proietto, J. (2011). Why is treating obesity so difficult? Justification for the role of bariatric surgery. *Medical Journal of Australia, 195*(3), 144–146.

Romero-Picó, A., Novelle, M. G., Folgueira, C., López, M., Nogueiras, R., & Diéguez, C. (2013). Central manipulation of dopamine receptors attenuates the orexigenic action of ghrelin. *Psychopharmacology, 229*(2), 275–283. https://doi.org/10.1007/s00213-013-3096-3097

Romero-Pico, A., Vázquez, M. J., González-Touceda, D., Folgueira, C., Skibicka, K. P., Alvarez-Crespo, M., et al. (2013). Hypothalamic κ-opioid receptor modulates the orexigenic effect of ghrelin. *Neuropsychopharmacology, 38*(7), 1296–1307. https://doi.org/10.1038/npp.2013.28

Saper, C. B., Chou, T. C., & Elmquist, J. K. (2002). The need to feed: Homeostatic and hedonic control of eating. *Neuron, 36*(2), 199–211. https://doi.org/10.1016/S0896-6273(02)009698

Slochower, J., Kapłan, S. P., & Mann, L. (1981). The effects of life stress and weight on mood and eating. *Appetite, 2*(2), 115–125.

Suzuki, K., Jayasena, C. N., & Bloom, S. R. (2012). Obesity and appetite control. *Experimental Diabetes Research, 2012*. https://doi.org/10.1155/2012/824305

Valbrun, L. P., & Zvonarev, V. (2020). The opioid system and food intake: Use of opiate antagonists in treatment of binge eating disorder and abnormal eating behavior. *Journal of Clinical Medicine Research, 12*(2), 41. https://doi.org/10.14740/jocmr4066

Volkow, N. D., Wang, G. J., & Baler, R. D. (2011). Reward, dopamine and the control of food intake: Implications for obesity. *Trends in Cognitive Sciences, 15*(1), 37–46. https://doi.org/10.1016/j.tics.2010.11.001

Vrang, N., Madsen, A. N., Tang-Christensen, M., Hansen, G., & Larsen, P. J. (2006). PYY (3–36) reduces food intake and body weight and improves insulin sensitivity in rodent models of diet-induced obesity. *American Journal of Physiology-Regulatory, Integrative and Comparative Physiology, 291*(2), R367–R375. https://doi.org/10.1152/ajpregu.00726.2005

Williams, D. L., Kaplan, J. M., & Grill, H. J. (2000). The role of the dorsal vagal complex and the vagus nerve in feeding effects of melanocortin-3/4 receptor stimulation. *Endocrinology, 141*(4), 1332–1337. https://doi.org/10.1210/endo.141.4.7410

Yau, S. W., Henry, B. A., Russo, V. C., McConell, G. K., Clarke, I. J., Werther, G. A., & Sabin, M. A. (2014). Leptin enhances insulin sensitivity by direct and sympathetic nervous system regulation of muscle IGFBP-2 expression: Evidence from nonrodent models. *Endocrinology, 155*(6), 2133–2143. https://doi.org/10.1210/en.2013-2099

Yu, J. H., & Kim, M. S. (2012). Molecular mechanisms of appetite regulation. *Diabetes & Metabolism Journal, 36*(6), 391. https://doi.org/10.4093/dmj.2012.36.6.391

Zwirska-Korczala, K., Konturek, S., Sodowski, M., Wylezol, M., Kuka, D., Sowa, P., et al. (2007). Basal and postprandial plasma levels of Pyy, Ghrelin. *Journal of Physiology and Pharmacology, 58*(1), 13–35.

Role of Endogenous Opioids in the Pathophysiology of Obesity and Eating Disorders

Sylvana Stephano Zuniga, Marcela Rodriguez Flores, and Adriana Albu

Abstract This second chapter in our trilogy reviews and critically appraises the scientific evidence for the role of endogenous opioid system (EOS) activity in the onset and progression of both obesity and eating disorders. Defining features of normative eating and maladaptive eating behaviors are discussed as a foundation. We review the scientific literature pertaining to the predisposing risk factors and pathophysiology for obesity and eating disorders. Research targeting the association between obesity, disordered eating, and psychiatric comorbidities is reviewed. We conclude by discussing the involvement of endogenous opioids in neurobiological and behavior traits, and the clinical evidence for the role of the EOS in obesity and eating disorders.

Keywords Endogenous opioids · Normative eating · Eating disorders · Obesity

Introduction

In the previous chapter, we explained the biological mechanisms of metabolic and hedonic eating behavior and the contribution of endogenous opioids and opioid receptors in the biology of metabolic needs, motivation, reward, and food-seeking behavior. This chapter reviews the evidence for the role of endogenous opioids in developing and maintaining both obesity and eating disorders (anorexia nervosa, binge eating disorder, and bulimia nervosa). We begin by addressing the differences between normative and maladaptive eating behaviors, followed by subsequent discussion of their predisposing risk factors, pathophysiology, relationship with

---

S. S. Zuniga (✉) · M. R. Flores
Obesity and Eating Disorders Clinic, Instituto Nacional de Ciencias Médicas y Nutrición Salvador Zubirán, Tlalpan, Mexico

A. Albu
2nd Department of Internal Medicine, University of Medicine and Pharmacy, Cluj-Napoca, Romania

P. L. Kerr et al. (eds.), *Endogenous Opioids*, Advances in Neurobiology 35,
https://doi.org/10.1007/978-3-031-45493-6_17

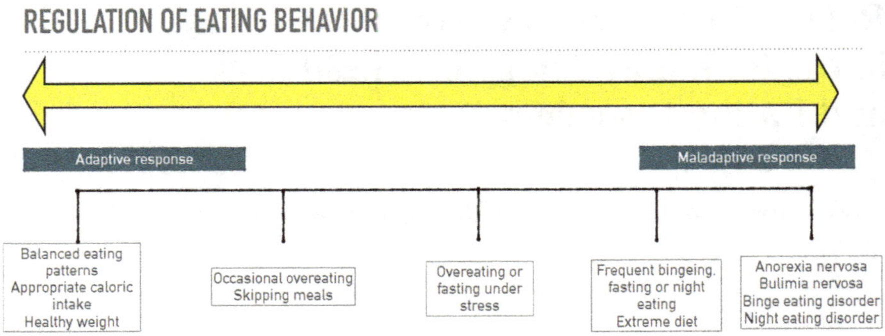

Fig. 1 Regulation of eating behavior. (Adapted from: Stuart, 2013)

psychiatric comorbidities, the involvement of endogenous opioids in neurobiological and behavior traits, and the clinical evidence of opioids in obesity and eating disorders.

Eating Behavior, Obesity, and Eating Disorders

Eating behavior is regulated by genetic, psychophysiological, cognitive, and developmental factors; these behaviors can evolve into a maladaptive response and promote overeating or become an eating disorder itself (Fig. 1). Eating disorders are psychiatric disorders characterized by abnormal eating and excessive body image concerns, including insufficient or excessive food consumption, concomitant damage to physical health, and impairments in psychosocial functioning (American Psychiatric Association, 2013). By contrast, due to the heterogeneous causes and manifestations of obesity, this condition cannot accurately be classified as any form of psychiatric disorder, and is not associated with an eating disorder in all individuals who experience it. Symptoms and associated behaviors overlap across the range of eating disorders and obesity, requiring diagnoses to comprise a complex clinical examination.

Neurobiological and Behavioral Traits in Obesity and Eating Disorders

In neuropsychiatric diseases, we use the term endophenotype to describe the observable signs and symptoms of behaviors that arise from underlying biological processes that could serve as targets for individualized treatments. With this approach, we integrate knowledge from genes, neurotransmitters, receptors, neural circuits, and behavior (Fig. 2).

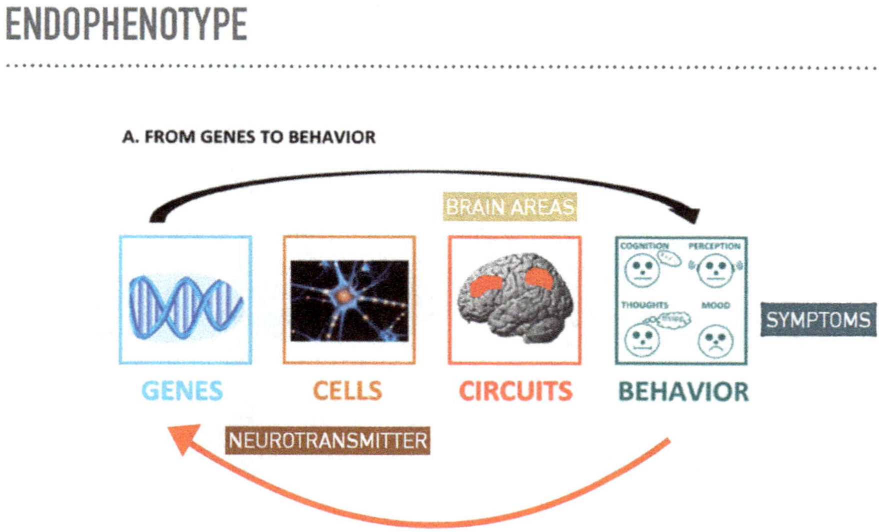

Fig. 2 Endophenotype concept

SYMP.One of the significant issues in eating disorders and obesity diagnosis and treatment is the heterogeneity within each diagnosis. There is also substantial overlap of different neurobiological and behavioral traits. In recent years, researchers have proposed a research domain criteria to approach eating disorders (Dunlop et al., 2016); these domains are negative valence systems, positive valence systems, cognitive systems, social processes, and arousal/regulatory systems (Fig. 3). Although obesity is not classified as a psychiatric disorder, several studies have tried to determine the neural circuits behind diverse cognitive and behavioral traits that increase susceptibility to the development of obesity, or that accompany individuals with obesity in the maintenance or relapse of behaviors related to obesity. These domains consist of broad spectrum of features that modulate such traits (Fig. 4). We will further approach the domains and subdomains with endogenous opioid involvement in eating disorders and obesity.

Negative Valence Systems

The negative valence system is a predominantly neurobiological construct. It is associated with the limbic system, insula, and prefrontal cortex. These systems are in charge of responding to aversive, harmful, and threatening situations or thoughts, generating a negative emotional valence of fear or anxiety.

In anorexia nervosa, the dorsolateral prefrontal cortex (dlPFC) is abnormally hyperactive to pain anticipation and the receipt of punishment. The anterior cingulate cortex (ACC) is hyperactive for aversive food stimuli, the receipt of

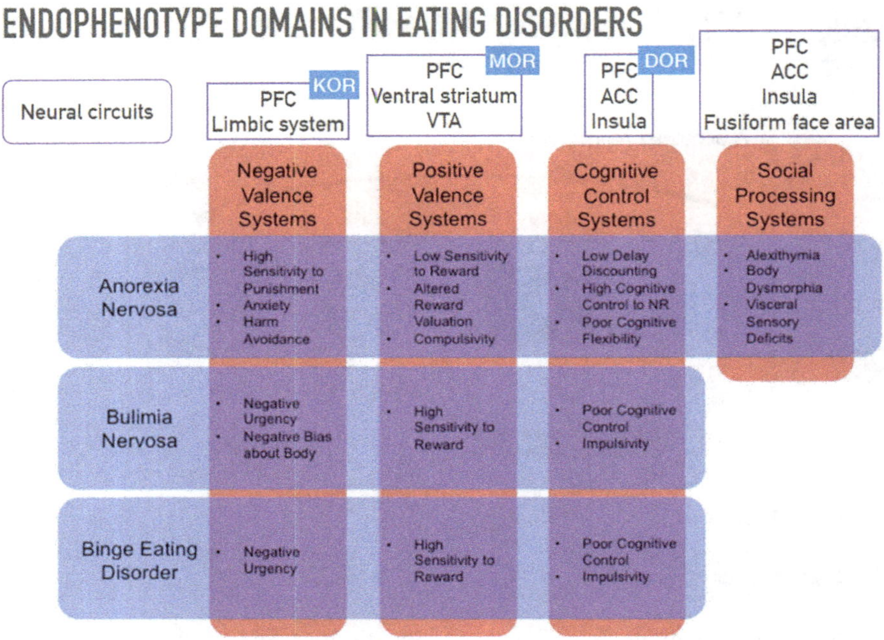

Fig. 3 Endophenotype domains in eating disorders and implicated neural circuits. (Adapted from: Dunlop et al., 2016)

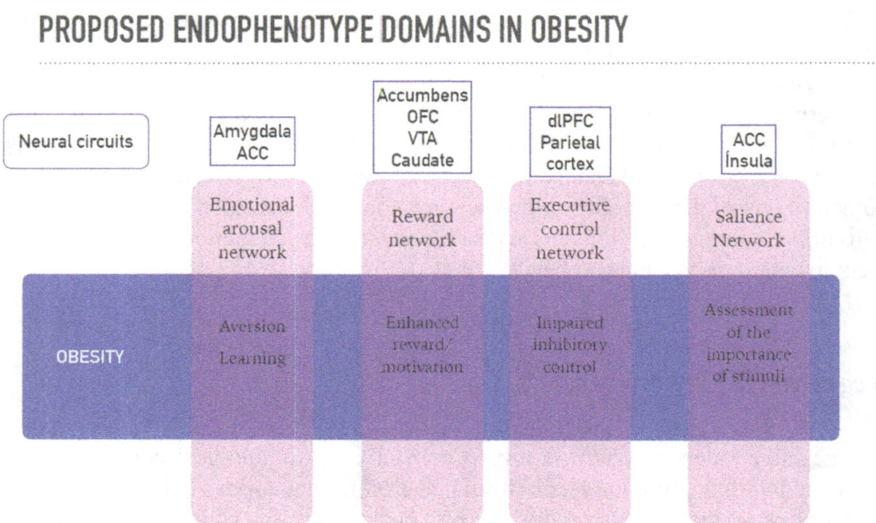

Fig. 4 Proposed endophenotype domains in obesity and implicated neural circuits. (Adapted from: Gupta et al., 2015)

punishment, and anxiety. The anterior insula is abnormally hyperactive during anxiety and the anticipation of pain (Dunlop et al., 2016). In bulimia nervosa and binge eating disorder, the ACC is abnormally activated for negative words about the body, and the insula is hyperactive during adverse effects (Dunlop et al., 2016).

There is evidence that amygdaloid lesions lead to hyperphagia and obesity, especially with lesions of the basolateral and lateral nuclei. In addition, research has found that patients with obesity have higher levels of anxiety compared with normal-weight controls (Baumeister & Härter, 2007); although the mechanism is not well established, the limbic system structures like the amygdala must be involved.

Positive Valence Systems

The positive valence system is related primarily to reward. This system is integrated by the dorsal prefrontal cortex (dlPFC), orbitofrontal cortex (OFC), and ventral striatum (Fig. 5). These circuits regulate reward, anticipation, and motivation, which form habits through repetition.

In anorexia nervosa, binge eating, and bulimia nervosa, the ventral striatum and ventromedial prefrontal cortex appear hypoactive while the orbitofrontal cortex and temporal cortex appear hyperactive, causing compulsivity in eating behaviors, such as restricting and purging (Dunlop et al., 2016).

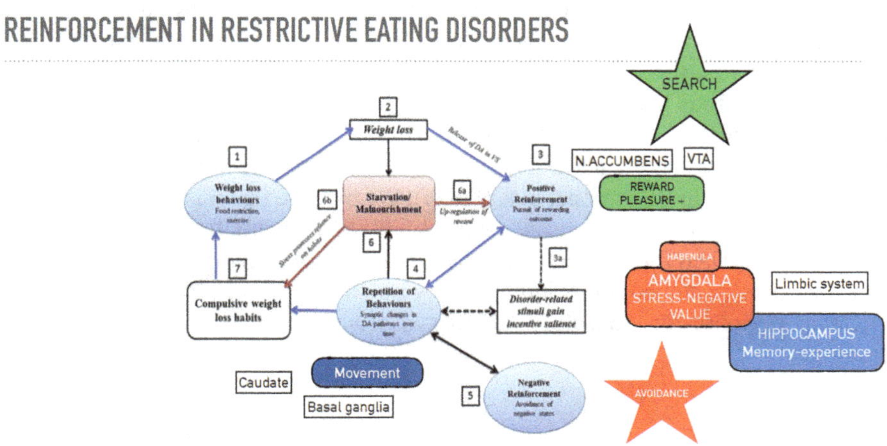

Fig. 5 Reinforcement in restrictive eating disorders. In restrictive eating disorders, all systems interact in the maintenance of the disease. Malnutrition and social reinforcement increase motivation and seeking to restrictive behaviors while the limbic system contribute to negative reinforcement and avoidance like social isolation and anxiety. The prefrontal cortex malfunction gives place to compulsive behaviors and executive functions disturbances

Cognitive Systems

The cognitive systems dimension refers to processes responsible for cognitive processing, including attention, perception, memory, language, and cognitive control. In healthy control studies, these behaviors are associated with activity in the DMPFC, DLPFC, and anterior insula. These networks tend to be responsible for response selection and inhibition.

Numerous cognitive biases have been studied in subjects with overweight and obesity. Impairments in learning and memory, poor decision-making associated with delay discounting and risk-taking under ambiguity, and loss of control can make people more susceptible to overeating, consumption of unhealthy foods, as well as to limit achieving behavior change in obesity control, – for example, resorting to "magical" treatments reduced motivation and weight internalization (D'Souza, 2019). In individuals with binge eating disorder, several cognitive factors are associated with the main behaviors of the disorder: attentional biases, enhanced behavioral inflexibility or compulsivity with habit formation and perseveration despite change in relevance, generalized cognitive interference in working memory, leading to variable risk seeking according to BMI and impulsivity (Voon, 2015). Both in obesity and eating disorders, food craving is a form of anticipatory reward that is regulated by endogenous opioids and mesolimbic dopaminergic systems. Reduction of dopamine 2 receptors (D2R) in the striatum in individuals with obesity has been associated with a decrease in the activity of prefrontal regions involved with neurobiological processes that underlie sensitivity to reward and that modulate inhibitory control, favoring overfeeding (Volkow et al., 2008).

Eating disorders commonly manifest with deficits on tasks related to cognitive control, including behavioral inhibition, working memory, selective attention, and cognitive flexibility. Generally, people with BED display more deficient response inhibition and lower activity in the inferior frontal gyrus and ventromedial PFC. Binge eating phenotypes show diminished activity in the inferior frontal gyrus, ACC, OFC, and DLPFC. Patients with obesity without binge eating can also present with disturbances in the dlPFC, which lead to impaired inhibitory control in feeding behavior and attentional bias (Fig. 6). We propose that this prefrontal hypofunction fails to regulate other systems like limbic and reward, leading to emotional dysregulation. Therefore, not all patients with obesity have features of major depressive disorders or anxiety disorders, but they commonly manifest disproportionate affective responses to stressful situations associated with inability to regulate emotions.

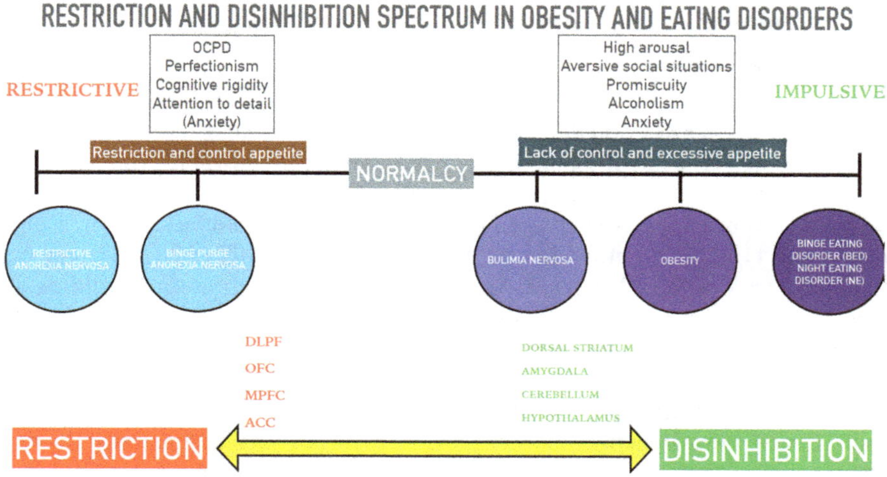

Fig. 6 Cognitive systems (mainly prefrontal cortex) involved in ED and obesity. DLPF dorsolateral prefrontal cortex, OFC orbitofrontal cortex, MPFC medial prefrontal cortex, ACC anterior cingulate cortex. (Adapted from Stuart, 2013)

Social Processing Systems: Social Cognition and Theory of Mind

Social processing systems refer to circuits involved in social communication and the perception and understanding of oneself and others. One key neurological structure involved in social processing systems is the insula, which is responsible for an interception of the temporoparietal junction, for the theory of mind-related processing, and other facial expressions.

Patients with eating disorders present with alexithymia, deficits in interoceptive awareness, and distortion in the perception of body shape. These disturbances are related to the hypofunction of the insula, thalamus, and ACC. Also, lower active activation of the right temporoparietal junction is shown during social decision-making tasks.

Social cognition and alexithymia are not well studied in obesity. However, there is some evidence of deficit in theory of mind tasks in people with obesity (Caldú et al., 2019). Research has found that people with obesity have difficulties in identifying emotions in themselves and others (Smith et al., 2011). Additionally, people with obesity exhibit difficulty identifying interoceptive sensations, which increases with weight (Smith et al., 2011).

Epidemiologic Aspects and Risk Factors for Obesity and Eating Disorders

Obesity has become a worldwide public health problem due to its increasing prevalence and wide-reaching morbidity, which reflects the diverse underlying environmental and biological risk factors (Jaacks et al., 2019). The current global burden of obesity is estimated to have increased from 1% to 6–8% in children since 1975, and from 3% to 11% in the same period, to 6% to 15% in adults (Blüher, 2019). However, some countries have had faster increases in prevalence, with up to two-thirds of their populations affected by overweight or obesity, such as the case of the USA (Centers for Disease Control and Prevention, 2018).

Obesity is a chronic disease with a multifactorial etiology that involves varying disturbances in a multitude of systems that regulate energy intake and expenditure, and central nervous system (CNS) functions (Heymsfield & Wadden, 2017). Its rising prevalence results from biological susceptibility resulting in complex physiological and psychosocial responses to a rapidly changing environment (Qasim et al., 2018). Genetic contributions account for up to 70% of the biological predisposition to obesity, characterized by a programming to store fat and motivation towards eating behaviors, consistent with when food was scarce. Many of these genes are expressed in the CNS (Table 1), with recent advances in genomics studying their role by screening the entire genome in large populations (Bray et al., 2016). This has led to the identification of polygenic markers of risk that convey varying influence according to the life cycle pattern, including the FTO and MCR4 genes as some of the most prominent traits (Sanz-de-Galdeano et al., 2020). Still, with over 700 loci identified, they have been found to individually account for less than 5% of

Table 1 Genes associated with obesity

Gene symbol	Gene name	Gene product's role in energy balance
ADIPOQ	Adipocyte-. C1q-, and collagen domain-containing	Produced by fat cells, adiponectin promotes energy expenditure
FTO	Fat mass- and obesity-associated gene	Promotes food intake
LEP	Leptin	Produced by adipocytes to induce satiety
LEPR	Leptin receptor	When bound by leptin, inhibits appetite
INSIG2	Insulin-induced gene 2	Regulation of cholesterol and fatty acid synthesis
MCR4	Melanocortin 4 receptor	When bound by alpha-melanocyte stimulating hormone, stimulates appetite
PCSK1	Proprotein convertase subtilisin/kexin type 1	Regulates insulin biosynthesis
PPARG	Peroxisome proliferator-activated receptor gamma	Stimulates lipid uptake and development of fat tissue
BDNF	Brain-derived neurotrophic factor	Promotes survival, growth, and maturation of nerve cells

Table 2 Genes associated with eating disorders

Genes in eating disorders			
Gene	Description	Polimorphism	Eating disorder
HTR1D	Serotonin receptor 1D	rs6300/ rs674386/rs856510	AN
HTR2A	Serotonin receptor 2A	rs6311	AN
HTR2C	Serotonin receptor 2C	1S6318	AN
SLC6A4	Serotonin transporter	5HTLPR	AN BN BED
DRD2	Dopamine receptor D2	rs1799742/rs2283265	BN BED
COMT	Catechol-O-methyltransferase	rs4680	AN BN
ANKK1	Ankyrin repeat and kinase domain containing 1	rs1800497	BED
GHRL	Ghrelin	rs4684677/rs2075356	AN BN
MC4R	Melanocortin 4 receptor	rs52820871	BED
AGRP	Agouti related protein	rs13338499	AN
ESR1	Estrogen receptor 1	rs726281	AN
BDNF	Brain derived neutrophil factor	rs6265/rs56164415	AN BN
CNR1	Cannabinoid receptor	AAT trinucleotide STR/rs1049353	AN BN
OPRD1	Opioid receptor delta 1	rs53706/rs760589/rs204081/ rs569356/rs4654327	AN
OPRM	Opioid receptor mu 1	1799971	BED
FTO	Fat mass and obesity associated	rs9939609	AN BN

From: Yilmaz et al. (2015)

AN Anorexia nervosa, *BN* Bulimia nervosa, *BED* binge eating disorder

individual variation in body mass index (BMI), and genome-wide estimates suggest that common variation accounts for >20% of BMI variation (Locke et al., 2015; Yengo et al., 2018). Therefore, not all individuals exposed to these variants develop obesity, just like not all individuals living in an obesogenic environment develop it either, highlighting the characteristic heterogeneity of obesity. In contrast, obesity (often severe) due to monogenic mutations of genes that codify for regulators of energy homeostasis, such as leptin and melanocortin-4 receptors, is highly rare (Pigeyre et al., 2016).

Eating disorders have a strong familial pattern, suggesting implications of genetic factors in their development and their psychopathology. The genes involved codify for several receptors from different neurotransmitters, hormones, and neuropeptides like serotonin, dopamine, norepinephrine, endogenous opioids, estrogens, ghrelin, and cannabinoids (Table 2).

In addition to genetic predisposition, many influences of modern society, along with metabolic mechanisms known to affect energy balance, such as periods of nutritional deficiency, the use of drugs that promote weight gain, and modification of the microbiota, contribute with the pathogenesis of obesity (Elshenawy & Simmons, 2016; Qasim et al., 2018). Some of the most recognized factors that have collided with our biology include a built environment with widespread availability of increasing portion sizes of palatable food that also hinders physical activity, and sociocultural forces that promote widespread use of technology favoring extraordinary screen time and consumerism (Kaila & Raman, 2008; Seidell & Halberstadt, 2015). All of these factors fluctuate according to diverse cultural contexts and socioeconomic status, resulting in higher predisposition of populations within the poorest environment (Hruby & Hu, 2015; Jaacks et al., 2019; Kanter & Caballero, 2012). On top of this, biological and psychological stress is recognized as a key driver of overeating, reduced self-care, and other behaviors of susceptibility for obesity (Centers for Disease Control and Prevention, 2018; Gubareva & Posokhina, 2014; Lopresti & Drummond, 2013; Fig. 7).

Obesity is the natural result of biological susceptibility responding to environmental changes. Importantly, some common behaviors in modern societies are known to strongly contribute to the obesity epidemic. These include disturbed eating and sleep timing, substitution of fruit and vegetable consumption for ultra-processed food, and high sedentariness (Asghari et al., 2017; Chaput, 2014; Jacquet

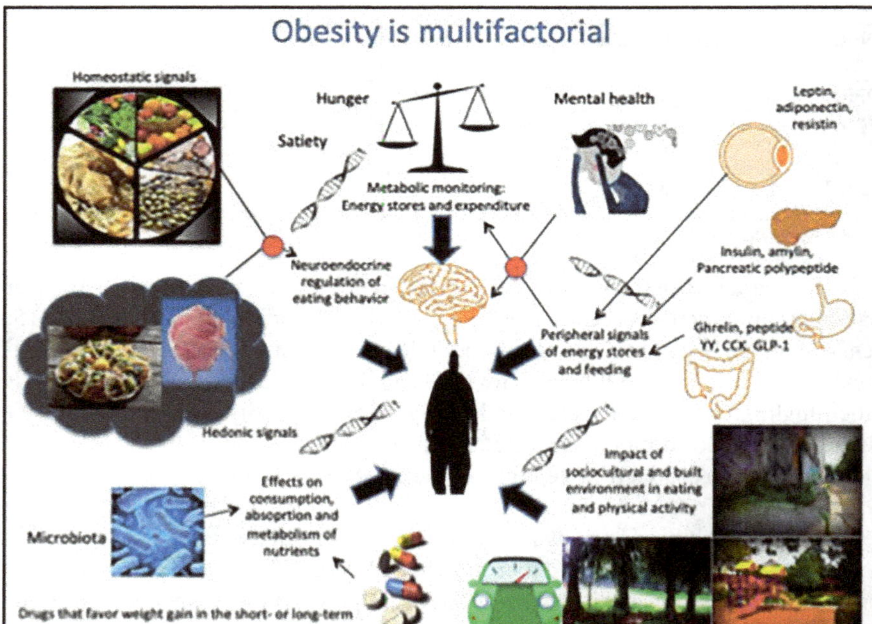

Fig. 7 Multifactorial etiology of obesity

et al., 2020; Pietiläinen et al., 2012). The result is activation of biological responses for energy conservation and immunological dysregulation (Darby et al., 2007; Depner et al., 2014; Miller et al., 2015). The development of processed and hyper-palatable foods, mostly with no or minimal nutritional value, enhances expression of reward responses that in our modern environment translate into disadvantageous and potentially addictive behaviors (Davis, 2014).

Patients with eating disorders have a prevalence of 50–70% of psychiatric comorbidities. In patients with obesity, psychiatric comorbidities increase with body mass index up to 50% (Berkowitz & Fabricatore, 2005). Mood disorders are the most frequent, followed by anxiety disorders, substance use disorders, and personality disorders.

Both obesity and eating disorders have been described in the medical literature for centuries. Obesity was first described in 1611 by Tobias. Anorexia nervosa was first described in the nineteenth century by Gull and Lasegue. Bulimia nervosa was described in the early twentieth century, and binge eating disorder was identified in 1959 by Albert Stunkard (Hay et al., 2014). Since then, the understanding of these diseases, along with their diagnosis and treatment, has increased progressively.

Eating disorders and obesity risk factors can overlap; for this discussion, we have divided them into restrictive, bulimic traits, and obesity for better understanding (Treasure et al., 2020; see Figs. 8, 9 and 10). All of these risk factors include

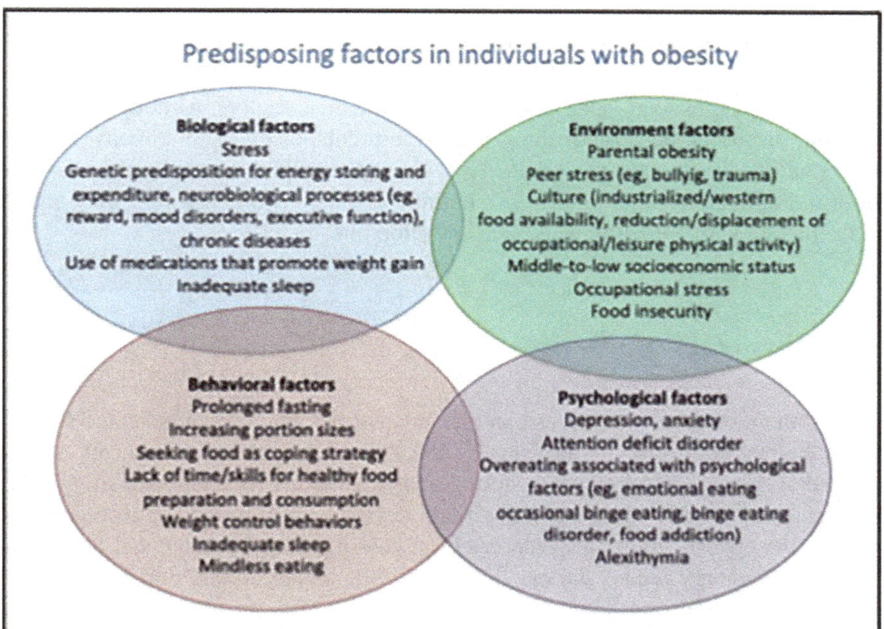

Fig. 8 Risk factors for obesity. (Adapted from: Treasure et al. (2020). Lancet Publishing Group. https://doi.org/10.1016/S0140-6736(20)30059-3)

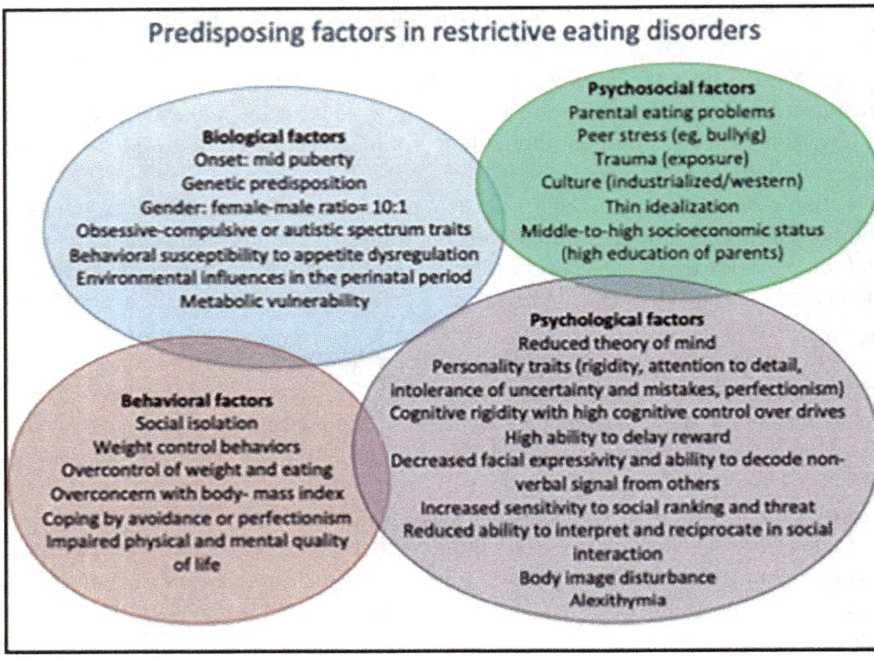

Fig. 9 Risk factors in restrictive eating disorders. (Adapted from: Treasure et al. (2020). Lancet Publishing Group. https://doi.org/10.1016/S0140-6736(20)30059-3)

biological, psychosocial, behavioral, and psychological factors. Although these are different disorders, because of the overlapping predisposing factors, many patients have multiple disorders across the life span, for example, anorexia nervosa may onset in adolescence, and later in life bulimia or binge eating disorder may onset; and someone with a healthy weight can develop obesity.

Pathophysiology of Obesity

Development of obesity involves complex adaptations of energy intake and expenditure that are controlled by three key CNS systems and their communication with peripheral organs of assimilation, storing and expenditure of energy substrates: the gastrointestinal tract, white and brown adipocytes, and muscle. The three central centers that govern autonomic responses, processing, interpretation, and behaviors related with energy intake and expenditure are the melanocortin, dopamine mesolimbic, opioid, and endocannabinoid systems. They are the networks behind the food preferences, eating- and health-related behaviors, as well as psychosocial and psychological factors previously mentioned to promote the development of obesity (Richard, 2015).

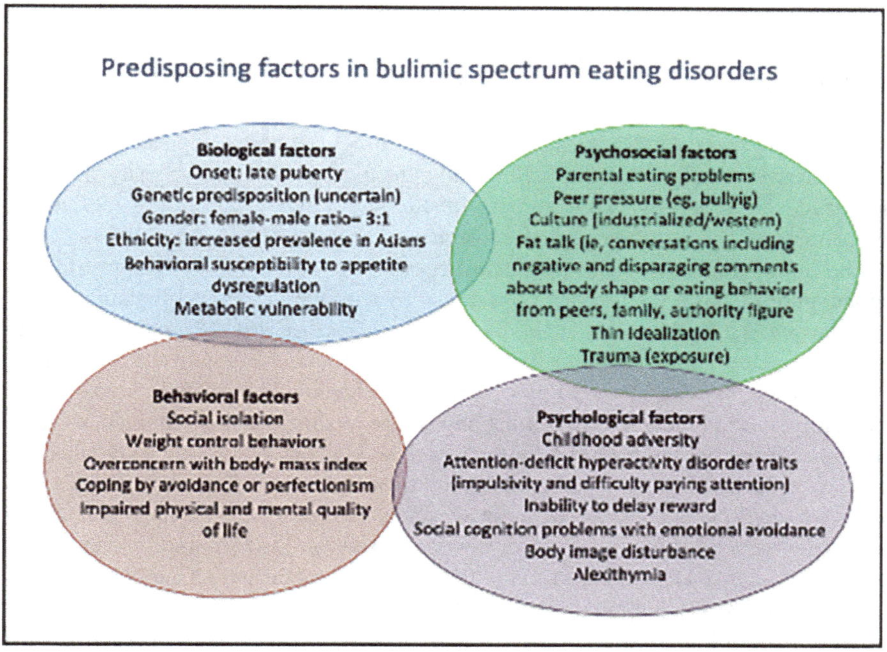

Fig. 10 Risk factors in bulimic spectrum eating disorders. (Adapted from: Treasure et al. (2020). Lancet Publishing Group. https://doi.org/10.1016/S0140-6736(20)30059-3)

The plasticity of the CNS allows the development of new circuits necessary for learning, which is more marked during gestation and the first years of life. Maternal malnutrition and obesity favor the development of obesity and related metabolic diseases, even more so if there is genetic predisposition (Marciniak et al., 2017). This fetal programming of cardiometabolic and neurodevelopmental outcomes occurs in a maternal environment that comprise hyperglycemia, micro- or macronutrient restriction, increased saturated fatty acids, inflammation, and dysbiosis (Desmet et al., 2020; Tounian, 2011).

Many neurobiological and metabolic mechanisms are able to promote weight gain, and many are triggered in particular stages of life. In the early years and during adolescence, diverse psychological stressors, including family dynamics and raising styles, can promote habits towards healthy activities and nutrition, as well as self-regulation (Hassan, 2017). On the other hand, high media use, unhealthy nutrition, and overvaluation of shape can contribute with unhealthy eating patterns and executive bias associated with obesity and binge eating disorder (Neumark-Sztainer et al., 2006; Pearl et al., 2014; Spengler et al., 2014).

Obesity and compulsive overeating are often associated with social, cognitive, and emotional impairments (Lacroix & von Ranson, 2021). Furthermore, distinctive patterns of brain structural connectivity have been found in subjects with overweight. Gupta et al. (2015) found greater connectivity between the reward network regions and regions of the executive control, emotional arousal, and somatosensory networks, accompanied by decreased nerve fiber density between the ventromedial prefrontal cortex and the anterior insula, and between the thalamus and executive control network regions. Regions of the reward, salience, executive control, and emotional arousal networks were associated with lower morphological values, while the opposite pattern was seen for regions of the somatosensory network (Gupta et al., 2015).

Altered metabolic-neural signaling, whether primary or once obesity is established, develops a network of ongoing functions throughout the life of individuals with obesity. The neurobiological substrate of appetite regulation consists of the anchoring of motivation with the reward circuits (Yu et al., 2015). The gastric hormone, ghrelin, has a vital role in food intake by stimulating its hypothalamic receptors; however, it has also been involved in food reward processes. Research has found that the manipulation of ghrelin and its receptors may interfere with the mesolimbic dopaminergic system, modulating food reward behavior. These effects are probably most important in inducing overeating and, subsequently, obesity mediated by ghrelin's specific receptor GHSA-1A, highly expressed in the CNS (Skibicka & Dickson, 2011). Peripheral sites for energy intake and expenditure can play varying roles during the natural history of obesity, some of which are not susceptible of modifications (e.g., age, genetic predisposition for energy storing and expenditure capacity, and autonomic neural circuits), and many that are affected during an individual's life, including sociocultural practices around food, physical activity, and body image, use of drugs that influence weight, modifications of the microbiota, and environment exposures (Heymsfield & Wadden, 2017; Pigeyre et al., 2016).

Once obesity develops, the function of the homeostatic system is to defend the body against starvation (Summermatter et al., 2009). The CNS is highly sensitive fat loss that is accompanied by lower leptin levels, and it is also highly susceptible to external cues generated through learning processes that alert feeding responses to the high-fat, high-sugar food (Chhabra et al., 2016). In addition, peripheral hormonal signals that stimulate orexigenic responses have been demonstrated several months after intentional weight loss, even after weight regain starts (Sumithran et al., 2011). In response, the regulation of food intake communicates with the executive function areas in the brain and the mesolimbic system to activate more motivation to eat through the appetite signals discussed previously (Blüher, 2019). In an environment where food is scarce, the brain is programmed to motivate us toward calorie consumption behaviors (Berridge et al., 2009). This is a biological drive to food that can translate into huger, desire, crave, impulsivity, and compulsion under different circumstances (Davis, 2014). Therefore, it is fair to say that executive and reward systems play as important roles as the autonomic regulation of energy homeostasis in our energy balance throughout life.

Obesity and Psychiatric Diseases

Hypothalamus-pituitary-adrenal (HPA) axis disturbances, imbalance of inflammatory pathways, increased oxidative stress, and reduced antioxidant defenses result in neurodegeneration, apoptosis, reduced neurogenesis, and mitochondrial disturbances, all of which have been hypothesized contribute to psychiatric symptoms in people with obesity, influenced intersectionally by biological, environmental, and genetic mechanisms (Lopresti & Drummond, 2013). Emotion dysregulation has a role in the development and maintenance of emotional eating (EE). Depression and anxiety have been proposed to mediate this phenomenon, with anxiety found to mediate EE in individuals with mild and moderate obesity, and depression found to mediate EE in those with severe obesity (Willem et al., 2020). Inflammation and neural progress must be related to the damage in connectivity. Also, there is increasing evidence of opioid receptors in glial cells in the central nervous system (Machelska & Celik, 2020), which can be dysregulated in inflammatory states. Suffering from both conditions is likely to have an additive influence on these pathways, which can influence food choice, activity patterns, hunger and satiety cues, cognition and emotion regulation.

Anorexia Nervosa

Anorexia nervosa (AN) is a psychiatric disorder characterized by food restriction, body shape concerns, and in some cases purging behaviors (Table 3). Heritability estimates for AN range from 0.28 to 0.74 (Bulik et al., 2019). Because of strict energy restriction of macronutrients and micronutrients, patients present with diverse endocrine disorders; increased of lipolysis, muscle breakdown,

Table 3 Diagnostic criteria anorexia nervosa DSM 5 (American Psychiatric Association, 2013)

DSM 5 criteria anorexia nervosa	
A.	Restriction of energy intake relative to requirements leading to a significantly low body weight in the context of age, sex, developmental trajectory, and physical health. Significantly low weight is defined as a weight that is less than minimally normal or, for children and adolescents, less than that minimally expected.
B.	Intense fear of gaining weight or becoming fat, or persistent behavior that interferes with weight gain, even though at a significantly low weight.
C.	Disturbance in the way in which one's body weight or shape is experienced, undue influence of body weight or shape on self-evaluation, or persistent lack of recognition of the seriousness of the current low body weight.
	Specify if: In partial remission: After full criteria for anorexia nervosa were previously met. Criterion A has not been met for a sustained period, but either Criterion B or C is still met. In full remission: After full criteria for anorexia nervosa were previously met, none of the criteria have been met for a sustained period of time. Specify current severity: The minimum level of severity is based, for adults, on current body mass index (BMI) (see below) or for children and adolescents, on BMI percentile. The ranges below are derived from World Health Organization categories for thinness in adults; for children and adolescents, corresponding BMI percentiles should be used. The level of severity may be increased to reflect clinical symptoms, the degree of functional disability, and the need for supervision. Mild: BMI >17 kg/m^2 Moderate: BMI 16–16.99 kg/m^2 Severe: BMI 15–15.99 kg/m^2 Extreme: BMI <15 kg/m^2

hypothyroidism, decreased estradiol and testosterone, as well as oxytocin and altered glucose uptake and lipogenesis (Misra & Klibanski, 2014)

AN affects women more than men, and its prevalence in western countries varies from 0.1% to 5.7%, and in non-western countries from 0.002% to 0.09% (Makino et al., 2004), suggesting a role of culture in onset and progression. It usually develops in adolescents and young adults. Patients with severe AN have the highest mortality rate of any mental health illness due to organ dysfunction caused by the underlying severe malnutrition and risk of suicide (Hay et al., 2012). AN represents a challenge because patients have many medical complications in multiple organ systems and decreased productivity.

Pathophysiology of AN

Patients present dysregulation in the anterior ventral striatal pathway that create vulnerability for dysregulated appetitive behaviors, the high level of self-control in these patients produced by an exaggerated dorsal cognitive functioning allows them to inhibit feeding behavior. Impairment of the dopamine and serotonin systems in AN lead to malfunction in satiety, impulse control, mood and aberrant rewarding effects of motivation and food (Kaye et al., 2013). Predisposing factors and genetics combined with dysfunctional social relationships since developmental stages (mainly family or primary relationships), onset of puberty (neural pruning) and trauma gives place to altered awareness and integration of interoceptive and exteroceptive stimuli in the insula and thalamus, hyperactivation of the limbic system (affective traits), hypoactivation of the dopaminergic system, and cognitive inflexibility.

Bulimia Nervosa

Bulimia nervosa (BN) starts in adolescents mainly women with body image altered perception, purging and binging episodes as clinical features (Table 4). Prevalence in European countries is 1% and incidence is up to 40 per 100,000 persons among women 12–19 years old. Although anorexia has the highest mortality rate of all eating disorders, bulimia has a mortality rate of 1.74 per 1000 person per year and suicide accounts for 23% of the deaths (Smink et al., 2012). The heritability of BN is estimated to be approximately 0.60 having multiple genetic polymorphisms involved in the genesis of the disease (Bulik et al., 2019; see Table 4).

Pathophysiology of BN

Individuals with BN have aberrant dopamine activity; low levels of dopamine possibly result from downregulation of an overstimulated dopamine (DA) system that leads to binge eating. Serotonin facilitates the release of dopamine in the nuclei

Table 4 Diagnostic criteria bulimia nervosa DSM 5 (American Psychiatric Association, 2013)

DSM 5 criteria bulimia nervosa	
A.	Recurrent episodes of binge eating. An episode of binge eating is characterized by BOTH of the following: 1. Eating in a discrete amount of time (ex: within a 2 h period) an amount of food that is definitely larger than what most individuals would eat in a similar period of time under similar circumstances. 2. Sense of lack of control over eating during an episode.
B.	Recurrent inappropriate compensatory behavior in order to prevent weight gain, such as self-induced vomiting: misuse of laxatives, diuretics, or other medications: fasting: or excessive exercise.
C.	The binge eating and inappropriate compensatory behaviors both occur, on average, at least once a week for three months
D.	Self-evaluation is unduly influenced by body shape and weight.
E.	The disturbance does not occur exclusively during episodes of anorexia nervosa.
	Specify if: In partial remission: After full criteria for bulimia nervosa were previously met, some, but not all of the criteria have been met for a sustained period of time. In full remission: After full criteria for bulimia nervosa were previously met. none of the criteria have been met for a sustained period of time. Specify current severity: The minimum level of severity is based on the frequency of inappropriate compensatory behaviors (see below). The level of severity may be increased to reflect other symptoms and the degree of functional disability. Mild: An average of 1–3 episodes of inappropriate compensatory behaviors per week Moderate: An average of 4–7 episodes of inappropriate compensatory behaviors per week Severe: An average of 8–13 episodes of inappropriate compensatory behaviors per week Extreme: An average of 14 or more episodes of inappropriate compensatory behaviors per week

accumbens (NAc) and stratum, and in patients with BN serotonin release disturbances may alter reward sensitivity through diminished effect of dopaminergic mesolimbic pathways. Also, serotonin receptors 2A in the medial prefrontal cortex have reduced function leading to disinhibition in patients with BN. In patients with chronic bulimia, levels of serotonin decrease progressively with frequent binge episodes (Umberg et al., 2012).

Binge Eating Disorder

Binge eating disorder (BED) is a recently recognized diagnostic category and is the most prevalent eating disorder. Patients with this diagnosis have recurrent episodes with eating an increased amount of food in short periods and sense of lack of control over food consumption during an episode (Table 5). Twin-based heritability estimates are between 0.39 and 0.45 (Buliket al., 2019). Twenty-five percent of patients with obesity searching medical attention have BED and 50–75% with morbid obesity. The lifetime prevalence is 1.4–4% in women and 0.8–2% in men. The age of onset is adolescents and young adults and with earlier onset more medical and psychiatric

Table 5 Diagnostic criteria for binge eating disorder DSM 5 (American Psychiatric Association, 2013)

DSM 5 criteria binge eating disorder	
A.	Recurrent episodes of binge eating. An episode of binge eating is characterized by BOTH of the following: 1. Eating in a discrete amount of time (ex: within a 2 h period) an amount of food that is definitely larger than what most individuals would eat in a similar period of time under similar circumstances. 2. Sense of lack of control over eating during an episode.
B.	The binge eating episodes are associated with: 1. Eating much more rapidly than normal 2. Eating until feeling uncomfortably full 3. Eating large amounts of food when not feeling physically hungry 4. Eating alone because of felling embarrassed by how much one is eating 5. Feeling disgusted with oneself, depressed, or very guilty afterwards.
C.	Marked distress regarding binge eating is present
D.	The binge eating occurs, on average, at least once a week for 3 months.
E.	The binge eating is not associated with the recurrent use of inappropriate compensatory behavior as in bulimia nervosa and does not occur exclusively during the course of bulimia nervosa or anorexia nervosa.
	Specify if: In partial remission: After full criteria for binge eating disorder were previously met, binge eating occurs at an average frequency of less than one episode per week for a sustained period of time. In full remission: After full criteria for binge eating disorder were previously met, none of the criteria have been met for a sustained period of time. Specify current severity: The minimum level of severity is based on the frequency of episodes of binge eating (see below). The level of severity may be increased to reflect other symptoms and the degree of functional disability. Mild: An average of 1–3 episodes binge eating episodes per week Moderate: An average of 4–7 episodes binge eating episodes per week Severe: An average of 8–13 episodes of binge eating episodes per week Extreme: An average of 14 or more episodes of binge eating episodes per week

comorbidities (Kessler et al., 2016). As stated previously, this disorder overlaps with both bulimia nervosa and obesity in their neurobiological and behavioral traits.

Pathophysiology of BED

The major neurobiological disruptions presented in BED are related to impulsivity, compulsivity, and reward processing along with attention biases directed toward food and impaired cognitive function. The ventral striatum, dorsal striatum, prefrontal cortex, and the insula have functional alterations.

Altered dopamine function is the best known contributor to BED; imbalance between the direct striatonigral output pathway (D1 receptors) and indirect striatopallidal pathway (D2 receptors) have reduced function and expression mediating impulsive and compulsive eating behaviors of highly palatable foods (Fig. 11).

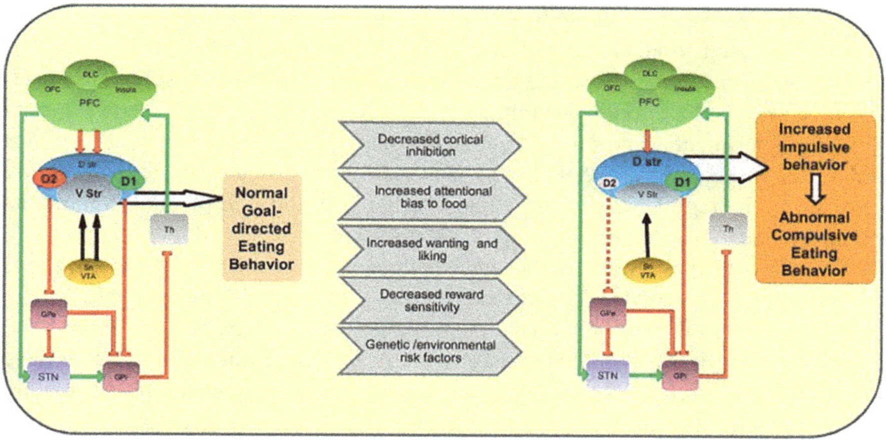

Fig. 11 Cortico-basal and thalamic pathways disturbances in reward response related of BED. PFC prefrontal cortex, OFC orbitofrontal cortex, DLC dorsolateral cortex, insula insular cortex; Dstr dorsal striatum, Vstr ventral striatum, D1 dopomamine D1 like receptor, D2 dopamine D2 like receptor, Th thalamus, Sn VTA substantia nigra, ventral tegmental area nuclei, GPe Globus pallidus external segment, GPi Globus pallidus internal segment, STN subthalamic nucleus. (Adapted from: Kessler et al., 2016)

Research Applications of Endogenous Opioid Involvement in Obesity and Eating Disorders

As we described in the previous chapter, many neurotransmitters are involved in the neural circuits that control eating behavior. In spite that endogenous opioids have been less studied than GABA, serotonin, and dopamine, there is abundant research in opioid receptors and opioid physiology involved in obesity and eating disorders. Because the opioids system has a role in reward, aversion, and cognition networks, modifying mood, stress and hedonic response, they participate in eating regulation of eating behavior and play a role in in hedonic eating and dysfunction manifestations such as impulsivity, attention, anxiety response, and social bias. We will discuss the evidence of many disturbances of this system in eating disorders and obesity.

Role of Opioids in the Pathogenesis of Obesity and Eating Disorders

Opioids in Obesity

Over the last 30 years, experimental and clinical research has focused on the involvement of endogenous opioids in feeding behavior, with a specific dual emphasis on both homeostatic and hedonic mechanisms of food intake (Zheng et al., 2007).

Stress and negative affect are reduced by food, and an inability to control stress may lead to increased food intake and obesity (van Strien et al., 2016). Research has found that the endogenous opioid system is involved in stress and affective responses and may also mediate food consumption (Zheng et al., 2007).

Both experimental and clinical studies suggest a relationship between opioids and obesity (Feng et al., 2012). As mentioned above, endogenous opioids are involved in the regulation of food intake, particularly through β-endorphin and its receptors, stimulating consumption of palatable foods and hedonic responses to feeding. This is thought to be the main pathway through which the endogenous opioid system contributes to the development of obesity (Zheng et al., 2010). Increased levels of endorphins have been reported in obesity, in both animal and human studies. Experimental studies in genetically obese mice and rats that have found elevated levels of β-endorphin in circulation and brain suggested hereditary disturbances of opioids in obesity (Nogueiras et al., 2012). Furthermore, ingestion of food with high sugar and fat content is associated with an increase in β-endorphin (Dum et al., 1983). Levels of endorphins found to be elevated in individuals with obesity did not demonstrate decrease after intentional weight loss induced by caloric restriction (Givens et al., 1980).

Regarding the metabolic abnormalities associated with obesity, assessment of plasma glucose and pancreatic hormones in an obesity-prone population (first-degree relatives of subjects with obesity) and healthy controls without family history of obesity showed that, under basal conditions, there were no significant differences in plasma glucose, insulin, C-peptide, and glucagon, between the two groups. In contrast, plasma β-endorphin levels were significantly higher in relatives of individuals with obesity. Both physiological and pharmacological levels of β-endorphin induced significant pancreatic secretion of insulin, C-peptide, and glucagon in the early phase of human obesity (Cozzolino et al., 1996).

Another study found a physiologic elevation of plasma β-endorphin levels associated with marked increase of glucose, insulin, C-peptide, and glucagon, a progressive decline in free fatty acids, and no change in glycerol plasma levels in patients with obesity. No significant difference was found in pancreatic hormones, but a considerable increase of free fatty acids and suppression of glycerol levels in normal-weight subjects was reported (Giugliano et al., 1988).

Positron emission tomography (PET) studies have found different traits of opioidergic system dysfunction associated with obesity. Evaluation of the central opioid receptor (μ-opioid receptors; MOR) function in subjects with obesity has found decreased availability compared to lean individuals in brain regions relevant for reward processing, including ventral striatum, insula, and thalamus (Karlsson et al., 2015). Individuals with obesity had lower activation of the opioid neurotransmitter system in response to a standard meal compared to lean subjects. After moderate weight loss, MOR availability and activation increased, but without reaching the levels found in lean subjects. Interestingly, a strong association between meal-related MOR system activation and decreases in negative affect was observed in lean subjects but not in those with obesity (Burghardt et al., 2015). Brain MOR availability has been assessed in patients with severe obesity before and 6 months after bariatric surgery. The results of this study indicate that bariatric surgery, whether through severe and sustained weight loss, or through the metabolic effects it creates, achieves significant changes in MOR not observed by nutritional interventions alone. The authors speculated that the low levels associated with increased body weight are most probably a consequence rather than a cause of obesity (Karlsson et al., 2015) (Fig. 12).

Endogenous Opioids in Anorexia Nervosa

Clinical studies about endogenous opioids in anorexia nervosa are limited. Since there are increased levels of opioids in the cerebrospinal fluid (CSF) of patients with an established diagnosis of AN, it has been proposed that opioid activity may be a maladjustment of the internal homeostatic adaptations to starvation in result of prolonged dieting. The behavioral effects of food restriction make endogenous opioid activity fluctuate. Starvation downregulates metabolism and activates food intake; hypothalamic structures sense a decrease in metabolic needs; therefore, opioid receptors increase their expression and release endorphins increasing palatability of food that leads to hyperphagia (Marrazzi & Luby, 1986). Also, this chronic increase in opioids underlie reproductive and metabolic changes observed in AN.

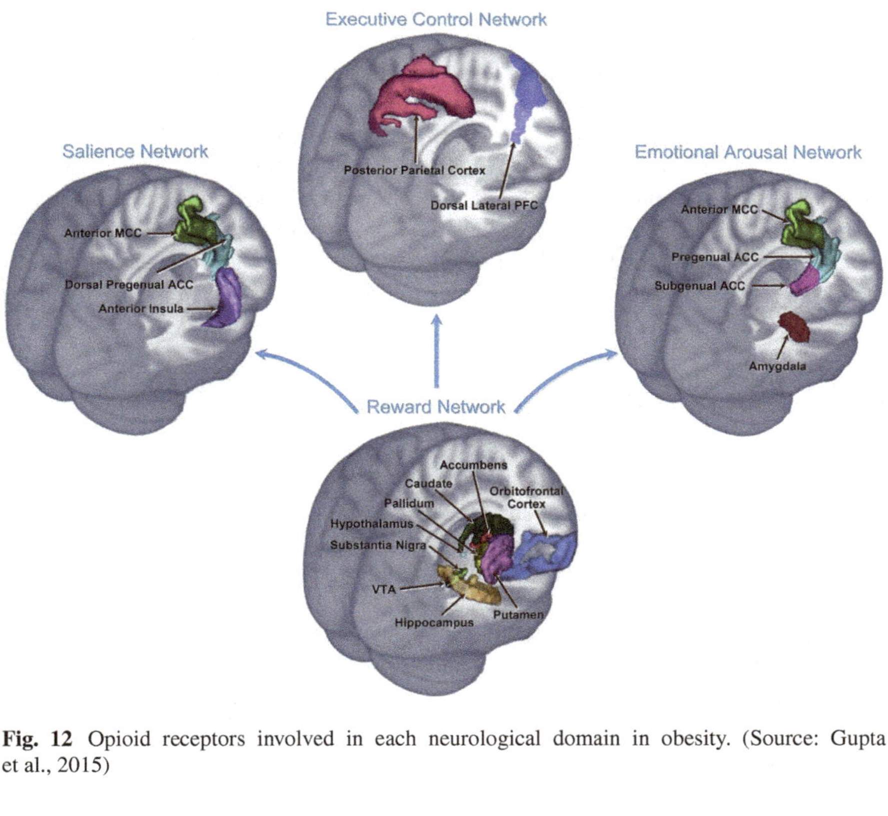

Fig. 12 Opioid receptors involved in each neurological domain in obesity. (Source: Gupta et al., 2015)

Because reward is enhanced by opioid release, another perspective of these findings is that release of endogenous opioids in response to starvation reinforce the restrictive and purging symptoms of anorexia nervosa. These observations have been applied to propose an auto-addictive model of behavior in AN (Marrazzi & Luby, 1986). However, this model lacks direct empirical evidence, and it is likely that endogenous opioids play a different role in the development and maintenance of AN.

Activity-based anorexia (ABA) rat model is an experimental model in which rodents with free access to voluntary exercise on a running wheel experience food restriction, many, but not all, become hyperactive running more than they ran prior to the onset of food restriction (Chowdhury et al., 2015). One hypothesis for this phenomenon is explained as a mechanism to cope with the stress of food restriction. Plasma and hypothalamic levels of β-endorphin and dynorphin-A have been reported to be elevated in an activity-based anorexia rat model (Avena & Bocarsly, 2012) meaning that levels of dynorphin may be elevated as a result of stress. Van Kuyck et al. (2007) noted low metabolism in the ventral striatum with micro-PET (positron emission tomography) in rats in the ABA model; also, a positive correlation was noted between body weight loss and brain metabolism in the cingulate cortex.

Gene polymorphism studies evidenced certain single nucleotide polymorphisms (SNP) in OPDR 1 gene (delta-1 receptor) mainly link to restricting type of anorexia nervosa (Bergen et al., 2003). This gene is associated with pain and stress response; if translated clinically, this system could be responsible for the altered stress response in anorexia nervosa which leads to abnormal eating behavior and the presentation of depression and anxiety comorbidities.

Research will be necessary to investigate the role of MOR and KOR in anorexia nervosa, but disturbances in other affected domains like positive valence, cognitive and social systems may have opioid involvement because of the expression of receptors (Fig. 13). Moore et al. used naloxone, a competitive opioid receptor antagonist, in 12 female patients with anorexia nervosa who were in an inpatient unit showing effect in weight gain and lipolysis (Valbrun & Zvonarev, 2020).

Endogenous Opioids in Bulimia Nervosa

Most of the evidence related to endogenous opioid receptor involvement in BN focuses on the bingeing component. Researchers have proposed that people with BN have lower levels of ß-endorphin and MOR binding. In between purging and bingeing episodes, patients have an upregulation of MOR binding that is proportional to food craving (Valbrun & Zvonarev, 2020). Also, opioid antagonists reduce binge eating in BN patients with selective suppression of ingestion of highly palatable foods. PET studies using radioligands have demonstrated decreased l-opioid binding in the temporo-insular cortex at rest in patients with BN and decreased 5-HT(2a) binding in the medial OFC in patients recovered of BN.

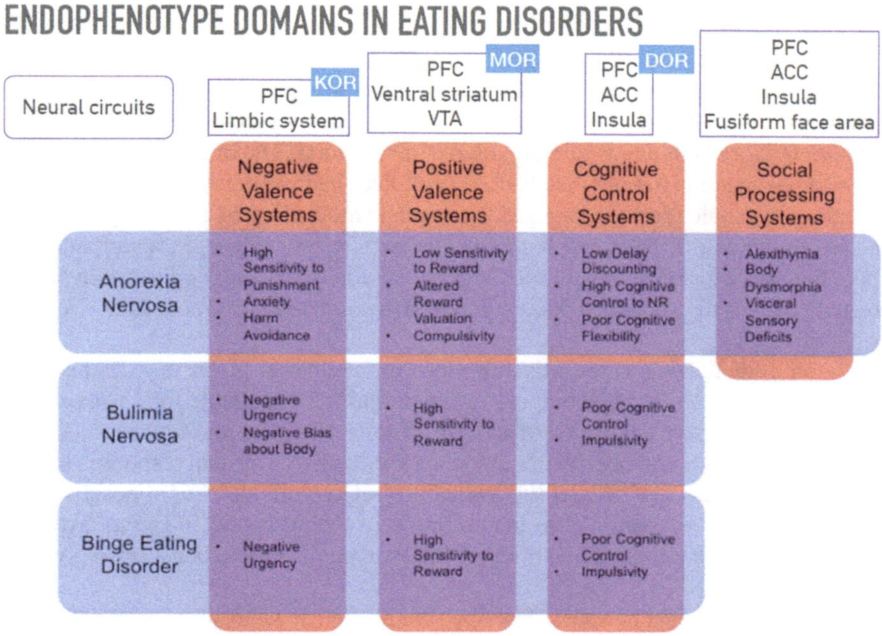

Fig. 13 Endophenotypes domains in eating disorders, implicated neural circuits, and opioid receptor expression. Adapted from (Dunlop et al., 2016). Patients with anorexia and bulimia nervosa exhibit antibodies against melanocortin peptide alpha-MSH, which is responsible for decreasing food intake and is influenced by pre- and postsynaptic action of endogenous opioids

Endogenous Opioids in Binge Eating Disorder

Animal models show that MOR antagonists suppress food bingeing. Dietary restraint for "forbidden" high-palatability foods in people who binge eat leads to relapse episodes (Nathan & Bullmore, 2009). Food restriction followed by access to food show increases in MOR binding in limbic regions, including the cingulate cortex, hippocampus, and the nucleus accumbens shell (Colantuoni et al., 2001). These findings suggest that repeated activation of MORs by food restriction and binge eating may help sustain behavior, including the desire to acquire a high-fat diet. People with obesity who engage binge eating show decrease motivational responses to palatable food, for example, chocolate with MOR agonist (Novelle & Diéguez, 2018).

MORs within the medial prefrontal cortex (mPFC) can suppress natural reward-related behaviors that have a role in binge-eating development. The DOR antagonist naltrindole blocks binge-eating behaviors (Valbrun & Zvonarev, 2020).

Food Addiction

As discussed in the previous chapter, eating behavior is regulated by metabolic and hedonic mechanisms (see Fig. 1, last chapter). Hedonic mechanisms are essential for anticipation, planning, and motivation to obtain food. The increasing availability of food has made metabolic needs easily fulfilled, and overstimulation of pleasure by eating palatable food has changed the way and motivation of our eating behavior, including the use of palatable food to cope with stress. Hence, in anticipation of food, the insula and caudate are activated; in these patients, these areas are hyperactivated, exhibiting enhanced motivation and reward towards food cues (Volkow et al., 2013a, b).

"Food addiction" is a recently introduced term that is given to a "behavioral addiction" characterized by compulsive consumption of palatable food with consequential activation of the reward system despite adverse consequences (Davis, 2014). Weight gain and obesity are associated with altered striatal responses to palatable food or cues that predict the availability of such food; also, hypoactivity of the prefrontal cortex can lead to disinhibition of eating behavior and an aversive response from the hyperactivation of the amygdala and other limbic system structures.

Combined, the disturbances described above can lead to overconsumption of food, but this does not mean that someone can become "addicted" to food in the same way that addiction is used in reference to substances. While there is emerging evidence that some specific types of food products (e.g., highly processed foods with added fats and sugars) may be associated with dysregulated eating, the validity of "food addiction" as a clinical construct remains debated (Hauck et al., 2020). This term is contentious and dubious on an empirical basis.

References

American Psychiatric Association. (2013). *Diagnostic and statistical manual of mental disorders: DSM-5* (Vol. 5). American Psychiatric Association.

Asghari, G., Mirmiran, P., Yuzbashian, E., & Azizi, F. (2017). A systematic review of diet quality indices in relation to obesity. *British Journal of Nutrition, 117*(8), 1055–1065.

Avena, N. M., & Bocarsly, M. E. (2012). Dysregulation of brain reward systems in eating disorders: neurochemical information from animal models of binge eating, bulimia nervosa, and anorexia nervosa. *Neuropharmacology, 63*(1), 87–96.

Baumeister, H., & Härter, M. (2007). Mental disorders in patients with obesity in comparison with healthy probands. *International Journal of Obesity, 31*(7), 1155–1164.

Bergen, A. W., Van Den Bree, M. B. M., Yeager, M., Welch, R., Ganjei, J. K., Haque, K., et al. (2003). Candidate genes for anorexia nervosa in the 1p33–36 linkage region: serotonin 1D and delta opioid receptor loci exhibit significant association to anorexia nervosa. *Molecular Psychiatry, 8*(4), 397–406.

Berkowitz, R. I., & Fabricatore, A. N. (2005). Obesity, psychiatric status, and psychiatric medications. *Psychiatric Clinics, 28*(1), 39–54.

Berridge, K. C., Robinson, T. E., & Aldridge, J. W. (2009). Dissecting components of reward: 'liking', 'wanting', and learning. *Current Opinion in Pharmacology, 9*(1), 65–73.

Blüher, M. (2019). Obesity: Global epidemiology and pathogenesis. *Nature Reviews Endocrinology, 15*(5), 288–298.

Bray, M. S., Loos, R. J., McCaffery, J. M., Ling, C., Franks, P. W., Weinstock, G. M., et al. (2016). NIH working group report—Using genomic information to guide weight management: From universal to precision treatment. *Obesity, 24*(1), 14–22.

Bulik, C. M., Flatt, R., Abbaspour, A., & Carroll, I. (2019). Reconceptualizing anorexia nervosa. *Psychiatry and Clinical Neurosciences, 73*(9), 518–525.

Burghardt, P. R., Rothberg, A. E., Dykhuis, K. E., Burant, C. F., & Zubieta, J. K. (2015). Endogenous opioid mechanisms are implicated in obesity and weight loss in humans. *The Journal of Clinical Endocrinology & Metabolism, 100*(8), 3193–3201.

Caldú, X., Ottino-González, J., Sánchez-Garre, C., Hernan, I., Tor, E., Sender-Palacios, M. J., et al. (2019). Effect of the catechol-O-methyltransferase Val 158Met polymorphism on theory of mind in obesity. *European Eating Disorders Review, 27*(4), 401–409.

Centers for Disease Control and Prevention. (2018). *Overweight & obesity: Adult obesity causes & consequences.*

Chaput, J. P. (2014). Sleep patterns, diet quality and energy balance. *Physiology & Behavior, 134*, 86–91.

Chhabra, K. H., Adams, J. M., Jones, G. L., Yamashita, M., Schlapschy, M., Skerra, A., et al. (2016). Reprogramming the body weight set point by a reciprocal interaction of hypothalamic leptin sensitivity and Pomc gene expression reverts extreme obesity. *Molecular Metabolism, 5*(10), 869–881.

Chowdhury, T. G., Chen, Y. W., & Aoki, C. (2015). Using the activity-based anorexia rodent model to study the neurobiological basis of anorexia nervosa. *JoVE (Journal of Visualized Experiments), 104*, e52927.

Colantuoni, C., Schwenker, J., McCarthy, J., Rada, P., Ladenheim, B., Cadet, J. L., et al. (2001). Excessive sugar intake alters binding to dopamine and mu-opioid receptors in the brain. *Neuroreport, 12*(16), 3549–3552.

Cozzolino, D., Sessa, G., Salvatore, T., Sasso, F. C., Giugliano, D., Lefebvre, P. J., & Torella, R. (1996). The involvement of the opioid system in human obesity: a study in normal weight relatives of obese people. *The Journal of Clinical Endocrinology & Metabolism, 81*(2), 713–718.

D'Souza, M. S. (2019). Brain and cognition for addiction medicine: From prevention to recovery neural substrates for treatment of psychostimulant-induced cognitive deficits. *Frontiers in Psychiatry, 509*, 1–27.

Darby, A., Hay, P., Mond, J., Rodgers, B., & Owen, C. (2007). Disordered eating behaviours and cognitions in young women with obesity: Relationship with psychological status. *International Journal of Obesity, 31*(5), 876–882.

Davis, C. (2014). Evolutionary and neuropsychological perspectives on addictive behaviors and addictive substances: Relevance to the "food addiction" construct. *Substance Abuse and Rehabilitation, 5*, 129–137. https://doi.org/10.2147/sar.s56835

Depner, C. M., Stothard, E. R., & Wright, K. P. (2014). Metabolic consequences of sleep and circadian disorders. *Current Diabetes Reports, 14*(7), 1–9.

Desmet, K. L., Marei, W. F., Richard, C., Sprangers, K., Beemster, G. T., Meysman, P., et al. (2020). Oocyte maturation under lipotoxic conditions induces carryover transcriptomic and functional alterations during post-hatching development of good- quality blastocysts: Novel insights from a bovine embryo-transfer model. *Human Reproduction, 35*(2), 293–307.

Dum, J., Gramsch, C. H., & Herz, A. (1983). Activation of hypothalamic β-endorphin pools by reward induced by highly palatable food. *Pharmacology Biochemistry and Behavior, 18*(3), 443–447.

Dunlop, K. A., Woodside, B., & Downar, J. (2016). Targeting neural endophenotypes of eating disorders with non-invasive brain stimulation. *Frontiers in Neuroscience, 10*(30), 1–20.

Elshenawy, S., & Simmons, R. (2016). Maternal obesity and prenatal programming. *Molecular and Cellular Endocrinology, 435*, 2–6.

Feng, Y., He, X., Yang, Y., Chao, D. H., Lazarus, L., & Xia, Y. (2012). Current research on opioid receptor function. *Current Drug Targets, 13*(2), 230–246.

Giugliano, D., Cozzolino, D., Salvatore, T., Torella, R., Franchimont, P., D'Onofrio, F., & Lefebvre, P. J. (1988). Altered metabolic and hormonal responses to epinephrine and β-endorphin in human obesity. *The Journal of Clinical Endocrinology & Metabolism, 67*(2), 238–244.

Givens, J. R., Wiedemann, E., Andersen, R. N., & Kitabchi, A. E. (1980). β-endorphin and β-lipotropin plasma levels in hirsute women: correlation with body weight. *The Journal of Clinical Endocrinology & Metabolism, 50*(5), 975–976.

Gubareva, L. I., & Posokhina, V. V. (2014). Obesity is one of the main reactions to stress. *International Journal of Psychophysiology, 92*(2), 258–258.

Gupta, A., Mayer, E. A., Sanmiguel, C. P., Van Horn, J. D., Woodworth, D., Ellingson, B. M., et al. (2015). Patterns of brain structural connectivity differentiate normal weight from overweight subjects. *NeuroImage: Clinical, 7*, 506–517.

Hassan, A. (Ed.). (2017). *Pediatric behavioral nutrition factors: Environment, education, and self-regulation*. CRC Press.

Hauck, C., Cook, B., & Ellrott, T. (2020). Food addiction, eating addiction and eating disorders. *Proceedings of the Nutrition Society, 79*(1), 103–112.

Hay, P. J., Touyz, S., & Sud, R. (2012). Treatment for severe and enduring anorexia nervosa: a review. *Australian & New Zealand Journal of Psychiatry, 46*(12), 1136–1144.

Hay, P., Chinn, D., Forbes, D., Madden, S., Newton, R., Sugenor, L., et al. (2014). Royal Australian and New Zealand College of Psychiatrists clinical practice guidelines for the treatment of eating disorders. *Australian & New Zealand Journal of Psychiatry, 48*(11), 977–1008.

Heymsfield, S. B., & Wadden, T. A. (2017). Mechanisms, pathophysiology, and management of obesity. *New England Journal of Medicine, 376*(3), 254–266.

Hruby, A., & Hu, F. B. (2015). The epidemiology of obesity: A big picture. *PharmacoEconomics, 33*(7), 673–689.

Jaacks, L. M., Vandevijvere, S., Pan, A., McGowan, C. J., Wallace, C., Imamura, F., et al. (2019). The obesity transition: Stages of the global epidemic. *The Lancet Diabetes & Endocrinology, 7*(3), 231–240.

Jacquet, P., Schutz, Y., Montani, J. P., & Dulloo, A. (2020). How dieting might make some fatter: Modeling weight cycling toward obesity from a perspective of body composition autoregulation. *International Journal of Obesity, 44*(6), 1243–1253.

Kaila, B., & Raman, M. (2008). Obesity: A review of pathogenesis and management strategies. *Canadian Journal of Gastroenterology, 22*(1), 61–68.

Kanter, R., & Caballero, B. (2012). Global gender disparities in obesity: A review. *Advances in Nutrition, 3*(4), 491–498.

Karlsson, H. K., Tuominen, L., Tuulari, J. J., Hirvonen, J., Parkkola, R., Helin, S., et al. (2015). Obesity is associated with decreased μ-opioid but unaltered dopamine D2 receptor availability in the brain. *Journal of Neuroscience, 35*(9), 3959–3965.

Kaye, W. H., Wierenga, C. E., Bailer, U. F., Simmons, A. N., & Bischoff-Grethe, A. (2013). Nothing tastes as good as skinny feels: the neurobiology of anorexia nervosa. *Trends in Neurosciences, 36*(2), 110–120.

Kessler, R. M., Hutson, P. H., Herman, B. K., & Potenza, M. N. (2016). The neurobiological basis of binge-eating disorder. *Neuroscience & Biobehavioral Reviews, 63*, 223–238.

Lacroix, E., & von Ranson, K. M. (2021). Prevalence of social, cognitive, and emotional impairment among individuals with food addiction. *Eating and Weight Disorders-Studies on Anorexia, Bulimia and Obesity, 26*(4), 1253–1258.

Locke, A. E., Kahali, B., Berndt, S. I., Justice, A. E., Pers, T. H., Day, F. R., et al. (2015). Genetic studies of body mass index yield new insights for obesity biology. *Nature, 518*(7538), 197–206.

Lopresti, A. L., & Drummond, P. D. (2013). Obesity and psychiatric disorders: Commonalities in dysregulated biological pathways and their implications for treatment. *Progress in*

Neuropsychopharmacology and Biological Psychiatry, 45, 92–99. https://doi.org/10.1016/j.pnpbp.2013.05.005

Machelska, H., & Celik, M. Ö. (2020). Opioid receptors in immune and glial cells—implications for pain control. *Frontiers inimmunology, 11*: 300, 472575.

Makino, M., Tsuboi, K., & Dennerstein, L. (2004). Prevalence of eating disorders: a comparison of Western and non-Western countries. *Medscape General Medicine, 6*(3), 49.

Marciniak, A., Patro-Małysza, J., Kimber-Trojnar, Ż., Marciniak, B., Oleszczuk, J., & Leszczyńska-Gorzelak, B. (2017). Fetal programming of the metabolic syndrome. *Taiwanese Journal of Obstetrics and Gynecology, 56*(2), 133–138.

Marrazzi, M. A., & Luby, E. D. (1986). An auto-addiction opioid model of chronic anorexia nervosa. *International Journal of Eating Disorders, 5*(2), 191–208.

Miller, A. L., Lumeng, J. C., & LeBourgeois, M. K. (2015). Sleep patterns and obesity in childhood. *Current Opinion in Endocrinology, Diabetes, and Obesity, 22*(1), 41–47.

Misra, M., & Klibanski, A. (2014). Endocrine consequences of anorexia nervosa. *The lancet Diabetes & endocrinology, 2*(7), 581–592.

Nathan, P. J., & Bullmore, E. T. (2009). From taste hedonics to motivational drive: central μ-opioid receptors and binge-eating behaviour. *International Journal of Neuropsychopharmacology, 12*(7), 995–1008.

Neumark-Sztainer, D., Paxton, S. J., Hannan, P. J., Haines, J., & Story, M. (2006). Does body satisfaction matter? Five-year longitudinal associations between body satisfaction and health behaviors in adolescent females and males. *Journal of Adolescent Health, 39*(2), 244–251.

Nogueiras, R., Romero-Picó, A., Vazquez, M. J., Novelle, M. G., López, M., & Diéguez, C. (2012). The opioid system and food intake: homeostatic and hedonic mechanisms. *Obesity Facts, 5*(2), 196–207.

Novelle, M. G., & Diéguez, C. (2018). Food addiction and binge eating: lessons learned from animal models. *Nutrients, 10*(1), 71.

Pearl, R. L., White, M. A., & Grilo, C. M. (2014). Overvaluation of shape and weight as a mediator between self-esteem and weight bias internalization among patients with binge eating disorder. *Eating Behaviors, 15*(2), 259–261.

Pietiläinen, K. H., Saarni, S. E., Kaprio, J., & Rissanen, A. (2012). Does dieting make you fat? A twin study. *International Journal of Obesity, 36*(3), 456–464.

Pigeyre, M., Yazdi, F. T., Kaur, Y., & Meyre, D. (2016). Recent progress in genetics, epigenetics and metagenomics unveils the pathophysiology of human obesity. *Clinical Science, 130*(12), 943–986.

Qasim, A., Turcotte, M., De Souza, R. J., Samaan, M. C., Champredon, D., Dushoff, J., et al. (2018). On the origin of obesity: Identifying the biological, environmental and cultural drivers of genetic risk among human populations. *Obesity Reviews, 19*(2), 121–149.

Richard, D. (2015). Cognitive and autonomic determinants of energy homeostasis in obesity. *Nature Reviews Endocrinology, 11*(8), 489–501.

Sanz-de-Galdeano, A., Terskaya, A., & Upegui, A. (2020). Association of a genetic risk score with BMI along the life-cycle: Evidence from several US cohorts. *PLoS One, 15*(9), e0239067.

Seidell, J. C., & Halberstadt, J. (2015). The global burden of obesity and the challenges of prevention. *Annals of Nutrition and Metabolism, 66*(Suppl. 2), 7–12.

Skibicka, K. P., & Dickson, S. L. (2011). Ghrelin and food reward: The story of potential underlying substrates. *Peptides, 32*(11), 2265–2273.

Smink, F. R., Van Hoeken, D., & Hoek, H. W. (2012). Epidemiology of eating disorders: incidence, prevalence and mortality rates. *Current Psychiatry Reports, 14*(4), 406–414.

Smith, E., Hay, P., Campbell, L., & Trollor, J. N. (2011). A review of the association between obesity and cognitive function across the lifespan: Implications for novel approaches to prevention and treatment. *Obesity Reviews, 12*(9), 740–755.

Spengler, S., Mess, F., Schmocker, E., & Woll, A. (2014). Longitudinal associations of health-related behavior patterns in adolescence with change of weight status and self-rated health over a period of 6 years: Results of the MoMo longitudinal study. *BMC Pediatrics, 14*(1), 1–11.

Stuart, G. W. (2013). *Principles and practice of psychiatric nursing* (10th ed., p. 478). Elsevier Mosby.

Sumithran, P., Prendergast, L. A., Delbridge, E., Purcell, K., Shulkes, A., Kriketos, A., & Proietto, J. (2011). Long-term persistence of hormonal adaptations to weight loss. *New England Journal of Medicine, 365*(17), 1597–1604.

Summermatter, S., Marcelino, H., Arsenijevic, D., Buchala, A., Aprikian, O., Assimacopoulos-Jeannet, F., et al. (2009). Adipose tissue plasticity during catch-up fat driven by thrifty metabolism: Relevance for muscle-adipose glucose redistribution during catch-up growth. *Diabetes, 58*(10), 2228–2237.

Tounian, P. (2011). Programming towards childhood obesity. *Annals of Nutrition and Metabolism, 58*(Suppl. 2), 30–41.

Treasure, J., Duarte, T. A., & Schmidt, U. (2020). Eating disorders. *Lancet, 395*, 899–911.

Umberg, E. N., Shader, R. I., Hsu, L. G., & Greenblatt, D. J. (2012). From disordered eating to addiction: the "food drug" in bulimia nervosa. *Journal of Clinical Psychopharmacology, 32*(3), 376–389.

Valbrun, L. P., & Zvonarev, V. (2020). The opioid system and food intake: use of opiate antagonists in treatment of binge eating disorder and abnormal eating behavior. *Journal of Clinical Medicine Research, 12*(2), 41–63.

van Kuyck, K., Casteels, C., Vermaelen, P., Bormans, G., Nuttin, B., & Van Laere, K. (2007). Motor-and foodrelated metabolic cerebral changes in the activity-based rat model for anorexia nervosa: a voxel-based microPET study. *Neuroimage, 35*(1), 214–221.

van Strien, T., Konttinen, H., Homberg, J. R., Engels, R. C., & Winkens, L. H. (2016). Emotional eating as a mediator between depression and weight gain. *Appetite, 100*, 216–224.

Volkow, N. D., Wang, G. J., Telang, F., Fowler, J. S., Thanos, P. K., Logan, J., et al. (2008). Low dopamine striatal D2 receptors are associated with prefrontal metabolism in obese subjects: Possible contributing factors. *NeuroImage, 42*(4), 1537–1543.

Volkow, N. D., Wang, G. J., Tomasi, D., & Baler, R. D. (2013a). Obesity and addiction: neurobiological overlaps. *Obesity Reviews, 14*(1), 2–18.

Volkow, N. D., Wang, G. J., Tomasi, D., & Baler, R. D. (2013b). The addictive dimensionality of obesity. *Biological Psychiatry, 73*(9), 811–818.

Voon, V. (2015). Cognitive biases in binge eating disorder: The hijacking of decision making. *CNS Spectrums, 20*(6), 566–573.

Willem, C., Gandolphe, M. C., Doba, K., Roussel, M., Verkindt, H., Pattou, F., & Nandrino, J. L. (2020). Eating in case of emotion dys-regulation, depression and anxiety: Different pathways to emotional eating in moderate and severe obesity. *Clinical Obesity, 10*(5), e12388.

Yengo, L., Sidorenko, J., Kemper, K. E., Zheng, Z., Wood, A. R., Weedon, M. N., et al. (2018). Meta-analysis of genome-wide association studies for height and body mass index in ~700000 individuals of European ancestry. *Human Molecular Genetics, 27*(20), 3641–3649.

Yilmaz, Z., Hardaway, J. A., & Bulik, C. M. (2015). Genetics and epigenetics of eating disorders. *Advances in Genomics and Genetics, 5*, 131–150.

Yu, Y. H., Vasselli, J. R., Zhang, Y., Mechanick, J. I., Korner, J., & Peterli, R. (2015). Metabolic vs. hedonic obesity: A conceptual distinction and its clinical implications. *Obesity Reviews, 16*(3), 234–247.

Zheng, H., Patterson, L. M., & Berthoud, H. R. (2007). Orexin signaling in the ventral tegmental area is required for high-fat appetite induced by opioid stimulation of the nucleus accumbens. *Journal of Neuroscience, 27*(41), 11075–11082.

Zheng, H., Townsend, R. L., Shin, A. C., Patterson, L. M., Phifer, C. B., & Berthoud, H. R. (2010). High-fat intake induced by mu-opioid activation of the nucleus accumbens is inhibited by Y1R-blockade and MC3/4R-stimulation. *Brain Research, 1350*, 131–138.

Integration of Endogenous Opioid System Research in the Interprofessional Diagnosis and Treatment of Obesity and Eating Disorders

Marcela Rodriguez Flores and Sylvana Stephano Zuniga

Abstract This third and final chapter in our trilogy introduces the clinical distinctions and phenotypical similarities between obesity and eating disorders. Research elaborating on the shared neurobiological substrates for obesity and eating disorders is discussed. We present an interprofessional model of treatment for both disordered eating and for obesity. Additionally, this chapter establishes the translational importance of research connecting endogenous opioid activity with both obesity and eating disorders, with an emphasis on clinical interventions. We conclude with a discussion of future directions for research.

Keywords Endogenous opioids · Interprofessional treatment · Obesity · Eating disorders

Introduction

Obesity and eating disorders are separate phenomena with distinct clinical features, which share an etiological link with food consumption for some people. Obesity itself is not classified as an eating disorder, nor any other form of psychiatric disorder. Clearly, not all people with obesity have an eating disorder, and not all people with eating disorders involving overeating develop obesity. However, there may be phenotypical similarities between people with obesity and those with eating disorders in relation to food consumption, which research has established activates

M. R. Flores (✉)
Department of Brain Sciences, Imperial College London, Hammersmith Hospital, London, UK

S. S. Zuniga
Obesity and Eating Disorders Clinic, Instituto Nacional de Ciencias Médicas y Nutrición Salvador Zubirán, Tlalpan, Mexico

P. L. Kerr et al. (eds.), *Endogenous Opioids*, Advances in Neurobiology 35,
https://doi.org/10.1007/978-3-031-45493-6_18

endogenous opioid activity. These commonalities may originate from the shared neurobiological substrates we have described, especially those associated with the endogenous opioid system. In this section, we delineate the translational importance of research connecting endogenous opioid activity with both obesity and eating disorders, with an emphasis on clinical interventions.

Assessment and Diagnosis of Obesities: Role of Endogenous Opioids

Obesity is defined as an increase in body fat, which can be identified by a body mass index (BMI) ≥30 kg/m^2. Abdominal obesity is defined as the presence of waist circumference ≥88 cm in women and ≥102 cm in men, with varying cut points according to ethnic group (Fitzpatrick et al., 2016). A large corpus of literature has found that both overweight (BMI ≥25 kg/m^2) and obesity, when associated with increased visceral adipose tissue, significantly increase risk for chronic diseases (e.g., type 2 diabetes mellitus; cardiovascular disease; osteoarticular, cardiopulmonary, gastrointestinal, genitourinary and neurological disorders; and several types of cancer), disability, and early mortality (Arnold et al., 2015; Blüher, 2019; Ryan & Kahan, 2018). Finally, there is a broad array of psychosocial conditions associated with obesity that might function as promoters, consequences, or barriers for treatment, such as lack of peer support, socioeconomic disadvantage, mood and anxiety disorders, and eating disorders (Mechanick et al., 2017).

There is empirical evidence that classification of obesity according to BMI alone is insufficient to predict health outcomes and to guide clinical decision-making in obesity management (Kuk et al., 2011). Moreover, BMI by itself fails to account for the complex intersection of neurobiological factors, including endogenous opioid signalling, involved in obesity. Thus, alternative systems are needed.

One example of an alternative approach is the Edmonton Obesity Staging System (EOSS). The EOSS was developed to provide a more holistic clinical assessment of the health burden induced by obesity, beyond the sole measurement of a person's size (Fig. 1; Blüher, 2019). The EOSS has demonstrated advantages in clinical settings. For example, research indicates that the EOSS is superior to BMI in predicting bariatric surgery complications, fertility treatment outcomes, and mortality (Chiappetta et al., 2016; Kuk et al., 2011; Paterson et al., 2016).

The EOSS is based on the three areas most often affected by obesity (medical aspects, functionality, and mental health) in each individual to direct clinical decision-making (Sharma & Kushner, 2009). EOSS classifies the health burden associated with overweight/obesity into four strata, as seen in Fig. 1. By obtaining this portrait of every patient in each clinical encounter, there is a health-based indication for weight loss actions along with an objective assessment of consequences and potential barriers for improved outcomes.

Multidimensional assessment plays a crucial part in the diagnosis of obesity, as it may highlight relevant targets for treatment. The EOSS was designed to be a

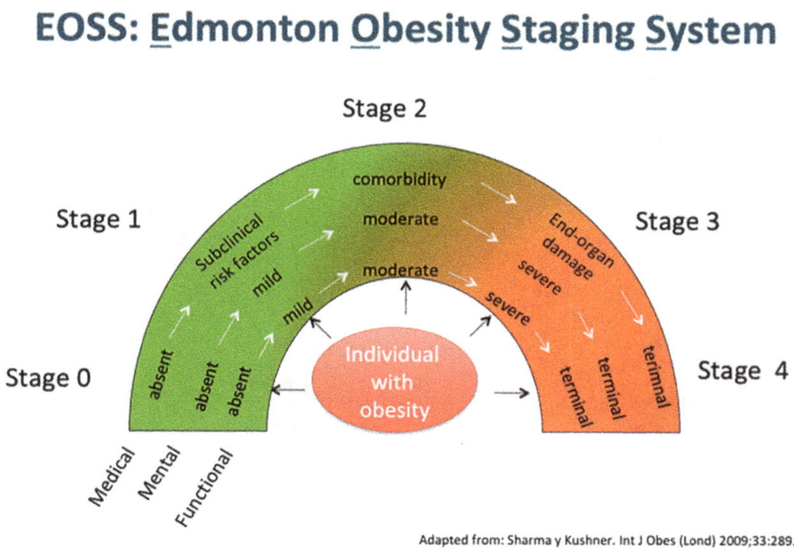

Fig. 1 Edmonton Obesity Staging System for diagnosis of individuals with overweight/ obesity

multidimensional assessment system. In the area of eating behavior, a practical approach is to assess the patterns that correspond to the three main neurological regulators of energy balance: autonomic circuits, mesolimbic system, and executive function. In combination with other neurobiological processes, endogenous opioid signaling is involved in multiple systems that may underpin the etiology and maintenance of obesity (see Table 1). Consonant with this, endogenous opioid receptor activity is predictive of response to some obesity treatments (e.g., bariatric surgery), as we will discuss later. Thus, a system that promotes multidimensional assessment has inherent value in clinical settings.

A spectrum of overeating behavior or loss-of-control eating might be present in individuals with overweight and obesity, including none at all. This continuum can range from "emotional eating," to using food as a coping strategy, and binge eating (Mobbs et al., 2010). Practitioners in obesity management should inquire about eating behaviors to differentiate between EOSS levels, including aspects of adequate nutrition and hydration, and "conscious" perspectives of the patient, in order to treat accordingly (Kushner & Sarwer, 2011).

Understanding the complexities of the pathophysiology of obesity recognizes the historical bias of health professionals towards it, with a more cosmetic and moralistic perspective than a scientific and clinical one. In response to pervasive misinformation around obesity, which have prioritized commercial interests and contributed to the obesity epidemic, international efforts are being undertaken towards ending obesity stigma and clinical inertia to increase patients' access to science-based treatments (Rubino et al., 2020).

Table 1 From neurobiology to diagnosis of eating behavior in patients with obesity

Regulating system	CNS structures	Relevant mechanism	Signaling	Clinical Assessment
Autonomic energy homeostasis	Hypothalamus (ARC, LH)	Hunger and satiety	Opioids, POMC, AGRP, leptin, ghrelin	Questioning of hunger (associated with prolonged fasting, dieting), hyperphagia (syndromes, hypothalamic lesions): *Hunger patterns, lack of satiety, compulsivity (visual analog scales for appetite, thirst)*
Autonomic endocrine	Anterior, medial, and posterior pituitary, hypothalamic SON and PVN, thyroid and adrenal glands	Endocrine regulation	GH, ACTH, LH, FSH, TSH, ADH, thyroid hormones, aldosterone, cortisol	Questioning of hunger and satiety, and fluid and electrolyte regulation: *Abnormal hunger with or without lack of satiety after a normal meal, associated stages of life, fluid intake, thirst*
Positive valence systems	Striatum, AN, VTA Caudate nucleus	Reward	Opioid, dopamine, serotonin, endocannabinoid	Questioning of normal/ abnormal reward responses to food: *Craving, dependence, tolerance (Yale Food Addiction Scale)*
Negative valence system	Limbic system: Amygdala, hippocampus, ACC	Avoidance Anxiety Stress reactivity	Opioid, serotonin Endocannabinoid	Questioning avoidance and mood (guilt, dislike, pain, anxiety): *Barriers for consumption of vegetables physical activity, eating as coping mechanism, triggers*
Mesolymbic	Ventromedial PFC, lateral OFC, cortico-limbic circuit	Emotional dysregulation Self-control	Dopamine, norepinephrine, serotonin	Questioning self-harm, binge eating, emotional regulation: *Have you ever hurt yourself without the intention of killing yourself? Do you experience loss of control while eating? Do you feel emotions deeper than others? Is it hard for you to change from one emotion to another? Do you find it difficult to reappraisal your emotions?*

(continued)

Table 1 (continued)

Regulating system	CNS structures	Relevant mechanism	Signaling	Clinical Assessment
Cognitive control system	Lateral PFC, DLC, thalamus, VTA, ACC, parietal cortex, visual cortex, hippocampus	Attention	Norepinephrine, dopamine	Trail making test Digit span, Go/no-Go test Attention subtest in mini-mental state examination (MMSE) or Montreal cognitive assessment (MOCA)
Cognitive control system	Hippocampus, rhinal cortex, cerebellum, basal ganglia	Memory: Working, short-term, long-term, explicit. Semantic, implicit, procedural, episodic	Serotonin, glutamate, GABA, ach	Memory subtest in mini-mental state examination (MMSE) or Montreal cognitive assessment (MOCA)
Cognitive control system	PFC, basal ganglia, ACC, posterior parietal cortex	Cognitive flexibility	Glutamate, GABA, serotonin	Stroop test, Wisconsin card sorting test, A-not-B task, set-shifting task
Social system	Temporo-parietal junction, medial parietal cortex, medial PFC, insula, fusiform gyrus	Social interaction Emotion recognition Attachment	Serotonin, oxytocin, vasopressin, dopamine	Emotion recognition of self and others Theory of mind task battery

ARC Arcuate nucleus, *POMC* Proopiomelanocortin, *AGRP* Agouti-Related-Peptide, *SON* Supraoptic nucleus, *PVN* Paraventricular nucleus, *GH* Growth hormone, *ACTH* Adrenocorticotrophic hormone, *LH* Luetinizing hormone, *FSH* Follicle-stimulating hormone, *TSH* Thyroid-stimulating hormone, *ADH* Antidiuretic hormone, *AN* Accumbens nucleus, *PFC* Prefrontal cortex, *OFC* Orbitofrontal cortex, *DLC* Dorsolateral cortex, *VTA* Ventral tegmental area nuclei, *ACC* Anterior cingulate cortex

Management of Obesity Through Multidisciplinary Interventions

A full description of all interventions for obesity is beyond this chapter's emphasis on the role of endogenous opioids. Nonetheless, an overview of a multidisciplinary framework for intervention, including pharmacotherapies that activate endogenous opioid activity, is a useful contribution to this discussion.

The global growth of obesity has led to a proliferation of treatments to achieve weight loss and reduce morbidity. However, reducing obesity's impact on comorbidities requires long-term treatment through evidence-based, multidisciplinary weight management interventions. Our current understanding of the complex pathophysiology of obesity has generated a shift away from the antiquated "eat less,

Table 2 Identification of patients eligible for obesity treatment according to body mass index

	BMI category (kg/m^2)				
Intervention	25–26.9	27–29.9	30–34.9	35–39.9	≥40
Diet, physical activity, and behavioral therapy	Without comorbidities	Without comorbidities	+	+	+
Pharmacotherapy		With comorbidities	+	+	+
Bariatric surgery			With comorbidities		+

move more" principles that were the pillars of obesity treatment in previous decades, to evidence-based principles of chronic disease management that validate patients' lived experiences and address the root drivers of obesity (Busetto et al., 2018; Wharton et al., 2020).

Several studies have shown benefits in the control of chronic diseases associated with obesity with a 5–10% reduction in weight (Fayh et al., 2013). Therefore, the goal of weight reduction in the medical treatment of obesity is to achieve this range, which may reverse some of the harmful mechanisms of weight gain. Guidelines recommend initial detection of individuals who would benefit from treatment according to their BMI as an initial step in obesity management and increase intensity of treatment with increasing BMI category (Table 2; Busetto et al., 2018; NICE, 2014).

The heterogeneous manifestations of obesity require individualization of interventions, which has proven challenging. Significant weight loss is challenging to achieve and sustain; and the most effective obesity interventions have included complex and intensive components that are difficult to enact from standard primary care practice, which has limited their implementation in health systems around the world (Luig et al., 2020). Therefore, most guidelines recommend utilization of the 5As framework to engage clinicians and patients in effective obesity treatments, and use of EOSS as a clinical decision scheme to guide utilization of clinical tools and in conversations with patients beyond BMI (Luig et al., 2020).

Endogenous Opioid Activity and Pharmacological Interventions for Obesity

As a result of the biological mechanisms favoring weight regain after intentional weight loss, and the pathophysiological mechanisms of obesity as a disease, significant weight loss and maintenance with lifestyle recommendations alone is hard to achieve, and risks for adverse outcomes are not sufficiently reduced (Gilis-Januszewska et al., 2018). The addition of pharmacotherapy to lifestyle changes can help patients achieve an additional weight loss, between 5% and 10%, with which different clinical benefits occur, including enhanced maintenance of weight loss and reduction of cardiovascular risk factors and incidence of diabetes in patients with prediabetes.

Table 3 Drugs approved for obesity treatment

	Mechanism of action	Mean weight loss	Observations
Orlistat	Pancreatic lipase inhibitor	−6.1%	Side effects limit adherence
Liraglutide	GLP-1 receptor agonist	−7.4 to −11%	Injectable, gastrointestinal side effects
Phentermine/ topiramate ER	Sympathomimetic/ anticonvulsant	−7.8 to −9.8%	Teratogen
Naltrexone SR/ bupropion SR	Opioid receptor antagonist/ dopamine and noradrenaline reuptake inhibitor	−5.4%	Reduces food craving, particular side effects

Medications for the treatment of obesity are indicated in people with a BMI ≥ 27 kg/m^2 and obesity comorbidities, and in those with a BMI ≥ 30 kg/m^2, seeking to enhance weight loss and control of comorbidities. Pharmacotherapy should be combined with lifestyle modifications. Pharmacological options are still scarce, with only five drugs approved for the long-term treatment of obesity (>12 months) by the world's major regulatory agencies, the US Food and Drug Administration (FDA) and the European Agency of Medications (EMA), for having demonstrated efficacy for weight loss and safety (see Table 3).

Development of medications for the treatment of obesity has focused on peripheral and central nervous system (CNS) mechanisms that can promote weight loss. The main peripheral mechanisms comprise increased thermogenesis, lipolysis, modifications on gut microbiota, bile acids, renal glucose excretion, malabsorption, and enhanced gut peptide action. Mechanisms acting on the CNS can support behavior change through emotion and impulse regulation, decrease overactivation of aversive and anxious traits and craving, and improvement of attention disturbances (Bray et al., 2016).

Building on the neurobiological findings described earlier, endogenous opioid system activity is also one target of adjunctive pharmacotherapy for obesity. Naltrexone/Bupropion has been approved in the USA since 2012 and in the European Union since 2015. Bupropion stimulates POMC neurons through inhibition of dopamine and norepinephrine reuptake. Naltrexone potentiates this stimulation through blockade of the inhibitory effects of opioid receptors on satiety, activated by β-endorphin release (Greenway et al., 2010). This allows the inhibitory effects of α-melanocyte stimulating hormone to reduce food intake, and it seems particularly useful in patients with compulsive eating patterns (Valbrun & Zvonarev, 2020). This combination helps achieve mild to moderate weight loss with the greatest weight loss attributed to bupropion.

A recent meta-analysis of unpublished clinical study data found that this medication is superior to placebo for weight loss. However, the high attrition rate in studies may underestimate the actual effects (Onakpoya et al., 2020). Related to this, it is important to note that the safety of naltrexone/bupropion remains in question: the sympathomimetic effects of bupropion can increase blood pressure, significantly higher risk of gastrointestinal symptoms have been found, and this drug is labeled with a warning of suicidality (Nissen et al., 2016). Thus, any proposed benefits must be carefully weighed against this risk profile.

Liraglutide is a glucagon-like peptide 1 (GLP-1) receptor agonist with both central and peripheral mechanisms for weight loss. Although it does not target endogenous opioid receptors specifically, this compound does act through related substrates, that is, proopiomelanocortin (POMC). Liraglutide stimulates thermogenesis through hypothalamic receptors, where it also stimulates satiety through (1) direct activation of POMC/cocaine and amphetamine-related transcript (CART) neurons; and (2) indirect suppression of activating neuropeptide Y (NPY)/Agouti-related peptide (AgRP) neurons. This activation occurs through GABA-dependent signals, which also suppress the reward system (Farr et al., 2016; Kanoski et al., 2011). This translates in reduced interest, or "wanting" in food, at increasing doses. Peripherally, liraglutide influences gastrointestinal motor function, delaying gastric emptying. The result is moderate weight loss, with cardiometabolic benefits and neuropsychiatric safety. One study found that liraglutide significantly improved weight loss when combined with intensive behavioral therapy (O'Neil et al., 2017; Pi-Sunyer et al., 2015; Wadden et al., 2019).

Pharmacotherapies with non-opioid signaling mechanisms also merit discussion, including orlistat and phentermine/topiramate combination. Orlistat is the only pharmacotherapy for obesity with peripheral actions (and minimal absorption) by irreversibly blocking gastric and pancreatic lipase, promoting malabsorption of dietary fat and resulting in caloric deficit (Drew et al., 2007). This mechanism is responsible for its main side effects (diarrhea, fecal incontinence, fat-soluble vitamin deficiency), which limits its tolerance in some patients, but induces reduction in fat consumption in others. In contrast to Naltrexone/Bupropion, which targets endogenous opioid receptors that affect eating behavior, orlistat targets metabolic processes (fat absorption). This medication has modest effects that can increase the benefits of lifestyle changes and improve metabolic parameters (Furman, 2007).

The phentermine/topiramate combination works by combining the anorectic sympathomimetic effects of phentermine (approved for short-term use alone) together with those of the anticonvulsant topiramate that increases stimulation of GABA receptors (Gadde et al., 2018) It is an extended release formulation that uses lower doses of both drugs than are usually prescribed alone (Aronne et al., 2013). It helps achieve moderate weight loss with side effects that include paresthesia, dizziness, dysgeusia, insomnia, constipation, and dry mouth. It is teratogenic and contraindicated in patients with glaucoma (Gadde et al., 2011).

Bariatric Surgery for Obesity

As noted in Table 2, bariatric surgery is indicated for people with severe obesity (BMI ≥40 kg/m^2) and Class II obesity (BMI = 35–39.9 kg/m^2) when accompanied by comorbidities (Mechanick et al., 2020). The main types of surgery are restrictive

Table 4 Types of bariatric surgery, associated mechanisms, and weight loss

	Mechanism of action	Mean 2–3 year weight loss	Observations
Adjustable gastric banding	Restriction of gastric capacity	−15.9%	Rapid weight regain
Sleeve gastrectomy	Restriction of gastric capacity, some metabolic effects	−25 to −30%	Same efficacy the first years, more weight regain afterwards
Roux-en-Y gastric bypass	Restriction of gastric capacity, some metabolic effects	−30 to −35%	Chronic malabsorption, risk for nutritional deficiencies
One-anastomosis gastric bypass	Restriction of gastric capacity, some metabolic effects	−35.4%	Chronic malabsorption, risk for nutritional deficiencies
Biliopancreatic diversion	Restriction of gastric capacity, some metabolic effects	−40%	Severe malabsorption, higher risk for nutritional deficiencies

(e.g., adjustable gastric band and sleeve gastrectomy) and mixed procedures (e.g., Roux-en-Y gastric bypass, biliopancreatic diversion, and one-anastomosis gastric bypass). They work through restriction of the gastric capacity with or without significant changes in food absorption and in the secretion of intestinal hormones (Nguyen & Varela, 2017; Table 4). These mechanisms reduce risk of type 2 diabetes mellitus, cardiovascular events, some types of cancer, and improve quality of life and survival (Ryan & Kahan, 2018).

Among the biological predictors of outcomes after bariatric surgery, μ-opioid receptor (MOR) availability before and after the procedure has been consistently shown to play a role for weight loss, feeding behavior, and mental health. Positron emission tomography (PET) studies have found significant association between MOR availability in the ventral striatum (VST), dorsal caudate (DCAUD), putamen (PUT), insula (INS), amygdala (AMYG), thalamus (THA), orbitofrontal cortex (OFC), anterior cingulate cortex (ACC), medial cingulate cortex (MCC), and posterior cingulate cortex (PCC), and the magnitude of weight loss after surgery (Karlsson et al., 2017).

Bariatric surgery procedures also often induce changes in food preferences, with increased acceptance of healthier foods accompanied by reduction of desire for sweet and fatty food. This has been shown to correspond with reduction in activation of brain reward centers that can lead to aversive responses to them, linking the enhanced gut peptide response characteristic of these procedures with recovering of downregulation of MOR availability and decreased availability of dopamine type 2 receptors (DA D2) in reward areas of the brain (Behary & Miras, 2015; Dunn et al., 2010; Karlsson et al., 2016; Miras et al., 2012; Scholtz et al., 2015). These pathways are also involved in the development of alcohol and opioid use disorders after bariatric procedures, beyond the changes in their pharmacokinetics but associated with previous use (Blackburn et al., 2017; Raebel et al., 2014). King et al. (2017) reported decrease in use of analgesic opioid treatment soon after bariatric surgery, followed by increases above presurgical levels, in association with public versus private health insurance, more presurgical pain, undergoing subsequent surgeries, worsening or less improvement in pain, and starting or continuing non-opioid analgesics post-surgery (King et al., 2017).

Bariatric surgery requires that patients participate in multidisciplinary teams with complex preparation protocols and longitudinal follow-up to obtain optimal results in the (Fried et al., 2014). Routine follow-up is crucial to reduce the risk of side effects from different types of surgery (e.g., weight regain, nutritional and metabolic disorders, disordered eating, substance use disorders, depression, and suicidality; de Heide et al., 2016; Green et al., 2014; Paul et al., 2017; Sarwer et al., 2011; Steffen et al., 2015; Tabesh et al., 2019). Conversely, there is emerging evidence that anxiety and depressive disorders may be reduced for some people following bariatric surgery (Gill et al., 2019). However, great heterogeneity of studies and populations do not allow establishing bariatric surgery as a stand-alone therapeutic treatment for these disorders and recognition of individuals at risk for these problems remains a crucial part of follow-up (Sarwer et al., 2011).

Endogenous and Exogenous Opioids in Obesity: Breakthroughs in Research

Research has established that exogenous opioids have profound effects on eating behavior, likely because these peptides exhibit opioid-like activity, in addition to a wide range of biological functions (i.e., neuromodulation, hormone-like function, analgesia, sedative effect, sleepiness, antihypertensive effects, promotion of alcohol consumption, inflammation and apnea; Kaur et al., 2020). Opioids have been shown to have orexigenic effects, especially for palatable foods, influencing food motivation, inducing stress-related eating, and in some individuals promoting binge eating leading to obesity (Bodnar, 2019).

In the last 50 years, many basic and clinical studies have explored the mechanisms and clinical applications of the opioid system in the treatment of obesity. Several experiments suggest the involvement of endogenous opioids and cannabinoids in the selection of a specific nutrient related to their rewarding mechanisms. Central administration of opioids or their antagonists, at the level of the hypothalamus or NAC has been used as a method to study these pathways and responses. In a study in which rats were allowed to choose the preferred macronutrient, the agonism of MOR through administration of morphine increased fat intake and decreased carbohydrate intake (Cota et al., 2006). Another study found that some rats preferred fat while others preferred carbohydrates. The administration of morphine stimulated the ingestion of carbohydrates in those rats that preferred this macronutrient and the ingestion of fat in rats that displayed a preference for fat (Gosnell et al., 1990). Furthermore, the administration of [D-Ala2, N-Me-Phe4, Gly5-ol]-Enkephalin (DAMGO), a MOR agonist, into the nucleus accumbens stimulated a preferential intake of fat diet and not a carbohydrate-rich diet, independent of baseline preferences (Zhang et al., 1998).

In contrast, opioid receptor antagonists, for example, naloxone, naltrexone, and nalmefene, inhibit eating in various conditions (e.g., administration of insulin, norepinephrine or benzodiazepine, electrical brain stimulation, social conflicts, and stress-induced hyperphagia; Levine et al., 1990; Mercer & Holder, 1997). In rats, both naloxone and naltrexone have been associated with significantly greater reduction in food intake for palatable nutrients versus standard chow-based diets (Apfelbaum & Mandenoff, 1981). This line of research has also found that opioids reduce feeding that is stimulated by sweet taste (Levine et al., 1995). Opioid antagonists have also been demonstrated to be most effective in reducing the consumption of sweet liquids containing glucose, fructose, or saccharine (Daniels et al., 2018; Fantino et al., 1986).

Naloxone has been shown to decrease fat or carbohydrate intake, depending upon baseline preferences (Gosnell & Levine, 2009; Gosnell & Majchrzak, 1989). Previous authors have proposed that the opioid receptors responsible for this interaction are μ-opioid receptors and κ-opioid receptors because the effect of AgRP is suppressed by the concomitant blockade of these receptors (Nogueiras et al., 2012). Administration of naloxone also decreases food intake through the NPY pathway (Zheng et al., 2010). Subsequent findings suggest that naloxone reduces the ingestion of preferred food at lower doses than those needed for the suppression of non-preferred nutrients (Glass et al., 2001). Data also indicate that opioids' sensory and metabolic effects on food intake vary according to their localization, with those in the amygdala involved in processing the emotional salience of foods, and those in the hypothalamus processing the value of food for meeting energy needs (Glass et al., 1999). Naltrexone, injected directly in the central nucleus, decreased the intake of preferred food, while the injection into the paraventricular nucleus reduced the ingestion of both preferred and non-preferred food. These results reflect a differential effect of naltrexone for homeostatic regulation of periventricular injection and for a hedonic mechanism when blocking the paraventricular nucleus (Glass et al., 2000).

Clinical studies investigating the endogenous opioid system in obesity have led to abundant evaluations of receptor antagonists, particularly those that inhibit MORs and KORs, as an option for the treatment of obesity (Fig. 2). In experimental studies, the systemic or intracerebroventricular administration of opioid agonists or antagonists increases and reduces intake of palatable food (Eikemo et al., 2016). Both naloxone and naltrexone, administrated in subjects with obesity, suppress food intake, and in some cases, reduce hunger (Drewnowski et al., 1992). Also in individuals with obesity, the administration of naloxone reduced glucose response and annulled the pancreatic secretion, suggesting the role of opioid receptors in these metabolic reactions (Giugliano et al., 1988). Early research by Fantino et al. (1986) found that naltrexone reduces pleasantness ratings of glucose solutions; and a subsequent study (Yeomans et al., 1990) found nalmefene selectively reduced the ingestion of preferred foods accompanied by increased alertness.

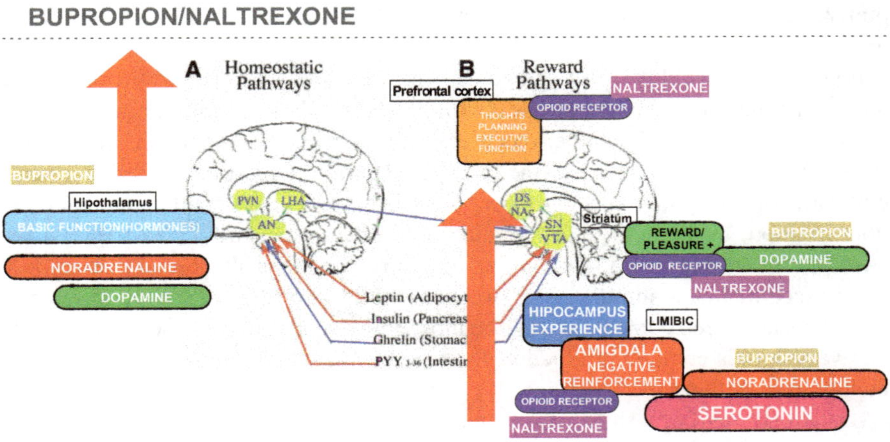

Fig. 2 Bupropion/naltrexone mechanisms within the central nervous system. This combination is the only opioid-based pharmacological treatment for obesity

Human studies in both normal-weight subjects and those with obesity have shown a significant reduction in the short term of food intake but without consistent weight loss (Yeomans & Gray, 2002). Metabolic and hormonal responses to β-endorphin infusion did not change in individuals with obesity after weight loss, supporting this neurobiological predisposition to weight gain in some (Giugliano et al., 1991).

These antagonists have shown reduction of binge episodes in people diagnosed with bulimia nervosa and or binge eating disorder (Nogueiras et al., 2012). Drewnowski et al. (1992) used sugar/fat mixtures to examine the effect of naloxone and butorphanol on food taste and consumption in a study that compared women reporting binge eating with normal-weight controls. The results indicated that naloxone significantly reduced taste preferences for sugar/fat sensory stimuli in all subjects and sugar/fat food ingestion in binge eaters, suggesting that abnormalities in endogenous opioids may be linked to preferences for sugar/fat food and compulsive overeating of sugar/fat mixtures. Hunger and fullness ratings were not affected by naloxone, the most critical effects being the reduction in episodes of binge eating. Consequently, the most important therapeutic role of opioid antagonists seems to be in reducing binge episodes and not in chronic control of body weight (Drewnowski et al., 1992).

Several synthetic opioid antagonists have demonstrated long-term effect in decreasing food intake. A dual MOR receptor and KOR antagonist LY255582 reduced food intake and body weight when injected intraventricularly, subcutaneously, or after chronic oral treatment in experimental studies. It also inhibited the consumption of palatable foods and blocked the activation of the mesolimbic dopamine system stimulated by highly palatable food (Czyzyk et al., 2010).

Another synthetic dual-action antagonist that blocks MOR and δ-opioid receptors, MZ-2 (H-Dmt-Tic-Lys-NH-CH_2-Ph), was found to reduce weight gain in sedentary obese mice. This compound also has some favorable metabolic effects, including a decrease of fat content, an increase of body mineral density, reduction of glucose, and insulin concentrations in obese mice (Marczak et al., 2009).

Opioid antagonists reduce food intake and body weight, and also ameliorate metabolic damage associated with obesity. Based on these findings, some medications aimed at reducing excessive food consumption and body weight, which act as opioid antagonists, have been developed and to date, only naltrexone has been approved as a long-term pharmacological treatment for obesity. Nevertheless, future research is needed to completely clarify the complex involvement of the endogenous opioid system in food intake and obesity that may lead to directed control of feeding behavior and to more efficient treatments for obesity.

Contextualizing the Diagnosis and Treatment of Eating Disorders Within the Endogenous Opioid System

As previously discussed, eating disorders are complex, multifactorial, and heterogeneous diseases (see diagnostic criteria in Tables 3, 4, and 5). The intersection of endogenous opioids and eating disorders treatment research is extremely limited. The data available at present consists only of pharmacological data, and primarily relate to bingeing and purging behaviors (Fig. 3).

Table 5 Review of studied pharmacological treatments in anorexia nervosa

PSYCHOPHARMACOLOGICAL TREATMENT OF ANOREXIA NERVOSA

Treatment (Grade of evidence)	Weight	Fear of gaining weight/body dissatisfaction	Re-feeding anxiety	Relapse	Depression	Obsessive symptoms
Olanzapine (B)	↑	↓	↓	NA	↓	NA
Amisulpride Risperidone Quetiapine (C)	↑	-	-	NA	Quetiapine	↓
Fluoxetine (E)	↑	NA	NA	Limited evidence after weight regain 60mg	↓	NA

Aigner et al. (2011), Hay and Claudino (2012), The National Institute for Health and Care Excellence (2009), Hay et al. (2014), and Wilson and Shafran (2005)

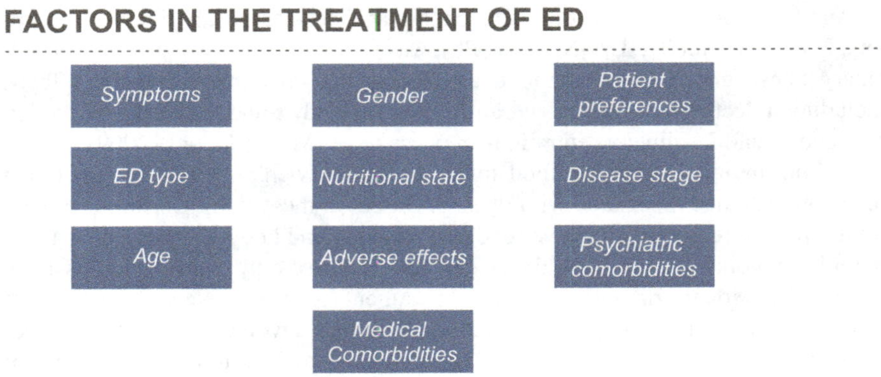

Fig. 3 Factors to consider in prescribing psychopharmacological treatment for eating disorders

Bulimia Nervosa

There is a limited set of studies that has found relatively inconsistent outcomes for the use of MOR antagonists in the treatment of people with BN. An early open label trial of high-dose (200–300 mg/day) versus standard-dose (50–100 mg/day) naltrexone for bulimia nervosa by Jonas et al. (1988) found that high-dose condition participants exhibited a significant reduction in bingeing and purging behaviors, while those in the standard dose condition did not. Similarly, a crossover study of naltrexone (50 mg/day) by Mitchell et al. (1989) in people with BN who were not obese found no changes in bingeing and purging behaviors over a six-week trial.

Consistent with Mitchell and colleagues, a study by Alger and colleagues (1991) compared naltrexone (100–150 mg/day), imipramine, and placebo in women with BN and people with obesity and binge eating, and found no between-groups or pre-post differences at 8 weeks. However, a within-groups difference was found for duration (but not frequency) of binge eating in participants with BN. Subsequently, Marrazzi et al.'s (1995) six-week crossover trial of naltrexone (200 mg/day) in people with BN yielded significant decreases in bingeing and purging behaviors, as well as reported decreases in intensity of binge and purge urges. In their study of naloxone, Drewnowski and colleagues (1995) found evidence that this drug reduced consumption of high-palatability foods (high-sweet/high-fat) in people with bulimia nervosa, regardless of weight status (i.e., obese, nonobese), but did not have this same effect for participants without BN.

Collectively, these studies suggest that MORs may be a fruitful target for pharmacotherapy, but higher doses may be required for efficacy. Despite the rigor of most of these trials, these data include a very limited number of participants overall, and substantially more research is needed in this area.

Anorexia Nervosa

As discussed previously, there is some data suggesting that neurobiological substrates underpinning AN symptomatology may intersect with or even include the endogenous opioid system. However, there is extremely limited research on the use of opioid receptor antagonists as pharmacotherapy for AN. Early data from Marrazzi and colleagues (1995) found that patients with AN binge-purge subtype responded favorably to naltrexone, with reduced bingeing and purging, and "ritualistic eating." However, this study included a small and highly heterogeneous group. A more recent study by Stancil et al. (2019) found that bingeing and purging behaviors were reduced in adolescents with either AN, BN, or OSFED in a treatment program for disordered eating. However, the study was retrospective, included a small total sample ($N = 33$), and an even smaller subgroup of patient with AN ($n = 12$). Additionally, differences between ED types were not evaluated, limiting the conclusions that can be drawn.

Researchers have tested the effects of endocannabinoid receptor agonists as a pharmacological intervention in AN. Early data from Gross et al. (1983) found no benefit for weight gain of Δ-9-tetrahydrocannabinol (THC) dosed up to 30 mg daily, but reported multiple adverse effects (e.g., sleep disturbance, interpersonal sensitivity). More recent results from a crossover RCT by Andries et al. (2014, 2015a) present a more nuanced picture. Patients receiving 5 mg of dronabinol (a synthetic cannabinoid analogue) daily for 4 weeks exhibited both significantly greater weight gain and increases in duration and intensity of physical activity. In their synthesis of this literature via systematic review, Rosager et al. (2021) note that the available data are quite limited (24 participants in one study, and 20 participants in another, all participants being women), the findings are mixed, and some aspects of the study designs lack adequate rigor. A more recently published case of a male with AN given 7.5 mg of dronabinol twice daily presented slightly different outcomes. Consistent with prior research, Graap et al. (2018) reported weight gain during use of dronabinol. However, in contrast to the study by Andries, this patient reported decreases in urges for physical activity, as well as reductions in AN-related cognitions and behaviors. His dose was nearly threefold higher than that used in the study by Andries et al. (2014, 2015a, b), and perhaps this accounts for the difference.

It is worth noting that some authors have also shown evidence that cannabis use may have an inverse relationship with BMI. This same line of research has also found that use of a cannabinoid receptor inverse agonist (rimonabant) was associated with weight loss in people with obesity (see Murphy & Le Foll, 2020). This is somewhat counterintuitive given the data on patients with AN. Thus, more research is clearly needed to comprehensively understand the transactions between the endocannabinoid system and body weight.

In sum, the etiological and intervention data pertaining to endogenous opioids and AN are limited in both quality and quantity. Despite speculation earlier in the literature, there is no direct evidence that AN functions as an "auto-addiction." Similar to the controversial concept of "food addiction," this remains an etiological hypothesis for which researchers have yet to find an empirical basis. This is important when considering pharmacotherapies for people with AN. While there may be some emerging evidence for MOR antagonists in people with BN, no such evidence exists for targeting endogenous opioid receptors in the treatment of AN (Tables 6, 7, and 8).

Table 6 Review of studied pharmacological treatments in bulimia

PSYCHOPHARMACOLOGICAL TREATMENT OF BULIMIA NERVOSA

Treatment (Grade of evidence)	Binges	Purging	Depression	Weight	Relapse	Abandonment
Imipramine (A)	Mild reduction	↓	↓	Insonsistent	NA	↑
Fluoxetine (A)	↓ 60 mg	↓	↓	Inconsistent	↑	-
Topiramate (A)	↓	↓	-	↓	NA	Inconsistent
Fluvoxamine (B)	150mg	↓	↓	NA	↓	-
Sertraline (B)	Slight reduction	Slight reduction	↓	-	NA	-
Citalopram (E)	↓	↓	↓	Inconsistent	NA	-

Aigner et al. (2011), Hay and Claudino (2012), The National Institute for Health and Care Excellence (2009), Hay et al. (2014), and Wilson and Shafran (2005)

Table 7 Review of studied pharmacological treatments in binge eating disorder

PSYCHOPHARMACOLOGICAL TREATMENT BINGE EATING DISORDER

Treatment (Grade of evidence)	Binges	Weight	Depression	ADHD
Imipramine (A)	↓	↓ With diet	↓	Inconsistent
Fluoxetine (B)	↓	Inconsistent	↓	-
Citalopram (A)	↓	↓	↓	-
Sertraline (B)	Improvement in mild symptoms	-	↓	-
Lisdexamfetamine (A)	↓	↓	↓	Improvement in symptoms
Escitalopram (A)	Slight reduction	↓	↓	-
Atomoxetine (B)	Slight reduction	↓	↓	Improvement in mild symptoms
Topiramate (B)	↓	↓	↓	Worsen symptoms

Aigner et al. (2011), Hay and Claudino (2012), The National Institute for Health and Care Excellence (2009), Hay et al. (2014), and Wilson and Shafran (2005)

Table 8 Proposed staging system in ED

Proposal of staging system in eating disorders		
Stage 0 (premorbid disease)	Overvaluation of weight/shape Fear of fatness or guilt when eating Psychiatric comorbidity	Social anxiety Alexithymia
Stage 1	Overvaluation of weight/shape Fear of fatness or behavior preventing weight gain Weekly binge eating Weekly compensatory behaviors Weekly restrictive behaviors Light physical complaints	Social anxiety Alexithymia Psychiatric comorbidity
Stage 2	Overvaluation of weight/shape Fear of fatness or behavior preventing weight gain Underweight/overweight Unmet nutritional and/or energy needs Physical factors requiring medical care Weekly binge eating Weekly compensatory behaviors Weekly restrictive behaviors	Psychiatric comorbidities Social isolation and anxiety
Stage 3	Overvaluation of weight/shape Fear of fatness or behavior preventing weight gain Severe malnutrition/obese Unmet nutritional and/or energy needs Severe physical factors requiring medical care Weekly binge eating Weekly compensatory behaviors Weekly restrictive behaviors	Psychiatric comorbidities Organ damage Complete Social isolation >3 years
Stage 4	Overvaluation of weight/shape Fear of fatness or behavior preventing weight gain Chronic severe malnutrition/obese Unmet nutritional and/or energy needs Physical factors requiring medical care	Several times a day binge eating Several times a day compensatory behaviors Several times a day restrictive behaviors Psychiatric comorbidities Organ damage Complete Social isolation >7 years

This proposal opens a field for individualized treatment and diagnosis in eating disorders. Clinical research is necessary to correlate the prognosis of each stage and treatment interventions. The current knowledge about the pathophysiology of obesity and eating disorders must be translated into best clinical interventions

Conclusions and Future Perspectives

Obesity and eating disorders are public health problems resulting from the collision of our biological predisposition with a changing environment characterized by widespread palatable food, decreasing settings for physical activity, and economic and psychosocial stress. Within our biological predisposition, the extremely complex and sophisticated neurobiological systems play a central role in the diverse expressions of these conditions. The three main neurobiological levels that manifest the mechanisms towards obesity and eating disorders are the areas of autonomic homeostasis regulating hunger and satiety, the subcortical limbic circuitry regulating the reward system, and the prefrontal cortex complex that regulate executive functions.

Endogenous opioids have proven to be crucial modulators of several pathways that influence dysfunctional eating behaviors. Evidence shows that opioids normally acting through dopamine release in reward systems in which MOR and DOR receptors become dysregulated and upregulated may lead to diverse endophenotypes, such as compulsive eating, as well as food-avoidant responses seen in anorexia nervosa and bulimia nervosa. These interactions, occurring through the activities of GABA prefrontal cortical neurons, appear to be in part responsible for the dysfunction of inhibitory processes that patients with eating disorders and obesity manifest.

The neurobiology of eating disorders is also shaped by different elements in our development; genetic and environmental factors, nurture, socioeconomic status, culture, food availability, social interaction, and individual education. There is no long-term evidence yet that could tell us how these disorders arise from all this combination of factors. Clinically, these diseases should be seen as complex model systems, aiming at individualized diagnosis and treatment by determining domains that can be present in different combinations for each patient. Research into how to best define combination of domains and construction of endophenotypes may provide clearer therapeutic pathways.

Finally, there are gaps in the prognostic and treatment approaches of eating disorders that many research groups have been trying to solve by segmenting the age, gender, chronicity and nutritional status of patients. Severity of eating disorders is evaluated by body mass index in anorexia nervosa and the number of binges and purges in bulimia nervosa and binge eating disorder. These criteria aim to detect physiological risk and guide the intensity of interventions, without reflecting the complexity of the disease. Here we propose a staging system in eating disorders based on clinical observation, Australian and New Zealand clinical guidelines for eating disorders, and FREED (first episode rapid early intervention for eating disorders) program of Maudsley Hospital (Hay et al., 2014).

References

Aigner, M., Treasure, J., Kaye, W., Kasper, S., & WFSBP Task Force on Eating Disorders. (2011). World Federation of Societies of Biological Psychiatry (WFSBP) guidelines for the pharmacological treatment of eating disorders. *The World Journal of Biological Psychiatry, 12*(6), 400–443. https://doi.org/10.3109/15622975.2011.602720

Andries, A., Frystyk, J., Flyvbjerg, A., & Støving, R. K. (2014). Dronabinol in severe, enduring anorexia nervosa: A randomized controlled trial. *International Journal of Eating Disorders, 47*(1), 18–23.

Andries, A., Frystyk, J., Flyvbjerg, A., & Støving, R. K. (2015a). Changes in IGF-I, urinary free cortisol and adipokines during dronabinol therapy in anorexia nervosa: Results from a randomised, controlled trial. *Growth Hormone & IGF Research, 25*(5), 247–252.

Andries, A., Gram, B., & Støving, R. K. (2015b). Effect of dronabinol therapy on physical activity in anorexia nervosa: A randomised, controlled trial. *Eating and Weight Disorders-Studies on Anorexia, Bulimia and Obesity, 20*(1), 13–21.

Apfelbaum, M., & Mandenoff, A. (1981). Naltrexone suppresses hyperphagia induced in the rat by a highly palatable diet. *Pharmacology Biochemistry and Behavior, 15*(1), 89–91. https://doi.org/10.1016/0091-3057(81)90344-0

Arnold, M., Pandeya, N., Byrnes, G., Renehan, A. G., Stevens, G. A., Ezzati, M., et al. (2015). Global burden of cancer attributable to high body-mass index in 2012: A population-based study. *The Lancet Oncology, 16*(1), 36–46. https://doi.org/10.1016/S1470-2045(14)71123-4

Aronne, L. J., Wadden, T. A., Peterson, C., Winslow, D., Odeh, S., & Gadde, K. M. (2013). Evaluation of phentermine and topiramate versus phentermine/topiramate extended-release in obese adults. *Obesity, 21*(11), 2163–2171. https://doi.org/10.1002/oby.20584

Behary, P., & Miras, A. D. (2015). Food preferences and underlying mechanisms after bariatric surgery. *Proceedings of the Nutrition Society, 74*(4), 419–425. https://doi.org/10.1017/S0029665115002074

Blackburn, A. N., Hajnal, A., & Leggio, L. (2017). The gut in the brain: The effects of bariatric surgery on alcohol consumption. *Addiction Biology, 22*(6), 1540–1553. https://doi.org/10.1111/adb.12436

Blüher, M. (2019). Obesity: Global epidemiology and pathogenesis. *Nature Reviews Endocrinology, 15*(5), 288–298. https://doi.org/10.1038/s41574-019-0176-8

Bodnar, R. J. (2019). Endogenous opioid modulation of food intake and body weight: Implications for opioid influences upon motivation and addiction. *Peptides, 116*, 42–62. https://doi.org/10.1016/j.peptides.2019.04.008

Bray, G. A., Frühbeck, G., Ryan, D. H., & Wilding, J. P. (2016). Management of obesity. *The Lancet, 387*(10031), 1947–1956. https://doi.org/10.1016/S0140-6736(16)00271-3

Busetto, L., Dicker, D., Azran, C., Batterham, R. L., Farpour-Lambert, N., Fried, M., et al. (2018). Obesity management task force of the European Association for the study of obesity released "practical recommendations for the post-bariatric surgery medical management". *Obesity Surgery, 28*(7), 2117–2121. https://doi.org/10.1159/000481825

Chiappetta, S., Stier, C., Squillante, S., Theodoridou, S., & Weiner, R. A. (2016). The importance of the Edmonton Obesity Staging System in predicting postoperative outcome and 30-day mortality after metabolic surgery. *Surgery for Obesity and Related Diseases, 12*(10), 1847–1855. https://doi.org/10.1016/j.soard.2016.02.042

Cota, D., Tschöp, M. H., Horvath, T. L., & Levine, A. S. (2006). Cannabinoids, opioids and eating behavior: The molecular face of hedonism? *Brain Research Reviews, 51*(1), 85–107. https://doi.org/10.1016/j.brainresrev.2005.10.004

Czyzyk, T. A., Nogueiras, R., Lockwood, J. F., McKinzie, J. H., Coskun, T., Pintar, J. E., et al. (2010). κ-Opioid receptors control the metabolic response to a high-energy diet in mice. *The FASEB Journal, 24*(4), 1151–1159. https://doi.org/10.1096/fj.09-143610

Daniels, S., Pratt, M., Zhou, Y., & Leri, F. (2018). Effect of steady-state methadone on high fructose corn syrup consumption in rats. *Journal of Psychopharmacology, 32*(2), 215–222. https://doi.org/10.1177/0269881117742116

de Heide, L. J. M., Emous, M., & van Beek, A. P. (2016). Postprandial Hyperinsulinemic Hypoglycemia in bariatric surgery. In V. R. Preedy, R. Rajendram, & C. R. Martin (Eds.), *Metabolism and pathophysiology of bariatric surgery: Nutrition, procedures, outcomes and adverse effects* (pp. 305–313). https://doi.org/10.1016/B978-0-12-804011-9.00054-6

Drew, B. S., Dixon, A. F., & Dixon, J. B. (2007). Obesity management: Update on orlistat. *Vascular Health and Risk Management, 3*(6), 817.

Drewnowski, A., Krahn, D. D., Demitrack, M. A., Nairn, K., & Gosnell, B. A. (1992). Taste responses and preferences for sweet high-fat foods: Evidence for opioid involvement. *Physiology & Behavior, 51*(2), 371–379. https://doi.org/10.1016/0031-9384(92)90155-U

Dunn, J. P., Cowan, R. L., Volkow, N. D., Feurer, I. D., Li, R., Williams, D. B., et al. (2010). Decreased dopamine type 2 receptor availability after bariatric surgery: Preliminary findings. *Brain Research, 1350*, 123–130. https://doi.org/10.1016/j.brainres.2010.03.064

Eikemo, M., Løseth, G. E., Johnstone, T., Gjerstad, J., Willoch, F., & Leknes, S. (2016). Sweet taste pleasantness is modulated by morphine and naltrexone. *Psychopharmacology, 233*(21), 3711–3723. https://doi.org/10.1007/s00213-016-4403-x

Fantino, M., Hosotte, J., & Apfelbaum, M. (1986). An opioid antagonist, naltrexone, reduces preference for sucrose in humans. *American Journal of Physiology – Regulatory, Integrative and Comparative Physiology, 251*(1), R91–R96. https://doi.org/10.1152/ajpregu.1986.251.1.r91

Farr, O. M., Sofopoulos, M., Tsoukas, M. A., Dincer, F., Thakkar, B., Sahin-Efe, A., et al. (2016). GLP-1 receptors exist in the parietal cortex, hypothalamus and medulla of human brains and the GLP-1 analogue liraglutide alters brain activity related to highly desirable food cues in individuals with diabetes: A crossover, randomised, placebo-controlled trial. *Diabetologia, 59*(5), 954–965. https://doi.org/10.1007/s00125-016-3874-y

Fayh, A. P. T., Lopes, A. L., da Silva, A. M. V., Reischak-Oliveira, A., & Friedman, R. (2013). Effects of 5% weight loss through diet or diet plus exercise on cardiovascular parameters of obese: A randomized clinical trial. *European Journal of Nutrition, 52*(5), 1443–1450. https://doi.org/10.1007/s00394-012-0450-1

Fitzpatrick, S. L., Wischenka, D., Appelhans, B. M., Pbert, L., Wang, M., Wilson, D. K., & Pagoto, S. L. (2016). An evidence-based guide for obesity treatment in primary care. *The American Journal of Medicine, 129*(1), 115-e1-115e7. https://doi.org/10.1016/j.amjmed.2015.07.015

Fried, M., Yumuk, V., Oppert, J. M., Scopinaro, N., Torres, A., Weiner, R., et al. (2014). Interdisciplinary European guidelines on metabolic and bariatric surgery. *Obesity Surgery, 24*(1), 42–55. https://doi.org/10.1007/s11695-013-1079-8

Furman, B. L. (2007). Orlistat. In S. J. Enna & D. B Bylund (Eds.), *xPharm: The comprehensive pharmacology reference* (pp 1–3). https://doi.org/10.1016/B978-008055232-3.62333-5

Gadde, K. M., Allison, D. B., Ryan, D. H., Peterson, C. A., Troupin, B., Schwiers, M. L., & Day, W. W. (2011). Effects of low-dose, controlled-release, phentermine plus topiramate combination on weight and associated comorbidities in overweight and obese adults (CONQUER): A randomised, placebo-controlled, phase 3 trial. *The Lancet, 377*(9774), 1341–1352. https://doi.org/10.1016/S0140-6736(11)60205-5

Gadde, K. M., Martin, C. K., Berthoud, H. R., & Heymsfield, S. B. (2018). Obesity: Pathophysiology and management. *Journal of the American College of Cardiology, 71*(1), 69–84. https://doi.org/10.1016/j.jacc.2017.11.011

Gilis-Januszewska, A., Barengo, N. C., Lindström, J., Wójtowicz, E., Acosta, T., Tuomilehto, J., et al. (2018). Predictors of long term weight loss maintenance in patients at high risk of type 2 diabetes participating in a lifestyle intervention program in primary health care: The DE-PLAN study. *PLoS One, 13*(3), e0194589. https://doi.org/10.1371/journal.pone.0194589

Gill, H., Kang, S., Lee, Y., Rosenblat, J. D., Brietzke, E., Zuckerman, H., & McIntyre, R. S. (2019). The long-term effect of bariatric surgery on depression and anxiety. *Journal of Affective Disorders, 246*, 886–894. https://doi.org/10.1016/j.jad.2018.12.113

Giugliano, D., Cozzolino, D., Salvatore, T., Torella, R., Franchimont, P., D'onofrio, F., & Lefebvre, P. J. (1988). Altered metabolic and hormonal responses to epinephrine and β-endorphin in human obesity. *The Journal of Clinical Endocrinology & Metabolism, 67*(2), 238–244. https://doi.org/10.1210/jcem-67-2-238

Giugliano, D., Cozzolino, D., Torella, R., Lefebvre, P. J., Franchimont, P., & D'Onofrio, F. (1991). Persistence of altered metabolic responses to beta-endorphin after normalization of body weight in human obesity. *European Journal of Endocrinology, 124*(2), 159–165. https://doi.org/10.1530/acta.0.1240159

Glass, M. J., Billington, C. J., & Levine, A. S. (1999). Opioids and food intake: Distributed functional neural pathways? *Neuropeptides, 33*(5), 360–368. https://doi.org/10.1054/npep.1999.0050

Glass, M. J., Billington, C. J., & Levine, A. S. (2000). Naltrexone administered to central nucleus of amygdala or PVN: Neural dissociation of diet and energy. *American Journal of Physiology – Regulatory, Integrative and Comparative Physiology, 279*(1), R86–R92. https://doi.org/10.1152/ajpregu.2000.279.1.r86

Glass, M. J., Grace, M. K., Cleary, J. P., Billington, C. J., & Levine, A. S. (2001). Naloxone's effect on meal microstructure of sucrose and cornstarch diets. *American Journal of Physiology – Regulatory, Integrative and Comparative Physiology, 281*(5), R1605–R1612. https://doi.org/10.1152/ajpregu.2001.281.5.r1605

Gosnell, B. A., & Levine, A. S. (2009). Reward systems and food intake: Role of opioids. *International Journal of Obesity, 33*(2), S54–S58. https://doi.org/10.1038/ijo.2009.73

Gosnell, B. A., & Majchrzak, M. J. (1989). Centrally administered opioid peptides stimulate saccharin intake in nondeprived rats. *Pharmacology Biochemistry and Behavior, 33*(4), 805–810. https://doi.org/10.1016/0091-3057(89)90474-7

Gosnell, B. A., Krahn, D. D., & Majchrzak, M. J. (1990). The effects of morphine on diet selection are dependent upon baseline diet preferences. *Pharmacology Biochemistry and Behavior, 37*(2), 207–212. https://doi.org/10.1016/0091-3057(90)90322-9

Graap, H., Erim, Y., & Paslakis, G. (2018). The effect of dronabinol in a male patient with anorexia nervosa suffering from severe acute urge to be physically active. *International Journal of Eating Disorders, 51*(2), 180–183.

Green, D. D., Engel, S. G., & Mitchell, J. E. (2014). Psychological aspects of bariatric surgery. *Current Opinion in Psychiatry, 27*(6), 448–452. https://doi.org/10.1097/YCO.0000000000000101

Greenway, F. L., Fujioka, K., Plodkowski, R. A., Mudaliar, S., Guttadauria, M., Erickson, J., et al. (2010). Effect of naltrexone plus bupropion on weight loss in overweight and obese adults (COR-I): A multicentre, randomised, double- blind, placebo-controlled, phase 3 trial. *The Lancet, 376*(9741), 595–605. https://doi.org/10.1016/S0140-6736(10)60888-4

Gross, H., Ebert, M. H., Faden, V. B., Goldberg, S. C., Kaye, W. H., Caine, E. D., et al. (1983). A double-blind trial of Δ9-tetrahydrocannabinol in primary anorexia nervosa. *Journal of Clinical Psychopharmacology, 3*(3), 165–171.

Hay, P. J., & Claudino, A. M. (2012). Clinical psychopharmacology of eating disorders: A research update. *International Journal of Neuropsychopharmacology, 15*(2), 209–222. https://doi.org/10.1017/S1461145711000460

Hay, P., Chinn, D., Forbes, D., Madden, S., Newton, R., Sugenor, L., Touyz, S., Ward, W., Australian, R., College, N. Z., & of Psychiatrists. (2014). Royal Australian and New Zealand College of Psychiatrists clinical practice guidelines for the treatment of eating disorders. *The Australian and New Zealand Journal of Psychiatry, 48*(11), 977–1008. https://doi.org/10.1177/0004867414555814

Jonas JM, Gold MS. The use of opiate antagonists in treating bulimia: a study of low-dose versus high-dose naltrexone. *Psychiatry Research,* 1988;*24*(2), 195–5.

Kanoski, S. E., Fortin, S. M., Arnold, M., Grill, H. J., & Hayes, M. R. (2011). Peripheral and central GLP-1 receptor populations mediate the anorectic effects of peripherally administered GLP-1 receptor agonists, liraglutide and exendin- 4. *Endocrinology, 152*(8), 3103–3112. https://doi.org/10.1210/en.2011-0174

Karlsson, H. K., Tuulari, J. J., Tuominen, L., Hirvonen, J., Honka, H., Parkkola, R., et al. (2016). Weight loss after bariatric surgery normalizes brain opioid receptors in morbid obesity. *Molecular Psychiatry, 21*(8), 1057–1062. https://doi.org/10.1038/mp.2015.153

Karlsson, H. K., Tuominen, L., Salminen, P., Nuutila, P., & Nummenmaa, L. (2017). Preoperative mu-opioid receptor availability predicts weight development following bariatric surgery. *Journal of Cerebral Blood Flow and Metabolism, 37*, 163–164.

Kaur, J., Kumar, V., Sharma, K., Kaur, S., Gat, Y., Goyal, A., & Tanwar, B. (2020). Opioid peptides: An overview of functional significance. *International Journal of Peptide Research and Therapeutics, 26*(1), 33–41. https://doi.org/10.1007/s10989-019-09813-7

King, W. C., Chen, J. Y., Belle, S. H., Courcoulas, A. P., Dakin, G. F., Flum, D. R., et al. (2017). Use of prescribed opioids before and after bariatric surgery: Prospective evidence from a US multicenter cohort study. *Surgery for Obesity and Related Diseases, 13*(8), 1337–1346. https://doi.org/10.1016/j.soard.2017.04.003

Kuk, J. L., Ardern, C. I., Church, T. S., Sharma, A. M., Padwal, R., Sui, X., & Blair, S. N. (2011). Edmonton obesity staging system: Association with weight history and mortality risk. *Applied Physiology, Nutrition, and Metabolism, 36*(4), 570–576. https://doi.org/10.1139/h11-058

Kushner, R. F., & Sarwer, D. B. (2011). Medical and behavioral evaluation of patients with obesity. *Psychiatric Clinics, 34*(4), 797–812. https://doi.org/10.1016/j.psc.2011.07.002

Levine, A. S., Grace, M., Billington, C. J., & Portoghese, P. S. (1990). Nor-binaltorphimine decreases deprivation and opioid-induced feeding. *Brain Research, 534*(1–2), 60–64. https://doi.org/10.1016/0006-8993(90)90112-O

Levine, A. S., Weldon, D. T., Grace, M., Cleary, J. P., & Billington, C. J. (1995). Naloxone blocks that portion of feeding driven by sweet taste in food-restricted rats. *American Journal of Physiology – Regulatory, Integrative and Comparative Physiology, 268*(1), R248–R252. https://doi.org/10.1152/ajpregu.1995.268.1.r248

Luig, T., Wicklum, S., Heatherington, M., Vu, A., Cameron, E., Klein, D., et al. (2020). Improving obesity management training in family medicine: Multi-methods evaluation of the 5AsT-MD pilot course. *BMC Medical Education, 20*(1), 1–11. https://doi.org/10.1186/s12909-019-1908-0

Marczak, E. D., Jinsmaa, Y., Myers, P. H., Blankenship, T., Wilson, R., Balboni, G., et al. (2009). Orally administered H-Dmt-Tic-Lys-NH-CH2-Ph (MZ-2), a potent μ-/δ-opioid receptor antagonist, regulates obese-related factors in mice. *European Journal of Pharmacology, 616*(1–3), 115–121. https://doi.org/10.1016/j.ejphar.2009.06.041

Mechanick, J. I., Hurley, D. L., & Garvey, W. T. (2017). Adiposity-based chronic disease as a new diagnostic term: The American Association of Clinical Endocrinologists and American College of Endocrinology position statement. *Endocrine Practice, 23*(3), 372–378. https://doi.org/10.4158/EP161688.PS

Mechanick, J. I., Apovian, C., Brethauer, S., Garvey, W. T., Joffe, A. M., Kim, J., et al. (2020). Clinical practice guidelines for the perioperative nutrition, metabolic, and nonsurgical support of patients undergoing bariatric procedures– 2019 update: Cosponsored by American Association of Clinical Endocrinologists/American College of Endocrinology, the Obesity Society, American Society for Metabolic & Bariatric Surgery, Obesity Medicine Association, and American Society of Anesthesiologists. *Surgery for Obesity and Related Diseases, 16*(2), 175–247. https://doi.org/10.1016/j.soard.2019.10.025

Mercer, M. E., & Holder, M. D. (1997). Food cravings, endogenous opioid peptides, and food intake: A review. *Appetite, 29*(3), 325–352. https://doi.org/10.1006/appe.1997.0100

Miras, A. D., Jackson, R. N., Jackson, S. N., Goldstone, A. P., Olbers, T., Hackenberg, T., et al. (2012). Gastric bypass surgery for obesity decreases the reward value of a sweet-fat stimulus as assessed in a progressive ratio task. *The American Journal of Clinical Nutrition, 96*(3), 467–473. https://doi.org/10.3945/ajcn.112.036921

Mitchell JE, Christenson G, Jennings J, Huber M, Thomas B, Pomeroy C, Morley J. A placebo-controlled, double blind crossover study of naltrexone hydrochloride in outpatients with normal weight bulimia. *Journal of Clinical Psychopharmacology*, 1989;*9*(2):94–7.

Mobbs, O., Crépin, C., Thiéry, C., Golay, A., & Van der Linden, M. (2010). Obesity and the four facets of impulsivity. *Patient Education and Counseling, 79*(3), 372–377. https://doi.org/10.1016/j.pec.2010.03.003

Murphy, T., & Le Foll, B. (2020). Targeting the endocannabinoid CB1 receptor to treat body weight disorders: A preclinical and clinical review of the therapeutic potential of past and present CB1 drugs. *Biomolecules, 10*(6), 855–898.

Nguyen, N. T., & Varela, J. E. (2017). Bariatric surgery for obesity and metabolic disorders: State of the art. *Nature Reviews Gastroenterology & Hepatology, 14*(3), 160–169. https://doi.org/10.1038/nrgastro.2016.170

NICE. (2014). *Obesity: Identification, assessment and management | Guidance and guidelines | NICE*. National Institute of Clinical Excellence Guideline.

Nissen, S. E., Wolski, K. E., Prcela, L., Wadden, T., Buse, J. B., Bakris, G., et al. (2016). Effect of naltrexone-bupropion on major adverse cardiovascular events in overweight and obese patients with cardiovascular risk factors: A randomized clinical trial. *JAMA, 315*(10), 990–1004. https://doi.org/10.1001/jama.2016.1558

Nogueiras, R., Romero-Picó, A., Vazquez, M. J., Novelle, M. G., López, M., & Diéguez, C. (2012). The opioid system and food intake: Homeostatic and hedonic mechanisms. *Obesity Facts, 5*(2), 196–207. https://doi.org/10.1159/000338163

O'Neil, P. M., Aroda, V. R., Astrup, A., Kushner, R., Lau, D. C., Wadden, T. A., et al. (2017). Neuropsychiatric safety with liraglutide 3.0 mg for weight management: Results from randomized controlled phase 2 and 3a trials. *Diabetes, Obesity and Metabolism, 19*(11), 1529–1536. https://doi.org/10.1111/dom.12963

Onakpoya IJ, Lee JJ, Mahtani KR, Aronson JK, Heneghan CJ. Naltrexone-bupropion (Mysimba) in management of obesity: A systematic review and meta-analysis of unpublished clinical study reports. *British Journal of Clinical Pharmacology*, 2020;*86*(4):646–667.

Paterson, N., Sharma, A. M., Maxwell, C., & Greenblatt, E. M. (2016). Obesity-related health status is a better predictor of pregnancy with fertility treatment than body mass index: A prospective study. *Clinical Obesity, 6*(4), 243–248. https://doi.org/10.1111/cob.12149

Paul, L., van der Heiden, C., & Hoek, H. W. (2017). Cognitive behavioral therapy and predictors of weight loss in bariatric surgery patients. *Current Opinion in Psychiatry, 30*(6), 474–479. https://doi.org/10.1097/YCO.0000000000000359

Pi-Sunyer, X., Astrup, A., Fujioka, K., Greenway, F., Halpern, A., Krempf, M., et al. (2015). A randomized, controlled trial of 3.0 mg of liraglutide in weight management. *New England Journal of Medicine, 373*(1), 11–22. https://doi.org/10.1056/NEJMoa1411892

Raebel, M. A., Newcomer, S. R., Bayliss, E. A., Boudreau, D., DeBar, L., Elliott, T. E., et al. (2014). Chronic opioid use emerging after bariatric surgery. *Pharmacoepidemiology and Drug Safety, 23*(12), 1247–1257. https://doi.org/10.1002/pds.3625

Rosager, E. V., Møller, C., & Sjögren, M. (2021). *Treatment studies with cannabinoids in anorexia nervosa: A systematic review* (pp. 1–9). Eating and Weight Disorders Studies on Anorexia, Bulimia and Obesity.

Rubino, F., Puhl, R. M., Cummings, D. E., Eckel, R. H., Ryan, D. H., Mechanick, J. I., et al. (2020). Joint international consensus statement for ending stigma of obesity. *Nature Medicine, 26*(4), 485–497. https://doi.org/10.1038/s41591-020-0803-x

Ryan, D. H., & Kahan, S. (2018). Guideline recommendations for obesity management. *The Medical Clinics of North America, 102*(1), 49–63. https://doi.org/10.1016/j.mcna.2017.08.006

Sarwer, D. B., Dilks, R. J., & West-Smith, L. (2011). Dietary intake and eating behavior after bariatric surgery: Threats to weight loss maintenance and strategies for success. *Surgery for Obesity and Related Diseases, 7*(5), 644–651. https://doi.org/10.1016/j.soard.2011.06.016

Scholtz, S., Goldstone, A. P., & le Roux, C. W. (2015). Changes in reward after gastric bypass: The advantages and disadvantages. *Current Atherosclerosis Reports, 17*(10), 1–7. https://doi.org/10.1007/s11883-015-0534-5

Sharma, A. M., & Kushner, R. F. (2009). A proposed clinical staging system for obesity. *International Journal of Obesity, 33*(3), 289–295. https://doi.org/10.1038/ijo.2009.2

Stancil SL, Adelman W, Dietz A, Abdel-Rahman S. Naltrexone Reduces Binge Eating and Purging in Adolescents in an Eating Disorder Program. *Journal of Child and Adolescent Psychopharmacology,* 2019;*29*(9):721–724.

Steffen, K. J., Engel, S. G., Wonderlich, J. A., Pollert, G. A., & Sondag, C. (2015). Alcohol and other addictive disorders following bariatric surgery: Prevalence, risk factors and possible etiologies. *European Eating Disorders Review, 23*(6), 442–450. https://doi.org/10.1002/erv.2399

Tabesh, M. R., Maleklou, F., Ejtehadi, F., & Alizadeh, Z. (2019). Nutrition, physical activity, and prescription of supplements in pre-and post-bariatric surgery patients: A practical guideline. *Obesity Surgery, 29*(10), 3385–3400. https://doi.org/10.1007/s11695-019-04112-y

The National Institute for Health and Care Excellence. (2009). *The NICE guideline on the treatment and depression the treatment and management of depression, NICE guidelines [CG90].* Available at: http://www.ncbi.nlm.nih.gov/books/NBK55364/

Valbrun, L. P., & Zvonarev, V. (2020). The opioid system and food intake: Use of opiate antagonists in treatment of binge eating disorder and abnormal eating behavior. *Journal of Clinical Medicine Research, 12*(2), 41–63. https://doi.org/10.14740/jocmr4066

Wadden, T. A., Walsh, O. A., Berkowitz, R. I., Chao, A. M., Alamuddin, N., Gruber, K., et al. (2019). Intensive behavioral therapy for obesity combined with liraglutide 3.0 mg: A randomized controlled trial. *Obesity, 27*(1), 75–86. https://doi.org/10.1002/oby.22359

Wharton, S., Lau, D. C., Vallis, M., Sharma, A. M., Biertho, L., Campbell-Scherer, D., et al. (2020). Obesity in adults: A clinical practice guideline. *Canadian Medical Association Journal, 192*(31), E875–E891. https://doi.org/10.1503/cmaj.191707

Wilson, G. T., & Shafran, R. (2005). Eating disorders guidelines from NICE. *The Lancet, 365*(9453), 79–81. https://doi.org/10.1016/S0140-6736(04)17669-1

Yeomans, M. R., & Gray, R. W. (2002). Opioid peptides and the control of human ingestive behaviour. *Neuroscience & Biobehavioral Reviews, 26*(6), 713–728. https://doi.org/10.1016/S0149-7634(02)00041-6

Yeomans, M. R., Wright, P., Macleod, H. A., & Critchley, J. A. J. H. (1990). Effects of nalmefene on feeding in humans. *Psychopharmacology, 100*(3), 426–432. https://doi.org/10.1007/BF02244618

Zhang, M., Gosnell, B. A., & Kelley, A. E. (1998). Intake of high-fat food is selectively enhanced by muopioid receptor stimulation within the nucleus accumbens. *Journal of Pharmacology and Experimental Therapeutics, 285*(2), 908–914.

Zheng, H., Townsend, R. L., Shin, A. C., Patterson, L. M., Phifer, C. B., & Berthoud, H. R. (2010). High-fat intake induced by mu-opioid activation of the nucleus accumbens is inhibited by Y1R-blockade and MC3/4R-stimulation. *Brain Research, 1350*, 131–138. https://doi.org/10.1016/j.brainres.2010.03.061

The Role of Endogenous Opioids in Cardioprotection

Cristina Sirbu

Abstract The opioid system involves opioid receptors (OPRs) and endogenous opioid peptides.

This chapter will focus on the distribution of OPRs in the cardiovascular system, the expression pattern in the heart, the activation by opioid peptides, and the effects of OPRs activation with potential relevance in cardiovascular performance. In the heart, OPRs are co-expressed with beta adrenergic receptors (β-ARs) in the G-protein-coupled receptor (GPCR) superfamily, functionally cross-talk with β-Ars and modify catecholamine-induced effects. They are involved in cardiac contractility, energy metabolism, myocyte survival or death, vascular resistance. The effects of the opioid system in the regulation of systemic circulation at both the central and peripheral level are presented. The pathways are discussed under physiological (i.e., aging) and pathological conditions (atherosclerosis, heart failure, essential hypertension, ischemic stress). Stimulation of OPRs not only inhibits cardiac excitation–contraction coupling, but also protects the heart against hypoxic and ischemic injury. An enhanced sensitivity to opioids of endocrine organs and neuronal systems is operative in hypertensive patients. The opioid system can be pharmacologically engaged to selectively mimic these responses via cardiac and nervous signaling. The clinical opportunities for the use of cardioprotective effects of opioids require future investigations to provide more specific details of the impact on cardiac performance and electrophysiological properties.

Keywords Opioid receptors · Cardioprotection · Endogenous opioids · Cardiovascular

C. Sirbu (✉)
Department of Cardiac Surgery and Transplantation, University Hospital Nancy-Brabois, Nancy, France

P. L. Kerr et al. (eds.), *Endogenous Opioids*, Advances in Neurobiology 35,
https://doi.org/10.1007/978-3-031-45493-6_19

Introduction

Four types of opioid receptors (OPRs) have been identified in the cardiovascular system: μ, δ, κ, and ORL1. Among these, δ, k, and ORL1 have the highest density in cardiomyocytes, μ and δ in endothelial cells, δ in vascular smooth muscle cells, and κ in the heart conduction system (Barron, 2000). OPRs are present in the sympathetic and parasympathetic nerve terminals in the heart, and their activation may affect the functional state of the heart. Each OPR type can be further subdivided, and, under pathological circumstances, the expression of OPR subtypes can be up- or downregulated (Barron, 2000; Maslov et al., 2016).

In the heart, OPRs and β-adrenergic receptors (β-ARs) are co-expressed and are members of G-protein-coupled receptor (GPCR) superfamily (Pepe et al., 2004). OPRs are Gi/Go-coupled receptors and are activated by endogenous peptides: endorphin, enkephalin, and dynorphin. β-ARs are Gs-coupled receptors, stimulated by catecholamines. OPRs functionally cross-talk with β-ARs and modify catecholamine-induced effects in the heart (Pepe et al., 2004; Weil et al., 2006).

The expression pattern of OPRs is different in the heart chambers, and their expression in this region may be developmentally dependent (Weil et al., 2006; Zimlichman et al., 1996). In adult rats, the κ-OPRs expression in the heart is the most abundant with equal levels in the atrial and ventricular myocardium. Prior studies have found that δ-OPRs are more abundant in atrial tissue, whereas μ-OPRs have a low abundance in the heart muscle in general. Moreover, there are specialized regions (e.g., AV-node region) with a much higher density of OPRs (Weil et al., 2006). All three OPRS (κ, δ, and μ) are present in the heart of neonatal rats. However, after day 7, μ-OPRs were not detected, and δ-OPRs reach adult levels at day 14 (Zimlichman et al., 1996). A nascent line of research in human heart tissue is worth consideration in this discussion. Bell et al. (2000) found evidence of both μ-OPR δ-OPR expression in adult human ventricular tissue. However, Peng et al. (2012) detected only δ-OPR and κ-OPR expression in heart tissue, despite finding substantial expression of all three OPRs throughout the central nervous system.

Stimulation of δ-OPRs directly modulates systemic vascular resistance (Saeed et al., 2000). Accordingly, activation of δ- and κ-OPRs exhibits an inhibitory effect on cardiac excitation-contraction coupling and causes a reduction in myocyte contractility by sarcoplasmic reticulum Ca^{2+} depletion (Ventura et al., 1992). This effect is mediated by Gi/o pathways and is reversed by the OPR antagonist naloxone (Ventura et al., 1992; Xiao et al., 1993). The "cross-talk" between OPRs and β-ARs antagonizes the β-AR-mediated positive inotropic effect, and increases in cAMP (Pepe et al., 1997). These anti-adrenergic effects are reversed by naloxone and prevented by inactivation of inhibitory G-proteins (Gi/o) by pertussis toxin (PTX) pretreatment (Pepe et al., 1997; Xiao et al., 1997). OPRs inhibit adenylyl cyclase via a PTX- sensitive Gi/o protein (Niroomand et al., 1996; Xiao et al., 1997). In regulation of myocardial contractility, δ-OPR signaling interacts selectively with β1-AR but not β2-AR (Pepe et al., 1997; Xiao et al., 1997) in relation with a cell architectural arrangement of β-AR subtypes (Rybin et al., 2000). The δ-OPR stimulation inhibits Gs and adenylate cyclase by activating Gi/o signaling pathways. This leads to a decrease in c-AMP production and protein kinase A (PKA) activation, a reduction in L-type Ca^{2+} current, depletion of Ca^{2+} from sarcoplasmic reticulum and reduced contractility (Schultz & Gross, 2001). The c-AMP independent inotropic effects of β2-AR agonism appear insensitive to δ-OPR activity (Pepe et al., 1997).

Thus, the interaction between OPRs and β-ARs might contribute to the inhibitory effects of OPRs stimulation on β-AR-mediated positive chronotropic and inotropic effects in the heart. The OPRs are activated by endogenous opioid peptides: enkephalins, endorphins and dynorphins, derived from distinct prohormones, which are encoded by three separate genes. Specifically, μ-OPRs are selectively sensitive to endorphins, the δ-OPRs have the highest affinity for enkephalins, and the κ-OPRs bind to dynorphins (Ventura et al., 1998; Baron, 2000).

Opioid peptides are co-released with catecholamines from peripheral neuronal terminals in the heart (Wilson et al., 1980). Moreover, the existence of a local enkephalin system in the heart has been recognized (Younès et al., 2000). The enkephalin peptide precursor (ppENK) has been found in mammalian ventricular tissue and in cultured myocytes (Springhorn & Claycomb, 1992; Ventura et al., 1998). In rats, the left ventricular myocardium has been shown to have four times higher levels of ppENK gene expression than in the right chamber (Howells et al., 1986; Weil et al., 1998). Thus, the ventricular myocardium supplies the heart and the body with enkephalins.

Opioid peptides of myocardial origin play important roles in local regulation of the heart. The ppENK synthesis may be triggered by increased hemodynamic load (Weil et al., 1998, 2006). The effects of locally synthesized and released opioids on contractility and dromotropy were investigated in isolated perfused rat hearts (Weil et al., 2006). In normal physiology, in vitro, endogenous opioids have no significant impact on contractility. Obviously, endogenous opioids modulate the positive dromotropic effect of noradrenaline. The δ- and κ-OPRs antagonists inhibited the noradrenaline positive dromotropic effect, whereas the endogenous enkephalins enhanced this effect (Weil et al., 2006). In the rat heart, a higher concentration of ppENK mRNA was found in the AV node region (Weil et al., 1998). Opioid system blockade resulted in a significant prolongation of the atrioventricular node conduction time without effects on the His-Purkinje system and the sinus node, whereas administration of pentazocine (opioid system stimulation) decreased atrioventricular conduction time (Markiewicz et al., 1991). These findings may have clinical implications; in intensive medical care patients, opioids may be administered while catecholamine levels are increased and may be involved in the pathogenesis of cardiac arrhythmia.

In aging heart tissue, opioid peptide gene expression is increased, and the cardiac response to β-AR stimulation decreases (Boluyt et al., 1993; Caffrey et al., 1994; Xiao et al., 1994; Lakatta & Levy, 2003). The suppression of cardiac response to β-AR stimulation is associated with a downregulation of β-AR density, and a decrease in agonist-stimulated adenylyl cyclase activity is associated with a decrease in cAMP and PKA. Thus, the antagonistic effects of stimulation of κ- and δ-OPRs on β-AR-mediated positive contractile response may explain the reduction in β-AR signaling in the aging heart related to the age-associated upregulation of OPR signaling (Pepe et al., 2004).

Role of Endogenous Opioids in Atherosclerosis

Psychosocial stress is involved in the progression of atherosclerosis, independent of traditional cardiac risk factors (Rozanski et al., 1999). Stress responses involving central and peripheral nervous system components lead to changes that improve the ability of the organism to adapt and increase its chances to survival. However, prolonged responses to stressors may be deleterious, resulting in increases in sympathetic pathways and activation of the hypothalamic-pituitary-adrenocortical axis (Pike et al., 1997). The early pathophysiology of atherosclerosis involves endothelial dysfunction, followed by the aggregation of platelets, smooth muscle cells, and oxidized lipids, that is, the "response-to-injury model" (Ross, 1999).

The mechanisms converting psychosocial stress into endothelial dysfunction involve opioid peptides, for example, β-endorphin which contributes to an imbalance between endothelin-1 (ET-1) and nitric oxide (NO) release, mediated by μ1-OPRs (Wilbert-Lampen et al., 2007). Under the influence of β-endorphin, the ET-1 release from endothelial and monocytic cells is increased, whereas NO release is decreased. The result is an enhanced production of ET-1, the most potent endogenous vasoconstrictor known (Yanagisawa et al., 1988), promptly activated under stress conditions (Noll et al., 1996) and a reduced bioavailability of NO, an endothelial vasodilator. Thus, β-endorphin has an important role in pathogenesis of stress-induced endothelial dysfunction leading to unopposed vasoconstriction (Wilbert-Lampen et al., 2007).

Interestingly, these effects were demonstrated with in vitro investigations of endothelial cells obtained from healthy volunteers. Contrary to these results, it was demonstrated that psychological stress induces an increase in β-endorphin, whereas ET-1 was decreased in patients with dilated cardiomyopathy (Fontana et al., 1998). The observed differences in β-endorphin and induced ET-1 secretion may be due to the observed species (healthy subjects versus people with cardiac dysfunction), the different local concentration of the stress hormone, as well as regional differences in μ1-OPRs distribution. Based on these data, it is possible that opioid system pathways are different under pathological conditions (i.e., heart failure, hypertension) in vivo, acting as cardiovascular protectors and modulators.

Role of Endogenous Opioids Heart Failure

Heart failure is associated with elevated levels of circulating catecholamines and a reduced cardiac contractile response to β-AR stimulation. The reduced β-AR inotropic effect results from reduction in β-AR density and sensitization of remaining receptors. The β1 and β2-AR may play opposing functional roles in the pathogenesis of heart failure.

Sustained β1-AR stimulation promotes cardiac hypertrophy (Morisco et al., 2001; Schäfer et al., 2000) and myocyte apoptosis (Communal et al., 1999; Zaugg et al., 2000) activating the classic c-AMP-PKA signaling pathway and PKA activation of calmodulin-dependent protein kinase II (Zhu et al., 2003; Schultz & Gross, 2001). Thus, in heart failure, the selective downregulation of β1-AR represents a cardioprotective mechanism, protecting myocytes against apoptosis and slowing the progression of cardiomyopathy and contractile dysfunction. The selective β1-AR blockers have beneficial effects in chronic heart failure patients (Bristow, 2000). The upregulation of β2-AR signaling is cardioprotective due to its contractile support and antiapoptotic effects.

As activation of δ-OPRs selectively inhibits β1-AR, but not β2-AR, in regulating myocardial contractility, they exhibit a cardioprotective effect in heart failure (Pepe et al., 1997; Xiao et al., 1997). The negative inotropic effect of OPRs may represent a cardioprotective mechanism in the early stage of heart failure; by reducing cardiac responsiveness to sympathetic stimulation they diminish oxygen demand by limiting work performance. However, exaggerated OPR signaling could contribute to the pathogenesis of heart failure. Increased levels of met-enkephalin are correlated with the degree of severity of the disease (Löwe, 1991; Fontana et al., 1993; Imai et al., 1994). Thus, endogenous enkephalins play an important role in mediating the myocardial depression that occurs in heart failure.

The endogenous opioid system is activated in heart failure. In patients with chronic heart failure, circulating β-endorphin is significantly increased and is correlated with New York Heart Association (NYHA) functional class (Kawashima et al., 1991).

β-endorphin improves cardiovascular function and exercise capacity in patients with dilated cardiomyopathy and II-III NYHA functional class (Cozzolino et al., 2004). Early work by Lord et al. (1977) found that β-endorphin engages in potent agonistic action on μ- and δ-OPRs. This leads to presynaptic inhibition of catecholamine release from neuronal endings and reduces neurohormonal activation (Gaddis & Dixon, 1982; Kienbaum et al., 2001), decreasing circulating vasoconstrictive endogenous substances: catecholamines, ET-1. β-endorphin could improve inotropic function via glucoside-like cardiac action, enacting an inhibitory effect on the sarcolemmal ouabain-sensitive Na^+-K^+ dependent ATP-aze activity (Ventura et al., 1987).

In accordance with this, Giugliano et al. (1987) found that a significant increase in β-endorphin mediated glucagon production. The glucagon increases left ventricular inotropic function by stimulating cardiomyocyte Ca^{2+} transiently via activation of adenylyl cyclase and inhibition of phosphodiesterase (Sauvadet et al., 1996). Activation of growth hormone (GH) and insulin-like growth factor-1 (IGF-1) production by β-endorphin was also reported and may explain a significant decrease in systemic vascular resistance by activating the endothelial NO pathway (Napoli et al., 2002, 2003). Thus, the improved left ventricular performance by β-endorphin in heart failure may occur by a direct positive inotropic effect, or indirectly by a reduction in afterload.

Clinical Implications

The preceding review of data regarding endogenous opioids and cardiac functioning can be synthesized, and some general clinical implications may be derived. In this section, some of the most promising therapeutic targets are described. These targets will require empirical investigation, but are encouraging nonetheless.

Clinical Implications for Cardiac Functioning

Our understanding of OPR agonist and antagonist mechanisms of action may contribute to better management of patients with cardiac dysfunction. For example, naloxone, an OPR antagonist, may improve cardiac performance by acting within the central nervous system (Sakamoto et al., 1989) and via direct inotropic effect (Llobel & Laorden, 1997). Additionally, there may be a role for targeting enkephalins in clinical therapeutic management. High doses of β-endorphin given for a short period could preserve cardiovascular function in patients with mild to moderate heart failure (Cozzolino et al., 2004). Beta-endorphin infusion in patients with chronic heart failure increases in oxygen consumption at peak effort and improves exercise duration soon after administration, indicating an improvement in exercise capacity (Cozzolino et al., 2004).

Clinical Implications for Management of Essential Hypertension

Sympathetic nervous system stimulation represents an early feature of the natural history of essential hypertension. Pressure overload and stimulation of the β-adrenergic signaling pathway have been identified as mechanisms leading to increased ppENK expression and may contribute to opioid system activation (Weil et al., 2006). Basal plasma levels of β-endorphin, norepinephrine, and ET-1 in hypertensive patients are significantly higher than in normotensive subjects (Fontana et al., 1994).

There is evidence in several animal species and in humans of involvement of the endogenous opioid system in essential hypertension. Opioid peptides exert hypotensive effects through central and peripheral mechanisms including sympathetic inhibition with subsequent vasodilation (Gayle et al., 1994). The stimulation of μ-OPRs decreases sympathetic nervous system activity in humans (Kienbaum, 2001). The δ-OPRs stimulation directly modulates systemic vascular resistance (Saeed et al., 2000).

Consequently, the enhanced-opioid mediated inhibition of blood pressure in people with hypertension could be a consequence of a hypersensitivity to opioid receptor agonists and/or an increased expression of central opioid receptors. An increased sensitivity of central and peripheral organs to β-c-endorphin might explain the exaggerated hypotensor and hormonal responses upon β-endorphin infusion in hypertensive patients (Cozzolino et al., 2005). The result is a decrease in vasoconstrictive neurohormones (norepinephrine and ET-1) and an increase in plasma concentrations of vasorelaxing peptides (atrial natriuretic factor (ANF), GH and IGF-1). These effects were abolished by naloxone, suggesting opioid receptor agonism as a mechanism of action (Cozzolino et al., 2005). Furthermore, increased hemodynamic load in hypertensives enhances ppENK gene expression and cardiomyocyte enkephalin production (Weil et al., 2006). This suggests a role in the pathogenesis of hypertensive cardiomyopathy. Considered together, these findings facilitate a more nuanced understanding of the relationship between cardiovascular functioning (especially hypertension) and OPRs, and suggest some novel targets for intervention in hypertension, as discussed below (see section "Treatment implications").

Treatment Implications

As noted earlier, β-endorphin administered at pharmacological doses for a short time period has hypotensive effects which are abolished by pretreatment with naloxone (Fontana et al., 1997; Cozzolino et al., 2005). Morphine also reduces blood pressure, and this effect is enhanced in people with essential hypertension. This should be considered to avoid an undesired drop of blood pressure, such as in the case of analgesic therapy before surgical interventions.

Implications for Management of Ischemic Preconditioning and Postconditioning

Ischemic preconditioning (IPC) is a phenomenon described in both humans (Murry et al., 1986; Abete et al., 1997) and animals (Karck et al., 2001; Laclau et al., 2001) to reduce infarct size, suppress ventricular arrhythmias, and improve functional recovery after a period of ischemia. Brief, nonlethal periods of ischemia protect the heart against a more sustained event in two phases: early preconditioning (immediately after the initial stimulus with a short duration of 1–2 h) and late preconditioning (occurring 24 h after stimulus with a duration of 72 h; Sommerschild & Kirkebøen, 2002).

The duration of ischemia is a critical factor in cellular survival. Cell death can be induced by necrosis and apoptosis. As the apoptotic component contributes to the extension of infarct size during reperfusion, inhibition of this component may improve contractile functioning in an ischemic heart (Zhao, 2003a, b). Apoptosis involves a mitochondrial permeability transition (MPT).The end effector is the mitochondrial permeability transition pore (PTP) complex, which demonstrates sensitivity to oxidative stress, and can determine mitochondrial and cell death via the consequences of MPT induction. The degree of this sensitivity to reactive oxygen species can be reduced by OPR activation. This results in an inhibition of apoptosis by increasing resistance to MPT induction and also cardioprotection. This process involves activation of a wide spectrum of kinases, which inhibit the activity of glycogen synthase kinase-3β (GSK-3β) (Juhaszova et al., 2005). δ- OPR stimulation by endogenous enkephalins enhances IPC, and may have a significant impact on cardiac protection and organ preservation for transplantation (Takasaki et al., 1999; Bell et al., 2000; Sigg et al., 2001; Bolling et al., 2001). Mediation of cardioprotection with sustained opioid agonism is β2-AR dependent (Peart & Gross, 2006). δ-OPR-mediated infarct reduction is negated by β2-AR blockade (Huang et al., 2007).

Stimulation of κ-OPRs may also play an important role in IPC-induced cardioprotection. In ischemic rat hearts, the inhibitory effect of κ-OPR stimulation on β-AR signaling is markedly enhanced, thus resulting in cardiac protection as evidenced by reduced arrhythmia (Yu et al., 1999). κ- OPR activation may also produce a reduction in myocardial infarct size (Wang et al., 2001). There is a cardioprotective cross-talk between OPRs and adenosine receptors (Peart & Gross, 2003). OPR stimulation in rats reduces the size of the infarct zone, and this effect is abolished by an adenosine A1 receptor antagonist. Inversely, OPR blocade attenuates cardioprotection in response to A1-AR activation. The protective effects of increased endogenous adenosine (following adenosine kinase inhibition) were found to be dependent upon δ-OPRs activity (Peart & Gross, 2005). Enkephalin and dynorphin that bind to δ- and κ- OPRs have been implicated in IPC, endogenous opioids induced IPC and/or reactions to myocardial ischemia (Romano et al., 2004; Schultz & Gross, 2001; Takasaki et al., 1999).

As noted earlier, IPC is a phenomenon in which brief episodes of ischemia/reperfusion occur at the time of reperfusion. Importantly, IPC can be as effective as preconditioning in reducing infarct size (Zhao et al., 2003a, b). Evidence also supports an essential role for OPRs in IPC. Beneficial effects of IPC are reproduced by OPR activation and countered by δ-OPR antagonism (Jang et al., 2008). Exogenous activation of κ- and δ-OPRs at reperfusion affords protective IPC and underlying mechanisms mirroring those for IPC responses (Gross et al., 2007; Peart et al., 2008; Tong et al., 2011). The agonists of κ2-OPRs may exacerbate the ischemic and reperfusion heart injury (Aitchison et al., 2000).

Treatment Implications of Cardioprotective Preconditioning and Postconditioning

OPR activation induces cardioprotection through endogenous release of opioid peptides during IPC or by exogenously administrated OPRs agonists. The nonselective opioid agonist morphine reduced infarct size when administered 10 min before ischemia or 5 min before reperfusion (Gross et al., 2004). Morphine mimics the cardioprotective effect of IPC; this effect was mediated by K-ATP channel activation (Schultz et al., 1996). IPC and morphine PC reduced infarct size to a similar degree (Schultz et al., 1996). The morphine-induced infarct size reduction is elicited via δ OPRs (Schultz et al., 1997). Morphine attenuated neutrophil and endothelial activation in patients with acute myocardial infarction and reduced the amount of adhesion molecules, suggesting diverse effects to reduce ischemia/reperfusion injury (Wang et al., 1997, 1998). While both central and peripheral OPRs mediated the reduction in infarct size by morphine-PC, only peripheral OPRs mediated IPC (Lu et al., 2011).

Fentanyl isothiocyanate, which binds selectively and irreversibly to δ-OPRs reduces infarct size when given prior to ischemia or prior to reperfusion, and this protection was extended to 10 seconds after reperfusion (Gross et al., 2005). Fentanyl protects the heart against myocardial ischemia/reperfusion injury via δ-OPR cross-talk with adenosine A1 receptors and through m-K-ATP channel opening and PKC-linked mechanisms (Henry et al., 1996; Kato & Foe, 2000).

Remifentanil is an ultrashort acting opioid that reduces infarct size by a mechanism involving both PKC activation and m K-ATP channel opening. Despite a high affinity for μ-OPRs, remifentanil mimics IPC via cardiac δ- and κ-OPRs and via extra-cardiac μ-OPRs (Dershwitz et al., 1995; Zhang et al., 2005).

OPR activation can increase the heart's tolerance to ischemia/reperfusion in a concentration-dependent manner. However, OPR stimulation can also exacerbate ischemia-reperfusion injury (Maslov et al., 2006; 2010). For example, morphine can have an adverse effect on the cardiovascular system at a dose many times exceeding the recommended therapeutic dose of 0.05–0.1 mg/Kg (Lappas et al., 1975; Meine et al., 2005). Thus, a therapeutic window dosage must be attained. By strategic modification of the reperfusion protocol, therapeutic interventions performed at the time of reflow may significantly limit infarct size. With appropriate attention to dosing, OPR agonists appear cardioprotective in different clinical cohorts (Headrick et al., 2015).

Conclusions

The endogenous opioid system represents a stress-sensitive cytoprotective system that is highly expressed in the heart; essential for cardioprotection; and significantly influenced by age, disease, and pharmacological agents. Different OPR ligands produce qualitatively different cardiovascular effects; more than one OPR subtype may be involved, and OPR subtypes may act at different anatomical sites. The cardiovascular effects of pharmacologically administered opioid peptides are complex and dependent upon species, doses, route and site of administration, presence or absence of anesthetics, and other interactions.

References

Abete, P., Ferrara, N., Cacciatore, F., Madrid, A., Bianco, S., Calabrese, C., et al. (1997). Angina-induced protection against myocardial infarction in adult and elderly patients: A loss of preconditioning mechanism in the aging heart? *Journal of the American College of Cardiology, 30*(4), 947–954.

Aitchison, K. A., Baxter, G. F., Awan, M. M., Smith, R. M., Yellon, D. M., & Opie, L. H. (2000). Opposing effects on infarction of delta and kappa opioid receptor activation in the isolated rat heart: Implications for ischemic preconditioning. *Basic Research in Cardiology, 95*(1), 1–10.

Baron, R. (2000). Neuropathic pain. The long path from mechanisms to mechanism-based treatment. *Der Anaesthesist, 49*(5), 373–386.

Barron, B. A. (2000). Cardiac Opioids (44507A). *Proceedings of the Society for Experimental Biology and Medicine, 224*(1), 1–7.

Bell, S. P., Sack, M. N., Patel, A., Opie, L. H., & Yellon, D. M. (2000). Delta opioid receptor stimulation mimics ischemic preconditioning in human heart muscle. *Journal of the American College of Cardiology, 36*(7), 2296–2302.

Bolling, S. F., Badhwar, V., Schwartz, C. F., Oeltgen, P. R., Kilgore, K., & Su, T. P. (2001). Opioids confer myocardial tolerance to ischemia: Interaction of delta opioid agonists and antagonists. *The Journal of Thoracic and Cardiovascular Surgery, 122*(3), 476–481.

Boluyt, M. O., Younes, A., Caffrey, J. L., O'neill, L., Barron, B. A., Crow, M. T., & Lakatta, E. G. (1993). Age-associated increase in rat cardiac opioid production. *American Journal of Physiology – Heart and Circulatory Physiology, 265*(1), H212–H218.

Bristow, M. R. (2000). Mechanistic and clinical rationales for using beta-blockers in heart failure. *Journal of Cardiac Failure, 6*(2 Suppl 1), 8–14.

Caffrey, J. L., Boluyt, M. O., Younes, A., Barron, B. A., O'Neill, L., Crow, M. T., & Lakatta, E. G. (1994). Aging, cardiac proenkephalin mRNA and enkephalin peptides in the Fisher 344 rat. *Journal of Molecular and Cellular Cardiology, 26*(6), 701–711.

Communal, C., Singh, K., Sawyer, D. B., & Colucci, W. S. (1999). Opposing effects of β1-and β2-adrenergic receptors on cardiac myocyte apoptosis: Role of a pertussis toxin–sensitive G protein. *Circulation, 100*(22), 2210–2212.

Cozzolino, D., Sasso, F. C., Salvatore, T., Torella, M., Cittadini, A., Gentile, S., et al. (2004). Acute effects of β-endorphin on cardiovascular function in patients with mild to moderate chronic heart failure. *American Heart Journal, 148*(3), 1–7.

Cozzolino, D., Sasso, F. C., Cataldo, D., Gruosso, D., Giammarco, A., Cavalli, A., et al. (2005). Acute pressor and hormonal effects of β-endorphin at high doses in healthy and hypertensive subjects: Role of opioid receptor agonism. *The Journal of Clinical Endocrinology & Metabolism, 90*(9), 5167–5174.

Dershwitz, M., Randel, G. I., Rosow, C. E., Fragen, R. J., Connors, P. M., Librojo, E. S., et al. (1995). Initial clinical experience with remifentanil, a new opioid metabolized by esterases. *Anesthesia & Analgesia, 81*(3), 619–623.

Fontana, F., Bernardi, P., Pich, E. M., Capelli, M., Bortoluzzi, L., Spampinato, S., & Canossa, M. (1993). Relationship between plasma atrial natriuretic factor and opioid peptide levels in healthy subjects and in patients with acute congestive heart failure. *European Heart Journal, 14*(2), 219–225.

Fontana, F., Bernardi, P., Boschi, S., De Iasio, R., Capelli, M., Carboni, L., & Spampinato, S. (1994). Endogenous opioid system and atrial natriuretic factor in normotensive offspring of hypertensive parents at rest and during exercise test. *Journal of Hypertension, 12*(11), 1285–1290.

Fontana, F., Bernardi, P., Spampinato, S., Boschi, S., De Iasio, R., & Grossi, G. (1997). Pressor effects of endogenous opioid system during acute episodes of blood pressure increases in hypertensive patients. *Hypertension, 29*(1), 105–110.

Fontana, F., Bernardi, P., Pich, E. M., Tartuferi, L., Boschi, S., De Iasio, R., & Spampinato, S. (1998). Opioid peptides in response to mental stress in asymptomatic dilated cardiomyopathy. *Peptides, 19*(7), 1147–1153.

Gaddis, R. R., & Dixon, W. R. (1982). Modulation of peripheral adrenergic neurotransmission by methionine-enkephalin. *Journal of Pharmacology and Experimental Therapeutics, 221*(2), 282–288.

Gayle, A. O., Olson, R. D., & Kastin, A. J. (1994). Endogenous opiates: 1993. *Peptides, 15*, 1513–1556.

Giugliano, D., Salvatore, T., Cozzolino, D., Ceriello, A., Torella, R., & D'Onofrio, F. (1987). Hyperglycemia and obesity as determinants of glucose, insulin, and glucagon responses to Beta-Endorphin in human diabetes mellitus. *The Journal of Clinical Endocrinology and Metabolism, 64*(6), 1122–1128.

Gross, E. R., Hsu, A. K., & Gross, G. J. (2004). Opioid-induced cardioprotection occurs via glycogen synthase kinase β inhibition during reperfusion in intact rat hearts. *Circulation Research, 94*(7), 960–966.

Gross, E. R., Peart, J. N., Hsu, A. K., Auchampach, J. A., & Gross, G. J. (2005). Extending the cardioprotective window using a novel δ-opioid agonist fentanyl isothiocyanate via the PI3-kinase pathway. *American Journal of Physiology – Heart and Circulatory Physiology, 288*(6), H2744–H2749.

Gross, E. R., Hsu, A. K., & Gross, G. J. (2007). GSK3β inhibition and K ATP channel opening mediate acute opioid-induced cardioprotection at reperfusion. *Basic Research in Cardiology, 102*(4), 341–349.

Headrick, J. P., See Hoe, L. E., Du Toit, E. F., & Peart, J. N. (2015). Opioid receptors and cardioprotection–'opioidergic conditioning'of the heart. *British Journal of Pharmacology, 172*(8), 2026–2050.

Henry, P., Demolombe, S., Pucéat, M., & Escande, D. (1996). Adenosine A1 stimulation activates δ-protein kinase C in rat ventricular myocytes. *Circulation Research, 78*(1), 161–165.

Howells, R. D., Kilpatrick, D. L., Bailey, L. C., Noe, M., & Udenfriend, S. (1986). Proenkephalin mRNA in rat heart. *Proceedings of the National Academy of Sciences, 83*(6), 1960–1963.

Huang, M. H., Wang, H. Q., Roeske, W. R., Birnbaum, Y., Wu, Y., Yang, N. P., et al. (2007). Mediating δ-opioid-initiated heart protection via the β2- adrenergic receptor: Role of the intrinsic cardiac adrenergic cell. *American Journal of Physiology – Heart and Circulatory Physiology, 293*(1), H376–H384.

Imai, N., Kashiki, M., Woolf, P. D., & Liang, C. S. (1994). Comparison of cardiovascular effects of Mu-and Delta-Opioid receptor antagonists in dogs with congestive heart failure. *The American Journal of Physiology, 267*(3 Pt 2), H912–H917.

Jang, Y., Xi, J., Wang, H., Mueller, R. A., Norfleet, E. A., & Xu, Z. (2008). Postconditioning prevents reperfusion injury by activating δ-opioid receptors. *The Journal of the American Society of Anesthesiologists, 108*(2), 243–250.

Juhaszova, M., Rabuel, C., Zorov, D. B., Lakatta, E. G., & Sollott, S. J. (2005). Protection in the aged heart: Preventing the heart-break of old age? *Cardiovascular Research, 66*(2), 233–244.

Karck, M., Tanaka, S., Bolling, S. F., Simon, A., Su, T. P., Oeltgen, P. R., & Haverich, A. (2001). Myocardial protection by ischemic preconditioning and δ-opioid receptor activation in the isolated working rat heart. *The Journal of Thoracic and Cardiovascular Surgery, 122*(5), 986–992.

Kato, R., & Foe, P. (2000). Fentanyl reduces infarction but not stunning via δ-opioid receptors and protein kinase C in rats. *British Journal of Anaesthesia, 84*(5), 608–614.

Kawashima, S., Fukutake, N., Nishian, K., Asakuma, S., & Iwasaki, T. (1991). Elevated plasma beta-endorphin levels in patients with congestive heart failure. *Journal of the American College of Cardiology, 17*(1), 53–58.

Kienbaum, P., Heuter, T., Michel, M. C., Scherbaum, N., Gastpar, M., & Peters, J. (2001). Chronic μ-opioid receptor stimulation in humans decreases muscle sympathetic nerve activity. *Circulation, 103*(6), 850–855.

Laclau, M. N., Boudina, S., Thambo, J. B., Tariosse, L., Gouverneur, G., Bonoron-Adèle, S., et al. (2001). Cardioprotection by ischemic preconditioning preserves mitochondrial function and functional coupling between adenine nucleotide translocase and creatine kinase. *Journal of Molecular and Cellular Cardiology, 33*(5), 947–956.

Lakatta, E. G., & Levy, D. (2003). Arterial and cardiac aging: Major shareholders in cardiovascular disease enterprises: Part III: Cellular and molecular clues to heart and arterial aging. *Circulation, 107*(3), 490–497.

Lappas, D. G., Buckley, M. J., Laver, M. B., Daggett, W. M., & Lowenstein, E. (1975). Left ventricular performance and pulmonary circulation following addition of nitrous oxide to morphine during coronary-artery surgery. *Anesthesiology, 43*(1), 61–69.

Llobel, F., & Laorden, M. L. (1997). Effects of mu-, delta-and kappa-opioid antagonists in atrial preparations from nonfailing and failing human hearts. *General Pharmacology, 28*(3), 371–374.

Lord, J. A., Waterfield, A. A., Hughes, J., & Kosterlitz, H. W. (1977). Endogenous opioid peptides: Multiple agonists and receptors. *Nature, 267*(5611), 495–499.

Löwe, H. (1991). Role of endogenous opioids in heart failure. *Zeitschrift fur Kardiologie, 80*, 47–51.

Lu, Y., Dong, C., Yu, J., & Li, L. (2011). Role of central and peripheral opioid receptors in the cardioprotection of intravenous morphine preconditioning. *Irish Journal of Medical Science, 180*(4), 881–885.

Markiewicz, K., Kuś, W., Cholewa, M., & Bubinński, R. (1991). Electrophysiologic effects of blocking and stimulating the opioid system in patients with unexplained heart palpitations. *Clinical Cardiology, 14*(10), 813–820.

Maslov, L. N., Platonov, A. A., Lasukova, T. V., IuB, L., Oeltgen, P., Nagase, H., et al. (2006). Delta-opioid receptor activation prevents appearance of irreversible damages of cardiomyocytes and exacerbates myocardial contractility dysfunction during ischemia and reperfusion. *Patologicheskaia fiziologiia i eksperimental'naia terapiia, 4*, 13–17.

Maslov, L. N., Lishmanov, Y. B., Oeltgen, P. R., Barzakh, E. I., Krylatov, A. V., Naryzhnaya, N. V., et al. (2010). Comparative analysis of the cardioprotective properties of opioid receptor agonists in a rat model of myocardial infarction. *Academic Emergency Medicine, 17*(11), 1239–1246.

Maslov, L. N., Khaliulin, I., Oeltgen, P. R., Naryzhnaya, N. V., Pei, J. M., Brown, S. A., et al. (2016). Prospects for creation of cardioprotective and antiarrhythmic drugs based on opioid receptor agonists. *Medicinal Research Reviews, 36*(5), 871–923.

Meine, T. J., Roe, M. T., Chen, A. Y., Patel, M. R., Washam, J. B., Ohman, E. M., et al. (2005). Association of intravenous morphine use and outcomes in acute coronary syndromes: Results from the CRUSADE Quality Improvement Initiative. *American Heart Journal, 149*(6), 1043–1049.

Morisco, C., Zebrowski, D. C., Vatner, D. E., Vatner, S. F., & Sadoshima, J. (2001). β- Adrenergic cardiac hypertrophy is mediated primarily by the β1-subtype in the rat heart. *Journal of Molecular and Cellular Cardiology, 33*(3), 561–573.

Murry, C. E., Jennings, R. B., & Reimer, K. A. (1986). Preconditioning with ischemia: A delay of lethal cell injury in ischemic myocardium. *Circulation, 74*(5), 1124–1136.

Napoli, R., Guardasole, V., Matarazzo, M., Palmieri, E. A., Oliviero, U., Fazio, S., & Saccà, L. (2002). Growth hormone corrects vascular dysfunction in patients with chronic heart failure. *Journal of the American College of Cardiology, 39*(1), 90–95.

Napoli, R., Guardasole, V., Angelini, V., D'Amico, F., Zarra, E., Matarazzo, M., & Saccà, L. (2003). Acute effects of growth hormone on vascular function in human subjects. *The Journal of Clinical Endocrinology & Metabolism, 88*(6), 2817–2820.

Niroomand, F., Mura, R. A., Piacentini, L., & Kübler, W. (1996). Opioid receptor agonists activate pertussis toxin-sensitive G proteins and inhibit adenylyl cyclase in canine cardiac sarcolemma. *Naunyn-Schmiedeberg's Archives of Pharmacology, 354*(5), 643–649.

Noll, G., Wenzel, R. R., Schneider, M., Oesch, V., Binggeli, C., Shaw, S., et al. (1996). Increased activation of sympathetic nervous system and endothelin by mental stress in normotensive offspring of hypertensive parents. *Circulation, 93*(5), 866–869.

Peart, J. N., & Gross, G. J. (2003). Adenosine and opioid receptor-mediated cardioprotection in the rat: Evidence for cross-talk between receptors. *American Journal of Physiology – Heart and Circulatory Physiology, 285*(1), H81–H89.

Peart, J. N., & Gross, G. J. (2005). Cardioprotection following adenosine kinase inhibition in rat hearts. *Basic Research in Cardiology, 100*(4), 328–336.

Peart, J. N., & Gross, G. J. (2006). Cardioprotective effects of acute and chronic opioid treatment are mediated via different signaling pathways. *American Journal of Physiology – Heart and Circulatory Physiology, 291*(4), H1746–H1753.

Peart, J. N., Gross, E. R., Reichelt, M. E., Hsu, A., Headrick, J. P., & Gross, G. J. (2008). Activation of kappa-opioid receptors at reperfusion affords cardioprotection in both rat and mouse hearts. *Basic Research in Cardiology, 103*(5), 454–463.

Peng, J., Sarkar, S., & Chang, S. L. (2012). Opioid receptor expression in human brain and peripheral tissues using absolute quantitative real-time RT-PCR. *Drug Alcohol Depend, 124, 223–228.*

Pepe, S., Xiao, R. P., Hohl, C., Altschuld, R., & Lakatta, E. G. (1997). 'Cross talk'between opioid peptide and adrenergic receptor signaling in isolated rat heart. *Circulation, 95*(8), 2122–2129.

Pepe, S., Van Den Brink, O. W., Lakatta, E. G., & Xiao, R. P. (2004). Cross-talk of opioid peptide receptor and β-adrenergic receptor signalling in the heart. *Cardiovascular Research, 63*(3), 414–422.

Pike, J. L., Smith, T. L., Hauger, R. L., Nicassio, P. M., Patterson, T. L., McClintick, J., et al. (1997). Chronic life stress alters sympathetic, neuroendocrine, and immune responsivity to an acute psychological stressor in humans. *Psychosomatic Medicine, 59*(4), 447–457.

Romano, M. A., Seymour, E. M., Berry, J. A., McNish, R. A., & Bolling, S. F. (2004). Relative contribution of endogenous opioids to myocardial ischemic tolerance1. *Journal of Surgical Research, 118*(1), 32–37.

Ross, R. (1999). Atherosclerosis—An inflammatory disease. *New England Journal of Medicine, 340*(2), 115–126.

Rozanski, A., Blumenthal, J. A., & Kaplan, J. (1999). Impact of psychological factors on the pathogenesis of cardiovascular disease and implications for therapy. *Circulation, 99*(16), 2192–2217.

Rybin, V. O., Xu, X., Lisanti, M. P., & Steinberg, S. F. (2000). Differential targeting of β- adrenergic receptor subtypes and adenylyl cyclase to cardiomyocyte caveolae: A mechanism to functionally regulate the cAMP signaling pathway. *Journal of Biological Chemistry, 275*(52), 41447–41457.

Saeed, R. W., Stefano, G. B., Murga, J. D., Short, T. W., Qi, F., Bilfinger, T. V., & Magazine, H. I. (2000). Expression of functional delta opioid receptors in vascular smooth muscle. *International Journal of Molecular Medicine, 6*(6), 673–680.

Sakamoto, S., Stone, C. K., Woolf, P. D., & Liang, C. S. (1989). Opiate receptor antagonism in right-sided congestive heart failure. Naloxone exerts salutary hemodynamic effects through its action on the central nervous system. *Circulation Research, 65*(1), 103–114.

Sauvadet, A., Rohn, T., Pecker, F., & Pavoine, C. (1996). Synergistic actions of glucagon and miniglucagon on Ca2+ mobilization in cardiac cells. *Circulation Research, 78*(1), 102–109.

Schäfer, M., Frischkopf, K., Taimor, G., Piper, H. M., & Schlüter, K. D. (2000). Hypertrophic effect of selective β1-adrenoceptor stimulation on ventricular cardiomyocytes from adult rat. *American Journal of Physiology-Cell Physiology, 279*(2), C495–C503.

Schultz, J. E. J., Hsu, A. K., & Gross, G. J. (1996). Morphine mimics the cardioprotective effect of ischemic preconditioning via a glibenclamide-sensitive mechanism in the rat heart. *Circulation Research, 78*(6), 1100–1104.

Schultz, J. J., Hsu, A. K., & Gross, G. J. (1997). Ischemic preconditioning and morphine-induced cardioprotection involve the delta (δ)-opioid receptor in the intact rat heart. *Journal of Molecular and Cellular Cardiology, 29*(8), 2187–2195.

Schultz, J. E. J., & Gross, G. J. (2001). Opioids and cardioprotection. *Pharmacology & Therapeutics, 89(2), 123–137.*

Sigg, D. C., Coles, J. A., Jr., Gallagher, W. J., Oeltgen, P. R., & Iaizzo, P. A. (2001). Opioid preconditioning: myocardial function and energy metabolism. *The Annals of Thoracic Surgery, 72*(5), 1576–1582.

Sommerschild, H. T., & Kirkebøen, K. A. (2002). Preconditioning – Endogenous defence mechanisms of the heart. *Acta Anaesthesiologica Scandinavica, 46*(2), 123–137.

Springhorn, J. P., & Claycomb, W. C. (1992). Translation of heart preproenkephalin mRNA and secretion of enkephalin peptides from cultured cardiac myocytes. *American Journal of Physiology – Heart and Circulatory Physiology, 263*(5), H1560–H1566.

Takasaki, Y., Wolff, R. A., Chien, G. L., & van Winkle, D. M. (1999). Met5-enkephalin protects isolated adult rabbit cardiomyocytes via δ-opioid receptors. *American Journal of Physiology – Heart and Circulatory Physiology, 277*(6), H2442–H2450.

Tong, G., Sun, Z., Wei, X., Gu, C., Kaye, A. D., Wang, Y., et al. (2011). U50, 488H postconditioning reduces apoptosis after myocardial ischemia and reperfusion. *Life Sciences, 88*(1–2), 31–38.

Ventura, C., Muscari, C., Spampinato, S., Bernardi, P., & Caldarera, C. M. (1987). Inhibitory action of opioid peptides on ouabain-sensitive Na+-K+ and Ca2+- dependent ATPase activities in bovine cardiac sarcolemma. *Peptides, 8*(4), 709–713.

Ventura, C., Spurgeon, H., Lakatta, E. G., Guarnieri, C., & Capogrossi, M. C. (1992). Kappa and delta opioid receptor stimulation affects cardiac myocyte function and Ca2+ release from an intracellular pool in myocytes and neurons. *Circulation Research, 70*(1), 66–81.

Ventura, C., Pintus, G., & Tadolini, B. (1998). Opioid peptide gene expression in the myocardial cell. *Trends in Cardiovascular Medicine, 8*(3), 102–110.

Wang, T. L., Hang Chang, M. D., Hung, C. R., & Yung-Zu Tseng, M. D. (1997). Attenuation of neutrophil and endothelial activation by intravenous morphine in patients with acute myocardial infarction. *The American Journal of Cardiology, 80*(12), 1532–1535.

Wang, T. L., Chang, H., Hung, C. R., & Tseng, Y. Z. (1998). Morphine preconditioning attenuates neutrophil activation in rat models of myocardial infarction. *Cardiovascular Research, 40*(3), 557–563.

Wang, G. Y., Wu, S., Pei, J. M., Yu, X. C., & Wong, T. M. (2001). κ-but not δ-opioid receptors mediate effects of ischemic preconditioning on both infarct and arrhythmia in rats. *American Journal of Physiology – Heart and Circulatory Physiology, 280*(1), H384–H391.

Weil, J., Eschenhagen, T., Fleige, G., Mittmann, C., Orthey, E., & Scholz, H. (1998). Localization of preproenkephalin mRNA in rat heart: Selective gene expression in left ventricular myocardium. *American Journal of Physiology – Heart and Circulatory Physiology, 275*(2), H378–H384.

Weil, J., Zolk, O., Griepentrog, J., Wenzel, U., Zimmermann, W. H., & Eschenhagen, T. (2006). Alterations of the preproenkephalin system in cardiac hypertrophy and its role in atrioventricular conduction. *Cardiovascular Research, 69*(2), 412–422.

Wilbert-Lampen, U., Trapp, A., Barth, S., Plasse, A., & Leistner, D. (2007). Effects of β- endorphin on endothelial/monocytic endothelin-1 and nitric oxide release mediated by μ1-opioid receptors: A potential link between stress and endothelial dysfunction? *Endothelium, 14*(2), 65–71.

Wilson, S. P., Klein, R. L., Chang, K. J., Gasparis, M. S., Viveros, O. H., & Yang, W. H. (1980). Are opioid peptides co-transmitters in noradrenergic vesicles of sympathetic nerves? *Nature, 288*(5792), 707–709.

Xiao, R. P., Spurgeon, H. A., Capogrossi, M. C., & Lakatta, E. G. (1993). Stimulation of opioid receptors on cardiac ventricular myocytes reduces L type Ca2+ channel current. *Journal of Molecular and Cellular Cardiology, 25*(6), 661–666.

Xiao, R. P., Spurgeon, H. A., O'Connor, F., & Lakatta, E. G. (1994). Age-associated changes in beta-adrenergic modulation on rat cardiac excitation-contraction coupling. *The Journal of Clinical Investigation, 94*(5), 2051–2059.

Xiao, R. P., Pepe, S., Spurgeon, H. A., Capogrossi, M. C., & Lakatta, E. G. (1997). Opioid peptide receptor stimulation reverses beta-adrenergic effects in rat heart cells. *American Journal of Physiology – Heart and Circulatory Physiology, 272*(2), H797–H805.

Yanagisawa, M., Kurihara, H., Kimura, S., Tomobe, Y., Kobayashi, M., Mitsui, Y., et al. (1988). A novel potent vasoconstrictor peptide produced by vascular endothelial cells. *Nature, 332*(6163), 411–415.

Younès, A., Pepe, S., Barron, B. A., Spurgeon, H. A., Lakatta, E. G., & Caffrey, J. L. (2000). Cardiac synthesis, processing, and coronary release of enkephalin-related peptides. *American Journal of Physiology – Heart and Circulatory Physiology, 279*(4), H1989–H1998.

Yu, X. C., Wang, H. X., Pei, J. M., & Wong, T. M. (1999). Anti-arrhythmic effect of κ- opioid receptor stimulation in the perfused rat heart: Involvement of a cAMP- dependent pathway. *Journal of Molecular and Cellular Cardiology, 31*(10), 1809–1819.

Zaugg, M., Xu, W., Lucchinetti, E., Shafiq, S. A., Jamali, N. Z., & Siddiqui, M. A. Q. (2000). β-Adrenergic receptor subtypes differentially affect apoptosis in adult rat ventricular myocytes. *Circulation, 102*(3), 344–350.

Zhang, Y., Chen, Z. W., Irwin, M. G., & Wong, T. M. (2005). Remifentanil mimics cardioprotective effect of ischemic preconditioning via protein kinase C activation in open chest of rats. *Acta Pharmacologica Sinica, 26*(5), 546–550.

Zhao, Z. Q., Corvera, J. S., Halkos, M. E., Kerendi, F., Wang, N. P., Guyton, R. A., & Vinten-Johansen, J. (2003a). Inhibition of myocardial injury by ischemic postconditioning during reperfusion: Comparison with ischemic preconditioning. *American Journal of Physiology – Heart and Circulatory Physiology, 285*(2), H579–H588.

Zhao, Z. Q. Z. Q., Morris, C. D., Budde, J. M., Wang, N. P., Muraki, S., Sun, H. Y., & Guyton, R. A. (2003b). Inhibition of myocardial apoptosis reduces infarct size and improves regional contractile dysfunction during reperfusion. *Cardiovascular Research, 59*(1), 132–142.

Zhu,W. Z., Wang, S. Q., Chakir, K., Kolbika, B. K., Cheng, H., & Xiao, R. P. (2003). Linkage of β1-adrenergic stimulation to apoptotic heart cell death through protein kinase A-independent activation of Ca^{2+}/calmodulin kinase II. *Journal of Clinical Investigation, 111, 617–25.*

Zimlichman, R., Gefel, D., Eliahou, H., Matas, Z., Rosen, B., Gass, S., et al. (1996). Expression of opioid receptors during heart ontogeny in normotensive and hypertensive rats. *Circulation, 93*(5), 1020–1025.

Endorphins, Sexuality, and Reproduction

Marjan Khajehei

Abstract Beta-endorphin is secreted from the hypothalamus and pituitary in both mother and newborn. The placenta produces numerous pituitary hormones from the third month of pregnancy, one of which is βE. It has been suggested that βE has a role in the appetitive and precopulatory phase of sexual behavior in animals. An increase in endorphin levels during sexual activity in humans may contribute to attachment and bonding between partners, but contradictory reports in the literature question the association between sexuality and βE levels. The level of βE also increases during pregnancy, rises in early labor, peaks in late labor, and drops in the postpartum period. This fluctuation provides natural analgesia, raises the pain threshold, decreases the sensation of pain, or suppresses pain, and decreases fear levels during labor and birth. Beta-endorphin also protects the fetus from hypoxia during labor and birth and potential neural damage by aiding blood flow to the brain under hypoxic conditions. It has been suggested that a variety of pharmacologic and nonpharmacologic complementary therapies, when used in pregnancy, labor, and birth, activate the opioid receptors in the CNS and alter the sensation of pain during labor and birth, affect the mother–child attachment and affect sexual function. These studies report contradictory results that will be discussed in this chapter.

Keywords Birth · Breastfeeding · Endorphins · Labor · Sexuality

Endorphins, Sexuality, and Reproduction

Endorphins are the body's natural opioids, or endogenous opioids, that are created and released by the central nervous system (CNS), hypothalamus, and pituitary gland. Endorphins have a reputation for pain reduction, enhancing excitement or satisfaction, boosting confidence, enabling control of emotions and generating

M. Khajehei (✉)
Westmead Hospital, Westmead, NSW, Australia
e-mail: Marjan.Khajehei@health.nsw.gov.au

P. L. Kerr et al. (eds.), *Endogenous Opioids*, Advances in Neurobiology 35,
https://doi.org/10.1007/978-3-031-45493-6_20

feelings of euphoria, and are involved in the natural reward cycle. It has been suggested that the release of endorphins in the human body is triggered by a variety of factors, including massage and bodywork (Bolbol-Haghighi et al., 2016; Hidayati et al., 2014), exercise (Hiramoto et al., 2013), active performance of music (Dunbar et al., 2012), consumption of certain foods such as dark chocolate (Parker et al., 2006), environmental factors such as ultraviolet light (Robinson & Fisher, 2014), and childbirth (Qu & Zhou, 2007). There is also evidence in the literature indicating the role of endorphins in sexuality. It is suggested that endorphins regulate the release of other hormones, such as sex hormones, prolactin, and growth hormone, that are involved in sexual function and love (Bancroft, 2005; Esch & Stefano, 2005).

Several endorphin-like substances have been identified in the human body, including βE (βE) proper, beta-lipotropin, N-terminal derivatives, and other related peptides (Harbach et al., 2008). βE was discovered in the 1970s. Since then, it has been the subject of substantial research attention, and several studies have been undertaken to explore its role in various aspects of reproduction (Dalayeun et al., 1993). βE is secreted from peripheral body tissues and the CNS and is a natural equivalent to pain-killing drugs such as pethidine and morphine (Mousa et al., 2003). From pregnancy through labor, birth, and the postpartum period, it has been suggested that endorphins, especially βE:

- Are involved in sexual function and attachment (Bancroft, 2005; Esch & Stefano, 2005)
- Have behavioral, thermoregulatory and neuroendocrine properties (Odent, 2001)
- Provide natural analgesia, increase pain threshold, decrease the sensation of pain or suppress pain and decrease fear levels, resulting in a sense of well-being and euphoria (Dabo et al., 2010)
- Promote a shift in a woman's consciousness, making her to communicate with her inner guidance accompanied by a sense of bliss and joy. This protects the laboring woman from external disturbances and directs her thoughts to sensations in her body and optimal positioning to facilitate birth (Baker, 2001)
- Protect the fetus from hypoxia (human studies) and neural damage (animal studies) during labor and birth (Espinoza et al., 1989; Lou et al., 1989)
- Activate the reward and pleasure center after birth, in relation to the birth and the newborn (Nelson & Panksepp, 1998; Sauriyal et al., 2011)
- Reach a peak immediately after birth (along with oxytocin), facilitating maternal excitement and stimulating the reward center, thus helping to establish a bond with newborn (Nelson & Panksepp, 1998)
- Reinforce breastfeeding in both mother and newborn (Ferrando et al., 1990; Franceschini et al., 1989; Grandison & Guidotti, 1977; Zanardo, Nicolussi, Carlo, et al., 2001a).

Lower levels of βE have been shown to occur in certain socioeconomic groups, in women receiving childbirth preparation and delivery assistance, in women who experience more stress during labor (Ombra et al., 2008), after caesarean section, in women with postpartum maternal anxiety (Zanardo et al., 2001b), in women with deteriorated mood and postpartum depression (Smith et al., 1990), and in women who receive analgesia during labor (Dabo et al., 2010; Smith et al., 1990; Thomas et al., 1982). While the roles of endorphins is well researched and understood in some areas of health, their potential impacts on reproduction and sexuality are only beginning to be understood. The purpose of this chapter is to review current understanding of endorphins, in particular βE, and how they relate to the field of reproduction and sexuality.

Beta-Endorphin and Sexuality

Sexual Response Cycle Models

Over the years, different models have been used to describe the human sexual response cycle. Masters and Johnson (1966, 1970) first described four main phases of the human sexual response cycle. Although this model, which included arousal, plateau, orgasm, and resolution, succeeded to explain human sexuality at that time, Katz (2009) has criticized it as a behavioral-based model, which failed to explain the role of the brain in organizing the human sexual response. Kaplan (1974) introduced a three-phase model (including desire, excitement, and orgasm) that focused on the role of brain in sexual emotion and cognition. Subsequently, Zilbergeld and Ellison (1980) also aimed to include both physiological and psychological aspects of the sexual response cycle, and they categorized the sexual response cycle into five phases including desire, arousal, plateau, orgasm, and resolution. Another four-phase circular model was suggested by Whipple and Brash-McGreer (1997) encompassing seduction (desire), sensations (excitement and plateau), surrender (orgasm) and reflection (resolution). According to Whipple and Brash-McGreer (1997), pleasant sexual experiences have substantial effects on an individual's sexual response.

During the last decade, many researchers have incorporated the role of the brain, emotion, and cognitive processes into the sexual response cycle and suggested models based on the biopsychosocial elements. A four-phase model has been suggested by Basson et al. (2000). This model is based on desire, excitement, orgasm, and resolution and explains the influence of the brain on corporal changes during sexual activity. The circular model for sexual response magnifies the importance of emotional intimacy and physical and emotional satisfaction during sexual activity (Basson, 2000; Kammerer-Doak & Rogers, 2008).

Effect of Beta-Endorphin on Sexual Response

Animal studies There is evidence in the literature indicating the role of endorphins in sexuality. Opioid peptides may have both excitatory and inhibitory effects on sexual performance and behaviors (Argiolas & Melis, 2013; Bancroft, 2005). According to animal studies (Melis et al., 1999; van Furth et al., 1995), when opioid peptides are released in response to stress, they impose their inhibitory effects by acting in the medial preoptic area and the paraventricular nucleus that, in turn, impairs sexual performance. It is suggested that endorphins regulate the release of other hormones, such as sex hormones, prolactin, and growth hormone, that are involved in sexual function and attachment (Bancroft, 2005; Esch & Stefano, 2005). It has also been suggested that this may be relevant to the low level of sexual desire in people with symptoms of depression (Dornan & Malsbury, 1989).

Preliminary studies have investigated the mechanisms of inhibition of sexual behavior by endogenous opioids. Myers and Baum (1979, 1980) showed that naloxone, the mu-opioid receptor (μ-OR) antagonist, has a facilitatory effect on masculine sexual performance in rats resulting in the release of gonadotropin releasing hormone (GnRH). A later study (Sirinathsinghji et al., 1983) indicated that infusion of μ-OR antagonists into the mesencephalic central gray matter increases neuronal GnRH output that in turn enhances the likelihood of lordosis behavior in estrogen-primed female rats. Other studies have shown that acute treatment with μ-OR antagonists augmented GnRH secretion followed by raised levels of serum LH and testosterone (Blank et al., 1986; Fraioli et al., 1985; Pfeiffer & Herz, 1984).

In a study by Csaba et al. (2003), administration of a single dose of endorphin to neonatal rats showed that sexual activity permanently decreased in females after 5 months and their tendency to refuse the male increased, in addition to male aggression increasing. Female rats showed a permanent increase in the density of uterine estrogen receptors, and male rats showed a decline in the serotonin content of the brain. Although little is known about the interaction of endorphin and other hormones or neurotransmitters in relation to human sexuality, results of the study by Csaba et al. (2003) suggest that there is a role for hormone imprinting at birth and that endorphin treatment influences sexual hormone production, which can affect sexual behaviors in later life.

During labor, the level of endorphin in the mother's blood increases and is dependent on the intensity of pain and the duration of labor (Bacigalupo et al., 1990; McLean et al., 1994; Weissberg et al., 1990). Therefore, it is presumed that neonatal endorphin imprinting affects later-life events such as sexual activity and aggression, because of the association between brain serotonin levels and aggressive behaviors (Sundblad & Eriksson, 1997). However, this hypothesis is based solely on data from rodent models, and its generalizability to other species, including primates (e.g., humans) is currently unclear.

The opioid peptides impose their excitatory effects by acting in the ventral tegmental area, increasing the activity of the mesolimbic dopaminergic system and promoting sexual arousal and motivation. There appears to be no research investigating the role of βE in human sexuality, making it impossible to determine whether this is a general effect of all opioid peptides or if it is specific for other peptides such as enkephalin, as reported in the literature (Argiolas & Melis, 2013).

Research in animal models has found βE affects brain activity and maintains a sense of balance and well-being by allowing the animals to perform feeding and drinking activities as well as social grooming (Keverne et al., 1989). A systematic review of animal studies (Veening & Barendregt, 2015) has also suggested that βE plays its main role in the appetitive and precopulatory phase of sexual behavior, in preparation for copulatory activities. Further, there is a relationship between βE and sex hormones.

Human studies Because this association is mutual and endogenous sex steroids can affect the neurobiology of sexual function, by directly influencing receptors at the nuclear and membrane level or by indirectly affecting the neurotransmitters of neuropeptides (oxytocin and βE) (Clayton, 2003; Frye, 2009), it has been suggested that βE may be involved in the regulation of sexual function in humans. Furthermore, early researchers proposed that a mild increase in the βE level creates a sense of well-being, and that a greater increase may lead to analgesia and euphoria (Henry, 1982). A variety of behavioral experiences can activate the release of βE (Izquierdo & Netto, 1985). For example, exercise stimulates secretion of corticotropin-releasing hormone (see Kerr et al., this volume), resulting in an increase in ACTH and endorphins that may enhance an individual's sexuality (Mastorakos & Pavlatov, 2004). In addition to aerobic exercise, discontinuation of tobacco use and illicit drug use and reduced alcohol consumption improve tissue oxygenation, promote metabolism, reduce body mass index, and stimulate endorphin release that may, in turn, boost sexual response (Basson et al., 2005).

An increase of endorphins during sexual activity in humans is presumed to contribute to attachment and bonding between partners, similar to that of a mother and her newborn (Esch & Stefano, 2005). However, contradictory reports in the literature question the association between sexuality and βE levels. For example, in a small study on ten healthy women, sexual arousal and orgasm resulted in a sharp increase in cardiovascular parameters and plasma catecholamine concentrations along with an increase in the concentration of plasma prolactin, but no changes were seen in the plasma concentrations of βE (Exton et al., 1999). Similar neuroendocrine response pattern to sexual arousal and orgasm in men was reported in an earlier study by Krüger et al. (1998). Although they showed a transient increase in heart rate and blood pressure as well as noradrenaline and prolactin plasma levels, no changes were seen in the plasma βE and other endocrine variables.

Less is known about the association between endogenous opioids and sexual function and behaviors in humans, but it is known that exogenous opiates negatively affect the sexuality of people with opioid use disorder (OUD), and contribute to their reduced sexual desire, impaired sexual arousal, decreased genital response, delayed or blocked ejaculation, orgasm dysfunction, and infertility (Katz & Mazer, 2009). Opioid use negatively affects sexual function through reducing the levels of sex hormones, and their effect on the endocrine system begins immediately after they are taken (Brennan, 2013). Although little is known about the exact mechanism of sexual dysfunction in people with OUD, and studies in this area are small, the available evidence shows a high prevalence of opioid-induced hypogonadism (up to 90%) in patients who use opioids, such as heroin (Rasheed & Tareen, 1995), methadone (Yee et al., 2014), intrathecal opioids (Roberts et al., 2002), and systemic (oral or transdermal) opioids (Fraser et al., 2009; Rhodin et al., 2010). According to a systematic review and meta-analysis of the testosterone levels in men and women while using opiates, regular drug use suppresses the testosterone level in men regardless of the type of opioid being ingested. Conversely, testosterone levels in women are not affected by opioids. This sex difference suggests that opioids may have differential mechanisms for endocrine disruption in men and women, and this should be taken into consideration when treating sexual problems in people who are opiate-dependent (Bawor et al., 2015). Because there may be different endocrine targets to aim for even in non-opioid-dependent men and women while trying to treat their sexual dysfunction using pharmaceutical drugs, any future drug development for sexual dysfunction needs to consider these differences.

The negative effects of opioids on male sexual function are reversible after opiate withdrawal (Pfaus & Gorzalka, 1987) or administration of μ-OR antagonists (Cicero et al., 1976). The positive effects of μ-OR antagonists are increased luteinizing hormone (LH) pulsatility, raised serum testosterone levels (Graves et al., 1993), increased in vitro sperm motility after administration of naloxone (Agirregoitia et al., 2012), recurrent spontaneous penile erections, frequent orgasms, and more intense sexual arousal and orgasm in healthy adult men who were not addicted to opiates, after administration of naltrexone (Mendelson et al., 1978; Sathe et al., 2001). However, these findings are not supported by other animal research indicating a lack of substantial influence of acute or chronic naloxone administration on different sexual activities of isolated and group-housed male rats (Decatanzaro et al., 1996).

The limited research in humans, especially in women, has created inconsistent but, in some cases, interesting results. For example, in the study by Goldstein and Hansteen (1977), a single male subject was recruited and the researchers prematurely concluded that there is no evidence of the involvement of endorphins in male sexual arousal. Another study by Gillman and Lichtigfeld (1983) found that administration of a 2 mg dose of naloxone on two separate occasions enhanced orgasm and pleasure in women, while a single 2 mg dose of naloxone inhibited arousal and orgasm for up to 10 minutes, suggesting that the relationship of naloxone to orgasm is dose-dependent and potentially parabolic. This is consistent with the notion that endogenous opioids, such as βE, have both inhibitory and excitatory effects, but the explanation for the dose–response effect remains obscure (Bancroft, 2005).

In conclusion, there is a lack of updated data on the mechanism of action of endorphins, especially βE, and their role in regulating human sexuality. While some animal studies report the effect of βE on GnRH, LH, and testosterone, these findings have not been supported by human research. Further studies are required to investigate the role of endorphins in human sexuality and their mechanism of action in men and women.

Role of Beta-Endorphin in Pregnancy, Labor, Birth, and Postpartum

Beta-Endorphin and Pregnancy

The increase in βE in the last weeks of pregnancy and during labor are part of the body's preparation for birth, and are also due to increases in steroid hormones, estrogen, and progesterone. Animal studies have shown that the steroid hormones of pregnancy activate non-βE opioid systems (κ- and δ-opioid receptors) in the spinal cord. They provide additional analgesia that lasts across labor, birth, and the early postpartum period (Gintzler & Liu, 2001). Progesterone is secreted from the placenta and levels continue to gradually increase in maternal serum throughout pregnancy. By the end of pregnancy, the levels of progesterone are 10–5000 times greater than that in nonpregnant women (Cunningham et al., 2010). Progesterone can cross the blood–brain barrier and enter the brain, but its distribution in the brain is not uniform, as the hypothalamus has been shown to contain the highest levels of progesterone (Wang et al., 1997). The synthesis of progesterone in brain cells results in the formation of neuro-steroids. These substances moderate the activity of hypothalamic and extra-hypothalamic noradrenergic, dopaminergic, and serotoninergic neurons and apply significant regulatory activities on neurons and glial cells (Baulieu, 1997). Experimental evidence suggests that progesterone plays a significant role in the release of hypothalamic βE into the bloodstream. This may be explained by the numerous progesterone receptors on βE neurons in the hypothalamus (Fox et al., 1990), and the overlap between the distribution of βE and progesterone receptors in the arcuate nucleus (Aleem & McIntosh, 1985).

Beta-Endorphin and Early Labor

The number of central μ-ORs (Dondi et al., 1991; Hammer et al., 1992) and central levels of βE increase during the last few weeks of pregnancy and around the physiological onset of labor (Sharon & Andrew, 1983). In a preliminary study, Cahill (1989) measured βE-like immunoreactivity to investigate the effect of βE on pain perception during labor. Results of the study showed that the level of plasma βE is higher in pregnant women compared to nonpregnant women at their mid-cycle. It also showed a steady increase in plasma βE-like immunoreactivity from second to third trimester of pregnancy and throughout labor increasing from early labor to full cervical dilatation. Although there was no significant correlation between βE levels and perceived labor pain, women in that study reported greater discomfort between contractions rather than during contraction. Cahill (1989) suggested that the increase of βE during labor may increase a woman's ability to tolerate acute pain. Other studies have supported these findings indicating that although the increased levels of βE, its receptors, and the non-βE opioid mechanisms do not generally eliminate pain entirely, these changes in βE levels increase maternal pain tolerance in late pregnancy and provide "pregnancy-induced analgesia" that reaches a peak when labor starts in both humans (Cogan & Spinnato, 1986; Whipple et al., 1990) and animals (Gintzler, 1980; Sander & Gintzler, 1987). This physiological analgesic system keeps the experience of labor within the levels for eustress (beneficial stress), without disrupting the progress of labor and helps the laboring woman tolerate labor pain (Tornetta, 1998).

Induction and Progress of Labor

During an undisturbed labor, endogenous oxytocin induces labor contractions and thus labor pain and stress. The oxytocin blocks beta-adrenergic receptors and decreases clearance of endorphins, especially βE (Kang & Park, 2000; Sweep et al., 1990). This process raises the level of plasma endorphins, increases their binding to opioid receptors on the uterine muscles and results in pain reduction, which can relieve some of the pain associated with uterine contractions and labor (Saisto et al., 2004). Subsequently, the release of βE when pain and stress levels are high prevents further release of endogenous oxytocin. This will, in turn, result in decelerating uterine contractions, less labor pain and more comfort for the laboring woman (Jowitt, 1993).

Induction or augmentation of labor contractions using oxytocics (i.e., synthetic oxytocin pharmaceutical compounds) may result in an increased sensation of pain (Hodnett, 2002). This has been suggested to be due to vigorous myometrial ischemia (Jurna, 1993; Lowe, 1996), or the inhibitory effects of oxytocic medication on the release of βEs in the brain, or both. Genazzani et al. (1985) compared βE levels in women in spontaneous labor with women who received an oxytocic medication to stimulate uterine contractions. They found that in women in spontaneous labor, plasma levels of βE increased significantly through labor until delivery. However, the βE levels remained constant or lower in women who received oxytocics. Genazzani et al. (1985) suggested that the lower levels of βE in women who received an oxytocic medication to stimulate uterine contractions could have been related mostly to the primary uterine hypocontractility, rather than to intravenous oxytocic administration. They also pointed out that using oxytocics to augment or induce labor can result in tocophobia (anxiety and fear of childbirth), which might be associated with lower βE levels in these women (Genazzani et al., 1985). The results of the study by Genazzani et al. (1985) were supported by a later study (Petraglia et al., 1986) suggesting that administration of synthetic oxytocin (Syntocinon 0·4 μg/min for 120 min) could inhibit the rise of βE in stressful situations in healthy humans.

Beta-Endorphin and Late Labor

Pain and stress management Childbirth has a reputation for intense pain and is known to be one of the most painful events in a woman's life (Lowe, 2002). The natural reflex response to labor pain is the release of endorphins from the pituitary gland. Endorphins are co-released into the peripheral bloodstream with adrenocorticotropic hormone (ACTH), oxytocin, epinephrine and norepinephrine, and play an important role in modulating pain during labor and birth (Saisto et al., 2004).

The level of βE increases during pregnancy, elevates in early labor, peaks in late labor, and drops in the postpartum period (Hofmeyr et al., 1995). These changes are positively correlated with the strength of labor contractions, progress of labor, self-reported pain, and rupture of membranes (Hofmeyr et al., 1995; Raisanen et al., 1984; Weissberg et al., 1990). During the pushing phase, the level of βE is four to eight times as high as its level before labor (Hoffman et al., 1984; Pancheri et al., 1985; Raisanen et al., 1984; Robin et al., 1981) and is comparable to that of athletes during an endurance event (Laatikainen, 1991). If a laboring woman does not use any pain relief medication, the level of βE continues to grow progressively. This helps the woman tolerate the increasingly painful uterine contractions, and helps her enter the reported "dream-like" or "on another planet" state (Odent, 2008; Reed et al., 2016).

Several studies have reported a correlation between levels of βE in pregnancy and later outcomes, although the reports may not be easy to interpret. Low levels of βE (Dabo et al., 2010) or dysregulation in relation to ACTH (Sandman et al., 1995) have been shown to be correlated with a greater need for analgesia during labor. In a longitudinal study by Dabo et al. (2010), a remarkable drop in the level of βE was shown from week 25 to week 34 of pregnancy. Subsequently, there was a significant increase in the plasma level of βE after 34 weeks, which continued to increase through the third trimester of pregnancy. No difference was found in the plasma concentration of βE between week 10 and week 37 of pregnancy. In their study, Dabo et al. (2010) followed up the women throughout labor and birth. Results of multivariate logistic regression analysis indicated that, compared to women with high levels of βE at 37 weeks, women who had the lowest levels of βE at 37 weeks of pregnancy required additional pain medication beyond nitrous oxide during labor, such as epidural analgesia, and this difference remained significant after adjusting for parity, body mass index, duration of labor, cervical dilatation on admission to birth unit, and the use of Syntocinon (synthetic oxytocin) during labor. Despite these intriguing findings, the researchers did not collect blood samples for the assessment of βE levels during labor, and any assumption about the association between the level of βE and the use of pain relief during labor was based on the last blood sample collected during pregnancy, but not during labor.

When opioids are systemically administered to women who are not in labor, they reduce pain and decrease the release of βE (Jowitt, 1993). It is suggested that they have the same effect in laboring women, but conflicting findings have been reported in the literature about the effect of opioids on the levels of βE and progress of labor.

Two studies reported no decrease in βE in laboring women after administration of meperidine (Brinsmead et al., 1985; McLean et al., 1994), while one study revealed a greater peak in βE (Riss & Bieglmayer, 1984) and other studies found lower levels of βE at each stage of labor following epidural analgesia (Scull et al., 1998; Thomas et al., 1982). These contradictory reports suggest that the effective dose of systemic opioids greatly depends on the individual variation in the βE system and opioid receptors that affect opioid sensitivity (Somogyi et al., 2007).

The chemical structure of endorphins is similar to opioids that are used to induce analgesia in labor (see Sessle, this volume); they activate opioid receptors in the brain and help diminish pain. Unlike exogenous opioids, endorphins are not addictive. When endorphins attach to the opioid receptors, they are broken down by enzymes. This allows recycling of the molecules so they can be reused later. However, when synthetic opioids attach to the opioid receptors, they resist the breakdown enzymes and continue to reactivate the same receptors, extending and increasing the resulting euphoria and increasing the probability of dependence (Waldhoer et al., 2004). It is therefore likely that opioids and analgesia in labor decrease a woman's βE levels in proportion to their efficacy at decreasing her stress and pain (Thomas et al., 1982), and that they prolong second-stage labor (Agrawal et al., 2014). There is controversy in the literature regarding an increased risk of caesarean section in women receiving epidural analgesia with opioids (Herrera-Gómez et al., 2017), but high-quality Cochrane reviews (Anim-Somuah et al., 2011; Sng et al., 2014) have shown that early or late initiation of epidural analgesia for labor does not affect labor and birth outcomes. These outcomes include the duration of the second stage of labor, incidence of caesarean section and instrumental birth, maternal satisfaction with pain relief, and occurrence of long-term backache.

The controversial reports in the literature could be due to the individual differences in βE levels during pregnancy, labor, and birth, as women with lower levels of βE are more likely to request and receive epidural analgesia during labor (Dabo et al., 2010). According to the American College of Obstetricians and Gynecologists (2002), a woman's request for labor pain relief is adequate medical indication for its provision. Therefore, obstetricians should take the woman's lead on when to implement epidural analgesia, rather than using other indicators. A suitable plan for pain relief needs to be developed in consultation with anesthesiologists, especially in women with risk factors such as marked obesity, severe edema, severe preeclampsia or bleeding disorders, to ensure the safety of both mother and her newborn (American Academy of Pediatrics and American College of Obstetricians and Gynecologists, 2012). Preliminary research has shown that the level of βE reaches a peak in maternal blood around the time of birth and drops during the first hour after birth (Hoffman et al., 1984), but shows another peak level during the delivery of the placenta (Facchinetti et al., 1982; Pancheri et al., 1985). In the first 72 hours after birth, the level of βE gradually decreases to those of late pregnancy (Bacigalupo et al., 1987; Pancheri et al., 1985). The human plasma half-life of βE has been estimated to be about half an hour. However, its central effects may last for 21 hours or longer (Foley et al., 1979).

Similar to other mammals, eustress in humans triggers the release of corticotropin-releasing hormone (CRH), also known as the stress hormone, from the hypothalamus. CRH, in turn, provokes the release of βE from the brain and pituitary gland into the bloodstream. This process also involves the release of other hormones such as epinephrine and norepinephrine, cortisol and prolactin. This well-coordinated general stress response is known as the "general adaptive syndrome" and enhances adaptation to extreme stress and restoration of homoeostasis during labor and birth (Selye & Horava, 1953; Voogt et al., 2001).

Studies on the association between βE, stress, and pain during labor are contradictory, which may reflect individual differences in subjective experiences of the analgesic effects of βE. Some studies report that higher levels of βE in labor are associated with less stress or pain (Benfield et al., 2010; Fajardo et al., 1994; Hartmann et al., 2005; Varrassi et al., 1989), but other studies have shown that women with elevated βE levels experience higher levels of stress and pain (Delke et al., 1985; Lechner et al., 1991; Punzio et al., 1994). The levels of βE have also been suggested to be higher in women with four or more previous deliveries (also known as "grand multiparous") than in women with fewer previous deliveries (Weissberg et al., 1990). This suggests that prior labor and birth experience may result in a more responsive βE system, and supports the results of an earlier study (Raisanen et al., 1984) in which unmedicated grand multiparous women had substantial increases in βE levels without severe stress or pain.

The adaptive midrange level of response, the balance between healthy and unhealthy stress, and the levels of stress hormones vary in different people and may be contingent on individual systems and the quantity and sensitivity of receptors (Anderson, 1976). Although short-term stress, along with moderate elevations of stress hormones, are part of the general adaptive syndrome with beneficial effects (Selye, 1976), excessive, long-term stress can result in extremely elevated stress hormone levels, reduced motor and social behaviors, insufficient adaptation, and detrimental effects on both mother and fetus or newborn (Garthus-Niegel et al., 2013; Shapiro et al., 2013). Animal studies involving rats (Douglas & Russell, 2001) and pigs (Jarvis et al., 2000) have shown that excessive stress results in over-release of βE to supraphysiological levels, followed by inhibited oxytocin levels, unbalanced epinephrine and norepinephrine levels, impaired transmission of visual and acoustic information and disturbed progress of labor (Hartwig, 1991). The supraphysiological increase in βE levels due to excessive stress, and its inhibiting effects on labor progress, suggests that the βE system is a physiological response that decelerates labor in situations of stress and potential danger, an effect mainly seen in animals (Jowitt, 1993). It can, therefore, be proposed that the relationship of stress and endogenous opioids has a parabolic curve that is similar to the parabolic association between naloxone and sexual arousal or orgasm, which was discussed earlier in this chapter.

Complementary and Alternative Therapies for Pain Management in Labor

A variety of non-pharmacological complementary approaches have been proposed to raise the level of βE and assist the progress of physiological labor through effective pain management. These include, but are not limited to, massage therapy (Bolbol-Haghighi et al., 2016; Hosseini et al., 2013), transcutaneous electrical nerve stimulation (TENS) (Sluka & Walsh, 2003), acupuncture (Tempfer et al., 1998), acupressure (Lee et al., 2004), yoga, and relaxation (Field, 2011), exercise, visualization, controlled breathing (Levett et al., 2016), reflexology (Smith et al., 2012), music therapy (Taghinejad & Delpisheh, 2010), water immersion (Benfield et al., 2010), heat therapy, and the birth ball (Taavoni et al., 2016). In addition to these interventions, research on nonpregnant healthy adults (Leonard et al., 2010) has shown that the use of diffuse noxious inhibitory control (DNIC) creates a second pain elsewhere in the body (Willer et al., 1990) and decreases the sensation of labor pain through opioid mechanisms, possibly involving βE. This could be the potential mechanism for the analgesic effects of other non-pharmacologic complementary approaches and pain control techniques in labor. They may increase the level of βE, bind βE to receptor sites in the brain, and result in a reduced sensation of pain and increased psychological well-being and euphoria during labor (Bonapace et al., 2013; Tempfer et al., 1998).

A meta-analysis of 57 studies published from January 1990 to December 2012 (Chaillet et al., 2014) showed that non-pharmacological methods to relieve labor pain can offer remarkable benefits to women and their newborns without additional harm. Varrassi et al. (1989) investigated changes in plasma βE levels after exercise during pregnancy and labor in 36 multiparous pregnant women. Their results showed that, compared to the control group, the level of plasma βE was significantly elevated in women who exercised during pregnancy. The high level of βE was maintained in exercise-conditioned women throughout labor and they reported less pain during labor (measured by visual analogue pain scales). The exercise also resulted in a decline in other hormones during labor including cortisol, human growth hormone, and prolactin. These hormones have been reported to increase in response to pain, and thus their decrease can be correlated with less pain and discomfort (Choi et al., 2012; Patil et al., 2013).

Despite limited evidence from the literature, the endorphin hypothesis remains one of the most common justifications for the benefits of non-pharmacological complementary therapies during pregnancy, labor, and birth. Another example is the use of TENS, which is suggested as an activator of opioid receptors in the CNS. According to animal studies, mu-opioid receptors are activated by low-frequency TENS and delta-opioid receptors are activated by high-frequency TENS (Kalra et al., 2001; Sluka et al., 1999). Opioids affect μ-opioid receptors and in turn βE levels. If opioid tolerance develops, such as tolerance to morphine (a mu-opioid receptor agonist), then low-frequency TENS would not work but high-frequency TENS would be still effective (Sluka et al., 2000). A similar pattern has been shown in human research suggesting the ineffectiveness of TENS in cases of opioid tolerance (Léonard et al., 2011). These findings highlight the significance of understanding the potential interactions between opioids and TENS and the need for future research to explore other pharmacological interactions.

TENS has been shown to inhibit pain-modulating circuits in animals, increase the local release of endorphins, and promote the concentration of βE in the bloodstream and cerebrospinal fluid (Hughes et al., 1984; Salar et al., 1981; Sluka et al., 2013). A meta-analysis and systematic review of the effectiveness and safety of TENS when administered by physicians to patients with acute pain in a prehospital setting has shown that TENS significantly reduces the severity of pain for patients with moderate-to-severe acute pain. TENS produces lower posttreatment mean pain scores than sham TENS and is effective in reducing acute anxiety secondary to pain (Simpson et al., 2014). Randomized clinical trials on the use of TENS to relieve back pain during pregnancy (Keskin et al., 2012) and general pain in labor (Santana et al., 2016) have shown a significant difference in pain scores after using TENS. The intervention did not have negative effects on other maternal and neonatal outcomes, and the women were satisfied with the TENS outcome. Nevertheless, there are contradictory reports in the literature about its efficacy and outcomes (Sluka & Walsh, 2003).

Beta-Endorphin and the Fetus

In addition to the production of βE in the hypothalamus and pituitary, and in normal, uncomplicated circumstances in pregnancy, the placenta produces numerous pituitary hormones from the third month of pregnancy, one of which is βE (Abboud, 1988; Goebelsmann et al., 1984). The level of βE is also high in the amniotic fluid (Raffin-Sanson et al., 2000). Placental βE, which increases with gestation, along with increased levels in the maternal blood, exerts immunosuppressive effects on the mother. This effect is vital for tolerance of the genetically and immunologically alien fetus throughout pregnancy (Apari & Rózsa, 2006; Nandhra & Carson, 2000). It may also manipulate the mother's pattern of resource allocation. The mother may become dependent on placental βE as a reward and may trade βE for extra nutrients (Apari & Rozsa, 2006).

Both human (Abboud, 1988) and animal studies (Wardlaw et al., 1981) have shown that being exposed to hypoxia during labor and birth can cause hypoxic stress and result in the release of additional βE from the fetal pituitary gland (Goebelsmann et al., 1984). According to animal studies (Espinoza et al., 1989; Lou et al., 1989), the elevated fetal βE preserves blood flow to the brain under hypoxic conditions and has neuroprotective effects on the fetal brain. A similar process may happen in humans, protecting the fetus against hypoxia (Abboud, 1988; Goebelsmann et al., 1984).

A positive correlation has been shown in human (Fajardo et al., 1994) and animal (Chiang & Rodway, 1997) studies between maternal peripheral plasma βE and the umbilical vein βE, suggesting a role for endorphins in the regulation of maternal and fetal stress. Research has shown that any intervention during pregnancy, labor, or birth can interfere with the normal secretion and effect of endogenous opioids and can affect the fetal opioid system. For example, a hormonal stress response and an increase in fetal plasma concentration of βE along with cortisol and norepinephrine have been reported in intrauterine needling procedures (Giannakoulopoulos et al., 1994, 1999). These studies have shown that the increase of plasma βE in the hepatic vein was greater than in the umbilical cord, suggesting an independent hormonal stress response to invasive procedures in the fetus (Giannakoulopoulos et al., 1994, 1999). Indeed, research has shown that the fetal endorphinergic cells emerge in the anterior lobes of the pituitary gland by 20 weeks of pregnancy (Li et al., 1979). The pituitary-adrenal, sympathoadrenal, and circulatory stress responses to physical stimulation and medical procedures in the human fetus have also been shown to exist from 18–20 weeks of pregnancy (Giannakoulopoulos et al., 1994, 1999; Gitau et al., 2000). Animal (Barg et al., 1992; Duckhyun & Barr, 1995) and human (Zagon et al., 1990) studies have reported that opioid receptor expression is dynamic during fetal life and increases as the pregnancy progresses, continuing to rise after birth. Therefore, it is suggested that the hormonal stress response in the fetus is independent of maternal hormonal reactions to invasive procedures (Gitau et al., 2000) and is mediated by the fetal autonomic nervous system and the hypothalamic–pituitary–adrenal axis without conscious cortical processing (Carrasco & Van de Kar, 2003). This is supported by a study by Radunovic et al. (1993) that showed an increase in fetal plasma βE concentrations after repeated cordocenteses between 18 and 28 weeks of pregnancy. In addition, the intrahepatic vein procedures involving needling through the fetal abdomen at these gestational ages has resulted in two to six times increase in the fetal plasma cortisol and βE (Giannakoulopoulos et al., 1994; Teixeira et al., 1999).

These findings suggest that the human fetus may possibly feel pain in utero. Evidence supporting this claim has shown that the administration of opioid analgesic into the fetal umbilical vein before potentially painful procedures prevents the fetal stress response to intrauterine needling (Fisk et al., 2001; Gutsche, 2002). Nevertheless, it is argued that the fetal neuroendocrine responses may not be evidence of fetal pain, because the hormonal stress response is not specific to painful stimuli (Lee et al., 2005).

Previous authors have proposed that perinatal exposure to opioids such as epidural analgesics (Mardirosoff et al., 2002) and administration of high-dose opiates (Sultan et al., 2013) have negative effects on the fetal heart rate and the newborn Apgar score at 1 minute, and also on long-term adult behaviors (Nyberg et al., 2000). However, these claims have not been supported by cohort studies, systematic reviews or meta-analyses (Pereira et al., 2012; Reynolds, 2010; Sng et al., 2014; Ullman et al., 2010), which suggest that use of epidural analgesia during labor may reduce the incidence of negative fetal and neonatal outcomes, such as fetal metabolic acidosis, low Apgar score after birth and impaired breastfeeding, and that their benefits may outweigh potential undesirable effects (Reynolds, 2010).

Synthetic Oxytocin: Risks Versus Benefits

In a meta-analysis of 30 randomized clinical trials, Wei et al. (2009) reported an increase in spontaneous vaginal deliveries and a modest reduction in caesarean section and instrumental/operative vaginal deliveries after early administration of oxytocic agents. However, the null effect was not excluded. Furthermore, Wei et al. (2009) showed that early administration of oxytocics did not affect the interval from admission to delivery, the use of epidural analgesia or narcotics, postpartum hemorrhage larger than 500 mL, maternal blood transfusion, fetal distress, Apgar score less than 7 at 5 minutes, admission to special care baby unit, and neonatal jaundice. Nevertheless, it decreased the risk of antibiotics in labor or postpartum and increased the likelihood of uterine hyperstimulation and labor pain. Wei et al. commented that although oxytocics can increase the uterine contractions intensity and the sensation of pain, this effect may be neutralized by an increased likelihood of spontaneous vaginal delivery. In their meta-analysis, Wei et al. were, however, unable to identify other maternal and neonatal adverse outcomes due to the small sample size of the included clinical trials (Wei et al., 2009). In a later systematic review, Wei et al. (2010) reported an association between high-dose oxytocics (an initial dose of ≥4 mU/min and dose increments of at least 4 mU/min) and decreased risk of caesarean section, as well as increased likelihood of uterine hyperstimulation, spontaneous vaginal delivery, and labor duration. They showed no evidence of increased maternal or neonatal morbidity. Wei et al. noted that high-dose oxytocics could possibly be associated with an increase in labor pain; however, none of the studies in their systematic review described pain scores and women's experience of labor pain as an outcome. The authors recommended further large trials to detect the safety, effectiveness, acceptability, and cost implications of this approach in obstetric care (Wei et al., 2010).

In a Cochrane review, Mori et al. (2011) investigated the impact of the dose of oxytocics for augmentation of labor contractions on both maternal and neonatal outcomes. They reported that higher starting and increment doses of oxytocics (4 mU per minute or more) were associated with a significant reduction in length of labor and rate of caesarean section, and an increase in the rate of spontaneous vaginal birth. The higher dose of oxytocics, however, did not affect the neonatal mortality, uterine hyperstimulation, chorioamnionitis, epidural analgesia, Apgar scores, umbilical cord pH, or admission to special care baby unit. Mori et al. did not evaluate the following outcomes because they were not investigated in the included studies: perinatal mortality, women's satisfaction, instrumental vaginal birth, uterine rupture, postpartum hemorrhage, abnormal cardiotocography, women's pyrexia, dystocia, and neonatal neurological morbidity. The authors suggested that further research is required to explore the effect of high-dose oxytocics on maternal and neonatal outcomes as well as any effect on women's birth experience (Mori et al., 2011).

Beta-Endorphin and Postpartum

Lactation and Breastfeeding

Human studies have shown that βE crosses the blood-brain barrier (Gerner et al., 1982) and stimulates the release of prolactin (Schulz et al., 1980), which suggests its role in regulating and enhancing lactation and breastfeeding. Alterations in plasma concentrations of βE has been shown to be correlated with changes in plasma concentrations of prolactin. In other words, no rise in plasma prolactin concentration has been identified in the absence of an increase in plasma βE concentrations (Risch et al., 1982).

Franceschini et al. (1989) proposed that suckling in lactating women after birth is a stimulus that induces the release of prolactin and βE. In a preliminary research, they reported a sharp increase in plasma prolactin concentrations 5 min after the commencement of suckling, with the highest levels reached at 20 min. Research by Franceschini et al. (1989) has also shown that the mean plasma βE level significantly increases after suckling and reaches a peak at 20 min, a rise that accords with that of prolactin. Although Franceschini et al. (1989) proposed that the increase in βE levels could be due to the influence of psychological factors or the mother–newborn contact, they reported that the timing of βE surge does not support this theory (Franceschini et al., 1989). Their results agree with previous experimental animal research on rats that indicated an increase in plasma βE induced by suckling (Riskind et al., 1984). However, the results are inconsistent with those of an earlier study by Genazzani et al. (1982) showing that suckling activated the release of prolactin due to stimulation of serotoninergic neurons with no significant changes in βE levels. Genazzani et al. (1982) suggested that the minimal changes in βE levels after suckling could be due to a lack of proopiocortin-related peptide response because breastfeeding may not be a stressful enough situation to induce the release of βE.

A study of 10-day postpartum lactating rats (Selmanoff & Gregerson, 1986) showed that suckling-induced prolactin release was suppressed after the administration of large dose of naloxone (the aforementioned μ-OR antagonist). The authors suggested that this neuroendocrine response is mediated by opiate neurons that are either located in the neuronal pathway or are positioned in a way that they can modulate this pathway's neuronal transmission. They also suggested that the inhibitory β endorphinergic input to the tuberoinfundibular dopaminergic (TIDA) neurons is blocked by naloxone that in turn results in the stimulation of TIDA neurones and overriding the suckling response. In their study, Selmanoff and Gregerson (1986) reported a prolactin response after intravenous bolus administration of βE whose latency of onset and duration was similar to suckling response. However, the βE-induced response was two times greater than the response induced by suckling. In addition, smaller dose of naloxone was required to block the βE-induced prolactin release compared to the dose required to impede suckling-induced prolactin release. These findings suggest that the suckling response is mediated by two neural mechanisms.

The level of βE in colostrum is twice as high as maternal plasma and remains high for at least 10 days after birth (Ferrando et al., 1990; Grandison & Guidotti, 1977; Zanardo et al., 2001a). The colostrum βE is transferred to the newborn through breastfeeding and results in an alert newborn (Lothian, 2005). It may also be important in several biological functions in the newborn, such as analgesia, steroidogenesis, cardiovascular and endocrine functions, neuro-immunomodulation and sleep–wake patterns (Zanardo et al., 2001a).

The concentrations of some bioactive proteins in breast milk differ between women, because of factors such as their mode of birth and the newborn's gestational age at birth (Mehta & Petrova, 2011). These bioactive proteins include βE, secretory immunoglobulin A, lysozyme, lactoferrin, osteoprotegerin, leptin, and adiponectin. Zanardo et al. (Zanardo et al., 2001a) hypothesized that an elevation in the concentration of βE in colostrum and transitional milk of women who experience vaginal birth may contribute to postnatal neonatal adaptation to the stress of natural labor and birth.

Zanardo et al. (2001a) suggested that any factor interrupting or decreasing βE release may eventually interfere with lactation and breastfeeding. Elective caesarean section without commencement of labor contractions, epidural analgesia, and some drugs markedly decrease the levels of βE in breast milk and in newborn (Vogl et al., 2006). Epidural analgesia using fentanyl and bupivacaine, which help decrease maternal pain, reduce the production of maternal βE. These interventions lead to lower levels of maternally acquired βE in the newborn and possibly higher than usual pain in the newborn from potential birth injuries (Zanardo, Nicolussi, Carlo, et al., 2001a). The affected newborn will also be less likely to suck well, more likely to be separated from the mother for advanced neonatal care, and less able to benefit from the analgesic effects of skin-to-skin contact and breastfeeding (Gray et al., 2002). Even if the newborn remains with the mother, suckles and maintains skin-to-skin contact, the breast milk of a mother who has had an elective caesarean section or epidural analgesia has a reduced analgesic effect compared with that of a mother having a natural vaginal birth (Smith, 2007).

In the case of natural labor, the increasing level of endogenous oxytocin throughout labor blocks the beta-adrenergic receptor (beta-2), which reduces the clearance of βE and raises levels of βE in the mother's plasma (Engstrom et al., 1999; Kang & Park, 2000; Kovács & Telegdy, 1987). This process is disturbed in women who receive oxytocics, which is associated with disruptions in normative breastfeeding (Abdoulahi et al., 2017; Brimdyr et al., 2015; Genazzani et al., 1985; Petraglia et al., 1986). Retrospective studies have shown higher rates of formula-feeding or partial breastfeeding during the first few days or weeks after birth in women who have received Syntocinon alone or in combination with ergometrine (Jordan et al., 2009) and epidural analgesia during labor (Dozier et al., 2013; Wiklund et al., 2009). These women are also more likely to report not having enough milk after the birth (Volmanen et al., 2004). The study by Garcia-Fortea et al. (2014) showed that the use of Syntocinon multiplied the risk of bottle-feeding by 1.4 and the risk of breastfeeding withdrawal at 3 months by 2.2. The association between the use of

Syntocinon and impaired breastfeeding may be due to lower levels of βE in these women and also to the negative impacts of Syntocinon on the newborn's health (Saisto et al., 2004).

The association between Syntocinon and unsuccessful breastfeeding may also be due to the impact of Syntocinon on the newborn's Apgar score and other adverse neonatal outcomes associated with induction or augmentation of labor, such as hyperbilirubinemia (Trotman & Henny-Harry, 2012) and admission to the neonatal intensive care unit (Clark et al., 2009). Successful breastfeeding requires the child to be alert with coordinated sucking, swallowing and breathing responses, which may be impaired because of hyperbilirubinemia. Furthermore, birth asphyxia-related complications and admission to neonatal intensive care unit (NICU) are more common after induction of labor and use of oxytocics (Sandström et al., 2017). This requires the affected child to be separated from the mother for resuscitation and post-birth care, which can impair breastfeeding (Romano & Lothian, 2008).

According to preliminary animal research, short-acting μ-OR antagonists such as naloxone bind to opioid receptors, blocking the activity of βE and reversing its inhibitory or excitatory actions (Jacquet, 1980). Injection of naloxone into female rats before they suckle their pups has been shown to inhibit the release of prolactin (Miki et al., 1981), which may in turn have a negative impact on lactation behavior. However, there is a lack of substantial evidence in the literature and a need for further human research relating to this theory.

Beta-Endorphin and Maternal and Neonatal Behaviors: Bonding and Emotion

Endogenous opioids are released simultaneously with endogenous oxytocin during labor and birth and reach a peak after birth. According to animal studies, increased numbers of central oxytocin receptors in the maternal brain immediately after birth (Hammer et al., 1992) enhance maternal attachment by activating dopamine-associated reward circuits (Panksepp et al., 1994), and inhibit tolerance to βE (Kovács & Telegdy, 1987). The increase in βE in mother and newborn during labor and after birth results in an alert mother ready to take care of her newborn, and induces a strong mother–child bond within the first minutes after the birth (Odent, 2001; Winberg, 2005). Most nonhuman mammalian females eat their placenta after birth, with the possible benefits of increasing interaction between the mother and newborn, promoting analgesia in the mother, enhancing maternal brain opioid circuits, and facilitating maternal caretaking behaviors (Kristal et al., 2012).

Results from animal studies suggest that the release of βE does not end after birth, as ongoing contact between mother and newborn, such as skin-to-skin contact and breastfeeding, can help increase the release of βE in both mother and newborn (Nelson & Panksepp, 1998). The release of βE after birth is believed to be a "reward" that the mother receives for performing reproductive functions such as becoming pregnant and giving birth. It also provides pleasure in relation to reproduction, triggers the reward response, and imprints the pleasure with newborn contact (Fisher et al., 2006; Odent, 1999). Maternal reward circuits may justify the devoted care required from the new mother necessary for offspring survival as seen in animal models (Nelson & Panksepp, 1998). Although there is a lack of high-quality research in humans, this also seems to be true in humans (see Barenz et al., this volume, for further description of positive emotions and endorphins).

Immediately after birth, the βE level increases for several hours in the newborn, possibly due to the stress-reduction effects of skin-to-skin contact with the mother. The level then gradually decreases during the first 24 h (Chidambaram et al., 2014; Weller & Feldman, 2003), and this diminishing level of βE may prompt the newborn to seek out maternal attention and touch, and to suckle. The resulting maternal contact then elevates the level of βE again in the newborn (Nelson & Panksepp, 1998). According to both animal (Nelson & Panksepp, 1998) and human studies (Weller & Feldman, 2003; Wu et al., 2016), endogenous opioids help the newborn recognize maternal odors and have a role in mediating the positive influence of skin-to-skin contact on the newborn's level of comfort, analgesia, and self-regulation.

In a systematic review of studies on some mammalian and avian species, Nelson and Panksepp (1998) found that separation of the mother and newborn immediately after birth, when levels of βE are expected to rise, may interfere with the activation of the reward center in the mother and the newborn. Repeated short-term separations in the first few weeks after birth result in deleterious effects on offspring opioid systems (probably through epigenetic programming). This can have negative effects on pain sensitivity (Panksepp, 1981, 1991) and behavior activation, such as increased physical activity (Schapiro & Salas, 1970), increased heart and respiration rates (Hofer, 1984), release of corticosterone (Stanton et al., 1987), and the production of ultrasonic vocalizations (Noirot, 1972). These behavioral patterns are found in humans and other mammals whose opioid system has been disturbed. Exhibition of these behaviors by the offspring may affect the mother's biopsychological response including increased activation of the hypothalamic–pituitary–adrenal axis, release of βE, but also corticotropin-releasing factor, adrenocorticotropic hormone, proopiomelanocortin and cortisol in the service of sympathetic nervous system activation, which in turn may impact her learning history with the baby (Nelson & Panksepp, 1998).

Disturbed maternal and newborn behaviors have been identified in women who have undergone labor and birth interventions. Although manipulation of and changes in peptides such as endogenous oxytocin and, in turn, βE, during labor and birth have become an accepted aspect of human development, the issue has been markedly understudied (Carter, 2003). During childbirth, women and potentially their newborns may be exposed to medical and nonmedical interventions such as labor analgesics and anesthesia (George et al., 2013), various doses of Syntocinon (Dudley, 1997), caesarean section (Zanardo et al., 2001a), and the adverse effects of the birth experience (Garthus-Niegel et al., 2013). Although these complex manipulations are routine in modern obstetrics, little is known about their effects on the opioid system and the possible behavioral consequences. For example, the use of oxytocics for induction or augmentation of labor may contribute to reduced βE release and impair maternal and newborn behaviors. Since Syntocinon does not cross the blood–brain barrier, it does not stimulate the release of βE and its subsequent desirable behavioral responses (Mens et al., 1983). In addition, the contractions induced by Syntocinon are stronger and more painful than the natural uterine contractions. This effect increases the incidence of labor and birth complications, inhibits the release of βE, makes the mother and newborn less responsive to each other, and impairs the mother–child attachment (Trotman & Henny-Harry, 2012).

The mode of birth is another factor that may affect maternal and neonatal behaviors. A magnetic resonance imaging study of maternal brain responsiveness to the newborn's cry has reported that at 2–4 weeks postpartum, mothers who give birth vaginally show increased brain responses and are more sensitive to their newborns' cry than mothers who gave birth by caesarean section (Swain et al., 2008). Swain et al. (2008) showed that particular brain circuits, including sensory processing, empathy, arousal, motivation, reward and habit-regulation circuits, are activated in women who have had vaginal births, which suggests that the mode of birth affects maternal behavior. However, these findings are not supported by more recent research (Olza Fernández et al., 2013) showing no differences in maternal behaviors according to the mode of birth. Olza Fernández et al. (2013) showed that newborns, when separated from their mothers, cry less if they were born by caesarean section than babies born via induced vaginal birth. Although Olza Fernández et al. (2013) attempted to rule out the role of mode of birth in maternal and neonatal behaviors, they failed to investigate maternal and neonatal behaviors in mothers and newborns who had undisturbed physiological birth; this aspect of their study might have yielded different outcomes. It is also unknown whether the decreased crying of newborns born by caesarean section was due to their decreased responsiveness to maternal separation and an early symptom of nonattachment, or whether it was because of an altered stress response, because newborns born by caesarean section did not experience the stress of induced vaginal birth.

Contrary to these findings, reports of other preliminary research has shown that while maternal stress systems are expected to be less activated when a woman is undergoing caesarean section than when she has a vaginal birth, the associated psychological stress of having an operation can cause an increase in βE that in turn helps the woman to cope with the stress of the operation (Hoffman et al., 1984). These findings support other research indicating that the use of epidural analgesia blocks the increase in βE, but caesarean section with a general anesthetic is associated with a rise in βE (Abboud et al., 1984).

The long-term effects of perinatal exposure to opioids on the βE system of human offspring are understudied. However, preliminary studies on the rhesus monkey have shown that meperidine and bupivacaine influence behavioral maturation during certain periods of infancy. It has been suggested that these effects occur due to the impact of opioids on vulnerable brain processes; that the drugs interfere with the programming of brain development or alter early experiences (Golub, 1996). It has also been proposed that the opiates alter offspring behaviors later in life. For example, two cohort studies have shown an association between exposure to opioids or barbiturates in labor and an increased risk of opioid addiction in adulthood, which suggests that opioid system imprinting is involved (Jacobson et al., 1990; Nyberg, 2000). Conversely, more recent research has found no association between the use of pethidine for labor pain, and substance misuse or smoking in later life (Pereira et al., 2012). The widespread use of analgesics for labor pain, the extent of perinatal exposure of newborns to opioids, the increasing rates of interventions in labor such as use of Syntocinon and caesarean section, and the inconclusive results from research to date, all mean that there is a need for further studies in this area.

Few studies have empirically addressed the possible predictive role of the βE level in postpartum depression, by examining the association between βE levels in late pregnancy and depressive symptoms during the first few days after birth. Research by Yim et al. (2010) has shown that, compared to women with low levels of βE (21.90–24.86 pg/mL) during pregnancy, women who have higher levels of βE (41.80–62.13 pg/mL) throughout pregnancy are three times as likely to develop postpartum depression. Although these findings support earlier research indicating that women whose mood deteriorated from 38 weeks of pregnancy to day 2 after birth had a larger drop in plasma βE after birth (Brinsmead et al., 1985; Smith et al., 1990), they are in contrast with other research reporting no correlation between levels of βE and symptoms of postpartum depression, after correcting for multiple comparisons (Newnham et al., 1984).

There are two reasons why the βE level could be used as an early marker of future postpartum depression. First, the synthesis of βE undergoes vigorous changes throughout pregnancy, with the highest levels occurring during labor, and a sharp decline occurring after birth. Women with higher levels of βE may experience a larger drop during the postpartum period. This may contribute to their higher risk of postpartum depression, especially in women with underlying dysregulation of the hypothalamic–pituitary–adrenal axis that is concealed during pregnancy (Browning et al., 1983; Chan & Smith, 1992; Chan et al., 1993; Lindow et al., 1996). Second, further to its well-known analgesic, rewarding and pleasure-inducing properties (Amalric et al., 1987; Foley et al., 1979), the hyperactivity of βE has been shown to be linked to non-puerperal depression in some studies (Goodwin et al., 1993; Kennedy et al., 2006; Navines et al., 2008; Risch et al., 1982), although other studies do not support this link (Cohen et al., 1984; Djurovic et al., 1999; Gerner & Sharp, 1982). One explanation for this controversy could be the elevation of βE in some subtypes of depression but not in others. Nevertheless, since this research is inconclusive, further studies are needed to support variations in maternal and newborn behaviors in response to changes in βE levels (see also Kerr et al., this volume for a description of the role of endorphin activity in the development and treatment mood and anxiety disorders).

Conclusion

Although there is a lack of research on the role of βE in human sexuality, its neuroendocrine properties have been documented in the literature. According to the evidence reviewed in this chapter, the promotion, support, and protection of the hormonal physiology of mothers and newborns during labor, birth, and the postpartum period have remarkable benefits for both of them. During a physiologic labor and birth, βE provides natural analgesia and stress relief during labor and birth, protects from potential neural damage to the fetus, activates the mother's reward and pleasure center after birth, helps establish a bond between mother and newborn, and enhances breastfeeding. While maternity care interventions may be useful in certain circumstances, they may disrupt the natural function of hormones and interfere with these beneficial processes.

The aim of this chapter is to encourage clinicians to consider the benefits and harms of maternity care practices and interventions. Findings from animal studies suggest that the maternal and newborn hormonal systems in the perinatal period may have lasting effects on offspring hormonal and other biologic systems. Thus, any disruption to this system may interfere with epigenetic programming and have long-term impacts on offspring development and behaviors.

Recommendations for Practice

Since physiologic childbirth has valuable benefits to women and offspring in both the short and long term, it is recommended that physiologic childbirth should be promoted, supported, and protected. To optimize hormonal physiology and achieve this, a proper maternity care system needs to be adapted by ensuring:

- Professional education to clinicians on the hormonal physiology of childbirth. This can be done in entry-level education, during advanced professional training, and in continuing professional development (CPD) modules.
- Performance-improvement initiatives to protect laboring women from unnecessary disruption.
- Protocols promoting physiologic childbirth and implementing evidence-based clinical practice guidelines to decrease the use of epidural analgesia or caesarean section and other consequent interventions.
- Performance measures to evaluate and improve quality of service and care.
- Continuous support in collaborative practice throughout pregnancy and the postpartum period, with increased access to models of care that foster physiologic childbirth, such as midwifery care and doula support. This will help promote privacy, decrease stress and anxiety, and limit the use of redundant maternity care interventions.
- Support for pregnant women and those who plan pregnancy, to improve their understanding of the hormonal physiology of childbirth. Related educational

resources should also be provided to these women to encourage them to engage in shared decision-making and to express their values, expectations, and preferences.

- Physiologic onset of labor at term and avoiding pre-labor induction or elective caesarean section, except where an indication for an individual woman is supported by evidence and informed decision-making, to protect women and newborns from unnecessary interventions and disruption of physiologic processes.
- Non-pharmacological approaches to comfort women during labor and assist them in coping with labor pain. This will help prevent unintended impacts on the hormonal physiology of the mother, fetus and newborn, enhance breastfeeding, and improve the mother–newborn attachment.
- Support skin-to-skin contact and initiation of breastfeeding immediately after birth to reduce anxiety and stress in both mother and newborn. To this aim, clinicians need to be educated about the pivotal role of early skin-to-skin contact in successful breastfeeding. A skin-to-skin midwife can also be employed to provide clinical care of the baby at time of vaginal birth and caesarean section, enable skin-to-skin contact, and facilitate woman-centered preferences for care and feeding of the baby. These recommendation do not exclude safe, appropriate, and timely use of maternity care procedures and interventions whenever required. Instead, they can help clinicians make the best use of hormonal physiology as far as possible, and safely support the implementation of natural childbirth.

Recommendations for Future Research

A thorough review of the literature in this chapter has identified many gaps in knowledge of the association between βE, sexuality, and reproduction. Inconclusive reports in the literature question the association between sexuality and βE, and studies on the association between βE, stress, and pain during labor are contradictory. There are conflicting reports on the impact of synthetic opioid drugs on labor pain and the release of βE, and on the increased risk of caesarean section in women receiving epidural analgesia with opioids. There is a lack of substantial evidence in the literature about the association between βE, lactation, and breastfeeding, and most studies on βE are not current and involve animals such as rats, monkeys, and rabbits. Thus, the potential short- and long-term impacts of maternity care interventions on mothers and newborns warrant further investigation. Although the endorphin hypothesis remains one of the most common justifications for the benefits of non-pharmacologic complementary therapies during pregnancy, labor, and birth, there are contradictory reports in the literature about their efficacy.

The importance of physiological birth and its role in the health and well-being of women and children, and the gaps in the literature on the topic, highlight the need for high-quality research to extend our understanding of the hormonal physiology of childbirth and the role of βE in human sexuality and reproduction.

References

Abboud, T. (1988). Maternal and fetal beta endorphin: Effects of pregnancy and labour. *Archives of Disease in Childhood, 63*(7 Spec No), 707–709.

Abboud, T., Noueihed, R., Khoo, S., Hoffman, D., Varakian, L., Henriksen, E., et al. (1984). Effects of induction of general and regional anesthesia for cesarean section on maternal plasma [beta] – endorphin levels. *Obstetric Anesthesia Digest, 4*(1), 9.

Abdoulahi, M., Mousavi Asl, F., Delaram, M., & Namnabati, M. (2017). Association of using oxytocin during labor and breastfeeding behaviors of infants within two hours after birth. *Iranian Journal of Neonatology IJN, 8*(3), 48–52.

Agirregoitia, E., Subiran, N., Valdivia, A., Gil, J., Zubero, J., & Irazusta, J. (2012). Regulation of human sperm motility by opioid receptors. *Andrologia, 44*(Suppl 1), 578–585. https://doi.org/10.1111/j.1439-0272.2011.01230.x

Agrawal, D., Makhija, B., Arora, M., Haritwal, A., & Gurha, P. (2014). The effect of epidural analgesia on labour, mode of delivery and neonatal outcome in nullipara of India, 2011–2014. *Journal of Clinical and Diagnostic Research : JCDR, 8*(10), OC03–OC06. https://doi.org/10.7860/JCDR/2014/9974.4930

Aleem, F., & McIntosh, T. (1985). Plasma levels of beta-endorphin in postmenopausal women using a specific radioimmunoassay. *Maturitas, 3*, 76–84.

Amalric, M., Cline, E., Martinez, J., Bloom, F., & Koob, G. (1987). Rewarding properties of βE as measured by conditioned place preference. *Psychopharmacology, 91*(1), 14–19.

American Academy of Pediatrics and American College of Obstetricians and Gynecologists. (2012). *Guidlines for perinatal care* (6th ed.).

American College of Obstetricians and Gynecologists. (2002). Obstetric analgesia and anesthesia. *Practice Bulletin, 36*.

Anderson, C. (1976). Coping behaviors as intervening mechanisms in the inverted-U stress-performance relationship. *The Journal of Applied Psychology, 61*(7), 30–34.

Anim-Somuah, M., Smyth, R., & Jones, L. (2011). Epidural versus non-epidural or no analgesia in labour. *Cochrane Database of Systematic Reviews*, (12), CD000331. https://doi.org/10.1002/14651858.CD000331.pub3

Apari, P., & Rozsa, L. (2006). Deal in the womb: Fetal opiates, parent-offspring conflict, and the future of midwifery. *Medical Hypotheses, 67*(5), 1189–1194. https://doi.org/10.1016/j.mehy.2006.03.053

Argiolas, A., & Melis, M. (2013). Neuropeptides and central control of sexual behaviour from the past to the present: A review. *Progress in Neurobiology, 108*, 80–107. https://doi.org/10.1016/j.pneurobio.2013.06.006

Bacigalupo, G., Langner, K., & Schmidt, S. (1987). Plasma immunoreactive beta-endorphin, ACTH and cortisol concentrations in mothers and their neonates immediately after delivery-their relationship to the duration of labor. *Journal of Perinatal Medicine, 15*(1), 45–52.

Bacigalupo, G., Riese, S., Rosendahl, H., & Saling, E. (1990). Quantitative relationships between pain intensities during labor and beta-endorphin and cortisol concentrations in plasma. Decline of the hormone concentrations in the early postpartum period. *Journal of Perinatal Medicine-Official Journal of the WAPM, 18*(4), 289–296.

Baker, J. (2001). *Prenatal yoga and natural childbirth* (3rd ed.). North Atlantic Books.

Bancroft, J. (2005). The endocrinology of sexual arousal. *The Journal of Endocrinology, 186*, 411–427.

Barg, J., Rius, R., Bem, W., Belcheva, M., Loh, Y., & Coscia, C. (1992). Differential development of beta-endorphin and mu opioid binding sites in mouse brain. *Brain Research Developmental Brain Research, 66*, 71–76.

Basson, R. (2000). The female sexual response: A different model. *Journal of Sex & Marital Therapy, 26*(1), 51–65.

Basson, R., Berman, J., Burnett, A., Derogatis, L., Ferguson, D., Fourcroy, J., et al. (2000). Report of the international consensus development conference on female sexual dysfunction: Definitions and classifications. *The Journal of Urology, 163*(3), 888–893. https://doi.org/10.1016/s0022-5347(05)67828-7

Basson, R., Brotto, L., Laan, E., Redmond, G., & Utian, W. (2005). Womne's sexual dysfunctions: Assessment and Management of Women's sexual dysfunctions: Problematic desire and arousal. *The Journal of Sexual Medicine, 2*(3), 291–300. https://doi.org/10.1111/j.1743-6109.2005.20346.x

Baulieu, E. (1997). Neurosteroids: Of the nervous system, by the nervous system, for the nervous system. *Recent Progress in Hormone Research, 52*, 1–32.

Bawor, M., Bami, H., Dennis, B., Plater, C., Worster, A., Varenbut, M., et al. (2015). Testosterone suppression in opioid users: A systematic review and meta-analysis. *Drug and Alcohol Dependence, 149*, 1–9. https://doi.org/10.1016/j.drugalcdep.2015.01.038

Benfield, R., Hortobágyi, T., Tanner, C., Swanson, M., Heitkemper, M., & Newton, E. (2010). The effects of hydrotherapy on anxiety, pain, neuroendocrine responses, and contraction dynamics during labor. *Biological Research For Nursing*. https://doi.org/10.1177/1099800410361535

Blank, M., Fabbri, A., Catt, K., & Dufau, M. (1986). Inhibition of luteinizing hormone release by morphine and endogenous opiates in cultured pituitary cells. *Endocrinology, 118*(5), 2097–2101.

Bolbol-Haghighi, N., Masoumi, S., & Kazemi, F. (2016). Effect of massage therapy on duration of labour: A randomized controlled trial. *Journal of Clinical and Diagnostic Research : JCDR, 10*(4), QC12–QC15. https://doi.org/10.7860/JCDR/2016/17447.7688

Bonapace, J., Chaillet, N., Gaumond, I., Paul-Savoie, É., & Marchand, S. (2013). Evaluation of the Bonapace method: A specific educational intervention to reduce pain during childbirth. *Journal of Pain Research, 6*, 653–661. https://doi.org/10.2147/JPR.S46693

Brennan, M. (2013). The effect of opioid therapy on endocrine function. *The American Journal of Medicine, 126*(3, Supplement 1), S12–S18. https://doi.org/10.1016/j.amjmed.2012.12.001

Brimdyr, K., Cadwell, K., Widström, A., Svensson, K., Neumann, M., Hart, E., et al. (2015). The association between common labor drugs and suckling when skin-to-skin during the first hour after birth. *Birth, 42*(4), 319–328. https://doi.org/10.1111/birt.12186

Brinsmead, M., Smith, R., Singh, B., Lewin, T., & Owens, P. (1985). Peripartum concentrations of Beta endorphin and cortisol and maternal mood states. *Australian and New Zealand Journal of Obstetrics and Gynaecology, 25*(3), 194–197. https://doi.org/10.1111/j.1479-828X.1985.tb00642.x

Browning, A., Butt, W., Lynch, S., & Shakespear, R. (1983). Maternal plasma concentrations of beta-lipotrophin, and gamma-lipotrophin throughout pregnancy. *British Journal of Obstetrics and Gynaecology, 90*(12), 1147–1151.

Cahill, C. (1989). Beta-endorphin levels during pregnancy and labor: A role in pain modulation? *Nursing Research, 38*(4), 200–203.

Carrasco, G., & Van de Kar, L. (2003). Neuroendocrine pharmacology of stress. *European Journal of Pharmacology, 463*(1–3), 235–272. https://doi.org/10.1016/S0014-2999(03)01285-8

Carter, C. S. (2003). Developmental consequences of oxytocin. *Physiology and Behavior, 79*(3), 383–397.

Chaillet, N., Belaid, L., Crochetière, C., Roy, L., Gagné, G., Moutquin, J., et al. (2014). Nonpharmacologic approaches for pain management during labor compared with usual care: A meta-analysis. *Birth, 41*(2), 122–137. https://doi.org/10.1111/birt.12103

Chan, E., & Smith, R. (1992). Beta-endrophin immunoreactivity during human pregnancy. *The Journal of Clinical Endocrinology and Metabolism, 75*(6), 1453–1458. https://doi.org/10.1210/jcem.75.6.1464647

Chan, E., Smith, R., Lewin, T., Brinsmead, M., Zhang, H., Cubis, J., et al. (1993). Plasma corticotropin-releasing hormone, beta-endorphin and cortisol inter-relationships during human pregnancy. *Acta Endocrinologica, 128*(4), 339–344.

Chiang, F., & Rodway, R. (1997). Determinations of umbilical cord β-endorphin concentration and blood gas parameters in newborn piglets. *Research in Veterinary Science, 63*(2), 107–111.

Chidambaram, A., Manjula, S., Adhisivam, B., & Vishnu Bhat, B. (2014). Effect of Kangaroo mother care in reducing pain due to heel prick among preterm neonates: A crossover trial. *The Journal of Maternal-Fetal & Neonatal Medicine, 27*(5), 488–490. https://doi.org/10.3109/14767058.2013.818974

Choi, J., Chung, M., & Lee, Y. (2012). Modulation of pain sensation by stress-related testosterone and cortisol. *Anaesthesia, 67*(10), 1146–1151. https://doi.org/10.1111/j.1365-2044.2012.07267.x

Cicero, T., Wilcox, C., Bell, R., & Meyer, E. (1976). Acute reductions in serum testosterone levels by narcotics in the male rat: Stereospecificity, blockade by naloxone and tolerance. *Journal of Pharmacology and Experimental Therapeutics, 198*(2), 340–346.

Clark, S., Miller, D., Belfort, M., Dildy, G., Frye, D., & Meyers, J. (2009). Neonatal and maternal outcomes associated with elective term delivery. *American Journal of Obstetrics and Gynecology, 200*(2), 156.e151–156.e154. https://doi.org/10.1016/j.ajog.2008.08.068

Clayton, A. (2003). Sexual function and dysfunction. *Psychiatric Clinics of North America, 26*(3), 673–682.

Cogan, R., & Spinnato, J. (1986). Pain and discomfort thresholds in late pregnancy. *Pain, 27*(1), 63–68. https://doi.org/10.1016/0304-3959(86)90223-X

Cohen, M., Pickar, D., Extein, I., Gold, M., & Sweeney, D. (1984). Plasma cortisol and beta-endorphin immunoreactivity in nonmajor and major depression. *The American Journal of Psychiatry, 141*(5), 628–632. https://doi.org/10.1176/ajp.141.5.628

Csaba, G., Knippel, B., Karabélyos, C., Inczefi-Gonda, Á., Hantos, M., Tóthfalusi, L., & Tekes, K. (2003). Effect of neonatal β-endorphin imprinting on sexual behavior and brain serotonin level in adult rats. *Life Sciences, 73*(1), 103–114. https://doi.org/10.1016/S0024-3205(03)00254-6

Cunningham, F., Leveno, K., Bloom, S., Hauth, J., Rouse, D., & Spong, C. (2010). *Williams obstetrics* (23rd ed.). McGrow Hill.

Dabo, F., Nyberg, F., Qin, Z., Sundström-Poromaa, I., & Akerud, H. (2010). Plasma levels of β-endorphin during pregnancy and use of labor analgesia. *Reproductive Sciences, 17*(8), 742–747. https://doi.org/10.1177/1933719110370059

Dalayeun, J., Norès, J., & Bergal, S. (1993). Physiology of β-endorphins. A close-up view and a review of the literature. *Biomedicine & Pharmacotherapy, 47*(8), 311–320. https://doi.org/10.1016/0753-3322(93)90080-5

Decatanzaro, D., Douglas, A., Griffiths, J., & Muir, C. (1996). Differential sexual activity of isolated and group-housed male mice: Lack of substantial influence of acute or chronic naloxone administration. *Pharmacology Biochemistry and Behavior, 55*(1), 169–174. https://doi.org/10.1016/0091-3057(95)02212-0

Delke, I., Minkoff, H., & Grunebaum, A. (1985). Effect of Lamaze childbirth preparation on maternal plasma beta-endorphin immunoreactivity in active labor. *American Journal of Perinatology, 2*(4), 317–319.

Djurovic, D., Milic-Askrabic, J., & Majkic-Singh, N. (1999). Serum beta-endorphin level in patients with depression on fluvoxamine. *Farmaco, 54*(3), 130–133.

Dondi, D., Maggi, R., Panerai, A., Piva, F., & Limonta, P. (1991). Hypothalamic opiatergic tone during pregnancy, parturition and lactation in the rat. *Neuroendocrinology, 53*(5), 460–466.

Dornan, W., & Malsbury, C. (1989). Neuropeptides and male sexual behavior. *Neuroscience & Biobehavioral Reviews, 13*(1), 1–15. https://doi.org/10.1016/S0149-7634(89)80046-6

Douglas, A., & Russell, J. (2001). Chapter 5 Endogenous opioid regulation of oxytocin and ACTH secretion during pregnancy and parturition. In *Progress in brain research* (Vol. 133, pp. 67–82). Elsevier.

Dozier, A., Howard, C., Brownell, E., Wissler, R., Glantz, J., Ternullo, S., et al. (2013). Labor epidural anesthesia, obstetric factors and breastfeeding cessation. *Maternal and Child Health Journal, 17*(4), 689–698. https://doi.org/10.1007/s10995-012-1045-4

Duckhyun, K., & Barr, G. (1995). The induction of Fos-like immunoreactivity by noxious thermal, mechanical and chemical stimuli in the lumbar spinal cord of infant rats. *Pain, 60*(3), 257–265.

Dudley, D. (1997). Oxytocin: Use and abuse, science and art. *Clinical Obstetrics and Gynecology, 40*, 516–524.

Dunbar, R., Kaskatis, K., MacDonald, I., & Barra, V. (2012). Performance of music elevates pain threshold and positive affect: Implications for the evolutionary function of music. *Evolutionary Psychology, 10*(4). https://doi.org/10.1177/147470491201000403

Engstrom, T., Bratholm, P., Vilhardt, H., & Christensen, N. (1999). Effect of oxytocin receptor and beta2-adrenoceptor blockade on myometrial oxytocin receptors in parturient rats. *Biology of Reproduction, 60*(2), 322–329.

Esch, T., & Stefano, G. (2005). The neurobiology of love. *Neuroendocrinology Letters, 26*(3), 175–192.

Espinoza, M., Riquelme, R., Germain, A., Tevah, J., Parer, J., & Llanos, A. (1989). Role of endogenous opioids in the cardiovascular responses to asphyxia in fetal sheep. *American Journal of Physiology - Regulatory, Integrative and Comparative Physiology, 256*(5), R1063–R1068.

Exton, M., Bindert, A., Kruger, T., Scheller, F., Hartmann, U., & Schedlowski, M. (1999). Cardiovascular and endocrine alterations after masturbation-induced orgasm in women. *Psychosomatic Medicine, 61*(3), 280–289.

Facchinetti, F., Bagnoli, F., Bracci, R., & Genazzani, A. (1982). Plasma opioids in the first hours of life. *Pediatric Research, 16*(2), 95–98.

Fajardo, M., Florido, J., Villaverde, C., Oltras, C., González-Ramirez, A., & González-Gómez, F. (1994). Plasma levels of β-endorphin and ACTH during labor and immediate puerperium. *European Journal of Obstetrics & Gynecology and Reproductive Biology, 55*(2), 105–108. https://doi.org/10.1016/0028-2243(94)90062-0

Ferrando, T., Rainero, I., De Gennaro, T., Oggero, R., Mostert, M., Dattola, P., & Pinessi, L. (1990). β-endorphin-like and α-MSH-like immunoreactivities in human milk. *Life Sciences, 47*(7), 633–635. https://doi.org/10.1016/0024-3205(90)90575-C

Field, T. (2011). Yoga clinical research review. *Complementary Therapies in Clinical Practice, 17*(1), 1–8. https://doi.org/10.1016/j.ctcp.2010.09.007

Fisher, H., Aron, A., & Brown, L. (2006). Romantic love: A mammalian brain system for mate choice. *Biological Sciences, 361*(1476), 2173–2186. https://doi.org/10.1098/rstb.2006.1938

Fisk, N., Gitau, R., Teixeira, J., Giannakoulopoulos, X., Cameron, A., & Glover, V. (2001). Effect of direct fetal opioid analgesia on fetal hormonal and hemodynamic stress response to intrauterine needling. *Anesthesiology, 95*(4), 828–835.

Foley, K., Kourides, I., Inturrisi, C., Kaiko, R., Zaroulis, C., Posner, J., et al. (1979). Beta-endorphin: Analgesic and hormonal effects in humans. *Proceedings of the National Academy of Sciences of the United States of America, 76*(10), 5377–5381.

Fox, S., Harlan, R., Shivers, B., & Pfaff, D. (1990). Chemical characterization of neuroendocrine targets for progesterone in the female rat brain and pituitary. *Neuroendocrinology, 51*(3), 276–283.

Fraioli, F., Fabbri, A., Gnessi, L., Moretti, C., Bonifacio, V., Isidori, A., & Dufau, M. (1985). Naloxone increases bioactive LH in man: Evidence for selective release of early LH pool. *Journal of Endocrinological Investigation, 8*(6), 513–517.

Franceschini, R., Venturini, P., Cataldi, A., Barreca, T., Ragni, N., & Rolandi, E. (1989). Plasma beta-endorphin concentrations during suckling in lactating women. *BJOG: An International Journal of Obstetrics & Gynaecology, 96*(6), 711–713. https://doi.org/10.1111/j.1471-0528.1989.tb03287.x

Fraser, L., Morrison, D., Morley-Forster, P., Paul, T., Tokmakejian, S., Nicholson, R., et al. (2009). Oral opioids for chronic non-cancer pain: Higher prevalence of hypogonadism in men than in women. *Experimental and Clinical Endocrinology & Diabetes, 117*(01), 38–43.

Frye, C. (2009). Neurosteroids' effects and mechanisms for social, cognitive, emotional, and physical functions. *Psychoneuroendocrinology, 34*, S143–S161.

van Furth, W., Wolterink, G., & van Ree, J. (1995). Regulation of masculine sexual behavior: Involvement of brain opioids and dopamine. *Brain Research Reviews, 21*(2), 162–184.

Garcia-Fortea, P., Gonzalez-Mesa, E., Blasco, M., Cazorla, O., Delgado-Rios, M., & Gonzalez-Valenzuela, M. (2014). Oxytocin administered during labor and breast-feeding: A retrospective

cohort study. *The Journal of Maternal-Fetal & Neonatal Medicine, 27*(15), 1598–1603. https://doi.org/10.3109/14767058.2013.871255
Garthus-Niegel, S., von Soest, T., Vollrath, M., & Eberhard-Gran, M. (2013). The impact of subjective birth experiences on post-traumatic stress symptoms: A longitudinal study. *Archives of Women's Mental Health, 16*(1), 1–10. https://doi.org/10.1007/s00737-012-0301-3
Genazzani, A., Facchinetti, F., Parrini, D., Petraglia, F., La Rosa, R., Puggioni, R., & D'Antona, N. (1982). Puerperal breast feeding does not stimulate circulating opioids in humans. *Journal of Endocrinological Investigation, 5*(6), 367–371. https://doi.org/10.1007/bf03350534
Genazzani, A., Petraglia, F., Facchinetti, F., Galli, P., & Volpe, A. (1985). Lack of beta-endorphin plasma level rise in oxytocin-induced labor. *Gynecologic and Obstetric Investigation, 19*(3), 130.
George, R., Allen, T., & Habib, A. (2013). Intermittent epidural bolus compared with continuous epidural infusions for labor analgesia: A systematic review and meta-analysis. *Anesthesia & Analgesia, 116*(1), 133–144. https://doi.org/10.1213/ANE.0b013e3182713b26
Gerner, R., & Sharp, B. (1982). CSF beta-endorphin-immunoreactivity in normal, schizophrenic, depressed, manic and anorexic subjects. *Brain Research, 237*(1), 244–247.
Gerner, R., Sharp, B., & Catlin, D. (1982). Peripherally administered β-endorphin increases cerebrospinal fluid endorphin immunoreactivity. *The Journal of Clinical Endocrinology & Metabolism, 55*(2), 358–360. https://doi.org/10.1210/jcem-55-2-358
Giannakoulopoulos, X., Glover, V., Sepulveda, W., Kourtis, P., & Fisk, N. (1994). Fetal plasma cortisol and β-endorphin response to intrauterine needling. *The Lancet, 344*(8915), 77–81. https://doi.org/10.1016/S0140-6736(94)91279-3
Giannakoulopoulos, X., Teixeira, J., Fisk, N., & Glover, V. (1999). Human Fetal and maternal noradrenaline responses to invasive procedures. *Pediatric Research, 45*(4, Part 1 of 2), 494–499.
Gillman, M. A., & Lichtigfeld, F. J. (1983). The effects of nitrous oxide and naloxone on orgasm in human females: A preliminary report. *The Journal of Sex Research, 19*(1), 49–57.
Gintzler, A. (1980). Endorphin-mediated increases in pain threshold during pregnancy. *Science, 210*(4466), 193–195. https://doi.org/10.1126/science.7414330
Gintzler, A., & Liu, N. (2001). Chapter 6 The maternal spinal cord: Biochemical and physiological correlates of steroid-activated antinociceptive processes. In *Progress in brain research* (Vol. 133, pp. 83–97). Elsevier.
Gitau, R., Fisk, N., Teixeira, J., Cameron, A., & Glover, V. (2000). Fetal hypothalamic-pituitary-adrenal stress responses to invasive procedures are independent of maternal responses. *The Journal of Clinical Endocrinology & Metabolism, 86*(1), 104–109. https://doi.org/10.1210/jcem.86.1.7090
Goebelsmann, U., Abboud, T., Hoffman, D., & Hung, T. (1984). Beta-endorphin in pregnancy. *European Journal of Obstetrics & Gynecology and Reproductive Biology, 17*(2), 77–89. https://doi.org/10.1016/0028-2243(84)90132-1
Goldstein, A., & Hansteen, R. (1977). Evidence against involvement of endorphins in sexual arousal and orgasm in man. *Archives of General Psychiatry, 34*(10), 1179–1180. https://doi.org/10.1001/archpsyc.1977.01770220061006
Golub, M. (1996). Labor analgesia and infant brain development. *Pharmacology Biochemistry and Behavior, 55*(4), 619–628. https://doi.org/10.1016/S0091-3057(96)00254-7
Goodwin, G., Austin, M., Curran, S., Ross, M., Murray, C., Prentice, N., et al. (1993). The elevation of plasma beta-endorphin levels in major depression. *Journal of Affective Disorders, 29*(4), 281–289.
Grandison, L., & Guidotti, A. (1977). Regulation of prolactin release by endogenous opiates. *Nature, 270*(5635), 357–359.
Graves, G., Kennedy, T., Weick, R., & Casper, R. (1993). Endocrinology: The effect of nalmefene on pulsatile secretion of luteinizing hormone and prolactin in men. *Human Reproduction, 8*(10), 1598–1603.
Gray, L., Miller, L., Philipp, B., & Blass, E. (2002). Breastfeeding is analgesic in healthy newborns. *Pediatrics, 109*(4), 590–593. https://doi.org/10.1542/peds.109.4.590

Gutsche, B. (2002). Effect of direct fetal opioid analgesia on fetal hormonal and hemodynamic stress response to intrauterine needling. *Survey of Anesthesiology, 46*(4), 215.

Hammer, R., Mateo, A., & Bridges, R. (1992). Hormonal regulation of medial preoptic mu-opiate receptor density before and after parturition. *Neuroendocrinology, 56*(1), 38–45.

Harbach, H., Antrecht, K., Boedeker, R., Hempelmann, G., Markart, P., Matejec, R., et al. (2008). Response to delivery stress is not mediated by beta-endorphin (1–31). *European Journal of Obstetrics & Gynecology and Reproductive Biology, 136*(1), 39–45. https://doi.org/10.1016/j.ejogrb.2007.02.004

Hartmann, S., Bung, P., Schlebusch, H., & Hollmann, W. (2005). The analgesic effect of exercise during labor. *Zeitschrift fur Geburtshilfe und Neonatologie, 209*(4), 144–150. https://doi.org/10.1055/s-2005-871307

Hartwig, A. (1991). Peripheral beta-endorphin and pain modulation. *Anesthesia Progress, 38*(3), 75–78.

Henry, J. (1982). Circulating opioids: Possible physiological roles in central nervous function. *Neuroscience and Biobehavioral Reviews, 6*(3), 229–245.

Herrera-Gómez, A., Luna-Bertos, E., Ramos-Torrecillas, J., Ocaña-Peinado, F., García-Martínez, O., & Ruiz, C. (2017). The effect of epidural analgesia alone and in association with other variables on the risk of cesarean section. *Biological Research for Nursing, 19*(4), 393–398. https://doi.org/10.1177/1099800417706023

Hidayati, H., Barlianto, W., Candra, S., & Baktiyani, W. (2014). Effects of endorphin massage on B-endorphin level and Edinburgh postnatal depression scale (EPDS) score in women with postpartum blues. *Cukurova Medical Journal, 39*(3), 512–516.

Hiramoto, K., Kobayashi, H., Sekiyama, A., Sato, F., & E., Tsuruta, D., & Ishii, M. (2013). Mild exercise suppresses exacerbation of dermatitis by increasing cleavage of the β-endorphin from proopiomelanocortin in NC/Nga mice. *Journal of Clinical Biochemistry and Nutrition, 52*(1), 58–63. https://doi.org/10.3164/jcbn.12-51

Hodnett, E. (2002). Pain and women's satisfaction with the experience of childbirth: A systematic review. *American Journal of Obstetrics and Gynecology, 186*(5), S160–S172.

Hofer, M. (1984). Relationships as regulators: A psycholobiologic perspective on bereavement. *Psychosomatic Medicine, 46*, 183–197.

Hoffman, D., Abboud, T., Haase, H., Hung, T., & Goebelsmann, U. (1984). Plasma β-endorphin concentrations prior to and during pregnancy, in labor, and after delivery. *American Journal of Obstetrics and Gynecology, 150*(5), 492–496. https://doi.org/10.1016/S0002-9378(84)90426-5

Hofmeyr, G., Gülmezoglu, A., & Nikodem, V. (1995). Labor experience and beta-endorphin levels. *International Journal of Gynaecology and Obstetrics, 50*(3), 299–300.

Hosseini, E., Asadi, N., & Zareei, F. (2013). Effect of massage therapy on labor Progress and plasma levels of cortisol in the active stage of first labor. *Zahedan Journal of Research in Medical Sciences, 15*(9), 35–38.

Hughes, G., Lichstein, P., Whitlock, D., & Harker, C. (1984). Response of plasma beta-endorphins to transcutaneous electrical nerve stimulation in healthy subjects. *Physical Therapy, 64*(7), 1062–1066.

Izquierdo, I., & Netto, C. (1985). Role of beta-endorphin in behavioral regulation. *Annals of the New York Academy of Sciences, 444*, 162–177.

Jacobson, B., Nyberg, K., Grönbladh, L., Eklund, G., Bygdeman, M., & Rydberg, U. (1990). Opiate addiction in adult offspring through possible imprinting after obstetric treatment. *British Medical Journal, 301*(6760), 1067–1070. https://doi.org/10.1136/bmj.301.6760.1067

Jacquet, Y. (1980). Excitatory and inhibitory effects of opiates in the rat vas deferens: A dual mechanism of opiate action. *Science, 210*(4465), 95–97.

Jarvis, S., Lawrence, A., McLean, K., Chirnside, J., Deans, L., Calvert, S., et al. (2000). The effect of opioid antagonism and environmental restriction on plasma oxytocin and vasopressin concentrations in parturient gilts. *Journal of Endocrinology, 166*(1), 39–44. https://doi.org/10.1677/joe.0.1660039

Jordan, S., Emery, S., Watkins, A., Evans, J., Storey, M., & Morgan, G. (2009). Associations of drugs routinely given in labour with breastfeeding at 48 hours: Analysis of the Cardiff Births Survey. *BJOG: An International Journal of Obstetrics & Gynaecology, 116*(12), 1622–1632. https://doi.org/10.1111/j.1471-0528.2009.02256.x

Jowitt, M. (1993). Beta-endorphin and stress in pregnancy and labour. *Midwifery Matters, 56*, 3–4.

Jurna, I. (1993). Labor pain-causes, pathways and issues. *Schmerz, 7*(2), 79–84. https://doi.org/10.1007/bf02527864

Kalra, A., Urban, M., & Sluka, K. (2001). Blockade of opioid receptors in rostral ventral medulla prevents antihyperalgesia produced by transcutaneous electrical nerve stimulation (TENS). *Journal of Pharmacology and Experimental Therapeutics, 298*(1), 257–263.

Kammerer-Doak, D., & Rogers, R. (2008). Female sexual function and dysfunction. *Obstetrics and Gynecology Clinics of North America, 35*(2), 169–183. https://doi.org/10.1016/j.ogc.2008.03.006

Kang, Y., & Park, J. (2000). Brain uptake and the analgesic effect of oxytocin— Its usefulness as an analgesic agent. *Archives of Pharmacal Research, 23*(4), 391. https://doi.org/10.1007/bf02975453

Kaplan, H. (1974). The new sex therapy: active treatment of sexual dysfunctions New York: Brunner-Routledge.

Katz, A. (2009). *Sex when you are sick*. Preager.

Katz, N., & Mazer, N. (2009). The impact of opioids on the endocrine system. *The Clinical Journal of Pain, 25*(2), 170–175.

Kennedy, S., Koeppe, R., Young, E., & Zubieta, J. (2006). Dysregulation of endogenous opioid emotion regulation circuitry in major depression in women. *Archives of General Psychiatry, 63*(11), 1199–1208. https://doi.org/10.1001/archpsyc.63.11.1199

Keskin, E., Onur, O., Keskin, H., Gumus, I., Kafali, H., & Turhan, N. (2012). Transcutaneous electrical nerve stimulation improves low Back pain during pregnancy. *Gynecologic and Obstetric Investigation, 74*(1), 76–83.

Keverne, E., Martensz, N., & Tuite, B. (1989). Beta-endorphin concentrations in cerebrospinal fluid of monkeys are influenced by grooming relationships. *Psychoneuroendocrinology, 14*(1–2), 155–161.

Kovács, G., & Telegdy, G. (1987). β-endorphin tolerance is inhibited by oxytocin. *Pharmacology Biochemistry and Behavior, 26*(1), 57–60. https://doi.org/10.1016/0091-3057(87)90533-8

Kristal, M., DiPirro, J., & Thompson, A. (2012). Placentophagia in humans and nonhuman mammals: Causes and consequences. *Ecology of Food and Nutrition, 51*(3), 177–197. https://doi.org/10.1080/03670244.2012.661325

Krüger, T., Exton, M., Pawlak, C., Mühlen, A., Hartmann, U., & Schedlowski, M. (1998). Neuroendocrine and cardiovascular response to sexual arousal and orgasm in men. *Psychoneuroendocrinology, 23*(4), 401–411. https://doi.org/10.1016/S0306-4530(98)00007-9

Laatikainen, T. (1991). Corticotropin-releasing hormone and opioid peptides in reproduction and stress. *Annals of Medicine, 23*(5), 489–496. https://doi.org/10.3109/07853899109150508

Lechner, W., Jarosch, E., Sölder, E., Waitz-Penz, A., & Mitterschiffthaler, G. (1991). Beta-endorphins during childbirth under transcutaneous electric nerve stimulation. *Zentralblatt fur Gynakologie, 113*(8), 439–442.

Lee, M., Chang, S., & Kang, D. (2004). Effects of SP6 acupressure on labor pain and length of delivery time in women during labor. *The Journal of Alternative and Complementary Medicine, 10*(6), 959–965. https://doi.org/10.1089/acm.2004.10.959

Lee, S., Ralston, H., Drey, E., Partridge, J., & Rosen, M. (2005). Fetal pain: A systematic multidisciplinary review of the evidence. *JAMA, 294*(8), 947–954. https://doi.org/10.1001/jama.294.8.947

Leonard, G., Goffaux, P., & Marchand, S. (2010). Deciphering the role of endogenous opioids in high-frequency TENS using low and high doses of naloxone. *Pain, 151*(1), 215–219. https://doi.org/10.1016/j.pain.2010.07.012

Léonard, G., Cloutier, C., & Marchand, S. (2011). Reduced analgesic effect of acupuncture-like TENS but not conventional TENS in opioid-treated patients. *The Journal of Pain, 12*(2), 213–221. https://doi.org/10.1016/j.jpain.2010.07.003

Levett, K., Smith, C., Bensoussan, A., & Dahlen, H. (2016). Complementary therapies for labour and birth study: A randomised controlled trial of antenatal integrative medicine for pain management in labour. *BMJ Open, 6*(7). https://doi.org/10.1136/bmjopen-2015-010691

Li, J., Dubois, M., & Dubois, P. (1979). Ultrastructural localization of immunoreactive corticotropin, β-lipotropin, α- and β-endorphin in cells of the human fetal anterior pituitary. *Cell and Tissue Research, 204*(1), 37–51. https://doi.org/10.1007/bf00235163

Lindow, S., Newham, A., Hendricks, M., Thompson, J., & van der Spuy, Z. (1996). The 24-hour rhythm of oxytocin and beta-endorphin secretion in human pregnancy. *Clinical Endocrinology, 45*(4), 443–446.

Lothian, J. (2005). The birth of a breastfeeding baby and mother. *The Journal of Perinatal Education, 14*(1), 42–45.

Lou, H., Tweed, W., & Davis, J. (1989). Endogenous opioids may protect the perinatal brain in hypoxia. *Developmental Pharmacology and Therapeutics, 13*(2–4), 129–133.

Lowe, N. (1996). The pain and discomfort of labor and birth. *Journal of Obstetric, Gynecologic & Neonatal Nursing, 25*(1), 82–92. https://doi.org/10.1111/j.1552-6909.1996.tb02517.x

Lowe, N. (2002). The nature of labor pain. *American Journal of Obstetrics and Gynecology, 186*(5, Supplement 1), S16–S24. https://doi.org/10.1016/s0002-9378(02)70179-8

Mardirosoff, C., Dumont, L., Boulvain, M., & Tramèr, M. (2002). Fetal bradycardia due to intrathecal opioids for labour analgesia: A systematic review. *BJOG: An International Journal of Obstetrics & Gynaecology, 109*(3), 274–281. https://doi.org/10.1111/j.1471-0528.2002.01380.x

Masters, W., & Johnson, V. (1966). *Human sexual response*. Little Brown.

Masters, W., & Johnson, V. (1970). *Human sexual inadequacy*. Little Brown.

Mastorakos, G., & Pavlatov, M. (2004). Exercise and the stress system. *Folia Clinica En Obstetricia y Ginecologia, 46*, 6–38.

McLean, M., Thompson, D., Zhang, H., Brinsmead, M., & Smith, R. (1994). Corticotrophin-releasing hormone and β-endorphin in labour. *European Journal of Endocrinology, 131*(2), 167–172. https://doi.org/10.1530/eje.0.1310167

Mehta, R., & Petrova, A. (2011). Biologically active breast milk proteins in association with very preterm delivery and stage of lactation. *Journal of Perinatology, 31*(1), 58–62.

Melis, M., Succu, S., Spano, M., & Argiolas, A. (1999). Morphine injected into the paraventricular nucleus of the hypothalamus prevents noncontact penile erections and impairs copulation: Involvement of nitric oxide. *European Journal of Neuroscience, 11*(6), 1857–1864.

Mendelson, J., Ellingboe, J., Keuhnle, J., & Mello, N. (1978). Effects of naltrexone on mood and neuroendocrine function in normal adult males. *Psychoneuroendocrinology, 3*(3–4), 231–236. https://doi.org/10.1016/0306-4530(78)90013-6

Mens, W., Witter, A., & Van Wimersma Greidanus, T. (1983). Penetration of neurohypophyseal hormones from plasma into cerebrospinal fluid (CSF): Half-times of disappearance of these neuropeptides from CSF. *Brain Research, 262*(1), 143–149. https://doi.org/10.1016/0006-8993(83)90478-X

Miki, N., Sonntag, W., Forman, L., & Meites, J. (1981). Suppression by naloxone of rise in plasma growth hormone and prolactin induced by suckling. *Experimental Biology and Medicine, 168*(3), 330–333. https://doi.org/10.3181/00379727-168-41281

Mori, R., Tokumasu, H., Pledge, D., & Kenyon, S. (2011). High dose versus low dose oxytocin for augmentation of delayed labour. *Cochrane Database of Systematic Reviews*, (10), CD007201.

Mousa, S. A., Bopaiah, C. P., Stein, C., & Schäfer, M. (2003). Involvement of corticotropin-releasing hormone receptor subtypes 1 and 2 in peripheral opioid-mediated inhibition of inflammatory pain. *Pain, 106*(3), 297–307. https://doi.org/10.1016/s0304-3959(03)00302-6

Myers, B., & Baum, M. (1979). Facilitation by opiate antagonists of sexual performance in the male rat. *Pharmacology Biochemistry and Behavior, 10*(4), 615–618.

Myers, B., & Baum, M. (1980). Facilitation of copulatory performance in male rats by naloxone: Effects of hypophysectomy, 17 α-estradiol, and luteinizing hormone releasing hormone. *Pharmacology Biochemistry and Behavior, 12*(3), 365–370.

Nandhra, T., & Carson, R. (2000). β-endorphin inhibits the production of interleukin-8 by human chorio–decidual cells in culture. *Molecular Human Reproduction, 6*(6), 555–560.

Navines, R., Martin-Santos, R., Gomez-Gil, E., Martinez de Osaba, M., & Gasto, C. (2008). Interaction between serotonin 5-HT1A receptors and beta-endorphins modulates antidepressant response. *Progress in Neuro-Psychopharmacology & Biological Psychiatry, 32*(8), 1804–1809. https://doi.org/10.1016/j.pnpbp.2008.07.021

Nelson, E., & Panksepp, J. (1998). Brain substrates of infant–mother attachment: Contributions of opioids, oxytocin, and norepinephrine. *Neuroscience & Biobehavioral Reviews, 22*(3), 437–452. https://doi.org/10.1016/S0149-7634(97)00052-3

Newnham, J., Dennett, P., Ferron, S., Tomlin, S., Legg, C., Bourne, G., & Rees, L. (1984). A study of the relationship between circulating beta-endorphin-like immunoreactivity and post partum 'blues'. *Clinical Endocrinology, 20*(2), 169–177.

Noirot, E. (1972). Ultrasounds and maternal behavior in small rodents. *Developmental Psychobiology, 5*(4), 371–387. https://doi.org/10.1002/dev.420050410

Nyberg, K. (2000). Long-term effects of labor analgesia. *Journal of Obstetric, Gynecologic, & Neonatal Nursing, 29*(3), 226–226. https://doi.org/10.1111/j.1552-6909.2000.tb02043.x

Nyberg, K., Buka, S., & Lipsitt, L. (2000). Perinatal medication as a potential risk factor for adult drug abuse in a North American Cohort. *Epidemiology, 11*(6), 715–716.

Odent, M. (1999). *The Scientification of love*. Free Association Books.

Odent, M. (2001). *The Scientification of love*. Free association Books.

Odent, M. (2008). Dispelling the disempowering birth vocabulary. *Journal of Prenatal & Perinatal Psychology & Health, 23*(1), 5.

Olza Fernández, I., Marín Gabriel, M., García Murillo, L., Malalana Martinez, A., Costarelli, V., & Millán Santos, I. (2013). Mode of delivery may influence neonatal responsiveness to maternal separation. *Early Human Development, 89*(5), 339–342. https://doi.org/10.1016/j.earlhumdev.2012.11.005

Ombra, M., Musumeci, M., Simpore, J., Palano, G., & Musumeci, S. (2008). β-Endorphin concentration in colostrums of Burkinabe and Sicilian women. *Nutrition, 24*(1), 31–36. https://doi.org/10.1016/j.nut.2007.09.004

Pancheri, P., Zichella, L., Fraioli, F., Carilli, L., Perrone, G., Biondi, M., et al. (1985). ACTH, beta-endorphin and met-enkephalin: Peripheral modifications during the stress of human labor. *Psychoneuroendocrinology, 10*(3), 289–301. https://doi.org/10.1016/0306-4530(85)90006-X

Panksepp, J. (1981). Brain opioids: A neurochemical substrate for narcotic and social dependence. In S. Cooper (Ed.), *Progress in theory in psychopharmacology* (pp. 149–175). Academic.

Panksepp, J. (1991). Affective neuroscience: A conceptual framework for the neurobiological study of emotions. In K. Strongman (Ed.), *International reviews of emotion research* (Vol. 1, pp. 59–99). Wiley.

Panksepp, J., Nelson, E., & Siviy, S. (1994). Brain opioids and mother—Infant social motivation. *Acta Paediatrica, 83*, 40–46. https://doi.org/10.1111/j.1651-2227.1994.tb13264.x

Parker, G., Parker, I., & Brotchie, H. (2006). Mood state effects of chocolate. *Journal of Affective Disorders, 92*(2–3), 149–159. https://doi.org/10.1016/j.jad.2006.02.007

Patil, M., Green, D., Henry, M., & Akopian, A. (2013). Sex-dependent roles of prolactin and prolactin receptor in postoperative pain and hyperalgesia in mice. *Neuroscience, 253*, 132–141. https://doi.org/10.1016/j.neuroscience.2013.08.035

Pereira, R., Kanhai, H., Rosendaal, F., van Dommelen, P., Swaab, D., Rodrigues Pereira, E., & van de Wetering, B. (2012). Parenteral pethidine for labour pain relief and substance use disorder: 20-year follow-up cohort study in offspring. *BMJ Open, 2*(3). https://doi.org/10.1136/bmjopen-2011-000719

Petraglia, F., Facchinetti, F., D'Ambrogio, G., Volpe, A., & Genazzani, A. (1986). Somatostatin and oxytocin infusion inhibits the rise of plasma β-endorphin, β-lipotrophin and cortisol induced by insulin hypoglycaemia. *Clinical Endocrinology, 24*(6), 609–616. https://doi.org/10.1111/j.1365-2265.1986.tb01656.x

Pfaus, J., & Gorzalka, B. (1987). Opioids and sexual behavior. *Neuroscience & Biobehavioral Reviews, 11*(1), 1–34. https://doi.org/10.1016/S0149-7634(87)80002-7

Pfeiffer, A., & Herz, A. (1984). Endocrine actions of opioids. *Hormone and Metabolic Research, 16*(08), 386–397.

Punzio, C., Neri, E., Metelli, P., Bianchi, M., Venticinque, M., Ferdeghini, M., & Fioretti, P. (1994). The relationship between maternal relaxation and plasma β-endorphin levels during parturition. *Journal of Psychosomatic Obstetrics and Gynecology, 15*(4), 205–210. https://doi.org/10.3109/01674829409025647

Qu, F., & Zhou, J. (2007). Electro-acupuncture in relieving labor pain. *eCAM, 4*(1), 125–130. https://doi.org/10.1093/ecam/nel053

Radunovic, N., Lockwood, C., Ghidini, A., Alvarez, M., & Berkowitz, R. (1993). Is fetal blood sampling associated with increased beta-endorphin release into the fetal circulation? *American Journal of Perinatology, 10*, 112–114.

Raffin-Sanson, M., Ferre, F., Coste, J., Oliver, C., Cabrol, D., & Bertagna, X. (2000). Proopiomelanocortin in human pregnancy: Evolution of maternal plasma levels, concentrations in cord blood, amniotic fluid and at the feto-maternal interface. *European Journal of Endocrinology, 142*(1), 53–59. https://doi.org/10.1530/eje.0.1420053

Raisanen, I., Paatero, H., Salminen, K., & Laatikainen, T. (1984). Pain and plasma [beta]-endorphin level during labor. *Obstetrics & Gynecology, 64*(6), 783–786.

Rasheed, A., & Tareen, I. (1995). Effects of heroin on thyroid function, cortisol and testosterone level in addicts. *Polish Journal of Pharmacology, 47*(5), 441–444.

Reed, R., Barnes, M., & Rowe, J. (2016). Women's experience of birth: Childbirth as a rite of passage. *International Journal of Childbirth, 6*(1), 46–56.

Reynolds, F. (2010). The effects of maternal labour analgesia on the fetus. *Best Practice & Research Clinical Obstetrics & Gynaecology, 24*(3), 289–302. https://doi.org/10.1016/j.bpobgyn.2009.11.003

Rhodin, A., Stridsberg, M., & Gordh, T. (2010). Opioid endocrinopathy: A clinical problem in patients with chronic pain and long-term oral opioid treatment. *The Clinical Journal of Pain, 26*(5), 374–380.

Risch, S., Janowsky, D., Siever, L., Judd, L., Rausch, J., Huey, L., et al. (1982). Correlated cholinomimetic-stimulated beta-endorphin and prolactin release in humans. *Peptides, 3*(3), 319–322. https://doi.org/10.1016/0196-9781(82)90093-6

Riskind, P., Millard, W., & Martin, J. (1984). Opiate modulation of the anterior pituitary hormone response during suckling in the rat. *Endocrinology, 114*(4), 1232–1237. https://doi.org/10.1210/endo-114-4-1232

Riss, P., & Bieglmayer, C. (1984). Obstetric analgesia and immunoreactive endorphin peptides in maternal plasma during labor. *Gynecologic and Obstetric Investigation, 17*(3), 127–130.

Roberts, L., Finch, P., Pullan, P., Bhagat, C., & Price, L. (2002). Sex hormone suppression by intrathecal opioids: A prospective study. *The Clinical Journal of Pain, 18*(3), 144–148.

Robin, S., Sharon, L., Raymond, I., & Andrew, G. (1981). Human plasma β-endorphin during pregnancy, labor, and delivery. *The Journal of Clinical Endocrinology & Metabolism, 52*(1), 74–78. https://doi.org/10.1210/jcem-52-1-74

Robinson, K., & Fisher, D. (2014). Tanning as a substance abuse. *Communicative & Integrative Biology, 7*(5), e971579. https://doi.org/10.4161/cib.29890

Romano, A., & Lothian, J. (2008). Promoting, protecting, and supporting normal birth: A look at the evidence. *Journal of Obstetric, Gynecologic, and Neonatal Nursing, 37*(1), 94–104; quiz 104-105. https://doi.org/10.1111/j.1552-6909.2007.00210.x

Saisto, T., Kaaja, R., Helske, S., Ylikorkala, O., & Halmesmäki, E. (2004). Norepinephrine, adrenocorticotropin, cortisol and beta-endorphin in women suffering from fear of labor: Responses to the cold pressor test during and after pregnancy. *Acta Obstetricia et Gynecologica Scandinavica, 83*(1), 19–26.

Salar, G., Job, I., Mingrino, S., Bosio, A., & Trabucchi, M. (1981). Effect of transcutaneous electrotherapy on CSF β-endorphin content in patients without pain problems. *Pain, 10*(2), 169–172. https://doi.org/10.1016/0304-3959(81)90192-5

Sander, H., & Gintzler, A. (1987). Spinal cord mediation of the opioid analgesia of pregnancy. *Brain Research, 408*(1–2), 389–393. https://doi.org/10.1016/0006-8993(87)90412-4

Sandman, C., Wadhwa, P., Chicz-DeMet, A., Porto, M., & Garite, T. (1995). Third trimester POMC disregulation predicts use of anesthesia at vaginal delivery. *Peptides, 16*(2), 187–190. https://doi.org/10.1016/0196-9781(94)00183-9

Sandström, A., Altman, M., Cnattingius, S., Johansson, S., Ahlberg, M., & Stephansson, O. (2017). Durations of second stage of labor and pushing, and adverse neonatal outcomes: A population-based cohort study. *Journal of Perinatology, 37*(3), 236–242. https://doi.org/10.1038/jp.2016.214

Santana, L., Gallo, R., Ferreira, C., Duarte, G., Quintana, S., & Marcolin, A. (2016). Transcutaneous electrical nerve stimulation (TENS) reduces pain and postpones the need for pharmacological analgesia during labour: A randomised trial. *Journal of Physiotherapy, 62*(1), 29–34. https://doi.org/10.1016/j.jphys.2015.11.002

Sathe, R., Komisaruk, B., Ladas, A., & Godbole, S. (2001). Naltrexone-induced augmentation of sexual response in men. *Archives of Medical Research, 32*(3), 221–226. https://doi.org/10.1016/S0188-4409(01)00279-X

Sauriyal, D., Jaggi, A., & Singh, N. (2011). Extending pharmacological spectrum of opioids beyond analgesia: Multifunctional aspects in different pathophysiological states. *Neuropeptides, 45*(3), 175–188. https://doi.org/10.1016/j.npep.2010.12.004

Schapiro, S., & Salas, M. (1970). Behavioral response of infant rats to maternal odor. *Physiology & Behavior, 5*(7), 815–817. https://doi.org/10.1016/0031-9384(70)90285-4

Schulz, S., Wagner, R., van Kammen, D., Rogol, A., Davis, G., Wyatt, R., et al. (1980). Prolactin response to beta-endorphin in man. *Life Sciences, 27*(19), 1735–1741. https://doi.org/10.1016/0024-3205(80)90440-3

Scull, T., Hemmings, G., Carli, F., Weeks, S., Mazza, L., & Zingg, H. (1998). Epidural analgesia in early labour blocks the stress response but uterine contractions remain unchanged. *Canadian Journal of Anaesthesia, 45*(7), 626–630. https://doi.org/10.1007/bf03012090

Selmanoff, M., & Gregerson, K. (1986). Suckling-induced prolactin release is suppressed by naloxone and simulated by β-endorphin. *Neuroendocrinology, 42*(3), 255–259.

Selye, H. (1976). *Stress in health and disease*. Butterworths.

Selye, H., & Horava, A. (1953). Stress and the adaptive hormones. *Postgraduate Medical Journal, 29*(330), 217.

Shapiro, G., Fraser, W., Frasch, M., & Séguin, J. (2013). Psychosocial stress in pregnancy and preterm birth: Associations and mechanisms. *Journal of Perinatal Medicine, 41*(6), 631–645.

Sharon, L., & Andrew, G. (1983). Brain β-endorphin during pregnancy, parturition, and the post-partum period. *Endocrinology, 113*(5), 1664–1668. https://doi.org/10.1210/endo-113-5-1664

Simpson, P., Fouche, P., Thomas, R., & Bendall, J. (2014). Transcutaneous electrical nerve stimulation for relieving acute pain in the prehospital setting: A systematic review and meta-analysis of randomized-controlled trials. *European Journal of Emergency Medicine, 21*(1), 10–17. https://doi.org/10.1097/MEJ.0b013e328363c9c1

Sirinathsinghji, D., Whittington, P., Audsley, A., & Fraser, H. (1983). β-Endorphin regulates lordosis in female rats by modulating LH-RH release. *Nature, 301*(5895), 62–64.

Sluka, K., & Walsh, D. (2003). Transcutaneous electrical nerve stimulation: Basic science mechanisms and clinical effectiveness. *Journal of Pain, 4*, 109–121.

Sluka, K., Deacon, M., Stibal, A., Strissel, S., & Terpstra, A. (1999). Spinal blockade of opioid receptors prevents the analgesia produced by TENS in arthritic rats. *Journal of Pharmacology and Experimental Therapeutics, 289*(2), 840–846.

Sluka, K., Judge, M., McColley, M., & Reveiz, P. (2000). Low frequency TENS is less effective than high frequency TENS at reducing inflammation-induced hyperalgesia in morphine-tolerant rats. *European Journal of Pain, 4*(2), 185–193. https://doi.org/10.1053/eujp.2000.0172

Sluka, K., Bjordal, J., Marchand, S., & Rakel, B. (2013). What makes transcutaneous electrical nerve stimulation work? Making sense of the mixed results in the clinical literature. *Physical Therapy, 93*(10), 1397–1402. https://doi.org/10.2522/ptj.20120281

Smith, L. (2007). Impact of birthing practices on the breastfeeding dyad. *Journal of Midwifery & Women's Health, 52*(6), 621–630.

Smith, R., Cubis, J., Brinsmead, M., Lewin, T., Singh, B., Owens, P., et al. (1990). Mood changes, obstetric experience and alterations in plasma cortisol, beta-endorphin and corticotrophin releasing hormone during pregnancy and the puerperium. *Journal of Psychosomatic Research, 34*(1), 53–69.

Smith, C., Levett, K., Collins, C., & Jones, L. (2012). Massage, reflexology and other manualmethods for painmanagement in labour. *Cochrane Database of Systematic Reviews*, (2), CD009290.

Sng, B., Leong, W., Zeng, Y., Siddiqui, F., Assam, P., Lim, Y., et al. (2014). Early versus late initiation of epidural analgesia for labour. *Cochrane Database of Systematic Reviews*, (10), CD007238. https://doi.org/10.1002/14651858.CD007238.pub2

Somogyi, A., Barratt, D., & Coller, J. (2007). Pharmacogenetics of opioids. *Clinical Pharmacology & Therapeutics, 81*(3), 429–444. https://doi.org/10.1038/sj.clpt.6100095

Stanton, M., Wallstrom, J., & Levine, S. (1987). Maternal contact inhibits pituitary-adrenal stress responses in preweanling rats. *Developmental Psychobiology, 20*(2), 131–145. https://doi.org/10.1002/dev.420200204

Sultan, P., Murphy, C., Halpern, S., & Carvalho, B. (2013). The effect of low concentrations versus high concentrations of local anesthetics for labour analgesia on obstetric and anesthetic outcomes: A meta-analysis. *Canadian Journal of Anesthesia/Journal canadien d'anesthésie, 60*(9), 840–854. https://doi.org/10.1007/s12630-013-9981-z

Sundblad, C., & Eriksson, E. (1997). Reduced extracellular levels of serotonin in the amygdala of androgenized female rats. *European Neuropsychopharmacology, 7*(4), 253–259.

Swain, J., Tasgin, E., Mayes, L., Feldman, R., Constable, R., & Leckman, J. (2008). Maternal brain response to own baby-cry is affected by cesarean section delivery. *Journal of Child Psychology and Psychiatry, and Allied Disciplines, 49*(10), 1042–1052. https://doi.org/10.1111/j.1469-7610.2008.01963.x

Sweep, C., Boomkamp, M., Barna, I., Logtenberg, A., & Wiegant, V. (1990). Vasopressin enhances the clearance of β-endorphin immunoreactivity from rat cerebrospinal fluid. *Acta Endocrinologica, 122*(2), 191–200. https://doi.org/10.1530/acta.0.1220191

Taavoni, S., Sheikhan, F., Abdolahian, S., & Ghavi, F. (2016). Birth ball or heat therapy? A randomized controlled trial to compare the effectiveness of birth ball usage with sacrum-perineal heat therapy in labor pain management. *Complementary Therapies in Clinical Practice, 24*, 99–102. https://doi.org/10.1016/j.ctcp.2016.04.001

Taghinejad, H., & Delpisheh, A. (2010). Comparison between massage and music therapies to relieve the severity of labor pain. *Women's Health Issues, 6*(3), 377–381. https://doi.org/10.2217/whe.10.15

Teixeira, J., Glover, V., & Fisk, N. (1999). Acute cerebral redistribution in response to invasive procedures in the human fetus. *American Journal of Obstetrics and Gynecology, 181*, 1018–1025.

Tempfer, C., Zeisler, H., Heinzl, H., Hefler, L., Husslein, P., & Kainz, C. (1998). Influence of acupuncture on maternal serum levels of interleukins, prostaglandin f2 alpha, and beta-endorphin: A matched pair study. *Obstetrics and Gynecology, 92*, 245–248.

Thomas, T., Fletcher, J., & Hill, R. (1982). Influence of medication, pain and progress in labour on plasma β-endorphin-like immunoreactivity. *British Journal of Anaesthesia, 54*(4), 401–408. https://doi.org/10.1093/bja/54.4.401

Tornetta, G. (1998). *Painless childbirth: An empowering journey through pregnancy and childbirth*. Cumberland House Publishing.

Trotman, H., & Henny-Harry, C. (2012). Factors associated with extreme hyperbilirubinaemia in neonates at the University Hospital of the West Indies. *Paediatrics and International Child Health, 32*(2), 97–101. https://doi.org/10.1179/2046905512Y.0000000014

Ullman, R., Smith, L., Burns, E., Mori, R., & Dowswell, T. (2010). Parenteral opioids for maternal pain management in labour. *The Cochrane Database of Systematic Reviews*, (9), CD007396. https://doi.org/10.1002/14651858.CD007396.pub2

Varrassi, G., Bazzano, C., & Edwards, W. T. (1989). Effects of physical activity on maternal plasma β-endorphin levels and perception of labor pain. *American Journal of Obstetrics and Gynecology, 160*(3), 707–712. https://doi.org/10.1016/S0002-9378(89)80065-1

Veening, J., & Barendregt, H. (2015). The effects of Beta-endorphin: State change modification. *Fluids and Barriers of the CNS, 12*, 3. https://doi.org/10.1186/2045-8118-12-3

Vogl, S., Worda, C., Egarter, C., Bieglmayer, C., Szekeres, T., Huber, J., & Husslein, P. (2006). Mode of delivery is associated with maternal and fetal endocrine stress response. *BJOG, 113*(4), 441–445. https://doi.org/10.1111/j.1471-0528.2006.00865.x

Volmanen, P., Valanne, J., & Alahuhta, S. (2004). Breast-feeding problems after epidural analgesia for labour: A retrospective cohort study of pain, obstetrical procedures and breast-feeding practices. *International Journal of Obstetric Anesthesia, 13*(1), 25–29. https://doi.org/10.1016/S0959-289X(03)00104-3

Voogt, J., Lee, Y., Yang, S., & Arbogast, L. (2001). Regulation of prolactin secretion during pregnancy and lactation. *Progress in Brain Research, 133*, 173–185.

Waldhoer, M., Bartlett, S., & Whistler, J. (2004). Opioid receptors. *Annual Review of Biochemistry, 73*(1), 953–990. https://doi.org/10.1146/annurev.biochem.73.011303.073940

Wang, M., Wahlström, G., & Bäckström, T. (1997). The regional brain distribution of the neurosteroids pregnenolone and pregnenolone sulfate following intravenous infusion. *The Journal of Steroid Biochemistry and Molecular Biology, 62*(4), 299–306. https://doi.org/10.1016/S0960-0760(97)00041-1

Wardlaw, S., Stark, R., Daniel, S., & Frantz, A. (1981). Effects of hypoxia on β-endorphin and β-lipotropin release in fetal, newborn, and maternal sheep. *Endocrinology, 108*(5), 1710–1715. https://doi.org/10.1210/endo-108-5-1710

Wei, S., Luo, Z., Xu, H., & Fraser, W. (2009). The effect of early oxytocin augmentation in labor: A meta-analysis. *Obstetrics & Gynecology, 114*(3), 641–649.

Wei, S., Luo, Z., Qi, H., Xu, H., & Fraser, W. (2010). High-dose vs low-dose oxytocin for labor augmentation: A systematic review. *American Journal of Obstetrics and Gynecology, 203*(4), 296–304. https://doi.org/10.1016/j.ajog.2010.03.007

Weissberg, N., Algur, N., Schwartz, G., Shemesh, O., Eylath, U., & Abraham, A. (1990). The relationship between beta-endorphin levels and uterine muscle contractions during labor. *International Journal of Gynecology & Obstetrics, 33*(4), 313–316. https://doi.org/10.1016/0020-7292(90)90516-N

Weller, A., & Feldman, R. (2003). Emotion regulation and touch in infants: The role of cholecystokinin and opioids. *Peptides, 24*(5), 779–788. https://doi.org/10.1016/S0196-9781(03)00118-9

Whipple, B., & Brash-McGreer, K. (1997). Management of female sexual dysfunction. In M. Sipski & C. Alexander (Eds.), *Sexual function in people with disability and chronic illness. A health professional's guide* (pp. 509–534). Aspen Publishers.

Whipple, B., Josimovich, J., & Komisaruk, B. (1990). Sensory thresholds during the antepartum, intrapartum and postpartum periods. *International Journal of Nursing Studies, 27*(3), 213–221. https://doi.org/10.1016/0020-7489(90)90036-I

Wiklund, I., Norman, M., Uvnäs-Moberg, K., Ransjö-Arvidson, A., & Andolf, E. (2009). Epidural analgesia: Breast-feeding success and related factors. *Midwifery, 25*(2), e31–e38. https://doi.org/10.1016/j.midw.2007.07.005

Willer, J., Le Bars, D., & De Broucker, T. (1990). Diffuse noxious inhibitory controls in man: Involvement of an opioidergic link. *European Journal of Pharmacology, 182*(2), 347–355. https://doi.org/10.1016/0014-2999(90)90293-F

Winberg, J. (2005). Mother and newborn baby: Mutual regulation of physiology and behavior – A selective review. *Developmental Psychobiology, 47*(3), 217–229.

Wu, Y., Hsieh, W., Hsu, C., Chang, J., Chou, H., Hsu, H., et al. (2016). Intervention effects on emotion regulation in preterm infants with very low birth weight: A randomize controlled trial. *Research in Developmental Disabilities, 48*, 1–12. https://doi.org/10.1016/j.ridd.2015.10.016

Yee, A., Loh, H., Hisham Hashim, H., & Ng, C. (2014). The prevalence of sexual dysfunction among male patients on methadone and buprenorphine treatments: A meta-analysis study. *The Journal of Sexual Medicine, 11*(1), 22–32. https://doi.org/10.1111/jsm.12352

Yim, I., Glynn, L., Schetter, C., Hobel, C., Chicz-DeMet, A., & Sandman, C. (2010). Prenatal β-endorphin as an early predictor of postpartum depressive symptoms in euthymic women. *Journal of Affective Disorders, 125*(1–3), 128–133. https://doi.org/10.1016/j.jad.2009.12.009

Zagon, I., Gibo, D., & McLaughlin, P. (1990). Adult and developing human cerebella exhibit different profiles of opioid binding sites. *Brain Research, 523*(1), 62–68.

Zanardo, V., Nicolussi, S., Carlo, G., Marzari, F., Faggian, D., Favaro, F., & Plebani, M. (2001a). Beta endorphin concentrations in human milk. *Journal of Pediatric Gastroenterology and Nutrition, 33*(2), 160–164.

Zanardo, V., Nicolussi, S., Giacomin, C., Faggian, D., Favaro, F., & Plebani, M. (2001b). Labor pain effects on colostral milk beta-endorphin concentrations of lactating mothers. *Neonatology, 79*(2), 87–90.

Zilbergeld, B., & Ellison, C. (1980). Desire discrepancies and arousal problems in sex therapy. In S. Leiblum & L. Pervin (Eds.), *Principles and practices of sex therapy*. Guilford Press.

Depression, Cancer, Inflammation, and Endogenous Opioids: Pathogenic Relationships and Therapeutic Options

Jennifer Hancock, Cristian Sirbu, and Patrick L. Kerr

Abstract Endogenous opioids and their associated receptors form a system that maintains survival by positively reinforcing behaviors that are vital to life. Cancer and cancer treatment side effects capitalize on this system pathogenically, leading to maladaptive biological responses (e.g., inflammation), as well as cognitive and emotional consequences, most notably depression. Psychologists who treat people with cancer frequently find depression to be a primary target for intervention. However, in people with cancer, the etiology of depression is unique and complex. This complexity necessitates that psycho-oncologists have a fundamental working knowledge of the biological substrates that underlie depression/cancer comorbidity. Building on other chapters in this volume pertaining to cancer and endogenous opioids, this chapter focuses on the clinical applications of basic scientific findings.

Keywords Cancer · Endogenous opioids · Psycho-oncology

Introduction: Endogenous Opioids and Cancer

The body has a natural reward system composed of endogenous opioids. This system helps to ensure survival by rewarding the behaviors that are necessary to life (e.g., eating, sleeping, sex), while also activating the body's coping mechanisms under short-term duress (e.g., acute pain, sickness). Cancer has a way of hijacking this system through both changes affected by the cancer itself as well as short- and long-term side effects of treatments designed to fight the cancer. These effects consist of an interrelated set of physiological, emotional, and cognitive consequences of cancer and cancer treatment. Psychologists specializing in caring for people with

J. Hancock (✉) · C. Sirbu
Center for Cancer Research, Charleston Area Medical Center, Charleston, WV, USA
e-mail: jennifer.j.hancock@wv.gov

P. L. Kerr
West Virginia University School of Medicine-Charleston, Charleston, WV, USA

P. L. Kerr et al. (eds.), *Endogenous Opioids*, Advances in Neurobiology 35,
https://doi.org/10.1007/978-3-031-45493-6_21

cancer are often called upon in comprehensive care settings and interprofessional teams to address the severe, complex emotional toll of both disease and intervention. Because of this complexity, a working knowledge of the biological underpinnings of presenting symptoms in people with cancer is essential. Thus, the principal aim of this clinically focused chapter is to serve as a translational bridge between the basic scientific research findings related to endogenous opioids and cancer, and the practical applications of these findings in clinical settings. This chapter complements others in this volume that have focused on cancer, oncology, and endogenous opioids (i.e., Hankins & Harris; Nguyen).

Stress and Depression

Historically, stress has been thought of as having a somewhat parabolic effect on motivation and mood. This concept has often been represented as a Yerkes-Dodson inverted U-shaped curve, with "good stress" (i.e., "eustress") as the goal to aim for, and "bad stress" (i.e., distress) to be avoided. It is now generally recognized that this simplistic model is antiquated because it fails to account for the nuanced and contextual effects of different forms, types, sources, and even definitions of stress (see Le Fevre et al., 2003). For example, in people diagnosed with cancer, it has long been recognized that stress derives from multiple correlated sources (e.g., biological stress from tumors, physiological stress from treatment, emotional stress from all aspects of cancer and cancer treatment), especially as cancer progresses or recurs (De Faye et al., 2006; Rinaldis et al., 2012; Yang et al., 2008). We would clearly not expect stress from cancer nor its treatment to improve performance for someone. However, how one copes with the symptoms of cancer (and side effects of treatment) may mediate the relationship between cancer-related stress and stress-related outcomes, including depression symptoms.

Depression is associated with psychosocial stressors as a precipitant, an exacerbating factor, and an outcome, all of which appear to be mediated biologically across multiple substrates and systems. One system implicated in depression is the hypothalamic-pituitary-adrenocortical (HPA) axis, commonly thought of as the "stress response system" (see De Kloet et al., 1998). HPA axis activation is mediated in part by the endogenous opioid systems, which are central to adaptive stress responses (Hebb et al., 2005). As described extensively elsewhere in this volume (see Kerr and Gregg, chapter "The Roles of Endogenous Opioids in Placebo and Nocebo Effects", Pettrey et al., chapter "Physical Exercise as an Intervention for Depression: Evidence for Efficacy and Mu-Opioid Receptors as a Mechanism of Action"), endogenous opioids also play a role in mood regulation, and in depressogenic responses to stressors.

Research on the link between depression and HPA axis functioning suggests an empirical association, with relatively consistent findings of HPA axis dysfunction in people who are depressed (Murri et al., 2014). An early meta-analysis by Burke and colleagues (2005) found that people experiencing depression appear to have stronger HPA axis activation, measured via cortisol responses, to stressors compared to nondepressed individuals. Subsequent data aggregations have found concordant results across age groups. A meta-analysis by Murri and colleagues (2014) found evidence of elevated cortisol levels in people 60 years old and older. Lopez-Duran et al.'s (2009) meta-analysis of studies ($k = 34$) on children and adolescents found significant differences in basal cortisol levels, cortisol secretion during dexamethasone suppression test studies, as well as those experiments with psychological stressor paradigms. Thus, HPA axis dysfunction seems to be a robust finding across the life span for people with depression.

Additional meta-analytic data have shed more nuanced light on the data described above. For example, Zorn et al. (2017) found important sex differences in cortisol responses. Women with major depressive disorder exhibited a blunted cortisol response to psychological stressors; conversely, men with major depressive disorder exhibited an elevated cortisol response to psychological stress (Zorn et al., 2017). Furthermore, Knorr et al. (2010) found that cortisol levels measured via a commonly used laboratory technique (salivary cortisol measurement) did not reliably differ between depressed and control participants across 20 studies. Thus, methodology may be a factor in some of the previous findings.

While HPA axis dysfunction is clearly involved in the pathogenesis of depression, data on this system as a therapeutic target are less conclusive. An initial meta-analysis on antidepressant treatment found little evidence that HPA axis functionality was either a predictor of treatment outcomes or a useful outcome measure in itself (McKay & Zakanis, 2010). Lombardo's et al. (2019) meta-analysis of pharmacotherapies targeting the HPA axis directly (i.e., cortisol synthesis inhibitors; glucocorticoid receptor antagonists) found that baseline cortisol levels predicted response to cortisol synthesis inhibitors, but did not predict response to glucocorticoid receptor antagonists. These data were derived from studies of MDD as well as bipolar disorder, which may also have affected the results. Finally, a recent meta-analysis by Fischer and colleagues (2017) focused on cortisol level as a predictor of psychotherapy outcomes in depression, and found that higher cortisol levels predicted smaller symptom reduction at posttreatment; cortisol levels were positively correlated with symptom severity when depression treatment concluded.

Endogenous Opioids and Stress Responses

As noted above and in Fig. 1, the endogenous opioid systems also play a role in mediating various aspects of the mammalian stress responses. Specifically, enkephalin functions to co-regulate neuronal activity in the locus coeruleus during stress, acting as an opposing influence to corticotropin-releasing factor (CRF) (Valentino & Van Bockstaele, 2015). Stressors activate afferent neuronal inputs for both enkephalin and corticotropin-releasing hormone (CRH) in the locus coeruleus. CRH potentiates tonic discharge and inhibits phasic discharge within the locus coeruleus to facilitate adaptive and beneficial arousal states. Conversely, enkephalin

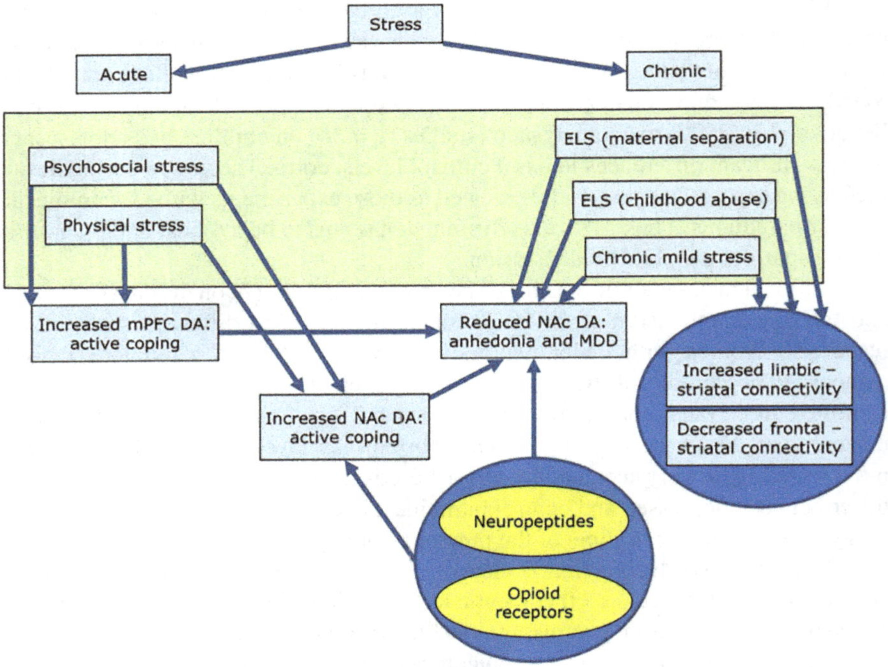

Fig. 1 Proposed stress-neuropeptide-opioid receptor relationships. From Ironside et al. (2018)

activation of μ-OR within the locus coeruleus leads to inhibition of tonic discharge and potentiation of phasic discharge, reducing arousal. This phasic activity is adaptive for responding to the removal or termination of the stressor. A balance in this activity is important, as an imbalance in favor of either CRH or enkephalin can lead to pathogenesis. Such imbalances may be more likely in people undergoing cancer treatment due to both cancer activity and the actions of cancer therapies.

Stress and Depression in Cancer

The immune system produces proteins that are either proinflammatory (liver-derived C-reactive protein, CRP) or anti-inflammatory (cell-derived IL-10 cytokine, IL-6). In acute infections, such as the flu, increases in these proteins combat the short-lived illness. However, in states of chronic stress and chronic infections, these proteins can remain chronically elevated, which may compromise the immune system, putting an individual at risk for disease, suboptimal or delayed healing, and mortality (Black & Slavich, 2016). Previous authors have posited a "sickness behavior" phenomenon induced by activation of proinflammatory cytokines, which is characterized by fatigue, anhedonia, low mood, social isolation, and irritability, and which is hypothesized to be adaptive in that it promotes healing from infection by decreasing energy expenditure (Dantzer & Kelley, 2007; De Berardis et al., 2010).

Cancer as a Source of Inflammation

Stress is associated both with cancer (as a consequence), and with depression (as an antecedent and a consequence). The link between this triad of phenomena appears to be inflammation, which is also observed in both depression and cancer. Much has been written regarding inflammation caused by tumorigenesis. We turn now to the intersecting literatures of oncology and inflammatory processes.

Cancer, broadly defined, is now understood to occur in the context of organ functioning, and is thus now thought of as an organ disease (Nisticò & Ciliberto, 2020). Certain conditions within organs are germane to tumorigenesis. The environments those conditions create are referred to as the "tumor microenvironment" (TME), which consists of heterogeneous variety of cells, for example, endothelial cells, monocytes, macrophages, T cells, NK cells, dendritic cells, as well as noncellular material (Galdiero et al., 2018). This biological material creates inflammation that is often progressive and chronic within the organ, and also is the basis of metastasis (Medzhitov, 2008). There is a reciprocal relationship between the TME and the tumor, such that the immune cells "select" the most viable of the cells that survive tumorigenesis, leaving the unselected cancer cells to replicate and progress (Nisticò & Ciliberto, 2020). At the same time, clinical phenomena such as deregulation of neurotransmitters, excessive and protracted secretion of cortisol, along with cellular toxicity, neurodegeneration, and cell death may also induce depression symptoms. In short, cancer cells develop in an inflamed environment (organ systems/cells with inflammation), and also physiologically fan the flames (immunological responses), leading to metastasis; and many these same processes and factors may also contribute to the development of depression symptoms (see Fig. 2).

While the tumor microenvironment plays a central role in the inflammatory process contributing to depression, epigenetic factors (which can be thought of as the "macroenvironment") may also play a role. Features of stress (chronicity, acuity, frequency, severity), factors contributing to stress (environmental, relational) and resources for responding to stress may all play intersecting roles in the creation of tumorigenic microenvironments (see Figs. 3 and 4). Epigenetic factors of macroenvironments may contribute to these features, forms, and sources of stress intersectionally, leading to increased risk for tumorigenesis and metastasis (Brower, 2011; Murata et al., 2019; Shanmugam & Sethi, 2013).

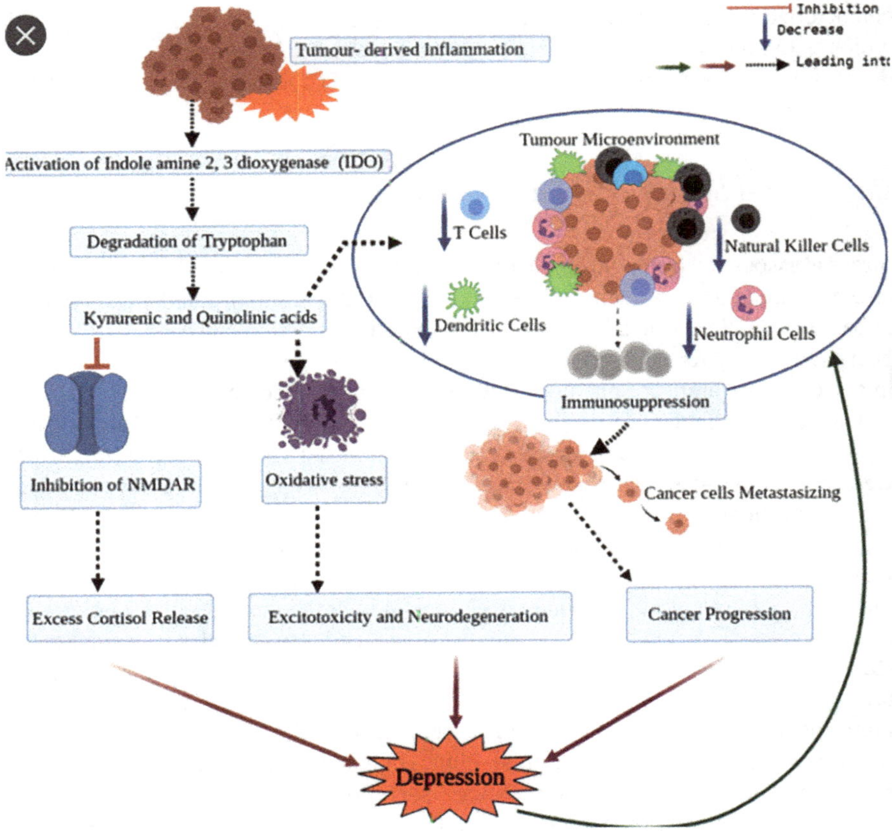

Fig. 2 Cancer as a source of inflammation leading to depression. (Ahmad et al., 2021)

Stress, Cancer, Depression, and Inflammation

In general, people exhibiting major depressive disorder have displayed higher levels of inflammatory markers, particularly IL-6 and C-reactive protein. They also exhibit increased production of other proinflammatory cytokines, including interleukin (IL)-1β and tumor necrosis factor (TNF)-a (Munzer et al., 2013). Hypothetically, the combined effects of cancer and depression may further increase inflammation in people with this comorbidity; however, the findings in this line of research are best characterized as mixed at this time. Data from a small number of studies suggest that people with cancer and depression exhibit higher levels of proinflammatory cytokines (Costanzo et al., 2005; Jehn et al., 2006; Musselman et al., 2001). Conversely, O'Connor et al. (2007) found that there were no differences in multiple inflammatory biomarkers (e.g., IL-6; TNF-α) in blood samples of people with and without depression in a family cancer registry. Adding more nuance, in a sample of patients with advanced stage cancers, Jacobson and colleagues (2008) found that

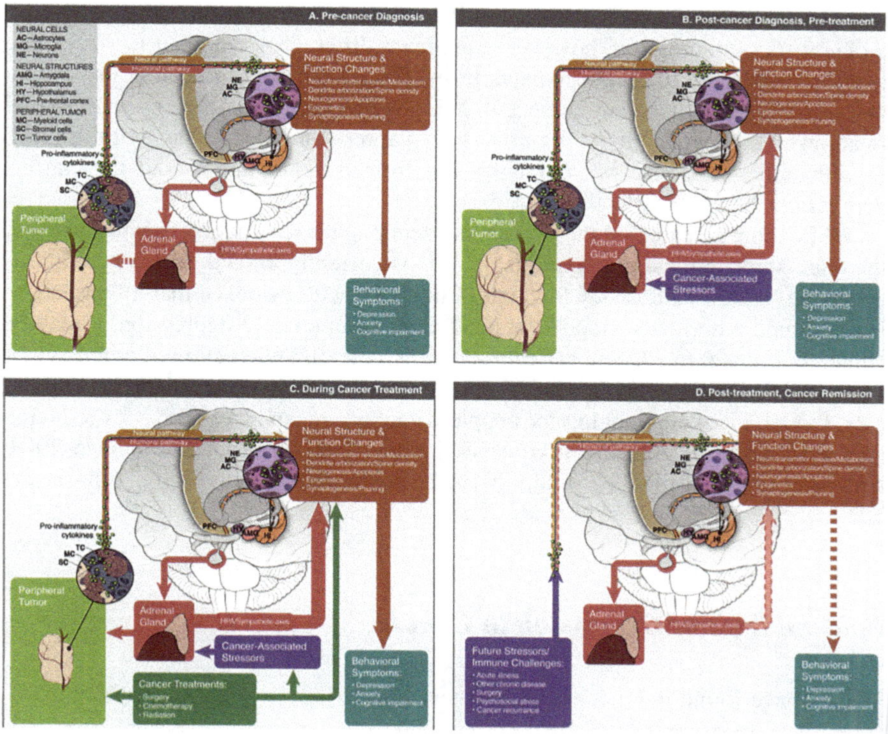

Fig. 3 Link between cancer and depression across the cancer continuum. (Schrepf et al., 2015)

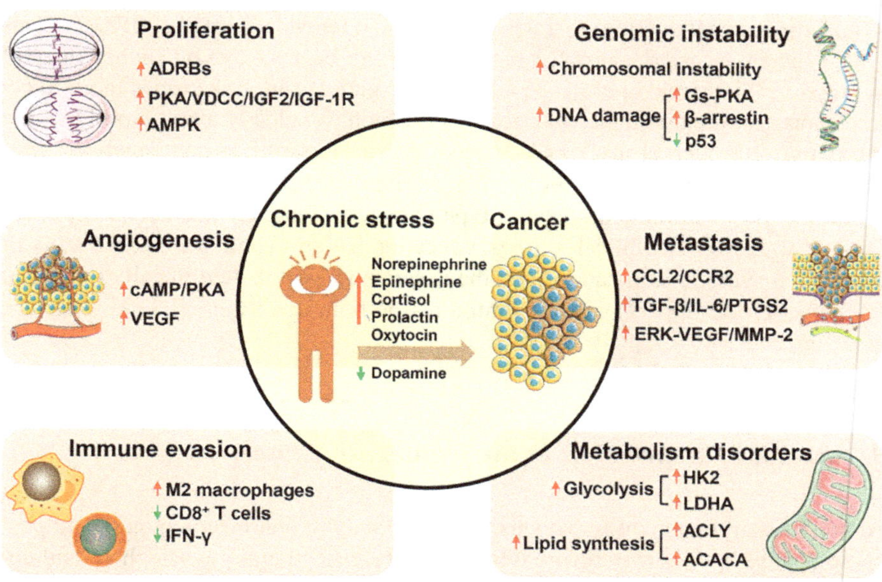

Fig. 4 Stress and cancer progression. Role of stress in cancer progression

there were no significant differences in IL-6 levels between those with and without a diagnosis of depression; however, significant differences emerged in subsample analyses with only those participants from whom samples had been drawn within 48 hours of the assessment interview. This may suggest a more temporally based relationship between inflammation and depression in people with cancer. Significantly more research is clearly required to better understand the cancer-depression-inflammation relationship.

While more research on the relationships among these variables remains needed, there are some clinical implications of what is currently known. First, there is evidence that inflammation may fluctuate. Combined with evidence that inflammation is associated with lower mood states and other symptoms of depression, it may be clinically prudent to educate patients about the potential fluctuation of mood symptoms, and the likely temporary of these changes. Second, stress management strategies, though often challenging for people with cancer, can be effective for reducing the acute and long-term effects of cancer-related stress (Antoni & Dhabhar, 2019; Antoni et al., 2009; Huang et al., 2016; Ledesma & Kumano, 2009; Tang et al., 2020; Xie et al., 2020; Zainal et al., 2013).

Epidemiology of Depression in Cancer

Studies have found that people with cancer have a higher risk of depression compared to those patients with other chronic illnesses, for example, chronic lung disease, heart disease (Eton & Lepore, 2002). People with cancer who have more advanced disease or greater physical symptoms display higher rates and more severe depressive symptoms. Approximately 16% of those with cancer experience depression, with approximately equal rates for males and females (Li et al., 2012). Patients with pancreatic, gastric, oropharyngeal, and lung cancers exhibit higher rates of depression (Fitzgerald et al., 2015). Depressed patients are three times more likely to be noncompliant with medical treatment, which can clearly affect morbidity and mortality (DiMatteo et al., 2000). Indeed, data from multiple meta-analyses have clearly indicated that depression in people with cancer is associated with a higher mortality risk, regardless of stage or type of cancer (Pinquart & Duberstein, 2010; Satin et al., 2009). In clinical settings, educating patients about the commonality of depression symptoms in cancer, and the itinerant risks, can be clinically important for reducing the impact of those symptoms on both quality of life and treatment engagement.

Treatment of Depression in the Context of Cancer

Antidepressant treatment in patients with cancer, both pharmacologically and psychologically, has been studied extensively. Pharmacotherapy research on antidepressant treatment of people with cancer has yielded hundreds of studies. While some analyses have suggested a robust benefit to treatment with antidepressant medications (e.g., Laoutidis & Mathiak, 2013), others have found mixed results (Riblet et al., 2014). Some authors have suggested that preventative SSRIs could help to mitigate depression comorbidity in cancer. However, a recent meta-analysis by Zahid et al. (2020) found limited and low-quality evidence supporting the prophylactic use of SSRIs in this population. Furthermore, SSRIs carry a risk of neurotoxicity, which must be weighed against any potential benefits. Whether prescribed proactively or reactively, the mediating and moderating mechanisms underlying the effects of antidepressant medications in people with cancer are currently unclear.

Meta-analyses of psychological interventions for depression in people with cancer generally support cognitive-behavioral interventions, as well as social support in reducing depression (e.g., Ekers et al., 2014; Hart et al., 2012; Li et al., 2020). The literature suggests psychological interventions in cancer patients should target specific symptoms and cancer type rather than general interventions (Daniels & Kissane, 2008). The biological mediators of the effects of psychological interventions in cancer-related depression remain unknown at this time.

One outcome of depression treatment that has generated interest in recent years is cancer survival rates. Despite some indications of effectiveness for depression symptoms, research has generally found little to no improvements in cancer survival resulting from treatment of depression (e.g., Oh et al., 2016; Xia et al., 2014; see also Mulick et al., 2018). This is likely due to the confluence of intersecting factors (e.g., later diagnosis, substandard treatment, limited access to optimal oncology services) that increase risk for cancer mortality in general, and are more prevalent among people with psychiatric disorders including depression (Mitchell, 2018). As noted by Mitchell (2018), with healthcare disparities at the center of this model, research trials of depression treatment in people with cancer are ill-equipped to address the root causes of higher cancer mortality.

Methodological factors in RCTs of depression treatment in people with cancer must also be considered. Because of the nature of both cancer and depression, longitudinal follow-up is often limited. Underpowering is a significant statistical issue that has yet to be solved in this literature. Thus, survival curves may be based on currently unavoidable biased samples.

Another methodological factor, also related to sampling, is generalizability. The term "cancer" refers to a class of over 100 types and subtypes of disease, each with unique etiologies, clinical idiosyncrasies, and practical effects on people's health. Populations of people with cancers are even more diverse than cancer itself. The intersection of such diversities arguably makes the endeavor to find a uniform and generalizable depression treatment protocol a fool's errand. Consider that a recent meta-analysis by Bravery et al. (2020) found substantial disproportionate representation of cancer types relative to cancer-type prevalence and prevalence of cancer-type/depression comorbidity. Nearly half (49.3%) of participants in depression treatment trials for people with cancer were diagnosed with breast cancer, yet this cancer type accounts for only approximately 11–12% of all cancers, with an approximate comorbidity rate of only about 10–11%. For comparison, cancers of the digestive tract represent approximately 27–28% of all cancers, but just over 11% of study participants, despite a depression comorbidity rate of 37–38%. A similar disparity can be seen for brain cancer: just 0.2% of research participants had brain cancers, despite representing 1.5% of cancer cases globally, and exhibiting a depression/cancer comorbidity rate of 2%. It is possible that such skewed data preclude an accurate understanding of the actual effects of depression treatment.

In the clinical context of providing psychological services to people with cancer, informing patients about realistic anticipated outcomes of such adjunctive treatment is essential. Based on the data thus far, mental health professionals should educate patients that psychological interventions may improve the quality of their life, but not necessarily improve the duration of their life. On the other hand, working to reduce stress associated with cancer and treatment, and resolving problems that interfere with treatment/management, can be parts of effective psychological treatments; these may also lead to improved cancer treatment participation, and potentially a greater opportunity for survival.

Treating Transdiagnostic Symptoms in Depression/Cancer Comorbidity

Somatic symptoms, for example, lack of appetite, fatigue, make diagnosing depression in cancer patients puzzling; it is difficult to attribute the causality of symptoms to cancer treatment side effects or depression. Some researchers argue that depression is often overlooked in people with patients, limiting behavioral health referrals (Reich et al., 2008; Williams & Dale, 2006).

Understanding the shared symptomatic presentation between symptoms common to both cancer and depression is important for monitoring and mitigation efforts. From a therapeutic perspective, targeting specific symptoms of cancer and cancer therapy side effects most closely associated with depression may yield better outcomes in this domain. We propose that a primary part of psychological intervention for depression symptoms in people with any form of cancer includes effective psychoeducation about the likely or potential sources of symptoms, expectations for duration and severity, and, when possible, strategies for mitigation (see Mitchell et al., 2007). In this section, we focus on two symptomatic presentations that are common to both depression and cancer: fatigue and sleep disturbance.

Fatigue in Cancer and Depression

Fatigue and sleep disturbance frequently co-occur. These symptoms are common in both cancer and depression. Cancer-related fatigue can include physical symptoms of tiredness, in addition to problems with sleeping and cognitive complaints. Fatigue in people with cancer is common. Al Maqbali et al.' (2021) recent meta-analysis of studies conducted between 1993 and 2020 ($k = 129$; $N = 71{,}568$) found an omnibus prevalence rate of 49%, with higher rates for advanced stage cancer (60.6%). Cancer fatigue appears to vary systemically, with gynecological (26.2%) and prostate (26.3%) cancers associated with relatively lower prevalence rates of fatigue, and gastrointestinal cancer (50%), breast cancer (48.9%), and lymphoma (43.3%) associated with higher prevalence rates of fatigue across studies (Al Maqbali et al., 2021). Cancer-related fatigue tends to peak in prevalence during active treatment (62%), and exhibits a progressive reduction thereafter, with prevalence rates of 50.1% during the first 3 months posttreatment, and 43% in the period beyond 3 months posttreatment (Al Maqbali et al., 2021). Bower and Lamkin (2013) note that for many people with cancer, fatigue will improve within 1 year of their last treatment. However, earlier work by Bower and colleagues (2006) also found that approximately 25–30% report persistent fatigue as long as 10 years posttreatment.

Tumors and treatment from radiation and chemotherapy can cause inflammation, which can produce diverse and systemically disruptive effects that may contribute to fatigue. The involvement of serotonergic receptors (5-HT) in depression pathogenesis has been identified for a subset of people with endogenous depression, and thus may be implicated in specific symptoms of depression, including fatigue. Earlier research found that TNF-α may dysregulate 5-HT metabolism by increasing both synaptic 5-HT secretion and clearance, which in turn decreases TNF-α synthesis, leading to further depletion of 5-HT (Morrow et al., 2002; Ryan et al., 2007).

Fatigue in breast cancer survivors treated with chemotherapy can lead to depressed mood and sleep disturbance, and previous authors have posited that inflammation (TNF-α) plays a role in post-chemotherapy fatigue (Bower et al., 2011). Consistent with this, people who have cancer-related fatigue may display higher levels of inflammatory activity, particularly sTNF-RII, IL-1RA, and CRP (e.g., Paulsen et al., 2017).

Prescribed and systematic physical activity shows promise as an intervention for improving fatigue in people with cancer. A meta-analysis of nine studies found that breast cancer patients undergoing radiotherapy post-surgery who engaged in combined aerobic-resistance exercise showed reduced levels of fatigue, and supervised exercise programs were more effective than at-home exercise programs (Lipsett et al., 2017).

The effect of exercise on fatigue in patients with advanced cancer is less clear. A review of patients with advanced cancer found that exercise improved quality of life in more than half of participants, while fatigue improved in less than half (Dittus et al., 2017). Stout et al. (2017) recommend exercise as an intervention for all cancer patients, assuming appropriate screening and precautions are taken as exercise can improve clinical and functional outcomes.

Another intervention that has shown promise for helping patients cope with fatigue include mindfulness-based stress reduction (MBSR). It is important to note that no gold standard exists on treatments for fatigue, and this research is still emerging. In a study of 35 cancer survivors with cancer-related fatigue, those who participated in a 7-week MBSR group, which included mindfulness meditation, yoga, and self-regulatory response for stress, showed significant improvement in fatigue interference, depression, and sleep disturbance after intervention and at 6 months post-intervention (Johns et al., 2015).

Although the mechanism of action for psychosocial interventions is unclear, one follow-up study of breast and prostate cancer patients who participated in a MBSR program found that participants had decreased cortisol levels and reduction in Th1 pro-inflammatory cytokines relative to baseline (Carlson et al., 2007).

While data are still emerging, and standards for psychological interventions in people with cancer have not been established, some clinical implications of the research findings described above can be distilled. First, similar to the discussion by Pettrey and colleagues (this volume), physical activity appears to have some potential for treating symptoms of depression for people with cancer, especially fatigue. Second, changing how one responds to stress by teaching specific therapeutic responses (as is found in MBSR) may further mitigate the stressful consequences of cancer. Indeed, the effects of such techniques may extend to immunological functioning and inflammatory biomarkers such as cytokines.

Sleep Disturbances in Cancer and Depression

Estimates of sleep disturbance prevalence in people with cancer have ranged from 30–75% (Ancoli-Israel et al., 2001; Berger et al., 2005; Fiorentino & Ancoli-Israel, 2006; Savard et al., 2001; Savard & Morin, 2001). Cancer and its treatments can negatively affect sleep through multiple pathways. Treatments (chemotherapy, radiation) can alter the levels of inflammatory cytokines or disrupt circadian rhythms or sleep-wake cycles, induce pain and fatigue, or trigger menopausal symptoms in women, including pain or fatigue, while medications used to treat side effects of treatment (e.g., corticosteroids) can cause insomnia (see Pasquini & Biondi, 2007).

One factor that may contribute to clinically significant sleep problems in people with cancer is pain and analgesic protocols. It is well understood that a reciprocal adversarial relationship exists between pain and sleep; pain disrupts sleep, while sleep deficiency dysregulates pain inhibition (Babiloni et al., 2021). Additionally, opioids are often part of analgesia protocols in oncology. As discussed extensively by Hankins and Harris (this volume), MOR agonists have been identified as promulgating factors for tumorigenesis. Essentially, MOR activity may create a cellular environment that is highly conducive to metastasis across several types of cancers. Dysregulation of the endogenous opioid systems may also disrupt sleep, as indicated by prior research findings of reduced pain inhibition in people experiencing sleep deprivation, including those with insomnia (Haack et al., 2012; Smith et al., 2007). In sum, we know that endogenous opioids are involved in cancer, chronic pain, and insomnia, as well as depression (see Pettrey et al., this volume). Educating patients that pain from cancer and intervention protocols may disrupt their sleep, which may, in turn, exacerbate their pain, can be important for managing both.

Cognitive behavioral therapy for insomnia (CBT-I) is an evidence-based treatment used to treat insomnia in young and older adults and is considered the "gold standard" treatment option (Arico et al., 2016). The brief intervention consists of stimulus control, sleep restriction, relaxation, cognitive and sleep hygiene components, allowing practitioners to address several dimensions of insomnia at once. Meta-analyses of CBT-I for insomnia in people with cancer have found that CBT-I improved sleep quality, mood, general and physical fatigue, and global and cognitive dimensions of quality of life (Garland et al., 2014; Johnson et al., 2016). Moreover, CBT-I also demonstrates effects on hot flashes and night sweats associated with menopausal symptoms as well as reduced frequency of medication use, and reduced levels of anxiety and depression. Improvements in insomnia and sleep quality generally persist for at least 12 months.

A promising direction for research on CBT-I and breast cancer patients pertains to research on the immune system. One study found that breast cancer patients receiving CBT-I showed higher secretion of interferon-y and a lower increase of lymphocytes at posttreatment. These levels of interferon-y and interleukin-1B increased significantly after treatment, and significant changes in white blood cells, lymphocytes, and interferon-y were found at follow-up (Savard et al., 2005).

Conclusions

This chapter has focused primarily on understanding and targeting symptoms of depression that are most likely to occur in cancer. In our summary of the literature, we sought to translate some of the most relevant basic scientific findings on cancer, depression, and endogenous opioids to practical clinical applications for clinicians practicing in oncology settings.

Research pertaining to physiological pathways of depression symptoms in cancer has been somewhat illuminating of the scope of the problem. We understand more now about this comorbidity than ever before. Nonetheless, cancer and people with cancer are both diverse. It is likely that no singular pathway for depressogenesis in cancer will be found. Similarly, while some key principles of psychological treatment have been identified, there is not a "one-size-fits-all" approach that can be applied broadly.

The intervention research we reviewed is promising. However, people with cancer present unique challenges for managing stress, fatigue, and sleep problems. Evidence-based interventions for problems with sleep and fatigue exist, but may require some adaptation to maintain efficacy. Understanding the idiopathic etiology of an individual patient's comorbid depression symptoms will guide clinicians to recommendations for the most effective psychological treatment options for their patients.

References

Ahmad, M. H., Rizvi, M. A., Fatima, M., & Mondal, A. C. (2021). Pathophysiological implications of neuroinflammation mediated HPA axis dysregulation in the prognosis of cancer and depression. *Molecular and Cellular Endocrinology, 520*, 111093.

Al Maqbali, M., Al Sinani, M., Al Naamani, Z., & Al Badi, K. (2021). Prevalence of fatigue in patients with cancer: A systematic review and meta-analysis. *Journal of Pain and Symptom Management, 61*(1), 167–189.

Ancoli-Israel, S., Moore, P. J., & Jones, V. (2001). The relationship between fatigue and sleep in cancer patients: A review. *European Journal of Cancer Care, 10*(4), 245–255.

Antoni, M. H., & Dhabhar, F. S. (2019). The impact of psychosocial stress and stress management on immune responses in patients with cancer. *Cancer, 125*(9), 1417–1431.

Antoni, M. H., Lechner, S., Diaz, A., Vargas, S., Holley, H., Phillips, K., et al. (2009). Cognitive behavioral stress management effects on psychosocial and physiological adaptation in women undergoing treatment for breast cancer. *Brain, Behavior, and Immunity, 23*(5), 580–591.

Arico, D., Raggi, A., & Raffaele, F. (2016). Cognitive behavioral therapy for insomnia in breast cancer survivors: A review of the literature. *Frontiers in Psychology, 7*, 1–9.

Babiloni, A. H., Beetz, G., Bruneau, A., Martel, M. O., Cistulli, P. A., Nixdorf, D. R., et al. (2021). Multitargeting the sleep-pain interaction with pharmacological approaches: A narrative review with suggestions on new avenues of investigation. *Sleep Medicine Reviews, 59*, 101459.

Berger, A. M., Parker, K. P., Young-McCaughan, S., Mallory, G. A., Barsevick, A. M., Beck, S. L., et al. (2005, November). Sleep/Wake disturbances in people with cancer and their caregivers: State of the science. *Oncology Nursing Forum, 32*(6), E98–E126.

Black, D. S., & Slavich, G. M. (2016). Mindfulness meditation and the immune system: a systematic review of randomizedcontrolled trials. *Annals of the New York Academy of Sciences, 1373*(1), 13–24.

Bower, J. E., & Lamkin, D. M. (2013). Inflammation and cancer-related fatigue: Mechanisms, contributing factors, and treatment implications. *Brain, Behavior, and Immunity, 30*, 548–557.

Bower, J. E., Ganz, P. A., Irwin, M. R., Kwan, L., Breen, E. C., & Cole, S. W. (2011). Inflammation and behavioral symptoms after breast cancer treatment: Do fatigue, depression, and sleep disturbance share a common underlying mechanism? *Journal of Clinical Oncology, 29*, 3517–3522.

Bravery, B., Loughnan, S., & Murphy, M. (2020). Depression treatment research in people with cancer does not reflect cancer prevalence: Findings from a systematic review. *Evidence-Based Mental Health, 23*(4), 155–160.

Brower, V. (2011). Epigenetics: Unravelling the cancer code. *Nature, 471*(7339), S12–S13.

Carlson, L. E., Speca, M., Faris, P., & Patel, K. D. (2007). One year pre-post intervention follow- up of psychological, immune, endocrine and blood pressures outcomes of mindfulness-based stress reduction (MBSR) in breast and prostate cancer outpatients. *Brain, Behavior, and Immunity, 21*, 1038–1049.

Costanzo, E. S., Lutgendorf, S. K., Bradley, S. L., Rose, S. L., & Anderson, B. (2005). Cancer attributions, distress, and health practices among gynecologic cancer survivors. *Psychosomatic Medicine, 67*(6), 972–980.

Daniels, J., & Kissane, D. W. (2008). Psychosocial interventions for cancer patients. *Current Opinion in Oncology, 20*, 367–371.

Dantzer, R., & Kelley, K. W. (2007). Twenty years of research on cytokine-induced sickness behavior. *Brain, Behavior, and Immunity, 21*, 153–160.

De Berardis, D., Conti, C. M. V., Serroni, N., Moschetta, F. S., Olivieri, L., Carano, A., Salerno, R. M., Cavuto, M., Farina, B., Alessandrini, M., Janiri, L., Pozzi, G., & Di Giannantonio, M. (2010). The effect of newer serotonin-noradrenalin antidepressants on cytokine production: A review of the current literature. *International Journal of Immunopathology and Pharmacology, 23*, 417–422.

De Faye, B. J., Wilson, K. G., Chater, S., Viola, R. A., & Hall, P. (2006). Stress and coping with advanced cancer. *Palliative & Supportive Care, 4*(3), 239–249.

De Kloet, E. R., Meijer, O. C., & van Haarst, A. D. (1998). Corticosteroid hormones and the organization of the stress response system. In *New frontiers in stress research. Modulation of brain function* (pp. 1–19). Harwood Academic Publishers.

DiMatteo, M. R., Lepper, H. S., & Croghan, T. W. (2000). Depression is a risk factor for noncompliance with medical treatment: Meta-analysis of the effects of anxiety and depression on patient adherence. *Archives of Internal Medicine, 160*(14), 2101–2107.

Dittus, K. L., Gramling, R. E., & Ades, P. A. (2017). Exercise interventions for individuals with advanced cancer. *Preventive Medicine, 104*, 124–132.

Ekers, D., Webster, L., Van Straten, A., Cuijpers, P., Richards, D., & Gilbody, S. (2014). Behavioural activation for depression; an update of meta-analysis of effectiveness and sub group analysis. *PLoS One, 9*(6), e100100.

Eton, D. T., & Lepore, S. J. (2002). Prostate cancer and health-related quality of life: A review of the literature. *Psycho-Oncology, 11*, 307–326.

Fiorentino, L., & Ancoli-Israel, S. (2006). Insomnia and its treatment in women with breast cancer. *Sleep Medicine Reviews, 10*(6), 419–429.

Fitzgerald, P., Miller, K., Li, M., & Rodin, G. (2015). Depressive disorders: Chapter 3. In J. C. Holland, W. S. Breitbart, P. B. Jacobsen, M. J. Loscalzo, R. McCorkle, & P. N. Butow (Eds.), *Psycho-oncology* (3rd ed.). Oxford University Press.

Galdiero, M. R., Marone, G., & Mantovani, A. (2018). Cancer inflammation and cytokines. *Cold Spring Harbor Perspectives in Biology, 10*(8), a028662.

Garland, S. N., Johnson, J. A., Savard, J., Gehrman, P., Perlis, M., Carlson, L., & Campbell, T. (2014). Sleeping well with cancer: A systematic review of cognitive behavioral therapy for insomnia in cancer patients. *Neuropsychiatric Disease and Treatment, 10*, 1113–1124.

Haack, M., Scott-Sutherland, J., Santangelo, G., Simpson, N. S., Sethna, N., & Mullington, J. M. (2012). Pain sensitivity and modulation in primary insomnia. *European Journal of Pain, 16*(4), 522–533.

Hart, S. L., Hoyt, M. A., Diefenbach, M., Anderson, D. R., Kilbourn, K. M., Craft, L. L., et al. (2012). Meta-analysis of efficacy of interventions for elevated depressive symptoms in adults diagnosed with cancer. *Journal of the National Cancer Institute, 104*(13), 990–1004.

Hebb, A. L., Laforest, S., & Drolet, G. (2005). Endogenous opioids, stress, and psychopathology. *Techniques in the Behavioral and Neural Sciences, 15*, 561–583.

Huang, H. P., He, M., Wang, H. Y., & Zhou, M. (2016). A meta-analysis of the benefits of mindfulness-based stress reduction (MBSR) on psychological function among breast cancer (BC) survivors. *Breast Cancer, 23*, 568–576.

Ironside, M., Kumar, P., Kang, M. S., & Pizzagalli, D. A. (2018). Brain mechanisms mediating effects of stress on reward sensitivity. *Current Opinion in Behavioral Sciences, 22*, 106–113.

Jehn, C. F., Kuehnhardt, D., Bartholomae, A., Pfeiffer, S., Krebs, M., Regierer, A. C., et al. (2006). Biomarkers of depression in cancer patients. *Cancer: Interdisciplinary International Journal of the American Cancer Society, 107*(11), 2723–2729.

Johns, S. A., Brown, L. F., Beck-Coon, K., Monahan, P. O., Tong, Y., & Kroenke, K. (2015). Randomized controlled pilot study of mindfulness-based stress reduction for persistently fatigued cancer survivors. *Psycho-Oncology, 24*(8), 885–893.

Johnson, J. A., Rash, J. A., Campbell, T. S., Savard, J., Gehrman, P. R., Perlis, M., et al. (2016). A systematic review and meta-analysis of randomized controlled trials of cognitive behavior therapy for insomnia (CBT-I) in cancer survivors. *Sleep Medicine Reviews, 27*, 20–28.

Knorr, U., Vinberg, M., Kessing, L. V., & Wetterslev, J. (2010). Salivary cortisol in depressed patients versus control persons: a systematic review and meta-analysis. *Psychoneuroendocrinology, 35*(9), 1275–1286.

Laoutidis, Z. G., & Mathiak, K. (2013). Antidepressants in the treatment of depression/depressive symptoms in cancer patients: A systematic review and meta-analysis. *BMC Psychiatry, 13*(1), 1–21.

Ledesma, D., & Kumano, H. (2009). Mindfulness-based stress reduction and cancer: a meta-analysis. *Psycho-Oncology: Journal of the Psychological, Social and Behavioral Dimensions of Cancer, 18*(6), 571–579.

Le Fevre, M., Matheny, J., & Kolt, G. S. (2003). Eustress, distress, and interpretation in occupational stress. *Journal of Managerial Psychology, 18*(7), 726–744.

Li, M., Fitzgerald, P., & Rodin, G. (2012). Evidence-based treatment of depression in patients with cancer. *Journal of Clinical Oncology, 30*, 1187–1196.

Li, J., Peng, X., Su, Y., He, Y., Zhang, S., & Hu, X. (2020). Effectiveness of psychosocial interventions for posttraumatic growth in patients with cancer: A meta-analysis of randomized controlled trials. *European Journal of Oncology Nursing, 48*, 101798.

Lipsett, A., Barrett, S., Haruna, F., Mustian, K., & O'Donovan, A. (2017). The impact of exercise during adjuvant radiotherapy for breast cancer on fatigue and quality of life: A systematic review and meta-analysis. *The Breast, 32*, 144–155.

Lombardo, G., Enache, D., Gianotti, L., Schatzberg, A. F., Young, A. H., Pariante, C. M., & Mondelli, V. (2019). Baseline cortisol and the efficacy of antiglucocorticoid treatment in mood disorders: A meta-analysis. *Psychoneuroendocrinology, 110*, 104420.

McKay, M. S., & Zakzanis, K. K. (2010). The impact of treatment on HPA axis activity in unipolar major depression. *Journal of Psychiatric Research, 44*(3), 183–192.

Medzhitov, R. (2008). Origin and physiological roles of inflammation. *Nature, 454*(7203), 428–435.

Mitchell, A. J. (2018). Why doesn't depression treatment improve cancer survival? *The Lancet Psychiatry, 5*(4), 289–291.

Mitchell, S. A., Beck, S. L., Hood, L. E., Moore, K., & Tanner, E. R. (2007). Putting evidence into practice: Evidence-based interventions for fatigue during and following cancer and its treatment. *Clinical Journal of Oncology Nursing, 11*(1), 99–113.

Morrow, G. R., Andrews, P. L., Hickok, J. T., Roscoe, J. A., & Matteson, S. (2002). Fatigue associated with cancer and its treatment. *Supportive Care in Cancer, 10*(5), 389–398.

Mulick, A., Walker, J., Puntis, S., Burke, K., Symeonides, S., Gourley, C., et al. (2018). Does depression treatment improve the survival of depressed patients with cancer? A long-term follow-up of participants in the SMaRT Oncology-2 and 3 trials. *The Lancet Psychiatry, 5*(4), 321–326.

Munzer, A., Sack, U., Mergl, R., Schönherr, J., Petersein, C., Bartsch, S., Kirkby, K. C., Bauer, K., & Himmerich, H. (2013). Impact of antidepressants on cytokine production of depressed patients in vitro. *Toxins, 5*, 2227–2240.

Murata, K., Saga, R., Monzen, S., Tsuruga, E., Hasegawa, K., & Hosokawa, Y. (2019). Understanding the mechanism underlying the acquisition of radioresistance in human prostate cancer cells. *Oncology Letters, 17*(6), 5830–5838.

Murri, M. B., Pariante, C., Mondelli, V., Masotti, M., Atti, A. R., Mellacqua, Z., et al. (2014). HPA axis and aging in depression: systematic review and meta-analysis. *Psychoneuroendocrinology, 41*, 46–62.

Musselman, D. L., Miller, A. H., Porter, M. R., Manatunga, A., Gao, F., Penna, S., et al. (2001). Higher than normal plasma interleukin-6 concentrations in cancer patients with depression: Preliminary findings. *American Journal of Psychiatry, 158*(8), 1252–1257.

Nisticò, P., & Ciliberto, G. (2020). Biological mechanisms linked to inflammation in cancer: Discovery of tumor microenvironment-related biomarkers and their clinical application in solid tumors. *The International Journal of Biological Markers, 35*(1_suppl), 8–11.

O'Connor, M. F., Irwin, M. R., Seldon, J., Kwan, L., & Ganz, P. A. (2007). Pro-inflammatory cytokines and depression in a familial cancer registry. *Psycho-Oncology, 16*(5), 499–501.

Oh, P. J., Shin, S. R., Ahn, H. S., & Kim, H. J. (2016). Meta-analysis of psychosocial interventions on survival time in patients with cancer. *Psychology & Health, 31*, 396–419.

Pasquini, M., & Biondi, M. (2007). Depression in cancer patients: A critical review. *Clinical Practice and Epidemiology in Mental Health, 3*(1), 1–9.

Paulsen, Ø., Laird, B., Aass, N., Lea, T., Fayers, P., Kaasa, S., & Klepstad, P. (2017). The relationship between pro-inflammatory cytokines and pain, appetite and fatigue in patients with advanced cancer. *PLoS One, 12*(5), e0177620.

Pinquart, M., & Duberstein, P. R. (2010). Associations of social networks with cancer mortality: A meta-analysis. *Critical Reviews in Oncology/Hematology, 75*, 122–137.

Reich, M., Lesur, A., & Perdrizet-Chevallier, C. (2008). Depression, quality of life and breast cancer: A review of the literature. *Breast Cancer Research and Treatment, 110*, 9–17.

Riblet, N., Larson, R., Watts, B. V., & Holtzheimer, P. (2014). Reevaluating the role of antidepressants in cancer-related depression: A systematic review and meta- analysis. *General Hospital Psychiatry, 36*(5), 466–473.

Rinaldis, M., Pakenham, K. I., & Lynch, B. M. (2012). A structural model of the relationships among stress, coping, benefit-finding and quality of life in persons diagnosed with colorectal cancer. *Psychology & Health, 27*(2), 159–177.

Ryan, J. L., Carroll, J. K., Ryan, E. P., Mustian, K. M., Fiscella, K., & Morrow, G. R. (2007). Mechanisms of cancer-related fatigue. *The Oncologist, 12*(S1), 22–34.

Satin, J. R., Linden, W., & Phillips, M. J. (2009). Depression as a predictor of disease progression and mortality in cancer patients: a meta-analysis. *Cancer, 115*(22), 5349–5361.

Savard, J., & Morin, C. M. (2001). Insomnia in the context of cancer: A review of a neglected problem. *Journal of Clinical Oncology, 19*(3), 895–908.

Savard, J., Simard, S., Blanchet, J., Ivers, H., & Morin, C. M. (2001). Prevalence, clinical characteristics, and risk factors for insomnia in the context of breast cancer. *Sleep, 24*(5), 583–590.

Savard, J., Simard, S., Ivers, H., & Morin, C. M. (2005). Randomized study on the efficacy of cognitive-behavioral therapy for insomnia secondary to breast cancer, part II: Immunologic effects. *Journal of Clinical Oncology, 23*(25), 6097–6106.

Schrepf, A., Lutgendorf, S. K., & Pyter, L. M. (2015). Pre-treatment effects of peripheral tumors on brain and behavior: neuroinflammatory mechanisms in humans and rodents. *Brain, Behavior, and Immunity, 49*, 1–17.

Shanmugam, M. K., & Sethi, G. (2013). Role of epigenetics in inflammation-associated diseases. In T. K. Kundu (Ed.), *Epigenetics: Development and disease* (pp. 627–657). Springer Nature.

Smith, M. T., Edwards, R. R., McCann, U. D., & Haythornthwaite, J. A. (2007). The effects of sleep deprivation on pain inhibition and spontaneous pain in women. *Sleep, 30*(4), 494–505.

Stout, N. L., Baima, J., Swisher, A. K., Winters-Stone, K. M., & Welsh, J. (2017). A systematic review of exercise systematic reviews in the cancer literature (2005–2017). *Contemporary Issues in Cancer Rehabilitation, 9*(9S2), S347–S384.

Tang, M., Liu, X., Wu, Q., & Shi, Y. (2020). The effects of cognitive-behavioral stress management for breast cancer patients: a systematic review and meta-analysis of randomized controlled trials. *Cancer Nursing, 43*(3), 222–237.

Valentino, R. J., & Van Bockstaele, E. (2015). Endogenous opioids: The downside of opposing stress. *Neurobiology of Stress, 1*, 23–32.

Williams, S., & Dale, J. (2006). The effectiveness of treatment for depression/depressive symptoms in adults with cancer: A systematic review. *British Journal of Cancer, 94*, 372–390.

Xia, Y., Tong, G., Feng, R., Chai, J., Cheng, J., & Wang, D. (2014). Psychosocial and behavioral interventions and cancer patient survival again: Hints of an adjusted meta-analysis. *Integrative Cancer Therapies, 13*(4), 301–309.

Xie, C., Dong, B., Wang, L., Jing, X., Wu, Y., Lin, L., & Tian, L. (2020). Mindfulness-based stress reduction can alleviate cancer-related fatigue: a meta-analysis. *Journal of Psychosomatic Research, 130*, 109916.

Yang, H. C., Brothers, B. M., & Andersen, B. L. (2008). Stress and quality of life in breast cancer recurrence: Moderation or mediation of coping? *Annals of Behavioral Medicine, 35*(2), 188–197.

Zahid, J. A., Grummedal, O., Madsen, M. T., & Gögenur, I. (2020). Prevention of depression in patients with cancer: A systematic review and meta-analysis of randomized controlled trials. *Journal of Psychiatric Research, 120*, 113–123.

Zainal, N. Z., Booth, S., & Huppert, F. A. (2013). The efficacy of mindfulness-based stress reduction on mental health of breast cancer patients: A meta-analysis. *Psycho-Oncology, 22*(7), 1457–1465.

Zorn, J. V., Schür, R. R., Boks, M. P., Kahn, R. S., Joëls, M., & Vinkers, C. H. (2017). Cortisol stress reactivity across psychiatric disorders: A systematic review and meta-analysis. *Psychoneuroendocrinology, 77*, 25–36.

Index

P. L. Kerr et al. (eds.), *Endogenous Opioids*, Advances in Neurobiology 35, https://doi.org/10.1007/978-3-031-45493-6

Printed in the USA
CPSIA information can be obtained
at www.ICGtesting.com
CBHW072049220624
10449CB00003BA/311

9 783031 454929